UPPER MAIN SEQUENCE STARS IN ANOMALOUS ABUNDANCES

ASTROPHYSICS AND SPACE SCIENCE LIBRARY

A SERIES OF BOOKS ON THE RECENT DEVELOPMENTS
OF SPACE SCIENCE AND OF GENERAL GEOPHYSICS AND ASTROPHYSICS
PUBLISHED IN CONNECTION WITH THE JOURNAL
SPACE SCIENCE REVIEWS

VOLUME 125
PROCEEDINGS

UPPER MAIN SEQUENCE STARS WITH ANOMALOUS ABUNDANCES

PROCEEDINGS OF THE 90TH COLLOQUIUM
OF THE INTERNATIONAL ASTRONOMICAL UNION,
HELD IN CRIMEA, U.S.S.R., MAY 13–19, 1985

Edited by

C. R. COWLEY
Department of Astronomy, The University of Michigan, Ann Arbor, Michigan, U.S.A.

M. M. DWORETSKY
Department of Physics and Astronomy, University College London, U.K.

and

C. MÉGESSIER
Observatoire de Paris, Meudon, France

D. REIDEL PUBLISHING COMPANY

A MEMBER OF THE KLUWER ACADEMIC PUBLISHERS GROUP

DORDRECHT / BOSTON / LANCASTER / TOKYO

Library of Congress Cataloging in Publication Data

CIP

IAU Colloquium (90th : 1985 : Krymskaia oblast; Ukraine)
Upper main sequence stars with anomalous abundances.

(Astrophysics and space science library; v. 125)
Includes index.
1. Stars—Congresses. 2. Stars—Magnetic fields—Congresses. 3. Photometry, Astronomical—Congresses. 4. Astrophysics—Congresses. I. Cowley, C. R. (Charles R.), 1934- . II. Dworetsky, M. M. (Michael M.) III. Mégessier, C. (Claud) IV. Title. V. Series.
QB799.I38 1985 523.8 86-13774
ISBN-13: 978-94-010-8597-7 e-ISBN-13: 978-94-009-4714-6
DOI: 10.1007/978-94-009-4714-6

Published by D. Reidel Publishing Company,
P.O. Box 17, 3300 AA Dordrecht, Holland.

Sold and distributed in the U.S.A. and Canada
by Kluwer Academic Publishers,
101 Philip Drive, Assinippi Park, Norwell, MA 02061, U.S.A.

In all other countries, sold and distributed
by Kluwer Academic Publishers Group,
P.O. Box 322, 3300 AH Dordrecht, Holland.

TABLE OF CONTENTS

SECTION II – SYSTEMATICS, PHOTOMETRY

SECTION IV – NON-MAGNETIC CP AND RELATED STARS

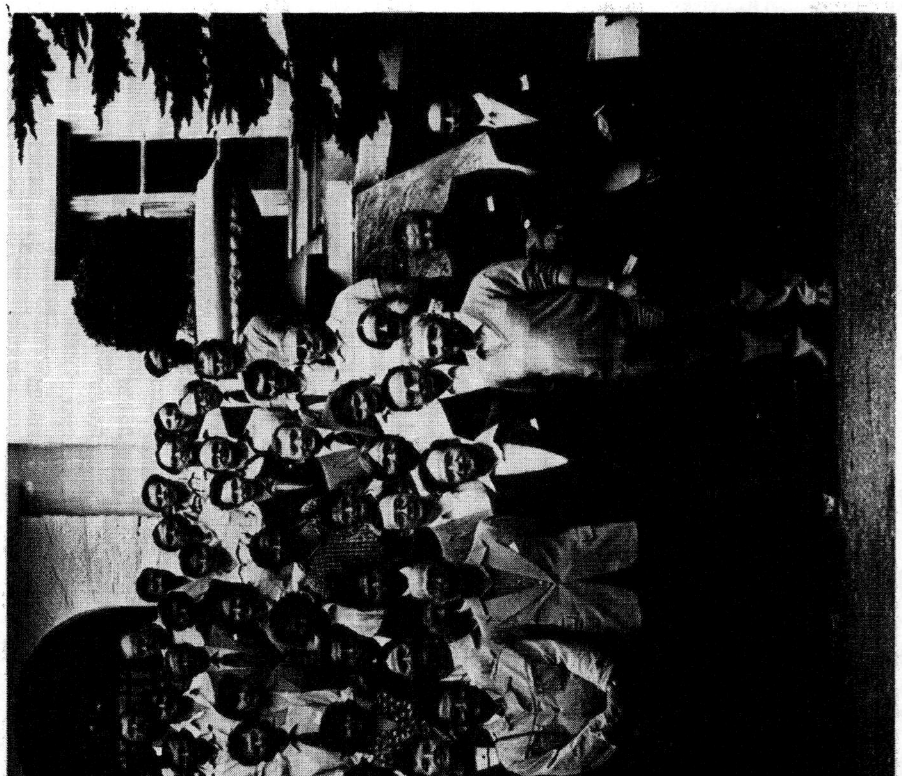

1. Cowley
2. Polosukhina
3. Engberg(son)
4. Mona Engberg
5. Catalano
6. Wehlau
7. Michaud
8. Krause
9. Mendoza
10. Schachovskoi
11. Ziznovsky
12. Dr. Engberg
13. Yagola
14. Khokhlova
15. Mégessier
16. Gertner
17. Rice
18. Hubený
19. Adelman
20. Pyper-Smith
21. Johansson
22. Vincze
23. Zverko
24. Dworetsky
25. unidentified
26. Takada-Hidai
27. Sadakane
28.
29. Muciek
30. Weiss
31. Dolginov
32. Silant'ev
33. Pogodin
34. Musielok
35. Boyarchuk
36. Kipper
37. Kumagarodskoya
38. North
39. Jugaku
40. Kuznetsov
41. Alecian
42. Iliev
43. Kuznetsova
44. Stepien
45. Van Rensbergen
46. Ruuzalepp
47. Zelwanowa
48. Ptitsyn
49. Kolev
50. Bitchkov
51. Pamyatnykh
52. Udovichenko
53. Skul'skij
54. Romanov
55. Miskevich
56. Ryabchikova
57. Kopylov
58. Savanov
59. Artru
60. Franzman
61. Drobyshevski
62. Vornt
62. Vorontsov
63. Kroll
64. Hensberge
65. Schneider
66. student of Crimea
67. Schöneich
68. Mikulashek
69. Hempelmann
70. Glagolevskij
71. Gershberg
72. Romanyuk
73. Plachinda

PREFACE

This volume contains papers presented at IAU **Colloquium** No. 90, at the Crimean Astrophysical Observatory in May of 1985. A few additional contributions are included from authors who for various reasons were unable to attend the meeting.

Four years have passed since the last major international conference on chemically peculiar stars of the upper main sequence was held in Liège, Belgium in 1981. Previous conferences were held in 1975 (Vienna, Austria) and in 1965 (Greenbelt, Maryland, USA). As the proceedings of this Colloquium show, the recent availability of ultraviolet spectra of large numbers of normal and chemically peculiar A and B stars is having a major impact on the way we study these objects, and has led to many new, exciting and unanticipated results. Simultaneously, the more traditional study of optical spectra has been advanced through the increasing use of very high spectral resolution with high signal-to-noise detectors.

The chemically peculiar (CP) stars on the upper main sequence belong in the standard framework within which we understand stellar evolution and the history of matter. Recent work has made it clear that the unusual chemistry and magnetic structure of these objects is of relevance across the broad domain of stellar astronomy, from the upper main sequence to horizontal branch stars and white dwarfs. Metal poor (λ Boo) as well as metal rich (Ap, Am) stars are an integral part of the picture. We do not know the fraction of A and B stars with significant ($>$ 0.3 dex) abundance anomalies. It is surely much larger than the fraction (~0.1) of *classified* peculiar stars. It may well be that the majority of A and B (and other?) stars have non-solar abundances for some elements. The Am's are recognized at classification dispersion because modest (~0.3 dex) deficiencies in calcium happen to be readily detectable in the K line. Much larger anomalies in yttrium or zirconium, for example, are known from high dispersion work, but they cannot be seen at survey dispersions. Chemically peculiar lower main sequence dwarfs have been recognized, but the relation to their upper main sequence congeners is unclear. Powerful observational selection effects prevent the ready detection of predicted mild abundance anomalies in stars slightly hotter than the sun.

Much of the chemistry of the classical Ap and Am stars is explicable in terms of diffusive fractionation. But it has become clear that the chemistry of neither the upper nor the lower main sequence can be understood without a consideration of a wide variety of hydrodynamic or hydromagnetic processes. For example, high mass loss rates are capable of explaining the λ Boo stars, while somewhat lower rates may control the magnitude of the anomalies that appear in Am stars. Additional factors that influence surficial abundances are meridional circulation, magnetic fields, and turbulence. Turbulent convection determines the spectral type at which the diffusion time scales approach the main sequence lifetimes (in late F or early G stars).

The surface chemistry of certain F and hotter stars is especially sensitive to two notorious, adjustable parameters of stellar hydrodynamics: the mixing length and the microturbulence. Abundances in these stars may therefore provide important observational tests of the theory that employs these parameters. Any insight that is gained in understanding the physical basis of these quantities has immediate and broad application. We mention especially, the burgeoning field of abundance work in faint, distant stars, where workers are forced to use features whose strengths depend upon the microturbulence.

In situ differentiation is perhaps the simplest and most promising of several mechanisms that can cause abundance anomalies in main sequence stars. But the observational complexity of the CP stars is such that it is reasonable to consider additional processes. For example, we must learn what abundance patterns to expect in the case of mass transfer, a question involving stellar evolution in binary systems. In violent events, the system may lose mass during the transfer process in a way that would greatly change the separation of the components, and account for the dearth of close binaries among magnetic A stars. We must attempt to identify mass-transfer candidates observationally, and theoreticians must consider the subsequent fractionations of the transferred material.

A variety of questions are now under investigation concerning the theory of stellar atmospheres and their spectra, in the presence of magnetic fields. Much of this work is being done in Eastern European countries. These studies have considered the generation of magnetic fields by chemical inhomogeneities, and the nature of prominence-type activity, in addition to the global effects of the general field on model atmospheres.

We now know that stellar magnetism extends from cool dwarfs, at least into the early B stars. Theoretical discussions of stellar magnetism now often take a global approach in which fossil and dynamo-generated fields are intercompared for both upper and lower main sequence stars. There is a great deal to be learned about hydromagnetic processes that are active during star formation and pre-main sequence phases, as well as during the hydrogen-burning lifetimes. These problems are not restricted to Ap stars, and in fact the most promising mechanisms of relevance to the magnetic and related CP stars are very general ones. It is frequently true that the *effects* of these mechanisms are most easily studied in the A stars, because of the overall simplicity of their spectra. It is therefore a challenge to study these matters in the observationally approachable domain of the upper main sequence stars.

Of course, not all CP stars have strong magnetic fields. Recent advances in Zeeman effect techniques demonstrate that certain of the mercury-manganese stars have no detectable fields, even at quite small sensitivity levels. An active area of current research centers around the origin of the distinct abundance patterns that manifest themselves in magnetic and non-magnetic CP stars.

Tape recordings were made of all of the discussions which followed the review papers as well as some of the contributed papers that were presented orally. Speakers were also requested to summarize their questions and responses in written form, and the edited discussions which appear in this volume are based on both the verbatim transcript and the written records. Although minor editing has been done in the interests of brevity and grammar, the Editors feel they are otherwise an accurate reflection of what was said. We hope readers will find them interesting, fascinating, and a valuable supplement to the papers themselves.

IAU Colloquium No. 90, at the Crimean Astrophysical Observatory, brought together workers with a mutual interest in CP stars who had not had the opportunity to talk with one another. The advantages of such personal contact cannot be overestimated. We are grateful to the International Astronomical Union for its support. We thank Drs. Severny and Boyarchuk and the Observatory Staff for their hospitality and hard work which made the meeting run very smoothly. Language barriers were overcome with the help of two excellent translators, N. Kuznetsov and A. Silina. Participants greatly enjoyed an extensive tour of the observatory and coach excursions to the palace at Bakhchisaray and the Southern Crimean coast, during the Colloquium.

C. R. Cowley, Ann Arbor, USA
M. M. Dworetsky, London, UK
C. Mégessier, Paris, France

THE ORIGIN AND STRUCTURE OF THE MAGNETIC FIELDS OF THE CHEMICALLY PECULIAR STARS

David Moss
Mathematics Department
The University
Manchester, M13 9PL
U.K.

ABSTRACT

Rival theories for the origin of the magnetic fields present in the CP stars are discussed, particular attention being paid to the claims of the 'contemporary dynamo' and 'fossil' theories. The internal structure of the field as predicted by calculations consistent with the fossil theory is discussed at length. It seems that current time dependent models can now give a coherent picture of the fields of the CP stars according to the fossil theory. Dynamo theory modelling has not been developed in such detail. As yet neither the theoretical predictions nor the observational material seem to be detailed enough to allow a decisive comparison between the theories.

1. INTRODUCTION

In recent years much attention has been paid to magnetic fields, activity cycles and related effects in stars of the lower main sequence - the "solar-stellar connection" - and at times those maintaining an interest in the classical magnetic stars of the middle main sequence have almost been made to feel rather old fashioned! Nevertheless there are many unsolved problems connected with such objects and in this review I will discuss some of those, limiting myself to the origin and large scale structure of the fields, and directly related topics. In the limited space available I do not intend to discuss topics such as the cause of the anomalously slow rotations of the magnetic CP stars, nor the short period oscillations or the origins of the anomalous chemical abundances.

The rigid rotator model will be adopted as a working hypothesis to interpret the observations, in common with the usual practice. In almost every case so far investigated field variations can be modelled by a dipole field, maybe with the dipole displaced from the stellar centre. An equivalent representation is as the sum of dipole and (smaller) quadrupole components. One star (HD37776, Thompson and Landstreet, 1985) appears to have a dominant quadrupolar component. The oblique rotator/displaced dipole model has dipole strength, inclinations i and χ

C. R. Cowley et al (eds.), Upper Main Sequence Stars with Anomalous Abundances, 1–10.

between the rotation axis and the line of sight and the magnetic axis respectively, and the fractional dipole displacement d as free parameters. The rather meagre observational evidence is consistent with a random distribution of χ (Hensberge et al 1979; Borra & Landstreet, 1980; Didelon, 1984).

An alternative interpretation of the rigid rotator model is the 'perpendicular rotator' in which χ is assumed to be 90°. The surface field is then modelled by a sum of dipolar and quadrupolar components with a common axis in the rotational equator. The ratio of B_s to B_e can now only be kept acceptably small by appealing to a markedly non uniform distribution of elements which have lines sensitive to magnetic fields. This results in a distribution with the apparent values of $\chi < 90°$ when effects of inhomogeneities are ignored (e.g. Oetken, 1977, 1979). Surface inhomogeneities undoubtably are present but fully self consistent models have not yet been calculated.

Theoretical problems in which magnetic fields are involved in CP stars include the following.

Origin of field.

Internal structure - relation between observed (surface) and interior fields.

Stability of field structure.

Explanation of the incidence of stellar magnetism.

Explanation of the χ distribution.

Evolutionary changes - secular effects.

Cause of the low angular velocities.

Cause of the chemical anomalies.

Rapid low amplitude oscillations.

This list is not exhaustive, nor are the topics independent. All but the last three will be touched upon under the main headings of 'Origin' and 'Structure'.

2. ORIGIN OF THE FIELD

Four types of theory have been advanced to explain the fields of the magnetic CP stars: the magnetic oscillator, battery, fossil and dynamo theories. Currently the first two have little favour (although a variant of a battery mechanism might provide an initial 'seed' field, or influence the toroidal field structure near the stellar surface, eg Dolginov, 1977; Mestel & Moss, 1983) and will not be discussed further here (but see, eg, Dolginov, 1984). The only satisfactory way to distinguish between the fossil and the dynamo theories is to develop the theories in detail and try to arrive at testable and distinct sets of predictions.

2.1 Contemporary dynamo theory

This proposes that the observed fields are the surface manifestations of a 'turbulent' dynamo operating in the convective core. Turbulent dynamo theory has been developed in great detail in the last twenty years, following the pioneering work of Krause et al collected in Roberts and Stix (1971). Briefly, dynamo action is possible where the net helicity of the fluid motions is non zero ($\langle \underline{u} \,.\, \nabla\times\, \underline{u}\rangle \neq 0$) and this condition is likely to be satisfied in rotating convection zones. In this case the normal MHD induction equation is supplemented by the "α-effect" term to give

$$\frac{\partial B}{\partial t} = \nabla \times (\underline{u}\times\underline{B} + \alpha\,\underline{B}) - \nabla \times (\eta_T \nabla\times \underline{B}),$$

where η_T is the turbulent resistivity. The additional source term invalidates the 'anti-dynamo' theorems and allows new poloidal field to be generated directly from the toroidal field. (Differential rotation and the α-effect are each capable of generating toroidal field from poloidal). There is a plethora of dynamo models, mostly however linear and kinematic, or limited to a parameterized non-linearity to represent the dynamical feed back of the field on the fluid motions. It is widely (but not universally) accepted that the solar field is generated by an unsteady dynamo, and it is tempting to try to explain the fields of the magnetic CP stars in a similar manner. There are, however, substantial differences between these cases. Any dynamo operating in the CP stars must be steady - both to agree with observations and, perhaps more stringently, to avoid being masked by the skin effect of the overlaying radiative zone. Some very young stars (age maybe of order 5×10^6 years) show kilogauss surface fields. If the dynamo turns on when nuclear reactions become important enough to set the core convecting then the field has to reach the surface within about 5×10^6 years. Classical diffusion times through the envelope are in excess of 10^9 years and even the accelerated 'forcing' discussed by Schüßler and Pähler (1978) only somewhat reduces this time. Many solar dynamo theorists want to have the dynamo just beneath the convection zone in order to prevent buoyancy in the convective region from removing flux too rapidly. There is a suggestion that at spectral type about M5 on the main sequence, where the complete star may become convective so that there is no 'bottom' to the convection zone, the dynamo efficiency declines rapidly. This might imply that a core dynamo is less efficient than predicted by simple turbulent dynamo theory.

If a core dynamo can operate satisfactorily in these stars then a crucial question is the relation between the surface and core fields. Presumably in a star of given structure the core is a function mainly of the angular velocity Ω, at least when a steady state has been reached. Non linear limiting effects must be crucial in determining the final field strength - eg. by modifying the small scale motions and so reducing the effective value of α, or by generating large scale meridional flows, or reducing differential rotation. Crudely it might be imagined that larger angular velocities would give rise to larger fields and this is certainly the usual assumption when modelling lower main sequence dynamos. It is conceivable that non-linear processes seen through the filter of a radiative envelope in which the hydrostatic and energy equations have to be satisfied might give a different result for the observable fields. The only detailed models so far calculated (Moss, 1983) lend some support to this conjecture, but the calculations were only for one, perhaps rather special case. More crudely the idea that for larger values of the angular velocity only oscillatory modes may be available, and so the skin effect will render the fields negligible at the surface, was discussed by Moss (1980).

To agree with the observations the theory must be capable of explaining why stars of the same angular velocity can have very different fields. Rädler (preprint) points out that differential rotation and a field component with symmetry axis perpendicular to the rotation axis cannot persist. Field lines of opposite sense will inevitably be brought close together, and ohmic dissipation of this perpendicular component will be enhanced. If $0 < \chi < 90°$ the reduction of the perpendicular component will reduce the observed value of χ. The Potsdam group have developed in detail the 'perpendicular rotator' dynamo model. They argue that dynamo theory demands that the field axes must be strictly parallel or perpendicular to the rotation axis (ie χ=0 or 90°). As discussed in section 1, they model the field by a sum of dipole and quadrupole components with axis perpendicular to the angular velocity, together with surface element inhomogeneities. Krause (1983) advances the 'fossil rotator' theory in the context of core dynamo theory with strictly perpendicular field and rotation axes. According to this theory, if a star initially has a strong differential rotation, the shear winds up and buries the field beneath the surface, whereas if the initial radial variation of angular velocity is small, or the field is strong enough to remove an initially strong angular velocity shear, the field continues to penetrate the surface and so to be visible. Such mechanisms would introduce some scatter into the relation between surface field and period in the dynamo model. Until calculations can be performed which calculate the inhomogeneities and the observed field in a self consistent manner it is difficult to assess the validity of this model. In general, in contrast to the fossil field model (below), the dynamo model still lacks detailed non-linear calculations which would enable a proper comparison to be made with the observational material.

2.2 Fossil theory

Given the long decay timescales (of order $\sim 10^{10}$ years) associated with large scale fields in a mainly radiative star it is possible that any field initially present when the star first settles on the main sequence could survive throughout the main sequence lifetime. In the simplest form of the theory, the initial field is a relic of the interstellar field which permeated the material from which the star was formed. The first question to be answered is whether such a large scale field can survive the global turbulence of the Hayashi phase (if these stars experience such a phase), or whether it will be mangled and expelled from the star. This is a difficult problem, to which there is no clearcut answer. Numerical calculations of the dynamical interaction between convection and magnetic fields (eg. Galloway, Proctor & Weiss 1978; Galloway & Weiss, 1981; Galloway & Proctor, 1983) do suggest the possibility that a significant amount of flux could survive this phase, having been concentrated into ropes by the turbulence, and so allowing convection to proceed freely in the more or less field free regions between the ropes. Current observations of the solar surface field do seem to indicate an intermittent structure.

After the convection dies away the field is assumed to diffuse back into a more uniform configuration. A variant of this picture appeals to a dynamo operating in the Hayashi phase to produce a large scale field. When the convection dies away the dynamo ceases to operate, but the relic field again survives over nuclear timescales. The consequences of this 'hybrid' theory are largely indistinguishable from the standard fossil theory, unless dynamo theory can make some definite statements about the initial distribution of the angle χ. For a fossil field to survive over nuclear timescales it must be able to resist a variety of instabilities which tend to reduce the total flux. It is worth emphasizing that, even for the full fossil theory, a certain amount of flux loss sometime in the history of a magnetic star is desirable, otherwise the high magnetic Reynolds number from the later stages of dynamical contraction onwards would ensure a significant flux for all stars. Field freezing may only be important after the 'molecular' phase of star formation - which in the absence of further flux loss would naively result in mean field strengths ($\bar{B}$) through the star of 10^4 to 10^5 gauss on the main sequence. This might be consistent with the observably magnetic stars, and it could be argued (see later) that rapid rotators have small ratios of surface to internal field and so are not seen as magnetic stars. However the slowly rotating HgMn stars do not have observable fields which suggests that further flux loss has occurred in some stars. Possible mechanisms of flux loss include the topological instabilities first investigated in detail by Wright (1973) and Markey and Tayler (1973). They noted that purely poloidal fields near a neutral line locally have the topology of the z-pinch which is known to be unstable, and detailed analysis confirms that purely poloidal (and also purely toroidal) fields are dynamically unstable. If the instability grows into the nonlinear region enhanced ohmic flux destruction can be expected. Linked poloidal and toroidal fields of comparable strength can be expected to be less vulnerable to such effects, although it is difficult to guarantee stability, as has been demonstrated in a series of papers by Goossens, Tayler and their collaborators (eg. Tayler (1982), Goossens et al (1981)). Even if dynamical stability is assured, in a stable subadiabatic radiative zone isolated flux tubes tend to rise to the surface at a rate determined by the diffusion of heat into the tube (Parker 1979, Acheson, 1979). This is a secular effect, timescales may well be in excess of 10^7 or 10^8 years, and its relevance to a complex geometry of continuously distributed interlinked poloidal and toroidal fields is hard to assess. Molecular weight gradients could plausibly aid stabilization. There is no evidence for the ratio $\epsilon \sim \frac{\bar{B}^2 R^4}{G M^2}$ of magnetic to gravitational energy ever being anything but a very small number. Even with the largest known surface field of $\sim 3 \times 10^4$g (HD 21544) and associated $\bar{B} \sim 10^6$g, $\epsilon \sim 10^{-5}$. It may well be that structures with larger ϵ are inevitably unstable, and that instabilities induce flux loss until a stable configuration with $\epsilon << 1$ is attained.

3. STRUCTURE OF THE FIELD

With the above considerations in mind we can discuss models for the large scale structure of the field throughout the star. Reflecting the detailed work done the discussion will be almost entirely concerned with models consistent with the fossil field theory. Early attempts at modelling, consistent with both the hydrostatic and energy equations, were "quasi-static" in that those which explicitly included meridional motions and the magnetohydrodynamic equation wrote the latter in a way equivalent to introducing a small source term to offset the overall decay of the field predicted by the simpler forms of the anti-dynamo theorems (eg. Cowling 1934). See Mestel & Moss (1977) and references therein. Models can be found for a variety of field geometries (dipole, even and odd multipole) and topologies (poloidal, linked poloidal and toroidal) some of which plausibly are stable against the more obvious dynamical instabilities. All these models predict an increasing ratio of internal to surface field for a given total flux as the angular velocity increases and for plausible parameters the field is strongly concentrated to the interior. At the time this seemed quite satisfactory, and was certainly in accord with what then seemed to be quite a strong anti-correlation between B_e and Ω. The thrust of the observational evidence then shifted somewhat, weakening (but not destroying) this anti-correlation, and also to suggest that the distribution of the angle χ was more nearly random. This was in contrast to what had appeared to be an approximately bimodal distribution, with a weaker peak near $\chi = 0$ and a stronger one near $\chi = 90°$. On the theoretical front Mestel et al (1981) worked out the details of what they called internal "ξ-motions", resulting from the rotation of a compressible body which is not symmetric about its rotation axis - for example an oblique rotator. In order to conserve angular momentum the rotation axis must precess about the axis of symmetry of the field with angular velocity $\omega \sim \epsilon\,\Omega$. These motions change the pressure-density field, and equilibrium can only be maintained by dynamically driven "ξ-motions", which are also of frequency ω. These distort the field and dissipate energy, predominantly by ohmic dissipation, and the body eventually adopts the minimum energy configuration for its fixed angular momentum, and so rotates about its axis of greatest moment of inertia. Thus $\chi \to 0$ if the star is dynamically oblate about the magnetic axis, and $\chi \to 90°$ if it is dynamically prolate. Poloidal fields give oblate configurations, toroidal prolate; the mixed fields so far calculated all gave oblate configurations, but the existence of prolate configurations was by no means ruled out. More importantly, the time scale for alignment or anti-alignment, τ_χ, is sensitive to the interior field concentration. Whereas for modest ratios of maximum interior radial field to surface field (0(10) say) $\tau_\chi \ll \tau_{ms}$, for significantly greater ratios such as found in the theoretical models discussed above $\tau_\chi \ll \tau_{ms}$ unless the period is greater than about 100 days. This implies that we should rarely see values of χ other than 0 or 90°, in contradiction to the simplest interpretation of the observations according to the oblique rotator/fossil field model which give a nearly random χ distribution (but note that if surface inhomogeneities can reconcile the perpendicular rotator version of dynamo theory with observations, then they can do the same for a fossil field in the asymptotic state $\chi = 90°$). Note also that ξ - motion theory is also relevant to dynamo models, with the additional complication that if the structure were such as to make χ tend to 0 from 90°, the new flux continuously being generated would tend to restore the original value of χ. Other mechanisms which might change χ include magnetic stellar wind torques (Mestel and Selley, 1970), and the purely kinematic effect of the rotationally driven circulation on the surface flux distribution in an initially oblique rotator (Moss, 1977).

Recently it has become possible to calculate improved, time dependent, magnetic models and to follow the evolution of the field over nuclear timscales (but, so far, ignoring the effects of stellar evolution on the underlying model). The calculations show that the previous quasi-static models are rather special cases selected from a much wider range of possible models (Moss, 1984, 1985). The evolution of models with fairly arbitrary initial flux distributions now can be followed over nuclear timescales. For models with $\chi = 0$ and relatively slow rotation ($P > P_c \sim 4^d$ for plausible values of mass and luminosity) the initial flux distribution survives with little change, except for slight ohmic decay, over a nuclear time scale. In particular a field that has small initial concentration to the interior never becomes concentrated below the surface. Stars with shorter periods and fields with initially low interior concentration will bury their fields, but on a time scale governed by the time scale of the of Eddington-Sweet circulation - typically of order 10^8 year or slightly more. The critical period is such that the magnetic Reynolds number of the Eddington-Sweet circulation is of order unity through the bulk of the radiative zone.

In the more general case of the oblique rotator, with respect to a spherical polar coordinate system (r,θ,λ) with axis the magnetic axis, the radial component of the Eddington-Sweet circulation can be written

$$v_r = V_{ES}P_2(\cos\theta)P_2(\cos\chi) - \frac{1}{4}V_{ES}\,\mathrm{Sin}^2\chi\, P_2^2(\cos\theta)\cos 2\lambda + \frac{1}{2}V_{ES}\sin 2\chi\, P_2^1(\cos\theta)\sin\lambda, \qquad (1)$$

where $V_{ES}(r)$ is the radial component of the Eddington-Sweet circulation along the rotation axis. When $\chi = 90°$ the λ independent part is in the opposite sense with respect to the field axis than when $\chi = 0$. Simple kinematic considerations (supported by detailed calculations) suggest that in this case the field will never be buried, but that if $P < P_c$ then the surface field will tend to be concentrated towards the magnetic poles. Further, equation (1) suggests that there is a critical value of χ, $\chi_c \simeq 55°$, such that for $\chi < \chi_c$ these models will behave quantitatively like aligned rotators and so bury their surface field if P is small enough (allowing for the reduction of V_r ($\propto \Omega^2$) by the factor $P_2(\cos\chi)$); and will never bury the surface field for any P if $\chi > \chi_c$. The existence of a critical value, $\chi_c \simeq 55°$, is also predicted from study of the hydrostatic equation by Galea and Wood (preprint). These conclusions also apply to displaced dipole models (Moss 1985). The interaction of a velocity field of the form (1) with an initially axisymmetric magnetic field when $\chi \neq 0$ can be expected to generate non axisymmetric components, although this aspect has not yet been investigated.

It is only quite recently that changes with time of the stellar fields have attracted serious study. Borra (1981) and North and Cramer (1984) find evidence for a decline in field strength on timescales of order 10^8 years. The calculations just discussed follow the evolution of the field under the influence of ohmic diffusion and the magneto-centrifugal circulation. A number of other effects can be mentioned. A linked poloidal - toroidal field topology seems to be essential for stability. If the field is a fossil, decay inevitably occurs. In the simplest axisymmetric case the torque free condition will hold approximately which implies $B_t = F(\psi)/(r\sin\theta)$, where the poloidal field $\underline{B}_p = \nabla \times (\psi/r \sin\theta \,\hat{\underline{t}})$, $\hat{\underline{t}}$ the unit toroidal vector. In general $\underline{B}_t$ and $\underline{B}_p$ will decay at different rates. Torques will arise, generating differential rotation, until a new torque free configuration is found. In reality continuing adjustment will occur. Similar effects can be expected to occur in non axisymmetric configurations. If a core dynamo operates and the diffusion of the core field to the surface has not reached a steady state, then the surface field may increase for some time (eg. Schüßler & Pähler, 1978). Correspondingly, both for dynamo and fossil fields, braking (eg. by a magnetic stellar wind) on the main sequence will cause a continuing adjustment of the magnetic structure. Finally the central condensation of the star will increase with time. This effect has been investigated in detail by Moss (1983), for initial field configurations corresponding to the quasi-static fields described above. Both these calculations, and the time dependent models of Moss (1984, 1985) suggest that in many cases surface fields will decline on roughly an evolutionary timescale, which is consistent with the very limited observational material.

4. CONCLUSIONS

The central issue remains the origin of the magnetic fields. It is not yet possible to give an unambiguous answer to this question. Fossil theory models have been worked out in considerable detail, which has not been matched by the dynamo models. Contemporary dynamo theory does appear to have problems in explaining the detailed distribution of field strength with period, in explaining the appearance of strong fields at the surface of very young stars, and the incidence of fields (eg. why don't the slowly rotating HgMn stars display fields?). The fossil theory has an extra degree of freedom in that the initial fluxes are fairly arbitrary, and it does now appear possible to begin to put together a coherent picture according to this theory. Suppose that stars lose nearly all their primeval flux before settling on the main sequence, and that these residual fluxes have, for example, a Maxwellian distribution. Assume that the initial fluxes are not strongly concentrated to the interior and that the high flux tail of this distribution contains the stars whose fields we see today. If $P > P_c$ then the initial configuration survives more or less unchanged, apart from some ohmic decay. If $P < P_c$ and the initial value of χ is less than χ_c then the field is buried, over a time scale roughly of order of the main sequence lifetime, depending on the period and initial χ value. If the initial $\chi > \chi_c$ the field is not buried, but may be concentrated towards the poles on a similar timescale. The observed value of χ may change because of kinematic effects of the predominantly rotationally driven circulation on the field, or because of wind torques. Magnetic braking may occur. This picture suggests that rapid rotators with strong fields should either be young or have large χ. This general scheme does not appear to be contradicted by the rather limited observational material, but clearly greater detail in both theory and observations is needed to provide a decisive test.

To finish on a note of caution: if the fossil field is a direct descendant of a large scale quasi-uniform primeval field which permeated the interstellar medium then, naively, the observed fields should be of odd parity - that is they should be represented approximately by the displaced dipole model (Moss, 1985). If the dynamo theory (either contemporary or hybrid) produces a combination of dipolar and quadrupolar models then, prima facie, it might be expected that we should sometimes see a predominantly quadrupolar field. Until recently the displaced dipole (ie. predominantly dipolar) model was adequate to approximate the field structure of all known stars. Thompson and Landstreet (1985) have recently modelled HD 37776 with a predominantly quadrupolar field. As yet this is an isolated case, and it is unclear how much weight should be given to it. Further evidence of this nature would be of great interest.

REFERENCES

Acheson, D.J., 1979. Solar Phys., 62, 23.

Borra, E.F., 1981. Astrophys. J. Lett., 249, L39.

Borra, E.F., & Landstreet, J.D., 1980. Astrophys. J. Suppl. 42, 421.

Cowling, T.G., 1934. Mon. Not. R. astr. Soc., 94, 39.

Didelon, P., 1984. Astron. Astrophys. Suppl. Ser., 55, 69.

Dolginov, A.Z., 1977. Astron. Astrophys., 54, 17.

Dolginov, A.Z., 1984. Astron. Astrophys., 136, 153.

Galloway, D.J., Proctor, M.R.E., & Weiss, N.O., 1978. J. Fluid Mech., 87, 243.

Galloway, D.J., & Proctor, M.R.E., 1983. In Planetary and Stellar Magnetism, ed. A.M. Soward; Gordon & Breach.

Galloway, D.J., & Weiss, N.O., 1981. Astrophys. J., 243, 945.

Goossens, M., Biront, D., & Tayler, R.J., 1981. Astrophys. Sp. Sci., 75, 521.

Hensberge, H., van Rensbergen, W., Goossens, M., & Deridder, G., 1979. Astron. Astrophys., 75, 83.

Krause, F., 1983. In Planetary and Stellar Magnetism, ed. A.M. Soward; Gordon & Breach.

Markey, P., & Tayler, R.J., 1973 Mon. Not. R. astr. Soc., 163, 77.

Mestel, L., & Selley, C.S., 1970. Mon. Not. R. astr. Soc., 149, 197.

Mestel, L., & Moss, D.L., 1977. Mon. Not. R. astr. Soc., 178, 27.

Mestel, L., & Moss, D.L., 1983. Mon. Not. R. astr. Soc., 204, 557.

Mestel, L., Nittmann, J., Wood, W.P., & Wright, G.A.E., 1981. Mon. Not. R. astr. Soc., 195, 979.
Moss, D.L., 1977. Mon. Not. R. astr. Soc., 178, 61.
Moss, D., 1980. Astron. Astrophys., 91, 319.
Moss, D., 1982. Mon. Not. R. astr. Soc., 201, 385.
Moss, D., 1983. Mon. Not. R. astr. Soc., 202, 1059.
Moss, D., 1984. Mon. Not. R. astr. Soc., 209, 607.
Moss, D., 1985. Mon. Not. R. astr. Soc., 213, 575.
North, P., & Cramer, N., 1984. Astron. Astrophys. Suppl. Ser., 58, 387.
Oetken, L., 1977. Astron. Nachr., 298, 197.
Oetken, L., 1979. Astron. Nachr., 300, 1.
Parker, E.N., 1979. Cosmical Magnetic Fields, Oxford, Oxford Univ. Press.
Roberts, P.H., & Stix, M., 1971. The Turbulent Dynamo, NCAR, Boulder, Co.
Schüßler, M., & Pähler, A., 1978. Astron. Astrophys., 68, 57.
Stift, M, 1980. Astron. Astrophys., 82, 142.
Tayler, R.J., 1982, Mon. Not. R. astr. Soc., 198, 811.
Thompson, I., & Landstreet, J.D., 1985. To appear in Astrophys. J.
Wolff, S.C., 1975. Astrophys. J., 202, 127.
Wright, G.A.E., 1973. Mon. Not. R. astr. Soc., 162, 339.

ORIGIN OF THE MAGNETIC FIELD

A. Z. Dolginov
A. F. Ioffe Physical-Technical Institute
Leningrad, USSR

Some progress has been achieved in recent years in the theory of the stellar magnetic field generation but a lot of questions remain unanswered. I would like to give here a brief critical review of the current state of the problem.

1. The Fossil Field Theory

The simplest is the suggestion that the stellar field is of a primordial nature. It needs, however, some special assumptions: 1) the conductivity of the stellar matter must be high during all period of the star formation. In the course of its evolution the star, apparently, passes the phase of intensive turbulent motions. The turbulent conductivity is low and the fossil field decays; 2) the field configuration must be stable against perturbations, leading to the decrease of the field scales. The stable configuration of the field must have both poloidal B_p (dipole, quadrupole, etc) and toroidal B_t components. The toroidal field B_t is maintained by the poloidal electric current j_p. If j_p has a component not parallel to B_p, then torques arise which produce motions suppressing this component, so that only $j_p = kB_p$ survives. The poloidal current, following magnetic lines of the poloidal field, cross outer stellar layers which have lower conductivity, provided the star does not possess a hot corona. In this case j_p and consequently B_t decay in a short time. The component B_p alone, without B_t, becomes unstable and all the field configuration decays (L. Mestel, D. Moss 1984[a]).

C. R. Cowley et al (eds.), Upper Main Sequence Stars with Anomalous Abundances, 11–21.

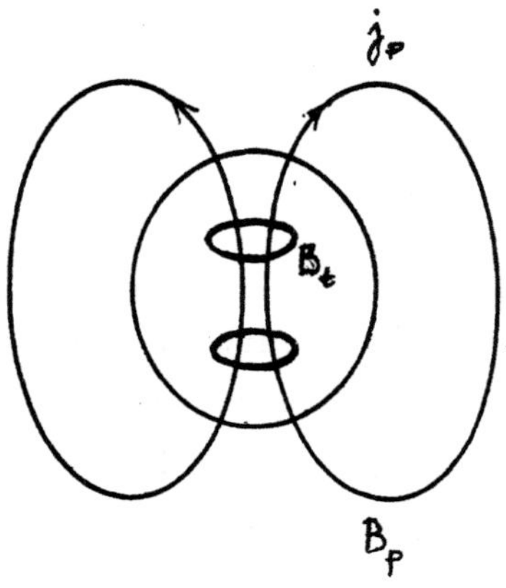

Fig. 1

In this case some external sources are needed to maintain the field. Hence, the fossil field theory in its classical form is rather inefficient.

2. The Dynamo Theory

The essence of the dynamo action in a conductive medium may be explained in a few words as follows. Consider a closed electric current loop which is frozen into plasma and, thus, follows all plasma motions. The current circuit may be expanded by motions and then twisted (e.g. by coriolis forces) into the geometrical form which looks like a number eight. After this the motions may put both parts of the twisted circuit together, restoring the initial single loop configuration, but with redoubled current. Motions of such kind are common in covective zones of rotating stars. The above process may be repeated many times, leading to the 2^n-fold amplification of the current and the field after the n-th cycle. The field configuration may be stable because both B_p and B_t are present. The convection developed in the star together with fast rotation create obviously favourable conditions for the dynamo to operate. However, Ap stars have either no outer convective zone or a very thin one, and possess, as a rule, slow rotation.

Appropriate conditions for the dynamo may be realized in the convective inner core (F. Krause, R. H. Rädler 1980). If the field is actually generated in the core, it must be transported upward, manifesting itself as the observed field. The time necessary for the field to penetrate through high conductive stellar layers to the surface exceeds the age of the star. However, the field may readily be lifted upward together with the hot plasma by buoyancy forces. Unfortunately this attractive possibility has not been considered quantitatively, and no evidence has been obtained that in Ap stars there exist some mechanism of the heat transport aside of that provided by the radiation. The field decreases toward the surface roughly proportional to some power of the density ($\sim\rho^{2/3}$). This means that the initial field near the core should be about several millions of gs. All up-to-date versions of the dynamo-theory make use of the linear approximation which does not permit calculations of the field value, but only the field growth rate at the very initial stage.

Thus, the dynamo origin of Ap star fields cannot be excluded but the current theory is far from being entirely conclusive.

3. The Biermann Effect

Magnetic fields in rotating bodies may also be generated by so called "Biermann effect" (L. Biermann 1955). This field is produced by the vortex electric field $E_B = \nabla P_e / eN_e$, P_e - being the electron pressure and N_e the electron number density. The Ohm law for a plasma may be written as

$$\vec{j} = \sigma \left[\vec{E} + \frac{\vec{v}}{c} \times \vec{B} - \frac{\vec{j} \times \vec{B}}{ecN_e} - \frac{\nabla P_e}{eN_e} \right] \quad (1)$$

where all the symbols have their standard meanings. For the fully ionized plasma $\nabla P_e/N_e$ is proportional to $\nabla P/\rho$ where P is the total pressure and ρ the plasma density.

From hydrostatic equilibrium, one has $\nabla P/\rho = \vec{g} + \vec{\Omega} \times (\vec{\Omega} \times \vec{r})$, where $\vec{g}$ is the gravity and $\vec{\Omega} \times (\vec{\Omega} \times \vec{r})$ is the centrifugal acceleration. If the rotation is differential, i.e. the angular velocity $\vec{\Omega}$ depends on coordinates, then $\nabla \times (\nabla P/\rho)$ 0. This means that the field E_B has vortex component and cannot be compensated by the electric field $E = -\nabla\phi$ produced by plasma polarization. The persistent electric current $j_B = -\sigma \nabla P_e/eN_e$ creates the magnetic field which may be large. However, the time of the field growth to the observed values exceeds, as a rule, the age of the star. That is why this mechanism has not be regarded as effective.

4. The Effect of Chemical Inhomogeneity

The following new mechanism of the field generation has been proposed by Dolginov (1977). If the mean molecular weight on a star is not distributed spherically symmetrically, then, in general, the partial pressure gradient per electron is not curl-free. In this case the equation which govern the magnetic field B has the form

$$\frac{\partial \vec{B}}{\partial t} = \nabla \times \left[\vec{V} \times \vec{B}\right] - \nabla \times \left[\frac{c^2}{4\pi\sigma} \nabla \times \vec{B}\right] - \nabla \times \frac{\vec{j} \times \vec{B}}{eN_e} - \frac{c}{e} \frac{\nabla N_e \times \nabla P_e}{N_e^2} \tag{2}$$

The last term on the right-hand side of eq. (2) describes the effect of molecular weight inhomogeneity. For Ap stars, the condition $\nabla N_e \nparallel \nabla P_e$ holds because of the presence of abundance patches on stellar surfaces and, first of all, of helium abundance patches. An axially symmetrical distribution of the patches leads only to the toroidal field generation. The poloidal field may be created in the following cases: (a) the helium distribution is not axially symmetric; (b) the surface temperature gradient ∇T has a tangential component nonparallel to that of ∇N; (c) a poloidal circulation of the stellar matter (e.g., meridional circulation) is present.

Calculations performed by Dolginov (1977), also by Dolginov and Urpin (1979) on the basis of observational data on chemical abundance patches of Ap stars yield the fields 10^3-10^4 gs, in agreement with observations. The time required to reach these values is about 10^5-10^6 years. Hence, the proposed mechanism is effective, if the lifetime of helium abundance patches exceeds 10^5-10^6 years. More than 10 year-long observations of Ap stars show no noticeable change in the surface abundance distribution. This means that the lifetime of patches exceeds 10^2-10^3 years. However, there exist no observational data concerning time scales $\sim 10^5$-10^6 years. This problem is closely connected with the problem to understand the nature of chemical anomalies.

The most elaborated model of abundance patches is based on the assumption that the patches are created by some diffusion process in the magnetic field (G. Megessier IAU Coll. 90, G. Alecian IAU Coll. 90). However, various instabilities which may lead to fast escape of the plasma from the magnetic trap have not been analysed and the lifetime of heavy elements in the traps has not been determined.

Dolginov (IAU Coll. 90) pointed out that the abundance patches may be created due to the thermal instability. Let us assume the existence of some initial local excess of atoms which can be easily excited by electron impacts. Then the blanketing effect of spectral lines may lead to the local cooling of the matter near the stellar surface. It may result in the local increase of the number density of the atoms which cause the cooling, i.e. in the abundance patches creation. This instability provides an example of a selforganizing process. The lifetime of the patches in this case may be rather long.

5. Combined Effect of Fossil Field and Chemical Inhomogeneity

L. Mestel and D. Moss (1983[b]) have made an attempt to use the effect of chemical inhomogeneity as a source for maintaining the fossil field. They have considered two possibilities: (a) The redistribution of mean molecular weight may be due to the meridional circulations which exist in rotating stars. In this case the redistribution is axially symmetric and may be responsible only for a toroidal field. If the fossil field is poloidal, then the generated toroidal component may provide the stability of the total field configuration. However, this effect is able to maintain observed fields only in rapidly rotating stars which possess hot coronae. Ap stars do not fit to these conditions: (b) As has been shown in previous section, helium abundance patches may be responsible for total Ap star fields. If the poloidal component is of a fossil nature, then the helium patches may be sufficienty effective to maintain the required toroidal component. This possibility has been considered in detail by L. Mestel and D. Moss (1983[b]). It has been shown that the effect of chemical inhomogeneities is important in the theory of Ap star fields without its having necessarily to be the explanation of the total surface field.

6. Role of Chemical Inhomogeneities in Various Cosmic Objects

Chemical inhomogeneities may be important not only for

such peculiar objects as Ap stars. They may be efficient also in the stars with developed convective zones. Near the boundary between the convective and radiative zones convective motions are slow, and the heat transport mechanism changes very sharply. Such conditions are favourable for the development of some instability which leads to chemical separation and produces, in this way, the field generation (Dolginov 1984).

It is highly likely that large scale chemical inhomogeneities at the core-mantle boundary inside the Earth play an important role for the terrestrial magnetic field generation. As has been shown by Dolginov (1984) even a 10% large scale tangential variation of the electron number density at the core-mantle boundary caused by chemical inhomogeneity is sufficient to generate the terrestrial field of the observed strength. Any theory of the Earth's field should take into account even a 1% large-scale variation of N_e.

7. Magnetic Field Generation in Binary Stellar Systems

(A) Tidal velocities in close binary stars whose rotational axes are non-parallel have been calculated by Dolginov (1974) and also by Dolginov and Yakovlev (1975). The tidal flow has a complicated geometry which is favourable for the dynamo action. In sufficiently close systems the time for the field growth by the dynamo mechanism driven by tidal forces is much shorter than the age of the star.

Tidal flows in close binary systems are capable of mixing the matter in surface layers and destroying chemical abundance patches on peculiar stars. That is why Ap stars are absent in close systems. Fast rotation may generate circulations which also can mix the matter. The absence of intensive mixing near the surface is apparently a necessary condition which enable the star to belong at Ap type.

(B) The inductive mechanism of field generation in close binary systems has been proposed by Dolginov (1973) and considered in detail by Dolginov and Urpin (1979). If one of the stars has some small initial field, then the stellar rotation and orbital motion induce electric currents in the counterpart. These currents produce a magnetic field which, in its turn, generates the current and additional field in the first component. This effect leads to the initial field amplification. The increment is proportional to the angular velocities of the orbital

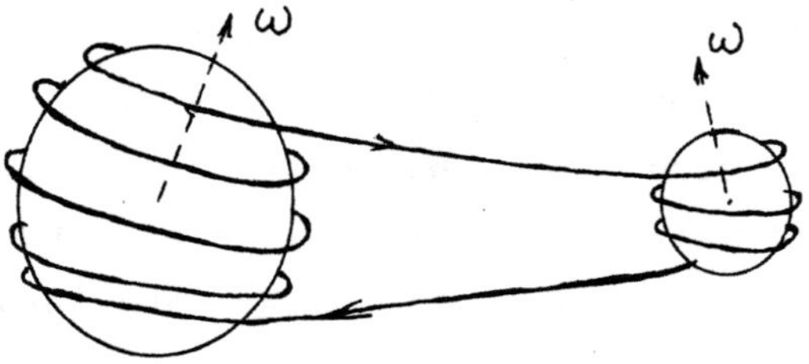

Figure 2

and axial rotation and inversely proportional to the fourth power of distance between the companions. If, for example, one component is a red dwarf with radius $\sim 10^{10}$ cm, the other one is a white dwarf with radius $\sim 10^{8}$ cm, the component sepration is 5.10^{10} cm, angular rotational velocities are equal to $\omega \sim 10^{-3} s^{-1}$, $\Omega \sim 3.10^{-4}\ s^{-1}$, and inclination of rotational axes is about 45°, then the time of field growth is about 10^{5} years. The field amplification is suppressed by nonlinear effects. It is quite possible that the observed strong fields $\sim 10^{5}$-10^{7} gs in some close binaries which contain two dwarfs are produced by the inductived mechanism.

(C) Let us consider the close binary system with the matter outflow from one star to the other. Let the first star possess the field, generated, for instance by the dynamo mechanism. Then, the accreting star may acquire the field together with the accumulated plasma (Drobyshevsky 1976). In this case the scale of such field seems to be small. The application of this mechanism to Ap stars is thought to be doubtful because there are no indications that Ap stars enter close binary systems in the course of their evolution.

8. The Thermomagnetic Instability

In the objects, where heat and electric current are provided by electrons, the field may be generated by the thermomagnetic instability (Dolginov and Urpin 1979, 1980, Blandford et al. 1983; Urpin et al. 1986). The instability operates as follows. Consider a thermodynamically open system with a temperature gradient ∇T_o (which is mainstained, for example, by nuclear reactions in the centre of the star). Any small initial magnetic field $\vec{B}_o$ induces the heat flow $\vec{q}_o$ perpendicular to both ∇T_o and $\vec{B}_o$. The flow $\vec{q}_o$ creates the thermoelectric field E_{th} and corresponding electric current. The current, in its turn, generates magnetic field $\vec{B}_1$, which amplifies the intial field $\vec{B}_o$. This instability has been investigated in laboratory. For example, a small grain ($\sim$0.01 cm) irradiated by an intensive laser beam acquires a field 10^6-10^7 gs in $\sim 10^{-9}$ s. It has been shown in the above mentioned papers that the thermomagnetic instability may very effectively generate magnetic fields in white dwarfs, neutron stars and hot stellar coronae. The considered instability does not operate in regions with radiative or convective heat transport.

Conclusion

We have made an attempt to show that large variations in physical conditions in cosmic objects may lead to quite different ways of the magnetic field generation. In some cases the fossil field may be most important, whereas in other cases the field may be generated by chemical inhomogeneities, by the dynamo-process, etc. One may hope that unambigous choices of the field generation mechanisms for certain objects is possible in the near future.

As for the case of the Ap stars, the following questions are most urgent: (a) What is the fine structure of the magnetic field? (b) What is the nature of an association between the magnetic field and chemical anomalies? (c) To what depths do the chemical anomalies penetrate and what is their lifetimes and space structure? (d) What is the relative role of diffusion and hydrodynamical motions in the process of chemical separation? (e) What is the role of various instabilities?

Rererences

1. Biermann, L., 1950, Zs. für Naturfosch., 5[a], 65.
2. Blandford, R. D., Applegate, J. H., Hernquist, L., 1983, Mon. Not. Roy. Astr. Soc., 204, 1025.
3. Dolginov, A. Z., 1977, Astron. Astroph., 54, 17; cf.IAU Coll. 32.
4. Dolginov, A. Z., Urpin, V. A., 1978, Sov. Astron. Journal, 55, 1015.
5. Dolginov, A. Z., 1984, Astron. Astrophy., 135, 61.
6. Dolginov, A. Z., 1974, Sov. Astron. Journal, 51, 388.
7. Dolginov, A. Z., Yakovlev, D. G., 1975, Astr. Space Sci., 36, 31.
8. Dolginov, A. Z., 1973, Pisma JTEP (Lett. JTEP), 18, 67.
9. Dolginov, A. Z., Urpin, V. A., 1979, Astron. Astroph., 79, 60.
10. Dolginov, A. Z., Urpin, V. A., 1979, Journ. Exp. Theor. Phys. (in Russian), 77, 1921.
11. Dolginov, A. Z., Urpin, V. A., 1980, Astr. Space Sci., 69, 259.
12. Drobyshevsky, E. M., 1973, Astrofisika (in Russian), 9, 119; Astr. Space Sci., 35, 403.
13. Krause, F., Rädler K.-H., "Mean-Field Magnetodynamics and Dynamo Theory" Acad. Verlag Berlin 1980; Krause, F., 1983, The Fluid Mech. Astr. Geoph., 2, 205.
14. Mestel, L., Moss, D., 1983[a], Mon. Not. Roy. Astr. Soc., 204, 575.
15. Mestel, L., Moss, D., 1983[b], Mon. Not. Roy. Astr. Soc., 204, 557.
16. Urpin, V. A., Yakovlev, D. G., Levshakow, S. A., 1986, Mon. Not. Roy Astr. Soc. in press.

DISCUSSION (Dolginov)

STĘPIEŃ: My question concerns the problem of the energy source for poloidal currents. If we assume a stationary situation, in the sense that the magnetic field will only decay very slowly, then we need a constant source of energy to support the poloidal currents. But, suppose that the poloidal currents decay faster than the internal currents, leading to an unstable field configuration. Then a kind of kink instability develops and a new stable configuration of the field is reached spontaneously with the poloidal currents having energy drawn from the field itself. Such an instability may occur many times during the star's lifetime. Is this picture possible?

DOLGINOV: For the fast rotating stars the rotation itself is sufficient to maintain the poloidal current. For the slow rotators the 'battery' current, produced by the helium spots, may be sufficient to compensate the energy losses of the stellar current system. As to the origin of the spotted surface, I have not discussed this today, but some ideas about this problem are given in a poster paper which I am presenting at this colloquium. Briefly, it is due to an instability not connected directly with the magnetic field.

ALECIAN: Is it possible that meridional circulation or mass loss bring toward the surface magnetic fields deep in the interior, produced by dynamo effects in the core?

DOLGINOV: No. Meridional circulation can not bring the large-scale dipole magnetic field toward the surface, but it is possible for the small-scale field. The circulation may be strongly suppressed by the effect of the molecular weight redistribution.

KRAUSE: It is true that a magnetic field which is excited in the convective core of a star would take far too long to come to the surface through the radiative zone, if field diffusion is the only thing we take into account. Instabilities, however, provide for a much quicker transport of the field to the surface. The most important one is the buoyancy instability. The density inside a magnetic tube is less than that in the surrounding material, because of the balance of the magnetic pressure. For a certain critical field strength the magnetic tubes become buoyant and rise to the surface. Sunspots are an excellent example. So, the magnetic field, combined with gravity, produces itself the motion which brings it very quickly to the surface. It is a possibility.

DOLGINOV: Yes, there exists a possibility that the magnetic field may be lifted by buoyancy from the inner core, but the magnetic field rising from the core toward the surface must be accompanied by the transportation of hot matter because the field is frozen in the plasma. The rate of transportation must be sufficiently high because of the field dissipation on the surface where the conductivity is low. Existing models of A stars do not include the possibility of such heat transfer. As far as I am aware, the theory of the lifting of the magnetic field from the inner core has not been elaborated quantitatively.

There is a second problem for the dynamo theory. Why do not all A stars, with convective cores and rotating faster than Ap stars, have

C. R. Cowley et al (eds.), Upper Main Sequence Stars with Anomalous Abundances, 23–24.

magnetic fields?

DROBYSHEVSKI: I would like to present a somewhat different approach to the origin of magnetic fields observed on CP stars, which exploits the semi-dynamo concept, i.e., MHD magnetic field generation process without self-excitation (see E. M. Drobyshevski, E. N. Kolesnikova & V. S. Yuferev, Geophys. Astrophys. Fluid Dyn., 23, p. 103, 1983). It is rather difficult to believe that no stars have relic fields. The most probable candidates to have such a field are He-stars and Si Ap stars. They are found in the youngest clusters.

The origin and many inherent peculiarities of the magnetic field in other CP stars may be understood and explained if one assumes a close binary nature and magnetic field generation in situ. Indeed, the duplicity of Am stars is well documented. The magnetic field may be generated by intensive convection concentrated in a very thin outer envelope of the A-star with $M_* \approx 1.5\text{-}2.0\ M_\odot$. If the secondary component is of solar type, say, then due to the continuous accretion of (nonmagnetized) stellar wind, the magnetic field generated in the outer layers will be carried into the interior of the A star and accumulated there. Thus, a first-generation magnetic star forms. It has a weak magnetic field (of order 10 - 100 gauss: see E. M. Drobyshevski & E. V. Ergma, Astron. Zh., 53, p. 1338, 1976).

The second-generation magnetic star, an intermediate or late type Ap star, is formed when the first-generation magnetic star becomes a red giant and overflows its critical Roche lobe. Then its magnetic field is transported, together with the overflowing matter, on to the secondary component and is strongly amplified by winding on the secondary. Eventually the secondary becomes a magnetic star with a highly disordered but strong magnetic field, and the original primary becomes a distant and virtually undetectable white dwarf. As an example of such a second-generation magnetic star one might consider α^2 CVn. The causes of chemical peculiarities may also be understood from the same model of a binary origin, and will be discussed in my paper tomorrow.

DOLGINOV: This seems qualitatively possible, but there are definitely single Ap stars, which can not be explained with such a picture. There are no known Ap stars in close binaries with mass exchange. The Am stars are binaries, but again, mass exchange is not observed. In any case it is not easy to get global large-scale dipole fields in the accretion process.

DWORETSKY: The observational detection of small-amplitude, long-period radial velocity variations is very difficult, especially when the star in question has intrinsic variations due to its spottedness. Drobyshevski is referring to systems in which mass exchange has already taken place, and usually the result of this is that the two components end up very much further apart than they began. Drobyshevski's model is not inconsistent with the radial velocity observations.

ON THE SPOTTEDNESS, MAGNETISM AND INTERNAL STRUCTURE OF STARS

R.E.Gershberg
Crimean Astrophysical Observatory
Crimea, Nauchny, 334413
U.S.S.R.

ABSTRACT. It is hypothesized that kinematical structures within stellar interiors that are results of a self-organization of these interiors as thermodynamically open nonlinear systems constitute the physical basis for stellar magnetism.

I. INTRODUCTION

Astrophysical observations give more and more evidences for heterogeneity of stellar surfaces. Using the obvious case of the Sun, one speaks about a spottedness of stars and connects it with stellar magnetism. As it is known, the contemporary theory of stellar magnetism is constructed on the principles of the magnetohydrodynamics whose basis has been formulated in the end of the fourties. However, in spite of strong efforts during decades, results of this theory are still rather modest. This fact becomes especially obvious if one compares these results and successes of the stellar evolution theory that was born also in the end of the fourties. Based on two simple hypotheses - stars are spherical gaseous bodies in a hydrostatical equilibrium and energy losses for radiation from most stars are compensated by thermonuclear reactions that are maintained with necessity within such bodies - the stellar evolution theory has permitted to order the vast volume of experimental data on physical features, chemical abundances, kinematics, spatial distributions and main non-stationarity events for such a diverse realm of stars, while the stellar magnetism theory is now a set of 'ad hoc' models for various manifestations of stellar magnetism. Such different intermediate finishes of two astrophysical disciplines - the both being very actively supported by the 'terrestrial' physics - have a simple explanation: magnetism of a star is determined completely by its internal motions which are known very badly, while these motions, if they

C. R. Cowley et al (eds.), Upper Main Sequence Stars with Anomalous Abundances, 25–28.

are not too strong, can be not essential for global observable characteristics and evolution of the star.

The fragmentarity of the stellar magnetism theory is clearly seen not only if one looks at such different - in respect of structure, age and evolution status - objects as the Sun, Ap stars and pulsars; even if we restrict ourselves with the Sun, we find that models of sunspot magnetic structures, of isolated magnetic flux tubes and of background magnetic fields are being developed independently also. This fragmentarity stimulates a search for some more general principles that could be used to represent the whole variety of stellar magnetism manifestations. It seems natural to look for such general principles in properties of motions within stellar interiors.

2. HYPOTHESIS

According to the main conclusion of the open system thermodynamics, an open non-linear system with metastable states has a high probability to be in one of such a state. It is quite clear that an emitting star is an open system. A non-linearity of equations of a stellar internal structure takes place even in approximation of simplest static models, and this non-linearity increases strongly if one includes into consideration the effects of stellar rotation, convective energy transfer, magnetic fields and gravitational interaction in close binary systems. The statement on existence of metastable states in equilibrium gaseous bodies with internal power energy sources is a main point of my hypothesis. It seems very likely: if in laboratory - even kitchen! - experiments during a heating of a plane layer of a fluid the effect of a self-organization occurs - the Bernard cells appear - all the more such effects of the open system thermodynamics should be expected in stars where energy fluxes are much stronger, ranges of physical parameter values are much wider, systems are essentially three-dimentional and their spherical symmetry is broken by the cited factors that increase their non-linearity. Then, it is very unlikely that a complex kinematical structure within non-homogeneous plasma body does not stimulate magnetic effects. Thus, see the abstract above.

3. DISCUSSION

3.I. If the proposed hypothesis is correct, the magnetism should be inherent in all stars while magnetic fields have been found not in all stars. The matter is that a direct discovery of stellar magnetic fields is extremely hard task, and most studied manifestations of stellar magnetism are determined now mainly with an observational selection. However, many indirect indications on stellar magnetism that have been obtained for last years (Vaughan, 1983; Vaiana, 1983) show much more wide-spreadness of stellar magnetic fields than it follows from the direct magnetometry of F, G, K and M stars. Then, there are many magnetic Ap stars and, finally, magnetism of pulsars and white dwarfs evidences for the existence of magnetic fields in rather massive stars which are predecessors of these degenerated objects. Consequently, the ubiquitness of stellar magnetism that follows from the hypothesis proposed does not contradict to observations.

3.2. What is a relation between suggested 'thermodynamical' stellar magnetism and the dynamo and battery effect theories? One may expect that the proposed conception will provide physically based initial conditions to integrate equations of these theories and if one uses such conditions many problems of stellar magnetism theory will have to be reduced to estimations of deformations of permanently existing kinematical structures and respective magnetic features of stars.

3.3. As it is known, investigations of stellar internal structures have lead to conclusion that meridional circulation velocities are small and the main sequence star evolution occurs without a noticeable mixing of stellar interiors. However, proposed kinematical structures can give rise to some enrichment of central regions of stars and, in particular, of the Sun by hydrogen that will decrease an expected neutrino flux, since for the solar luminosity observed the higher the hydrogen abundance the less the temperature in its central region (Roxbugh, 1983).

3.4. The conception proposed may have direct relation to the problem of a radiation deficit from sunspots and starspots: as any metastable states, proposed kinematical structures can be some energy reservoirs, and changes in these structures can be connected with variations of an energy flux emergent from a stellar surface while its internal energy sources remain constant. While discussing the conclusion by Hartmann and Rosner (1979) that in the BY Dra case the flux deficit problem cannot be resolved with a redistribution of radiating energy over frequencies or over the stellar surface and therefore a real time variability of the stellar luminosity is most likely, I have suggested the necessity to use the open system thermodynamics' ideas to find out internal energy reservoirs within the star (Gershberg, 1983). Since in the ACRIM experiment (Willson et al., 1981) the correlation between sunspot visibility and the solar luminosity has been discovered, we need to have a variable reservoir of an internal energy for the Sun also. Finally, the qualitative model by Gershberg and Petrov (Gershberg, I982) and the quantitative model by Appenzeller and Dearborn (I984) have been proposed to explain significant brightness variations and other activity manifestations of the T Tau type stars; in the both models irregular in time and rather small in amplitude magnetic field strength variations have been supposed, and the cause of such variations of the magnetic field on a stellar surface can be some reconstructions of kinematical structures within the stellar interior.

3.5. In the solar research a more direct way to investigate the internal kinematical structures appears: it is the analysis of differential rotation rate variations with the phase of the solar cycle (Howard and LaBonte, I983).

In our days the astrophysics only begins to assimilate ideas of the open system thermodynamics. Laborious investigations of a self-organization of stellar interiors must be fulfilled. However, we may hope that such investigations will lead us to deeper and more complete understanding of the internal structure, magnetism and spottedness of stars.

My deep thanks to Drs V.A.Zamkov and E.I.Mogilevskij for stimulating discussions.

REFERENCES

Appenzeller I. and Dearborn D.S.F., I984 - Astrophys. J. v. 278, p. 689.
Gershberg R.E., I982 - Astron.Nachr. v. 303, p. 25I.
Gershberg R.E., I983 - in P.B.Byrne and M.Rodonò (eds)'Activity in red dwarf stars'. Reidel.Dordrecht. p. 487.
Hartmann L. and Rosner R., I979 - Astrophys.J. v. 230, p. 802.
Howard R. and LaBonte B.J., I983 - in J.O.Stenflo (ed)'Solar and stellar magnetic fields: origins and coronal effects'. Reidel. Dordrecht. p. IOI.
Roxburgh I.W., I983 - in J.O.Stenflo (ibid), p. 449.
Vaiana G.S., I983 - in J.O.Stenflo (ibid), p. I65.
Vaughan A.H., I983 - in J.O.Stenflo (ibid), p. II3.
Willson R.C., Gulkis S., Janssen M., Hudson H.S., Chapman G.A., I98I - Science v. 2II, p. 700.

INVESTIGATIONS OF THE MAGNETIC FIELDS OF CHEMICALLY PECULIAR STARS OF DIFFERENT AGE

Yu. V. Glagolevskij, V. G. Klochkova,
I. M. Kopylov
Special Astrophysical Observatory of USSR
Academy of Sciences, Nizhnij Arkhyz

1. The program of comprehensive investigations of stellar magnetism in the Special Astrophysical Observatory of the USSR AS includes the study of origin and evolution of the stellar magnetic fields in open clusters and associations of different age. In the papers by Borra (1981), Brown et al. (1981) and North and Cramer (1984) have been found some indications on the evolutionary decay of the fossil stellar fields.

2. To get more valid answer on this question and for the quantitative determination of decay time τ , it is necessary to investigate the magnetic fields for rather large number of CP stars in a lot of the star groups within a maximum age range. We have studied 13 open clusters and associations. The magnetic fields of the stars were taken from the literature and for 19 CP stars we measured the fields from the zeeman spectra obtained on the 6-m telescope and with the hydrogen line magnitometer. We have found the mean values of the root-mean-square field $\langle Be \rangle$ for 61 group stars.

3. The initial stellar field, due to the large plasma conductivity, has the ohmic decay time τ evaluated by exponential law:

$$B = B_0 \cdot e^{-t/\tau} .$$

For the stars of spectral type B5 - A0 $\tau \sim 10^{10}$ years. The fig.1a shows a dependence $\langle Be \rangle$ (lg t) for the open clusters and associations. The mean value $\langle Be \rangle$ for the groups are marked with filled circles, the vertical lines indicate the dispersion of $\langle Be \rangle$, the open circles show the data for the clusters which have the values $\langle Be \rangle$ for small number of stars ($N \leq 2$). The curve on the fig.1a shows the exponential law of the field decay. It was constructed by the methd of least squares allowing for the weights of poinrs equal to the number of stars belonging to each group. The decay time $\tau = 6 \cdot 10^8$ years, B_0 = 1200 Gs, and this result does not contradict the

C. R. Cowley et al (eds.), Upper Main Sequence Stars with Anomalous Abundances, 29–32.

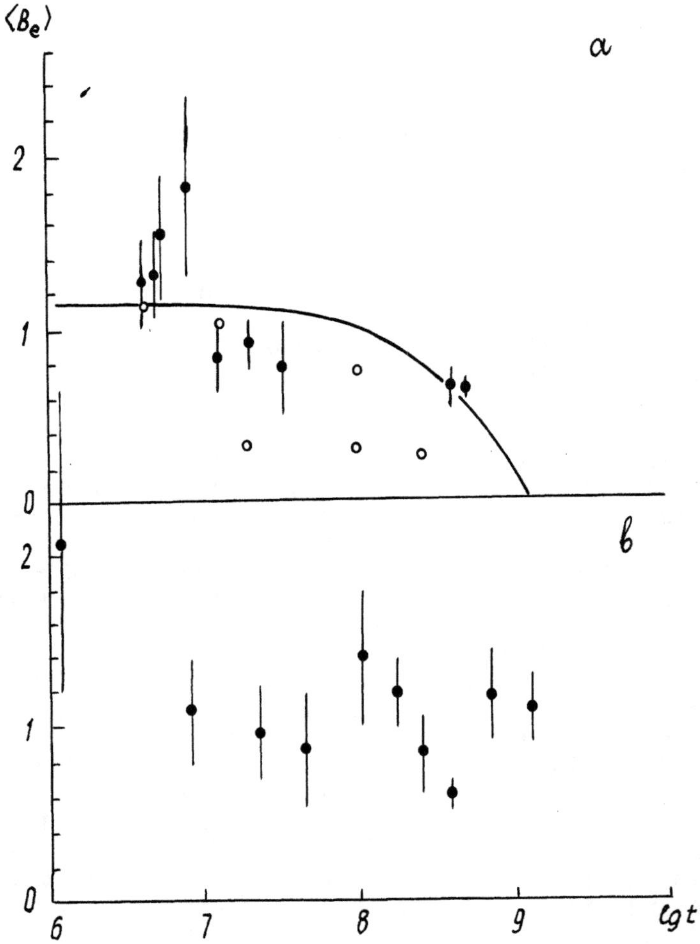

Fig. 1.

assumption about the magnetic field decrease with age, made by Borra, Brown et al. However a large scattering of points in fig.1a does not allow us to speak with a confidence about the field decay or the decay law.

4. To get the additional information about the behaiviour of the magnetic fields of CP stars with age, we have used the data for 71 stars of the galactic field. For each star the age was determined according to their position on the evolution track. The stellar effective temperatures were determined from (B - V) colors, corrected for the interstellar reddening. The absolute magnitudes Mv for the largest part of stars were determined with β-indices. Due to the fact that for cool A stars the β-indices can not be reliable indicators of luminosity, we de-

termined their Mv with help of Mv(Sp) dependences (Straizis, Kuriliene, 1981). The quantitative spectral classes for these stars have been determined from $(B - V)_0$ according to calibration Sp(B - V) (Страйжис, 1977). The dependence <Be> (lg t) for field stars presented in fig.1b shows, that the field dissipation is not discovered.

5. All the selected magnetic stars, both in the galactic field and in clusters (132 stars), we divided into 10 age groups with the equal number of stars in each ones. These data are plotted in fig.2. One can see that the mean values <Be> do not change in the wide range of age. This conclusion is supported by the dispersion analysis of <Be> for the whole number of stars of different age both in the galactic field and in the clusters.

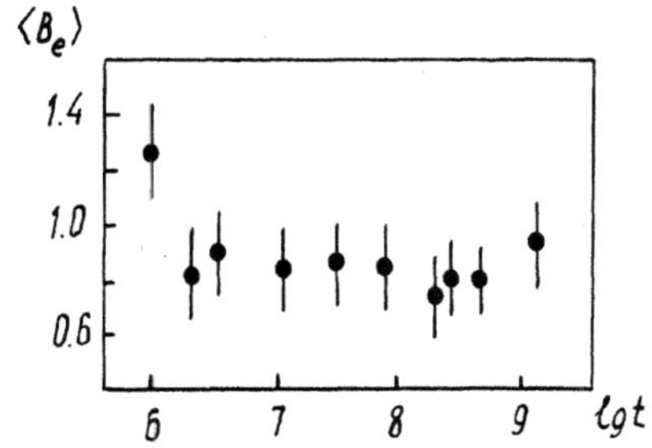

Fig. 2.

6. One can attempt to expose the probable changes of <Be> with age for the stars of different masses. For this purpose we estimated the mass for each star on the basis of its position on the evolution track on Mbol (lg Te) diagram. All the stars are divided into 3 groups depending on its mass:

1. massive stars: $\mathfrak{M} > 5\,\mathfrak{M}_\odot$ (22 stars),
2. stars with intermediate mass: $\mathfrak{M} \approx 3 - 5\,\mathfrak{M}_\odot$ (46 stars),
3. stars with small mass: $\mathfrak{M} < 3\,\mathfrak{M}_\odot$ (64 stars).

After that procedure all the stars in each group were divided into 4 equal subgroups a, b, c, d depending on the degree of their deviation from the zero age line. Inside each of the groups there appeared the stars being at close evolution stages. For each subgroup the mean values of the field <Be> were determined depending on the mean subgroup age. They are plotted in fig.3. This figure shows that values <Be> of the massive stars are systematically higher than those of both the middle mass stars and small mass stars, but the significance of differences is nonimportant and does not exceed 80%. The mean magnetic field Be for each of the three groups 1,2,3, decreases ~1,5 times during the evolution of CP stars from the zero age line up to the upper boundary of main sequence, but the significance of these differences is low.

Thus on the basis our results it is impossible to make a final conclusion about the magnetic field decay of CP stars with age during their evolution across the main sequence.

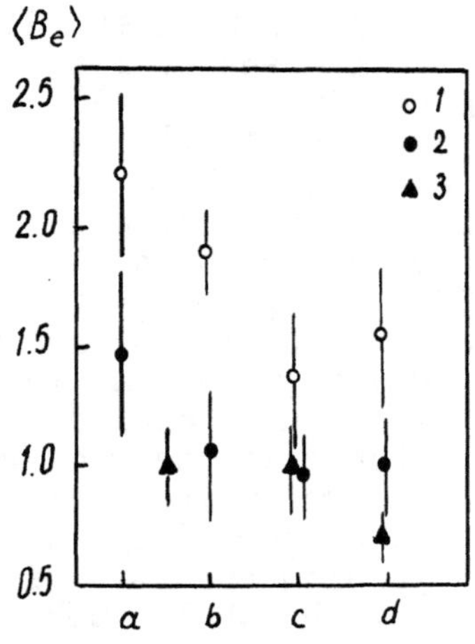

Fig.3.

References:

Borra E., 1981. Astrophys. J., Letters, 249, L39 - L42.
Brown D. N., Landstreet J. D., Thompson I., 1981. 23-d Liege Coll, p. 195 - 198.
North P., Cramer N., 1984. Astron. Astrophys., Suppl. Ser., 58, 387 - 403.
Straizys V., Kuriliene G., 1981. Astr. Sp. Sci., 80, 353 - - 368.
Страйжис В. Многоцветная фотометрия звёзд. Вильнюс, Мокслас, 1977.

THE MAGNETIC FIELD AND OTHER PARAMETERS OF THE CHEMICALLY PECULIAR STARS

Glagolevskij Yu.V., Romanyuk I.I., Chunakova N.M.
Special Astrophysical Observatory
Stavropolskij Kraj
Niznij Arkhyz 357147
USSR

1. The surface field B_s (but not the effective B_e one) is responsible for different processes occuring in the magnetic stars atmospheres. For this reason it is natural that different investigators are interested in working out and improving methods of determination of B_s. Especially great attention attracts the method based on using the multicolor Geneva photometry [1] . Cramer and Maeder have found a dependence between B_s and parameter Z of multicolor photometry wich they use for the estimation of the surface field. But it is necessary to investigate this method before application. Due to this reason we have put the following two problems:

A - The search for stars with a maximal predicted field

B - Examination of calibration of B_s (Z).

On the 6-m telescope we have received the spectrograms of 20 Cp stars with the maximal predicted magnetic field. Indeed, in 14 of them we discovered at first the magnetic fields and HD 147010 appeared to have an unique large B_e 4000 Gs in maximum (independently, Brown et al. made one estimation of B_e= 5050 Gs [2]). We calculated B_s for 70 magnetic stars on the basis of our own measurements and literature data using Stibbs-Preston method (dipolar field). A part of these stars are given in the Cramer-Maeder list. A comparison of B_s from these two lists showed that a half of them corresponds to the calibration of Cramer-Maeder and another half does not correspond to (Fig.1). The main parameters: T_e, pecularity type, mean period P, declination angle i, declination angle of the pole axe are absolutly equal in both groups. There is a difference only between v sini because of distinction of $V/R \sim \Omega$. It means, that the mean angle velocity of stars which does not correspond to the Cramer-Maeder calibration, is three times smaller. The reason of this effect is unclear.

C. R. Cowley et al (eds.), Upper Main Sequence Stars with Anomalous Abundances, 33–36.

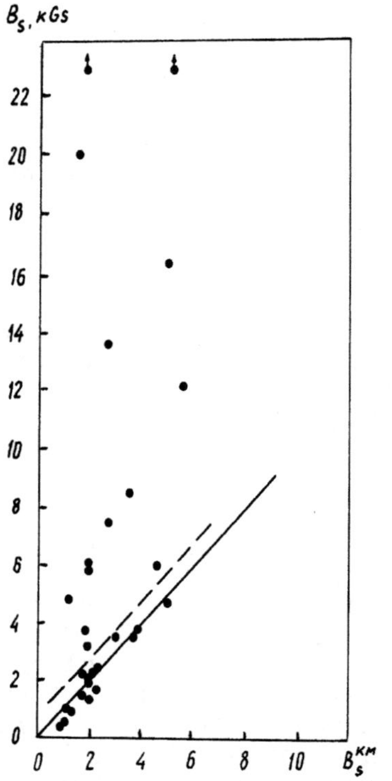

B_s, кGs
22
20
18
16
14
12
10
8
6
4
2
0
0
2
4
6
8
10
B_s^{KM}

Fig. 1

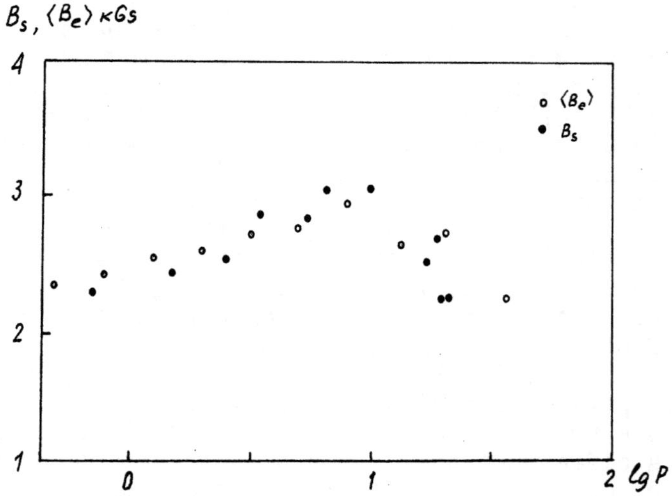

B_s, ⟨B_e⟩ кGs
4
3
2
1
0
1
2
lg P
⟨B_e⟩
B_s

Fig. 2

TABLE 1

	$\bar{P}$	N
$\beta < 45^0$	5.2	I3
$\beta > 45^0$	4.I	28

TABLE 2

	ΔU	σ	ΔB	σ	ΔV	σ	N
A	0.057	0.007	0.035	0.005	0.032	0.003	3I
B	0.046	0.009	0.035	0.0I2	0.035	0.0I0	I8

2. We investigated the dependence of B and $\langle B_e \rangle$ apon phase P on the basis of the mentioned 70 stars and 200 values of $\langle B_e \rangle$ calculated from our own measurements on the 6-m telescope and the literature data. The periods P are taken from the Didelon cataloque. It turned out that both B and $\langle B_e \rangle$ increase up to $P=8^d$ and then having passed the maximum they decrease. Thus we confirm the reverse correlation of the field with the angle speed discovered by Borra and Landstreet, but we found the existence of maximum and the right correlation $B_e \propto \Omega$ for the stars with $P > 8^d$. The existence of such a complicated dependence contradicts to the fossil hypothesis and supports the dynamo model. However, to explain the reverse correlation we support the idea of the influence of the meridional circulation which decreases the surface field for the rapid rotating stars.

3. In 1981 Fleck [3] has shown that the slow rotation of CP-stars can occur due to the interaction of the star magnetosphere with the interstellar matter. In this case the breaking will be proportional to surface fiel B_s. But on the other hand it must depend on the orientation of dipole relative to the axis of rotation. If $\beta \cong 0$ the breaking must be minimal, and if $\beta \approx 90°$the breaking is significant. To test this assumption we have found the mean rotational period for 13 stars with $\beta < 45°$ and for

28 stars with $\beta > 45°$. The mean period for stars of the first group $P = 5.2$ and for those of the second one $P = 4^d.1$. So it is clear that either the Fleck's theory is wrong or the breaking absent. (see Table 1).

4. With the purpose of finding other effects connecting with the rotating we conducted a comparative analysis of the amplitude variability ΔU, ΔB, ΔV of two stellar groups with $P < 2$ days (30 stars) and 10 days $< P < 100$ days (18 stars). These data are shown in the Table 2. We see that there are differences only between the parameters U, but within the limit 3σ. If we consider that the photometric variability appears as a result of inhomogenous distribution of chemical elements on the surface, we can draw a conclusion that the rotation does not influence on the concentration degree of chemical elements. The obtained data give a new base for theoretical investigations of stellar magnetism and phenomena connected with it.

REFERENCES

1. Cramer N.,Maeder A.,Astron.Astrophys.Suppl.Ser.,1980, v.41,p.111 - 115.
2. Brown D.N.,Landstreet J.D.,Thompson I. Upper main sequence,23-d Liege Coll.,Liege,1981,p.195 - 198.
3. Fleck R.C. Upper main sequence, 23-d Liege Coll., Liege,1981,p.341 - 342.

ON THE PERTURBATION TECHNIQUE FOR THE INFLUENCE OF MAGNETIC FIELD ON STELLAR OSCILLATIONS

S. V. Vorontsov
Department of Theoretical Physics
Institute of Physics of the Earth
B. Gruzinskaya IO
I23810 Moscow
USSR

ABSTRACT. The modification of the Rayleigh's principle is developed for the determination of the perturbations of the eigenfrequencies due to a weak magnetic field. The perturbational procedure is outlined, including its generalization for a slowly rotating magnetic star.

INTRODUCTION

The problem of adiabatic oscillations of a magnetic star has been of considerable interest in many years. The variational principle and the method of tensor virial equations have been constructed for the study of the simplest modes of oscillations (Kovetz 1966; Goossens 1979; and references therein). For a weak magnetic field, its effect on the oscillation frequencies can be determined with the use of the perturbation technique, which has been constructed by Ledoux and Simon (1957) and in much more developed form by Goossens (1972) and Goossens et al. (1976). In the present paper somewhat different approach is developed, which may become more tractable for certain magnetic configurations. It is based on the modification of Rayleigh's principle, which is customary to use in the theory of free oscillations of the Earth (e.g. Backus and Gilbert 1967; Zharkov et al. 1968; Woodhouse 1976).

THE MODIFIED RAYLEIGH'S PRINCIPLE

We consider here the oscillations of a magnetic configuration with equilibrium field continuous across the surface S, with pressure and density vanishing on S, so that the equilibrium configuration is in a state of "true equilibrium" (Roberts 1955). It is assumed that the internal magnetic field tends to force-free field near the surface.

C. R. Cowley et al (eds.), Upper Main Sequence Stars with Anomalous Abundances, 37–40.

The usual form of the equation of adiabatic oscillations is

$$\rho_0\omega^2\vec{u} = \nabla p_1 + \rho_1\nabla\psi_0 + \rho_0\nabla\psi_1 - \frac{1}{4\pi}\left[(\nabla\times\vec{H}_0)\times\vec{H}_1 + (\nabla\times\vec{H}_1)\times\vec{H}_0\right] , \qquad (1)$$

where the common time-dependent factor $\exp(i\omega t)$ is omitted, $\vec{u}$ is displacement field, ψ is gravitational potential, and

$$p_1 = -\Gamma_1 p_0 \nabla\cdot\vec{u} - \vec{u}\cdot\nabla p_0 , \qquad (2)$$

$$\rho_1 = -\nabla\cdot(\rho_0\vec{u}) , \qquad (3)$$

$$\nabla^2\psi_1 = 4\pi G\rho_1 , \qquad (4)$$

$$\vec{H}_1 = \nabla\times(\vec{u}\times\vec{H}_0) . \qquad (5)$$

Subscript 0 denotes the equilibrium value of the corresponding quantity, subscript 1 - its Eulerian perturbation. In the exterior vacuum $\vec{H}_1$ is described by its vector potential

$$\vec{H}_1^V = \nabla\times\vec{A}^V , \quad \nabla\times(\nabla\times\vec{A}^V) = 0 , \quad \nabla\cdot\vec{A}^V = 0 , \qquad (6)$$

$$\hat{n}\times\vec{A}^V = \hat{n}\times(\vec{u}\times\vec{H}_0) \quad \text{on } S \qquad (7)$$

(e.g. Kovetz 1966), where $\hat{n}$ is the unit normal to the surface. Denoting by $[\,]_-^+$ the discontinuity of the enclosed quantity across S , the boundary conditions for the eigenvalue problem (1) are known to be

$$[\psi_1]_-^+ = 0 \quad \text{and} \quad \left[\hat{n}\cdot\nabla\psi_1 + 4\pi G\rho_0\hat{n}\cdot\vec{u}\right]_-^+ = 0 \quad \text{on } S , \qquad (8)$$

$$\vec{H}_0\cdot\left[\vec{H}_1 + (\vec{u}\cdot\nabla)\vec{H}_0\right]_-^+ = 0 \text{ and } (\hat{n}\cdot\vec{H}_0)\left[\vec{H}_1 + (\vec{u}\cdot\nabla)\vec{H}_0\right]_-^+ = 0 \quad \text{on } S . \qquad (9^a)$$

If $\vec{H}_0$ is zero, the boundary condition (9^a) is replaced by the regularity condition

$$\Gamma_1 p_0 \nabla\cdot\vec{u} = 0 \quad \text{on } S . \qquad (9^b)$$

After the scalar multiplication of eq. (1) by $\vec{u}^*$, where the asterisk denotes complex conjugate, and integration over the volume of the star, the somewhat lengthy manipulations lead to

$$\int_E L\,dv + \frac{1}{4\pi}\int_S \hat{n}\times(\vec{u}^*\times\vec{H}_0)\cdot\left[\vec{H}_1\right]_-^+ ds = 0 , \qquad (10)$$

$$\begin{aligned} L = {} & \Gamma_1 p_0(\nabla\cdot\vec{u}^*)(\nabla\cdot\vec{u}) + \frac{1}{2}\Big[\rho_0(\vec{u}\cdot\nabla)(\vec{u}^*\cdot\nabla\psi_0) + \rho_0(\vec{u}^*\cdot\nabla)(\vec{u}\cdot\nabla\psi_0) + \\ & + (\nabla\cdot\vec{u}^*)(\vec{u}\cdot\nabla p_0) + (\nabla\cdot\vec{u})(\vec{u}^*\cdot\nabla p_0)\Big] + \rho_0(\vec{u}\cdot\nabla\psi_1^* + \vec{u}^*\cdot\nabla\psi_1) + \\ & + (1/4\pi G)\nabla\psi_1^*\cdot\nabla\psi_1 - \rho_0\omega^2\vec{u}^*\cdot\vec{u} + \\ & + (1/4\pi)\vec{H}_1^*\cdot\vec{H}_1 - (1/8\pi)\left[(\nabla\times\vec{H}_0)\times\vec{H}_1\cdot\vec{u}^* + (\nabla\times\vec{H}_0)\times\vec{H}_1^*\cdot\vec{u}\right] , \end{aligned} \qquad (11)$$

where E is all of space. The eq. (10) was obtained without using the boundary condition (9^a). If the eigenfunctions satisfy (9^a), the surface integral in (10) vanish. In the external vacuum p_0, ρ_0 and $\nabla\times\vec{H}_0$ are identically zero, so that all but two terms in the Lagrangian (11) vanish there. The Lagrangian (11) has the symmetric form and differ from those corresponding to the non-magnetic problem (e.g. Backus and Gilbert 1967) only in that the last line in (11) has appared. It contains the terms corresponding to the work done against the Lorentz forces.

Now we can wright the variational equation

$$\int_E L(\psi_j, \psi_{j,i}, p_k, \lambda)\, dv = 0 . \tag{I2}$$

The Lagrangian is homogeneous function of "fields" ψ_j and their space derivatives $\partial\psi_j/\partial x_i$ denoted by $\psi_{j,i}$. By "fields" in stellar interior we mean the components of displacement and the perturbation in the gravitational potential. In exterior vacuum, by "fields" we denote the perturbation in the gravitational potential and the components of the magnetic vector potential. Equilibrium ρ_0, p_0, ψ_0, Γ_1, the components of $\vec{H}_0$ and their space derivatives are denoted by "parameters" p_k. The eigenvalue λ is ω^2.

The perturbation of the integral in (I2) due to the perturbation in the fields alone is (to first order)

$$- \int_E \left[\frac{\partial}{\partial x_i}\frac{\partial L}{\partial \psi_{j,i}} - \frac{\partial L}{\partial \psi_j} \right] \delta\psi_j\, dv - \int_S n_i \left[\frac{\partial L}{\partial \psi_{j,i}} \delta\psi_j \right]_-^+ ds . \tag{I3}$$

The summation convention is used here and will be used later. This perturbation may be shown to be zero, if the fields (but not necessary their perturbations) satisfy the eqs. (I,6) and boundary conditions (8,9). The volume integral in (I3) is zero because the eqs. (I,6) are equivalent to Euler-Lagrange equations

$$\frac{\partial}{\partial x_i}\frac{\partial L}{\partial \psi_{j,i}} - \frac{\partial L}{\partial \psi_j} = 0 , \qquad j = I,2,\ldots \tag{I4}$$

The surface integral in (I3) is zero due to the boundary conditions (8,9). The perturbation in "fields" is not restricted by boundary conditions, so that the equation

$$\int_E L\, dv = 0 \tag{I5}$$

provides the basis for "unrestricted" variational principle, similar to those proposed by Kovetz (I966).

Now let us perturb the stellar model from non-magnetic to slowly magnetized configuration. We are left with the perturbation of the integral in (I2) due to the perturbation in the parameters and the eigenvalue, so that

$$\delta\lambda \int_E \frac{\partial L}{\partial \lambda}\, dv = - \int_E \frac{\partial L}{\partial p_k} \delta p_k\, dv , \tag{I6}$$

from which the perturbation of the eigenfrequency may be calculated when the eigenfunctions of non-magnetic spherically symmetric star are known. It is the form of Rayleigh's principle, generalized to include the effect of modification of boundary conditions.

Eq. (I6) gives the perturbation of the eigenfrequency as the sum of volume integrals over the spherical volume occupied by the non-magnetic star and one integral with $\vec{H}_1^* \cdot \vec{H}_1$ over the external vacuum. The latter is redused to the surface integral and determined with the use of (7).

When the perturbations in the parameters (δp_o , $\delta \rho_o$, $\delta \psi_o$, $\delta \nabla p_o$, $\vec{H}_o$, $\nabla\times\vec{H}_o$) are given in terms of scalar and vector spherical harmonics, the volume integrals may be reduced to radial integrals and angular integrals containing the triplets of spherical harmonics. These angular integrals may be computed with the use of Vigner's 3-j symbols (e.g. Luh 1973).

The first-order effects of rotation may be easily included into the perturbation technique. In the co-rotating coordinate system in which $\vec{H}_o$ is stationary, these effects will be given by the additional term $2\rho_o i\omega\Omega\,\hat{z}\times\vec{u}$ in the right-hand side of eq. (I). This term describes the Coriolis force, $\hat{z}$ is unit vector along the axis of rotation. The corresponding additional term in the Lagrangian is $2\rho_o\omega\Omega\,\vec{u}^*\cdot(i\hat{z}\times\vec{u})$ and results in the additional integral in the right-hand side of eq. (I6), which is

$$-\int_{V_o} 2\rho_o\omega\Omega\,\vec{u}^*\cdot(i\hat{z}\times\vec{u})\,dv\ .$$

If the axial symmetry is violated, the true zero-order eigenfunctions become linear combinations of $2l+1$ eigenfunctions with different m values. The coefficients may be found from the requirement that $\delta\lambda$ must be stationary to their variations. This requirement leads to the linear system of algebraic equations. If this situation includes rotation (e.g. an oblique magnetic rotator), then the field $i\hat{z}\times\vec{u}$ should be decomposed on to vector spherical harmonics (Vorontsov and Zharkov 1981).

The perturbational approach presented in this paper is restricted to the study of the perturbation of the eigenfrequencies. Quite different methods should be used to determine the perturbation of the eigenfunctions (Biront et al. 1982; Roberts and Soward 1983).

Author thanks Professor V.N.Zharkov for useful discussions.

REFERENCES

Backus, G.E. and Gilbert, J.F. 1967, Geophys.J.R.A.S., 13, 247.
Biront, D., Goossens, M., Cousens, A. and Mestel, J. 1982, M.N.R.A.S., 201, 619.
Goossens, M. 1972, Ap.Space Sci., 16, 386.
Goossens, M. 1979, Ap.Space Sci., 60, 401.
Goossens, M., Smeyers, P. and Denis, J. 1976, Ap.Space Sci., 39, 257.
Kovetz, A. 1966, Ap.J., 146, 462.
Ledoux, P. and Simon, R. 1957, Ann.d'Ap., 20, 185.
Luh, P.C. 1973, Geophys.J.R.A.S., 32, 187.
Roberts, P.H. 1955, Ap.J., 122, 508.
Roberts, P.H. and Soward, A.M. 1983, M.N.R.A.S., 205, 1171.
Vorontsov, S.V. and Zharkov, V.N. 1981, Soviet Astron.AJ., 58, 1101.
Woodhouse, J.H. 1976, Geophys.J.R.A.S., 46, 11.
Zharkov, V.N., Lubimov, V.M. and Osnach, A.I. 1968, Izv.Acad.Nauk USSR Fizika Zemli, N°10, 3.

ON THE MAGNETIC FIELD OF β CrB

S.I.Plachinda

Crimean Astrophysical Observatory
Nauchny, Crimea, SU - 334413, USSR

ABSTRACT. We have discussed the oscillations of the magnetic field (H_e) of β CrB with a period $P_2=P_1/5=3^d.6974$, where P_1 is stimulated by a rotation of the star. The analysis of other author's measurements has been confirmed the reality of P_2.

The mean value of the longitudinal component of the β CrB magnetic field was measured in a separate spectral line λ 4520.2 FeII (Borra and Vaughan 1977) and in λ 4923.9 FeII line, single measurement (Landstreet 1982). But since β CrB is a magnetic Ap-star of SrCrEu type, it seems mostly reasonable to measure the H_e in one of the peculiar elements line. Such a kind of measurements were carried out for the first time by A.B.Severny (Severny 1970; Severny et al. 1974) in the photoelectric observations of the magnetic stars in λ4254.33 CrI line.

We have measured the H_e of β CrB in λ4254.33 CrI line. The observations were realized using the magnetometer ZTSh of the 2.6-m telescope of the Crimean Astrophysical Observatory. For three nights the measurements were carried out on the 6-m telescope of SAO. The data were reduced by the formula (Plachinda in press):

$$\Delta\lambda_H = \int_{\Delta\lambda_1}^{\Delta\lambda_2} (N_1 - N_2)\, d(\Delta\lambda) \Big/ \int_{\Delta\lambda_1}^{\Delta\lambda_2} \partial N/\partial\lambda\, d(\Delta\lambda),$$

C. R. Cowley et al (eds.), Upper Main Sequence Stars with Anomalous Abundances, 41–44.

where $\Delta\lambda_H$ is a magnetic splitting, N_1 and N_2 are the intensities for the wavelength in question of the orthogonal polarized light fluxes, $\partial N/\partial\lambda$ is the weight function of V-parameter.

For 23 dates of observations, we have made 126 estimates of H_e for β CrB. Fig.1 shows the value of H_e averaged according to dates. To compute the phases we used the ephemeris published by Preston and Sturch (1967). The sinusoid calculated by the least square method is indicated here by a solid line. The deviations of H_e were calculated with respect to this line and the rest range of values was subjected to the analysis in order to search for the presence of periodicity. Fig.2a shows the behaviour of ΔH_e with the period

Fig.1

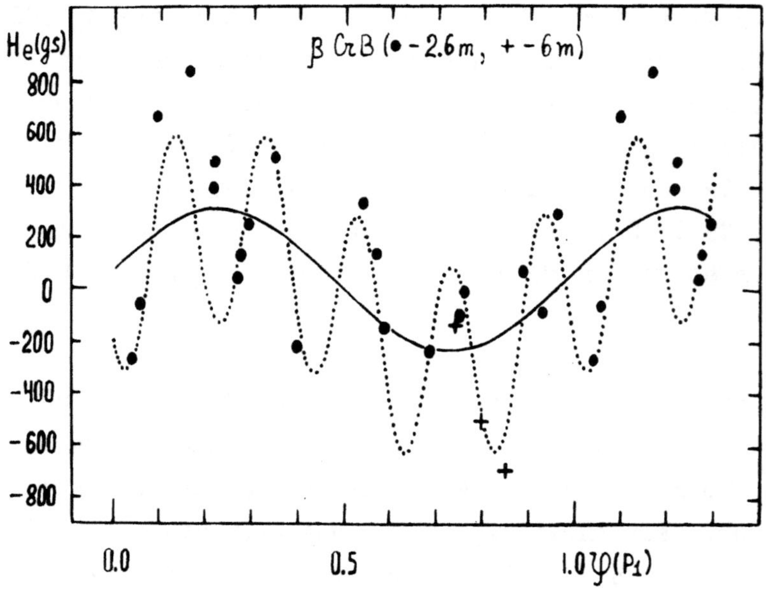

$P_2=P_1/5=3^d.6974$, for which $2A/\delta \simeq 3.1$ ("A" is the amplitude of a period). The zero epoch is the same as for the 18-d period (Preston and Sturch 1967). The superposition of periods P_1 and P_2 is presented in Fig.1 by a dotted line.

The same type of analysis was applied for the H_e measurements in λ4254.33 CrI line (Severny et al. 1974) and in λ4520.2 FeII line (Borra and Vaughan 1977). The approximation curve for P_2 according to the data published by Severny et al. (1974) is shown in Fig.2b and according to data published by Borra and Vaughan (1977) - in Fig.2c. The phase correlation of the three curves (see Fig.2a,b,c) supplies a completing evidence for the reality of the period $P_2 = 3^d.6974$ existing alongside with $P_1=18^d.487$ for the star β CrB.

Fig.2

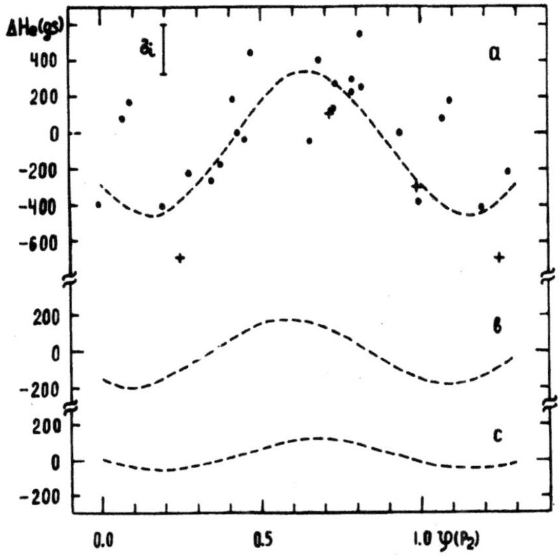

REFERENCES

Borra, E.F., and Vaughan, A.H. 1977, Ap.J., 216, 462.

Landstreet, J.D. 1982, Ap. J., 258, 639.

Plachinda, S.I. Izv. Krimsk. Astrofiz. Obs., 74, (in press).

Preston, G.W., and Sturch, C. 1967 , in The Magnetic and Related Stars, ed. R.C. Cameron (Baltimore: Mono Book Corp.), p. 111.

Severny, A.B. 1970, Ap.J. (Letters), 159, L73.

Severny, A.B., Kuvshinov, V.M., and Nikulin, N.S. 1974, Izv. Krimsk. Astrofiz. Obs., 50, 3.

Discussion appears after the following paper.

ESTIMATION OF STELLAR SURFACE MAGNETIC FIELDS BY THE CURVE-OF-GROWTH METHOD

T. A. Ryabchikova and N. E. Piskunov
The Astronomical Council
of the U.S.S.R. Academy of Sciences
Pyatnitskaya Str. 48, Moscow
109017 U.S.S.R.

ABSTRACT. The curve-of-growth method of estimating the stellar surface magnetic fields is applied to six Ap stars of different temperatures and one normal (or metallic-line) star σ Aqr. The values of H_s determined by the curve-of-growth technique are in good agreement with those obtained by the photometric method.

The curve-of-growth method of estimating the surface magnetic fields in Ap stars is examined using the 8 to 10 A/mm dispersion spectra. The spectra of faint Ap stars and Ap stars in clusters are usually obtained with such a dispersion. According to the basic assumption of the method proposed by Hensberge and De Loore (1974) the line broadening on the flat part of the curve of growth may be represented as a sum of Doppler's and Zeeman's broadening:

$$\Delta\lambda = (\Delta\lambda_D^2 + \Delta\lambda_H^2)^{1/2} \quad , \text{ where}$$

$$\Delta\lambda_H = k\cdot\lambda\cdot z\cdot H_S \qquad (k=4.66\times 10^{-13}).$$

Since $W \sim (\mathrm{Log}\,\eta_0)^{1/2}\Delta\lambda$, the equivalent width of a line formed in the presence of magnetic field is enhanced by a factor of $(1 + \Delta\lambda_H^2/\Delta\lambda_D^2)^{1/2}$, i.e.,

$$W_H = W_0[1 + (k\cdot\lambda\cdot c\cdot z\cdot H_S)^2/V_D^2]^{1/2}.$$

On minimizing the standard deviation of W_0 with respect to the computed curves of growth for various microturbulent velocities, we obtain H_s along with the abundance of the given element and the microturbulent velocity V_t.

This procedure was applied to six Ap stars and a normal star σ Aqr. The spectra of these stars with the dispersion of 8 to 9 A/mm were obtained at the Crimean Observato-

C. R. Cowley et al (eds.), Upper Main Sequence Stars with Anomalous Abundances, 45–47.

Table. Derived parameters of Ap stars.

Star	T_{eff}/Log(g)	Element	H_s kgauss	Log$\frac{Fe}{H}$	V_t km/s	H_s (phot)
σ Aqr	10150 K/4.0	FeI	0.35	-4.01	1	0.10
		FeII	0.36	-4.01	1	
HD 2453	9000 K/3.75	FeI	2.80	-3.80	2	3.13
		FeII	2.32	-3.64	2	
HD 8441	9200 K/3.5	FeI	1.23	-4.00	0	0.65
		FeII	0.70	-3.78	0	
HD 110066	9300 K/3.75	FeI	2.46	-3.62	2	3.12
		FeII	2.16	-3.69	2	
HD 118022	9650 K/4.0	FeI	2.30	-3.39	1	2.42
(78 Vir)		FeII	1.63	-3.28	1	
HD 168733	14500 K/ 3.6	FeII	0.86	-3.68	2	1.18
HD 192913	11000 K/3.5	FeI	2.26	-3.68	1	1.67
		FeII	1.00	-3.43	1	

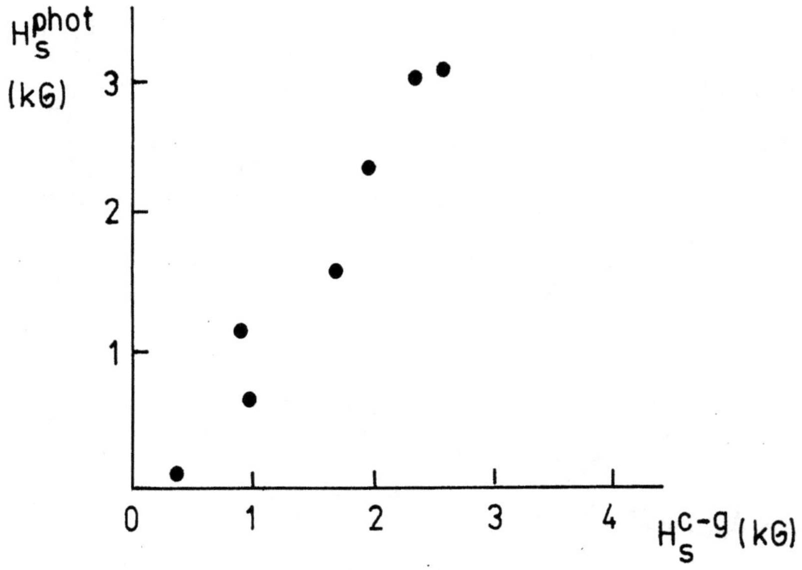

Fig.1. Comparison between estimates of H_s obtained by the photometric method and the curve-of-growth method.

ry (HD 118022), the Bulgarian National Astronomical Observatory (HD 2453, HD 8441 and HD 192913) and the Special Astrophysical Observatory of the USSR Acedemy of Sciences (HD 110066). Data for σ Aqr and HD 168733 were taken from Adelman and Nasson (1980) and Little and Aller (1970). The results are summarized in the Table. The rightmost column of this table contains the H_s values which were obtained using the photometric calibration (see Cramer and Maeder (1980)) and the temperatures from our table. Comparison between our mean estimates of H_s and the photometric ones results in Figure 1.

The following conclusions can be drawn from these results.

1. The curve-of-growth method provides the possibility of estimating the Ap star surface magnetic fields using moderate dispersion spectra.
2. One can simultaneously obtain the more accurate mean abundance of an element under investigation and the value of the microturbulent velocity.
3. The values of H_s obtained from FeII lines are smaller than the respective values obtained from FeI lines for all Ap stars in question. Due to the method error ($\sim$0.5kgs) it is still unclear whether there actually is the difference in reality.
4. The estimates of H_s yielded by the curve-of-growth method are systematically lower than those yielded by the photometric method. This may partly be explained by the fact that the calibration of the photometric method is insufficiently accurate for the surface magnetic fields smaller than 5 kgs.

REFERENCES

Adelman, S.J., Nasson, M.A. 1980, Publs. Astr. Soc. Pacif., 92, 348.

Cramer, N., Maeder, A. 1980, Astr. Ap., 88, 135.

Hensberge, H., De Loore, C. 1974, Astr. Ap., 37, 367.

Little, S.J., Aller, L.H. 1970, Ap. J. Suppl., 22, 157.

DISCUSSION (Plachinda)

ADELMAN: What is the wavelength of the Cr I line used? Is it 4254 Å?

PLACHINDA: Yes.

ADELMAN: That line is blended on the longward side by some line(s) of unknown origin in β CrB and other Ap stars. I believe there are better lines on which to make such measurements. I am concerned about the effects of the blending component(s) on all measurements of the magnetic field which use Cr I 4254 Å.

PLACHINDA: The Cr I line is significantly stronger than the blends.

TUTUKOV: Have you a model explaining your results?

PLACHINDA: Not yet. We do not intend to make an interpretation.

WEISS: [to Adelman] I agree that the Cr I line is blended and therefore the interpretation of the Zeeman effect will be complicated, but how would this influence the extraordinary effect of the superimposed $P_1/5$ variation?

ADELMAN: The blending line may have the same variability as Cr I 4254 Å, or it may be quite different. This would certainly complicate the analysis. I was simply pointing out that I think there are better lines to use for this star.

WEISS: Yes, correct.

DWORETSKY: I find the result most extraordinary. If it is true, an interpretation is extremely difficult. Could you please comment on the statistical significance of the $P_1/5$ periodicity which you have found?

PLACHINDA: We did not rely on statistics. Our argument is the phase correlation of periods obtained by three authors.

C. R. Cowley et al (eds.), Upper Main Sequence Stars with Anomalous Abundances, 49.

Does ν Cep indicate a non-axisymmetric dynamo?

F. Krause and G. Scholz
Zentralinstitut für Astrophysik der AdW der DDR, Potsdam, DDR

Abstract

According to observations of Scholz and Gerth the supergiant ν Cep has a magnetic field with a maximum field strength up to 2500 Gauss. This field shows a period of about 5 years. It is unplausible that this magnetic field is a relic since ν Cep was formed by expansion of a B-star. We claim here that ν Cep represents a dynamo exciting a magnetic field which in the average strongly deviates from symmetry about the rotation axis.

We know for some cosmical objects with certainty that their magnetic fields are dynamo excited: that are some planets (Earth, Jupiter, Saturn) and the Sun. The average magnetic fields of this objects are mainly axisymmetric with respect to the axis of rotation. Deviations from this symmetry are secondary effects.

Theoretical considerations show that under certain circumstance those magnetic fields are most easily excited which have no symmetry with respect to the axis of rotation. This is the case for α^2-dynamos with sufficiently strong anisotropy, which is due either to the influence of rotation (Rüdiger, 1978, 1980) or to the radial stratification (Rädler 1980, 1985). Dynamo models of that kind excite fields where the leading term is a dipol with its moment lying in the equatorial plane.

All magnetic stars possess highly non-axisymmetric fields. But they cannot, at present, used as examples for non-axisymmetric mean-field dynamos since the question whether these fields are relics or excited by dynamo action is still open. For a clarification the supergiant ν Cep could be a suitable case as we know that a short time ago it was in quite a different evolutionary state, probably an early B-star. That is why ν Cep is worth a closer analysis.

The supergiant ν Cep is of spectral type A2Ia. It is assumed that it is an evolved B-star with radius $R \approx 90\ R_{\odot}$. A magnetic field was first found by Scholz and Gerth (1980, 1981), meanwhile, this field shows a period of about 5 years (Scholz, Gerth, Glagolevskij and Romanjuk

C. R. Cowley et al (eds.), Upper Main Sequence Stars with Anomalous Abundances, 51–54.

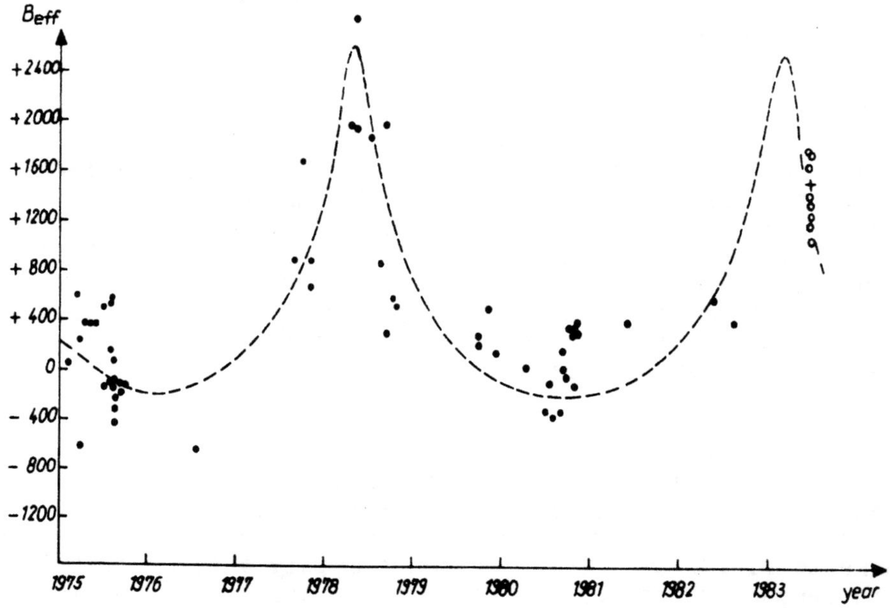

Magnetic field of ν Cep from measurements at Tautenburg (•), Selentschuk (o) and with the magnetograph of Selentschuk (+).

(1984)). It has a maximum value of about 2500 Gauss, the minimum is rather flat and seemed to be negativ of about -300 Gauss. What is the origin of this magnetic field?

It is generally assumed that a supergiant like ν Cep was formed by expansion of a B-star, say a star of about 6 $R_\odot$, i.e. it was enlarged by a factor 15. In case the present magnetic field is the relic of that of the main sequence B-star we find by extrapolation that this star had a field of at least some 10^5 Gauss strength. But this contradicts our knowledge about B-stars. Consequently, the possibility of a relic is rather unplausible, the magnetic field must be formed in a later evolutionary state.

If the magnetic field is formed in a later state the only possibility is that it is excited by a working dynamo. The period of 5y opens the possibility of an oscillatory dynamo.

From the Sun and the solar type stars we know periods of activity cycles down to about 7 years, i.e. periods of the magnetic field down to 14 years. The period of ν Cep, five years, is rather short. That is the more of importance since the period to some extend increases with the thickness of the convective layer (skin effect). Consequently, we would expect a much larger period for the supergiant ν Cep than for the (rather small) solar type stars.

Another argument against an oscillatory dynamo is the high degree of anharmonicity which the observed curve of B_{eff} shows. We know from the Sun and the (periodic) solar type stars that activity cycles do not so much differ.

So we have good reasons for excluding the possibility of a working oscillatory dynamo.

The most plausible explanation, which is now left, is that the 5 year period of the magnetic field is the period of rotation and the magnetic field strongly deviates from the symmetry with respect to the rotational axis. This view of the matter is supported by the following argumentation.

If we, as above, assume that ν Cep was formed by expansion of a B-star by a factor 15, and assume, in addition, a rotational period of this B-star of, say, 2 days, we find by taking into account the conservation of angular momentum a rotational period for ν Cep of about 500 days, i.e. about 1 1/2 years.

But mass loss is ubiquitous among highly luminous OBA stars. Therefore, even if the mass loss rate is unknown for ν Cep, the assumption of conservation of angular momentum is unlikely to be a correct one. For the increase of the rotational period up to about 5 years, mass loss in the presence of a magnetic field can presumably explain the braking of the rotational velocity. Thus ν Cep is well fitting in the picture we have from magnetic stars. Also the strongly anharmonic time variation of B_{eff} , one extrema narrow, the other one broad, is a typical behaviour of such stars like α_2CVn, 53 Cam and others.

So we have good reasons to assume that ν Cep is a realisation of a non-axisymmetric mean-field dynamo. Following this idea we have to conclude that ν Cep has an extended convection zone. This does not contradict the generally accepted view. In addition, a statement concerning the structure of the convection is possible:
It is well known that dynamos with remarkable differential rotation preferably excite fields showing symmetry with respect to the axis of rotation. Hence we have to conclude that ν Cep has a convection of such a structure, which does not cause differential rotation. This is possible for convection where the turnover time is large compared

with the rotational period (Rüdiger 1983, Hathaway 1984). Hence, according to our view, the supergiant ν Cep shows a convection with long living large cells.

References:

Hathaway, D.H., 1984, Astrophys. J. 276, 316.
Rädler, K.-H., 1980, Thesis Potsdam.
Rädler, K.-H., 1985, Astron. Nachr. (in press).
Rüdiger, G., 1978, Astron. Nachr. 299, 217.
Rüdiger, G., 1980, Astron. Nachr. 301, 181.
Rüdiger, G., 1983, GAFD 25, 213.
Scholz, G. and E. Gerth, 1980, Astron. Nachr. 301, 211.
Scholz, G. and E. Gerth, 1981, Mon.Not.R.astr.Soc. 195,853.
Scholz, G., E. Gerth, J.V. Glagolevskij, and I.I. Romanjuk, 1984, 6-th. sci. conf. on. magn. stars, Riga, 69.

DISCUSSION (Krause and Scholz)

DWORETSKY: What is the spectral type of ν Cep?
OETKEN: A2 Ia.
DOLGINOV: What is the evidence that in this case, in the presence of convection there is no differential rotation?
KRAUSE: Convection with no differential rotation seems, indeed, to be controversial, since it is generally accepted that the differential rotation of the Sun is caused by convection. This theory, however, also reveals a certain dependence of the differential rotation on the characteristic parameters of the convection; this is also true for the dynamo effect. These results (published by Rüdiger and Hathaway) state: with growing angular velocity, the convection takes such a structure that differential rotation becomes weak, and the dynamo effect favours a magnetic field with no symmetry with respect to the rotation axis.
SEVERNY: How sensitive are your conclusions to the rather large scatter of points in the observed magnetic curve?
KRAUSE: The crucial point is the existence of the field, and its origin. Our data do not allow us to define the magnetic geometry very well, but the existence of the field is established.
MÉGESSIER: Is there any photometry of ν Cep that would indicate quasi-periodic or periodic light variations? The B supergiants are known to exhibit such variations; what is the case for A supergiants?
KRAUSE: As far as we are aware, there is no light curve for this star. We have found a number of possible periods from the radial velocity data. Perhaps Dr. Gerth can comment.
GERTH: The significance of the magnetic field and the period was established by superimposing individual plates and examining the linear dependence of the z-value of $\Delta\lambda/\lambda^2$. The search for a period was statistical. We also measured radial velocities, and for this star the radial-velocity period is half the magnetic period.
KROLL: The rotation is slow, yet the magnetic field is fairly strong. Did you make quantitative estimates to see if the dynamo can support the field?
KRAUSE: Whether or not the rotation is slow depends on the time scales relevant to dynamo excitation. These are the decay time of the magnetic field in a convective cell, and the lifetime of that convective cell, or the turnover time. Because of the large geometric dimensions, the first time scale is surely large compared with the rotation period (for a granule in the solar convection zone, this decay time is about 30 years). I do not know whether there is reliable information on the lifetime of a convective cell in a supergiant. For other reasons (given in my reply to Dr. Dolginov earlier) we expect a large turnover time. The dynamo effect will be weak if both these time scales are very small compared with the rotation period.
HENSBERGE: How many supergiants were included in your search for magnetic fields? Have any others shown signs of a magnetic field?
KRAUSE: May I ask Dr. Oetken to answer this question?
OETKEN: We observed many stars, but our first inspection of the spectrograms showed that ν Cep has some remarkable features, so we

C. R. Cowley et al (eds.), Upper Main Sequence Stars with Anomalous Abundances, 55–56.

investigated this star first. So far no one else has found magnetic fields in supergiants, and we are wondering if this is a special star. We do not yet know whether magnetic fields are characteristic of other supergiants.

MODEL ATMOSPHERES AND RADIATIVE TRANSFER IN CHEMICALLY PECULIAR STARS:
INTERPRETATIONAL SIGNIFICANCE OF NON-LTE

I. Hubený
Astronomical Institute
Czechoslovak Academy of Sciences
251 65 Ondřejov, Czechoslovakia

1. INTRODUCTION

One of the main problems in studying chemically peculiar (CP) stars is the question of the extent to which the conceptual framework of the contemporary spectroscopic diagnostics is reliable. As the first step, it should be clarified whether the traditional assumption of local thermodynamic equilibrium (LTE) provides an adequate approximation to reality, or whether a more general non-LTE approach should be employed.

The period of rapid development of computational methods and extensive calculations of NLTE model atmospheres is now past its culmination point. The importance of a relevant NLTE description is viewed as unquestionable for hot stars (B and earlier). Consequently, the hot stars attract the attention of most "NLTE theoreticians," while considerably less attention is being devoted to later types (late B and A). Moreover, the abundance anomalies found in the CP stars are no longer expected to be a spurious result of an inadequate (LTE) analysis (Cowley 1981). It is not commonly accepted that other phenomena such as diffusion, magnetic field, inhomogeneous abundance distribution, etc., are quite essential ingredients of the general atmospheric pattern of CP stars. Therefore, a large part of the astronomers investigating the CP stars now consider the NLTE effects to be rather minor perturbations of the LTE predictions.

Is LTE indeed a satisfactory description of the CP stellar atmospheres, and if so, why? For what purposes? On the other hand, is an NLTE approach really necessary, and if so, why? For what diagnostic purposes? For what spectral types and/or spectral features? These are the most urgent questions which are of interest to both observers and theoreticians.

It is the aim of the present paper to discuss the questions raised above. We stress at the very outset that we shall only consider the so-called classical stellar atmospheres, i. e. those characterized by the assumptions of plane parallel horizontally homogeneous stratification, hydrostatic and radiative equilibrium, but not generally local thermodynamic equilibrium (LTE). Primary attention will be devoted to examining the changes in atmospheric structure and emergent spectrum that arise due to relaxing the assumption of LTE. Excellent reviews of the classical model stellar atmospheres, as well as those that relax one or more of the above assumptions can be found in Mihalas (1978) and in the series of NASA-CNRS monographs "Non-thermal Phenomena in Stellar Atmospheres" (Underhill and Doazan, 1982; Wolff, 1983; Thomas, 1983; and references therein). Further, we shall not consider more involved treatments of the classical NLTE radiative transfer as, for instance, departures from complete frequency redistribution in spectral lines (for recent reviews of this topic, see Linsky, 1985 and Hubený, 1985).

We shall primarily concentrate on the methodological problems, leaving aside the more technical aspects of the theory (for details, refer to the excellent textbook of Mihalas, 1978) as well as the discussion of the results of actual diagnostic studies (for a review see Wolff, 1983; and some other papers of this conference). An illuminating discussion of some general aspects

C. R. Cowley et al (eds.), Upper Main Sequence Stars with Anomalous Abundances, 57–75.

of spectroscopic diagnostics of CP stars may be found in interesting papers by Cowley (1980, 1981), Praderie (1982), and Cowley and Adelman (1983).

2. PRELIMINARY CONSIDERATIONS

2.1 LTE versus NLTE

Generally speaking, NLTE is a state of radiation and matter characterized by any kind of departure from the LTE. However, this term is often used in a limited sense. In the usual astrophysical context, the difference between LTE and NLTE consists only in the manner of calculating the atomic level populations.

Briefly, under the assumptions of LTE, the occupation numbers of bound and free states of atoms (level populations) at a specified point in the atmosphere are functions of only two thermodynamic variables - electron temperature and density - and are given by the well known Saha-Boltzmann equations. This is no longer generally true under the assumption of NLTE; instead, atomic level populations have to be determined by solving the statistical equilibrium equations which express the balance of all microscopic processes populating and depopulating the various atomic states. No other departures from LTE distribution functions are allowed in this restricted NLTE approach; this led some authors to propose more appropriate names, such as the statistical equilibrium (Wolff 1983) or kinetic equilibrium (Athay 1972) approach. However, we retain the name NLTE throughout the present paper.

It is customary to express NLTE level populations in terms of the LTE populations that correspond to the same local values of temperature and electron density as

$$n_i = b_i n_i^*,$$

where n_i is the actual (NLTE) population, n_i^* the corresponding LTE population, and b_i is the so-called departure coefficient (b-factor) of a given level i.

The solution of the statistical equilibrium equations approaches the LTE distribution ($b \to 1$) if either the collisional rates dominate over the radiative rates, or the radiation field approaches the equilibrium—Planckian—distribution. Taking into account the typical atmospheric conditions, the physical state of a stellar atmosphere may be characterized by a continuous transition from deep layers, where LTE prevails, to more superficial layers, where NLTE is more and more important.

Finally, it should be stressed that the (astrophysical) (LTE) already represents a non-equilibrium state of the radiation matter complex. The approximation of LTE allows for the most pronounced non-equilibrium property of stellar atmospheres: the radiation escapes the atmosphere (we see the star !) and, therefore, cannot generally possess a Planckian distribution. The actual radiation field is then determined by solving the radiative transfer equation; the opacity and the source function are, however, assumed independent of the radiation field. The LTE modeling also allows for an indirect influence of radiation on the thermodynamic parameters via the equation of radiative equilibrium. The only feature which is neglected by the LTE approach is the influence of radiation on the atomic level populations (and thus on the opacity and source function). Anyway, the LTE approach accounts for many essential features of stellar atmospheres; this explains its relative success in atmospheric modelling (see below).

2.2 Coupling versus Decoupling

To take into account all the microscopic processes populating and depopulating the atomic levels, thereby allowing for the coupling between all the populations and state parameters, is clearly beyond the state of the art. It is thus necessary to disregard, or crudely simplify, a vast majority of individual microscopic processes, believing that those which were neglected are indeed negligible.

Thus, while the coupling between processes (state parameters) is q quite essential NLTE feature, the idea of decoupling is the actual gist of any application. The usual decoupling, which is virtually always made, is a two-step calculation of predicted spectral features (e.g. line profiles): i) calculation of the model atmosphere, adopting LTE, etc., and ii) having been given the model atmosphere, the radiative transfer and the statistical equilibrium equations are solved for an individual atom or ion, taking into account as much detail as feasible (and reasonable). The assumption underlying this approach is that the detailed behavior of the level populations and the radiation field in the lines and continua of this atom (ion) do not affect the overall atmospheric structure significantly.

Notice that a decoupling is inevitably involved in the second step as well. Only a limited number of atomic levels and transitions between them is accounted for explicitly; the remaining ones are treated either approximately (e.g. in LTE) or are ignored completely. Consequently, careful test calculations are, in principle, always required to verify that nothing important has been overlooked.

However, there is one problem that makes the safe applicability of the two-step procedure doubtful. This problem is usually referred to as line blanketing, and represents a task of accounting for a large number of spectral lines which influence both the atmospheric structure and the emergent spectrum. Since every single spectral line is, at least in principle, affected by departures from LTE, and since we are interested in the cumulative effect of many lines, the very nature of the decoupling of the atmospheric structure and the detailed line transfer is violated.

On the other hand, in judging the efficiency and suitability of a modelling approach, one should be aware of an intimate link between the degree of sophistication of the theoretical calculations, on the one hand, and the intended interpretational goal on the other. Obviously, less accurate observations may be interpreted by means of simpler models; various spectral features in various stars often require different kinds of models.

Specifically, the following diagnostic scheme, which is parallel to the two-step atmospheric modelling, is commonly accepted: Step (i) - model atmospheres - is viewed as a procedure that yields the continuum flux, while step (ii) - detailed transfer solutions - is viewed as a procedure needed to predict detailed line profiles. Once again, line blanketing complicates the matter significantly, yet some features of the two-step procedure are found here as well. Schematically, less accurate treatments of numerous lines are used to compute the atmospheric structure and predicted continua, while more accurate treatments are used to compute detailed synthetic spectra.

There is another source of difficulties that may hamper or even inhibit accurate modelling. This is the necessity of knowing precisely an enormous amount of atomic data, such as oscillator strengths, photo-ionization cross-sections, collisional rates, line broadening parameters, etc. A particularly useful review of this problem in the context of CP stars is given by Cowley and Adelman (1983).

We have tried to summarize the above considerations in Figure 1. Further on, we shall discuss the three basic steps of modelling—model atmospheres, detailed line transfer, and line blanketing— in turn. Before doing this, we advance a simplified picture which will prove useful in the subsequent analysis.

2.3 Simplest NLTE Situations

Although the nature of the NLTE effects is generally quite complex, it may often be visualized as the result of superposition of or competition among some elementary phenomena. However, this idea must be used with great caution. Although useful, such a simplistic view may be quite misleading in certain cases.

Let us restrict ourselves to the case of a semi-infinite atmosphere with no incident

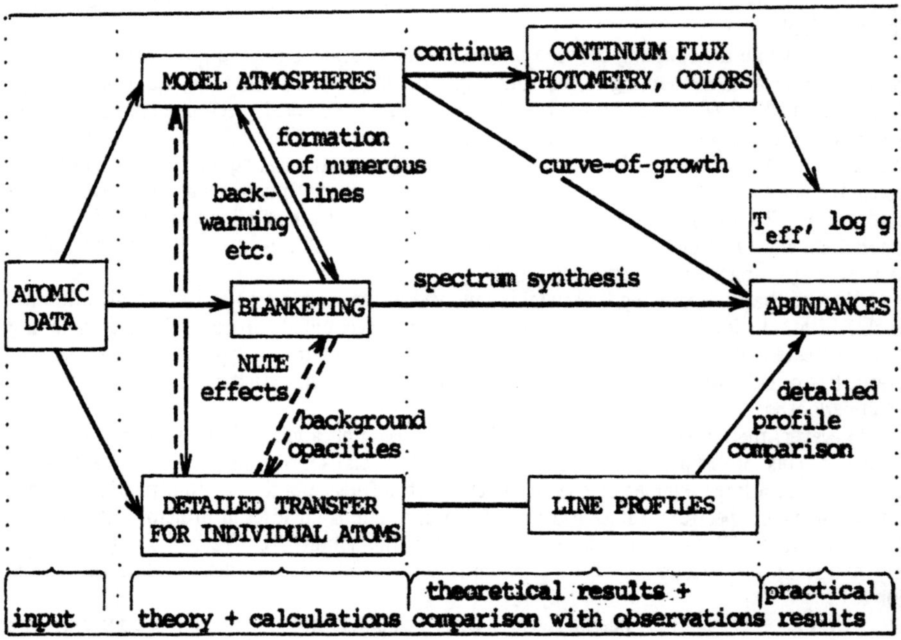

Figure 1. A Schematic diagram suggesting the logical connections between the different steps of atmospheric modelling and spectroscopic diagnostics discussed in the text. The full lines represent the standard procedures that have already been (at least partly) elaborated; the broken lines represent the procedures to be worked out.

radiation at the surface. There are now in principle two kinds of NLTE effects: 1) those which arise due to the presence of an outer boundary (a lack of photons that have escaped through the boundary affects the balance of photo-excitations and de-excitations or photo-ionizations and recombinations), irrespective of the presence of gradients of the thermodynamic parameters (temperature, density); the gradients may still modify the magnitude of these effects; and ii) effects that arise due to the presence of gradients. Three different cases can thus be distinguished which are displayed schematically ion Figure 2.

Case A, a typical NLTE effect, represents the behavior of a strong resonance line (or a strong ground state continuum, e.g. the hydrogen Lyman continuum). Cases B and C are typical of continua. In both these cases, the mean intensity J_ν decreases towards the surface, where it roughly equals $(1/2)S_\nu(\tau_\nu=1) \simeq (1/2)B_\nu$, S_ν being the source function and B_ν the Planck function. Consider a bound state i. The photo-ionization rate is proportional to $\int J_\nu ... d\nu$, while the recombination rate is proportional to $\int B_\nu ... d\nu$. Now if the statistical equilibrium from level i is dominated by radiative ionizations and recombinations (which occurs at depths where the lines are saturated while the continua are not) then $b_i > 1$ if $J_\nu < B_\nu$ (case B), and $b_i < 1$ if $J_\nu > B_\nu$ (case C).

Case C represents the characteristic behavior of the continua formed at wavelengths for which the gradient of the Planck function is sufficiently steep (roughly, shortward of the maximum of the Planck function for a temperature equal to the effective temperature). Actually for A and late B stars, the type C behavior is typical of the continua formed in the Balmer continuum.

Obviously, our scheme is neither exhaustive nor complete. Yet, while other similar schemes (see, e.g., Mihalas 1978; Freire and Praderie 1974) are devised primarily to understand the behavior of NLTE line profiles, our scheme is meant to be helpful in understanding the global NLTE multilevel, multitransition radiative transfer in lines and continua of complex atoms or ions.

3. MODEL ATMOSPHERES

3.1 LTE Models

Let us consider separately model atmospheres of normal and "peculiar" stars, the distinction between both groups being roughly given by the adopted chemical composition, or, sometimes, by including magnetic fields.

3.1.1. Normal Stars. By far the most extensive grid of LTE model atmospheres available at present is Kurucz's (1979). Line blanketing of about one million atomic lines is accounted for statistically by means of the opacity distribution functions. This grid, supplemented by the corresponding theoretical ubvy and UBV photometric indices, calculated by Relyea and Kurucz (1978) and Buser and Kurucz (1978), respectively, provides a widely used norm to which observed stellar spectra and colors are compared.

Concerning the spectroscopic diagnostics of A and late B stars, Kurucz himself showed that there is an excellent agreement between his model predictions and the spectroscopic observations of Vega. Afterwards, many investigators employed Kurucz's models and compared the predicted and observed flux distributions. Systematic studies of a sample of stars have been carried out, e.g. by Underhill et al. (1979), Böhm-Vitense (1981), and Malagnini et al. (1982, 1983). Generally, these authors have demonstrated that Kurucz's models reproduce the observed visual distribution fairly well, while the UV flux is reproduced somewhat worse, but still satisfactorily. However, the present observational uncertainties in the ultraviolet do not allow definite conclusions concerning model deficiencies to be drawn.

Kurucz's grid meets most of the practical requirements, hence only a few other LTE model calculations of normal stars have recently been carried out. We mention the study of Dreiling and Bell (1980), who have calculated a detailed line blanketed model atmosphere for Vega, and compared the theoretical predictions with the observations in the wavelength range $1250 \leq \lambda < 10800\mathring{A}$. They again found satisfactory agreement, except in the far UV region ($\lambda < 1500\mathring{A}$) where, however, the observational uncertainties are the largest.

3.1.2. LTE model atmospheres for peculiar stars. The published model atmospheres can be divided into two categories which mimic the two most pronounced characteristics of the peculiar stars, namely i) models with enhanced metal abundances, with otherwise the same assumptions as adopted for modelling the atmospheres of normal stars; and ii) models that account for the effects of magnetic fields.

The models of the first category basically aim at describing quantitatively the phenomena expected on simple theoretical grounds: an enhanced metal opacity in the ultraviolet decreases the UV flux and increases the visible flux via increased backwarming in the deep atmospheric layers. Strom and Strom (1969) and Peterson (1970) suggested that continua of common elements, e.g. Si, may supply such an enhanced opacity. (However, subsequent NLTE calculations showed that continua alone are insufficient - see below). A more realistic study has been carried out by Leckrone et al. (1974) who were able to demonstrate convincingly that

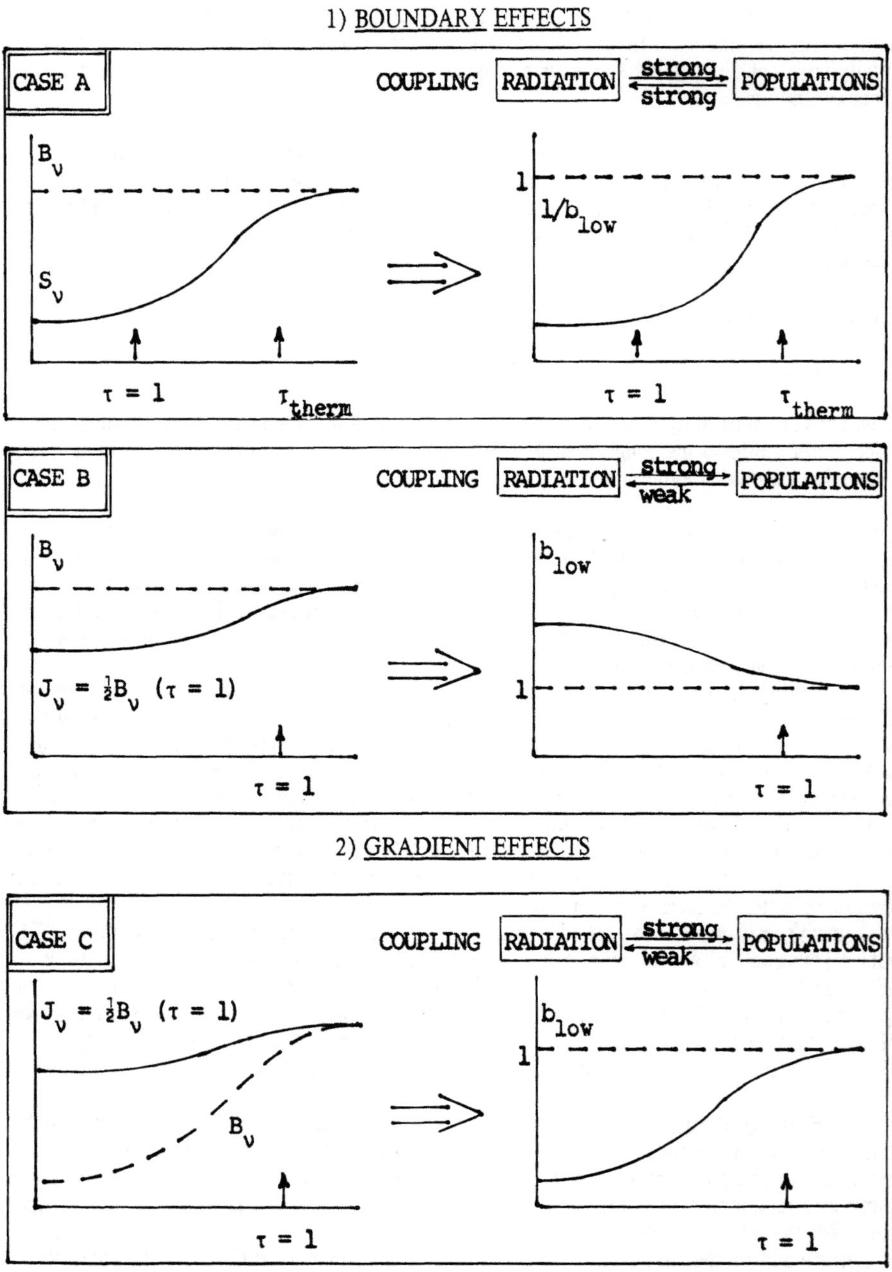

Figure 2. A highly schematic diagram illustrating the simplest NLTE situations. Left-hand graphs – see text; right-hand graphs – the corresponding b-factors for the lower level of a given transition.

an enhanced line opacity model simulates a hotter atmosphere in the visible and a cooler atmosphere is the UV, similar to what is actually observed.

Subsequently, a similar but more extensive study was undertaken by Muthsam and co-workers. Line blanketing of about $x10^5$ spectral lines has been accounted for by means of the opacity sampling procedure suggested by Peytremann (1974); representative abundances for CP stars have been adopted. Magnetic fields were not considered in the equation of hydrostatic equilibrium, but their effects on the formation of spectral lines (by Zeeman broadening) have been mimicked by introducing a pseudo-microturbulent velocity (Muthsam 1979). Besides calculating a grid of models (Muthsam 1978), they presented a comparison of the theoretical predictions with observations, both for a sample of Ap stars (Stepien and Muthsam 1980) and for individual stars (Muthsam and Stepien 1980; Stepien and Muthsam 1981; Muthsam and Cowley 1984). Some individual features present in the Ap-star spectra have also been considered by the spectral synthesis technique (Muthsam and Weiss 1978; Maitzen and Muthsam 1980).

There is a number of further studies that either employ Kurucz's (1979) models or calculate their own model atmospheres using a methodology similar to Kurucz's, and compare the theoretical predictions with observations for various types of CP stars [e.g. Kurucz and Furenlid 1979; Bell and Dreiling 1981 - Sirius; Lane and Lester 1984 - Am stars; Baschek et al. 1984 - λ Boo stars, etc.]. Attention has also been devoted to the search, by means of the LTE spectral synthesis, for extra opacity sources in the atmospheres of Ap stars (autoionization lines - Jamar et al. 1978; Jamar 1980; Artru et al. 1981; Artru and Lanz - this conference).

Generally, the LTE models have been proven to provide a good explanation of the many observed features of CP stars, thereby demonstrating that the classical methodology of computing line blanketed LTE model stellar atmospheres is indeed an efficient approach to modelling CP stellar atmospheres, which is worth pursuing further. On the other hand, many individual features, particularly in the ultraviolet but also in the visible, remain unexplained.

Finally, we shall briefly mention the models of the second category. An earlier paper by Stepien (1978) concluded that the effects of magnetic fields should be rather small. However, subsequent, more complete calculations by Madej (1983) indicate that the magnetic pressure can significantly influence the atmospheric structure and the emergent spectrum of an Ap star. Yet, these models do not consider metal line blanketing.

Carpenter (1985) has recently taken a major step forward by calculating LTE model atmospheres that take into account both the magnetic pressure and the line blanketing effects. He demonstrated that the structure changes (via alteration of the gas pressure distribution due to magnetic forces) in combination with enhanced blanketing (via Zeeman broadening of lines) cause the emergent spectrum to vary with the viewing inclination and to differ from the non-magnetic case, and concluded that these model atmospheres of magnetic Ap stars are consistent with the observations.

3.2 NLTE Model Atmospheres

3.2.1 Why calculate NLTE models? As discussed in the preceding section, LTE model atmospheres have been shown to provide an excellent basis for the interpretation of the observed spectra of normal stars, and a somewhat poorer but still satisfactory basis for CP stars. Moreover, numerous studies seem to indicate that the discrepancies between observations and theory are likely to be caused by observational errors and/or other physical mechanisms (inhomogeneous abundances, magnetic fields, etc.) rather than by departures from LTE. The natural question then arises whether to calculate NLTE model atmospheres for A and late B type stars.

In our opinion, the answer is affirmative due to the following reasons: i) The NLTE description of the interaction between the radiation and matter is principally superior to the LTE one. If for nothing else, the NLTE models can serve to justify the applicability of the LTE models. ii) NLTE is an inherent part of radiative transfer physics. While phenomena like inhomogeneous abundances or magnetic fields may or need not be actually present, the NLTE effects are unequivocally, to a higher or lesser degree, present always. Hence, NLTE phenomena should be understood well, otherwise a misinterpretation of the observations in terms of spurious phenomena may easily occur.

From a more practical point of view, NLTE models are necessary iii) in order to decide for which spectral features the departures from LTE are substantial; iv) to treat the uppermost layers of an atmosphere and related spectral features (cores of the strongest lines, X-radiation, etc.). Here, and NLTE treatment is obligatory, yet further refinements of the theory are still required.

3.2.2 Overview of NLTE model calculations ($T_{eff} \leq 15000K$). In spite of the great conceptual importance of NLTE model calculations, only a few studies have been undertaken so far. The first NLTE, pure hydrogen model atmosphere in this temperature range (T_{eff} = 12500K) was calculated by Auer and Mihalas (1970). They demonstrate that this model exhibits a behavior similar to that found for higher temperatures as, for instance, an increase of the electron temperature in the uppermost layers (due to the indirect effect of the Balmer lines on the continuum energy balance.

Analogous calculations for cooler atmospheres were carried out by Frandsen (1974) and Kudritzki (1973). Frandsen calculated the NLTE models for T_{eff} = 10000K, log(g) = 4 (and also 2, and 1); Kudritzki considered hydrogen-helium atmospheres and calculated NLTE models for T_{eff} = 10000K, log(g) = 1, for various ratios N(He)/N(H). Finally, Borsenberger and Gros (1978) have extended Mihalas's (1972) grid of NLTE model atmospheres towards cooler stars. They presented a grid of NLTE model atmospheres for $10000 \leq T_{eff} \leq 15000K$, log(g) = 4. The atmospheres were assumed to be composed of hydrogen, helium [with N(He)/N(H) = 0.1] and an average of C, N, O, with N(CNO)/ N(H) = 10^{-3}.

In all of the above models, departures from LTE were allowed for the first five levels of hydrogen; the hydrogen Hα, Hβ, Pα, Pβ, and Bα lines (except for Kudritzki's models) were taken into account explicitly in the statistical equilibrium equations. The Lyman lines were taken to be in detailed balance. The most important conclusions drawn from these studies are: i) the energy distribution in the continua is little affected by departures from LTE, except at very low gravities and in the (yet unobserved) Lyman continuum; and ii) significant departures from LTE occur in the cores of the Balmer and Paschen lines, the wings being described rather well by LTE.

On the other hand, it was recognized during the seventies that the NLTE effects in hydrogen are not the most important NLTE effects to be expected for A and late B stars. For instance Snijders (1977a,b), who did not actually calculate model atmospheres, but used Kurucz's models and recalculated the emergent radiation, demonstrated that the NLTE effects in neutral carbon (actually C type effects in our terminology) affect the emergent far UV radiation considerably. Consequently, it became clear that the NLTE modelling of the A and late B atmospheres should represent more than a straightforward extension of the methodology used for B and O stars.

More realistic model atmospheres of early A type stars ($9400 \leq T_{eff} \leq 9700K$, log(g) = 4), allowing for departures from LTE in H, H^-, C I, Si I, N I, S I and Mg II, were calculated by Hubený (1981). The effects of line wing opacities of the Lα to Lδ lines were included in the calculations, allowing for the departures from complete frequency redistribution (Hubený 1980). In agreement with the previous NLTE studies, the effects of departures from LTE have been found to be negligible in the visible and near UV continua ($\lambda > 1700$Å) while, in contrast, NLTE effects were found to be important for the far UV region ($\lambda < 1500$Å). In particular, for the region $1100 < \lambda < 1200$Å, the LTE flux is too low by a factor of as much as 50! Good agreement has been found between the predicted flux for the NLTE model with

T_{eff} = 9660, log(g) = 4, and the flux observed for Vega in which the UV and visible, and even in the region 1100 < λ 1200Å, which was previously a point of controversy (see Praderie et al 1975; Snijders 1977b; Praderie 1981).

All the above studies ignore the effects of metal line blanketing. The only attempt to mimic these effects in NLTE model atmospheres is due to Borsenberger and Jamar (1980) who studied the influence of an enhanced UV opacity on atmospheric structure. They found the behavior to be analogous to that described by Leckrone et al. (1974) for LTE models. However, their approach is questionable since they considered departures from LTE only in hydrogen and not in the most important UV absorbers (C, Si, etc.).

3.3 Discussion

The main interpretational conclusions following from the above-mentioned studies are summarized in Table 1. Although very schematic, this table illustrates several important methodological considerations outlined earlier.

Table 1: Spectroscopic Diagnostics of Continua

Region	Diagnostic method	Main problems
IR continue	LTE	circumstellar matter (?)
Vis continua		
-colors	LTE	(blanketing)
-special photometry (Geneva, Maitzen)	LTE ?	blanketing,magnetic fields NLTE ionization shifts
• near UV continua	LTE ?	blanketing, magnetic fields, atomic parameters, NLTE ionization shifts
far continua	NLTE	blanketing, atomic parameters, details of NLTE treatment
EUV, X continua	more exact physics	practically not tackled yet

From the practical point of view, the diagnostic method indeed depends on the spectral region and/or the precision of observations. Roughly speaking, LTE model atmospheres appear to be a satisfactory diagnostic tool for continua of A and late B stars. On the other hand, more detailed observations and/or the far UV region require an NLTE treatment.

From the methodological point of view, NLTE model atmospheres have already proven their usefulness. However, it should be kept in mind that their true value still lies in the philosophy of the problem rather than in practical applications. As an example, consider the far UV region in early A stars - see Fig. 3. The LTE and NLTE predictions for T_{eff} = 9600K are quite different; the NLTE prediction yields an almost perfect agreement with observations for Vega (Hubený 1981). Moreover, the NLTE predictions indicate that the violet wing of Lα is very sensitive to the effective temperature. Consequently, this region might seem to be an excellent indicator of T_{eff}. However, this is not quite so, since various uncertainties and approximations in the relevant atomic data and the treatment of the far UV opacity sources yield large differences in the calculated emergent flux. The corresponding scatter in the calculated departure coefficients for the C I ground state is given in Fig. 4. (for details refer to Hubený 1981).

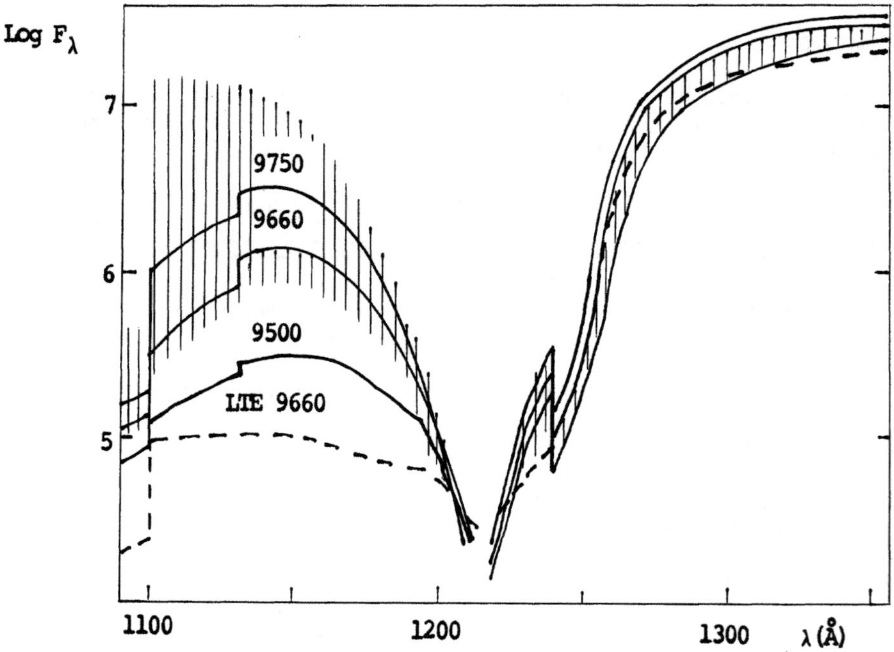

Figure 3. Predicted far UV flux for NLTE (full lines) and LTE (broken lines) model atmospheres. The curves are labeled with the corresponding value of T_{eff}. The shaded area represents the range of the emergent flux for T_{eff} = 9660 calculated under various approximations in the treatment of the far UV opacity sources (after Hubený 1981).

In summing up, NLTE models, although certainly important, are not yet capable of providing a reliable and accurate norm to which the observations can be compared in detail. The crucial problem to be solved is the inclusion of metal line blanketing in the NLTE model atmospheres. We shall return to this question in Section 5.

4. DETAILED TRANSFER SOLUTIONS FOR INDIVIDUAL ATOMS

4.1 Overview

NLTE studies of the individual atoms/ions are more common than those of model atmospheres. However, not all such studies have been motivated solely by reasons explained in

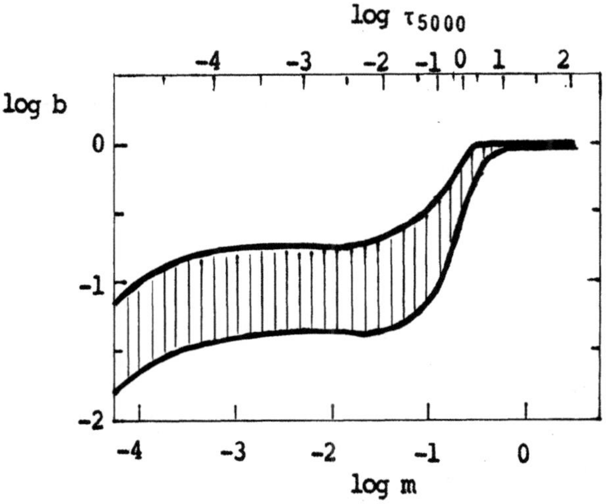

Figure 4. The corresponding range of the departure coefficient for the C I ground state calculated under various approximations in the treatment of the far UV opacity sources (after Hubený 1981).

Sect. 2.2. In fact, many NLTE studies have been carried out to examine numerically the parameter-free model of diffusion of chemical elements in CP stars (Michaud 1970, 1981, this conference; for computational details refer to Borsenberger 1979).

The calculations performed so far for A and late B stars are summarized in Table 2. For completeness, the solutions for atoms/ions, which were calculated self-consistently with the global atmospheric structure (and hence, strictly speaking, do not belong to the present category of studies), are also included.

From the computational point of view, the majority of studies use a "direct" approach: several lowest atomic levels are treated explicitly; higher levels are either disregarded or treated approximately. An entirely different approach has been adopted by Alecian and Michaud (1981). To cope with the complicated energy level structure of Mn III, they did not consider the individual energy levels separately; instead, they lumped several levels with similar energies together, assuming that all levels within each group have equal b-factors. Such an approach represents a promising way of examining NLTE effects in other "complicated" atoms, such as the iron group elements.

The main purpose of Table 2 is to extract the general features of the individual solutions. An inspection of Table 2 reveals two important conclusions which were, in fact, already suggested earlier (cf. Praderie 1976; Snijders 1977a), namely: i) For ions with the ionization limit in the Lyman continuum and with no low-lying levels collisionally coupled with the ground state, there are small continuum effects; line effects dominate the behavior of level populations; ii) For ions with the ionization limit in the Balmer continuum, or in the Lyman continuum but with at least one collisionally coupled low-lying level, the ionization energy of which lies in the Balmer continuum, there are important C-type continuum effects (see Sect. 2.3).

In the latter case, the departure coefficients of all levels considered explicitly exhibit the behavior displayed schematically in Figs 2(C) and 4. Higher up in the atmosphere, the line effects start to influence the populations. Despite the complex nature of these phenomena, two types of behavior may again be distinguished:

a) for abundant atoms (C, Mg) – the A-type effects completely dominate the formation of resonance lines, while the subordinate lines are still more or less affected by the C-type continuum effects;
b) for less abundant atoms (Be, B, Ca, Sr, Ba) – the resonance lines are now weaker; the result of the competition between the A-type effects for resonance lines and C-type for continua is very sensitive to the adopted abundance as well as to the actual values of the relevant atomic parameters. The C-type continuum effects often dominate the formation of the subordinate lines.

4.2 Ionization Shifts

The latter conclusion is particularly important from the point of view of abundance determinations and line blanketing studies. It shows that the C-type effects may dominate the formation of many observable lines which are thus formed as if the ***abundance of a given ion were lower than the corresponding LTE abundance***—the so-called ionization shift. However, this does not necessarily mean that the total abundance of the ion is actually lower than the LTE abundance. For instance, if the population of the ground state significantly exceeds the total population of the excited states, and if the A-type effects dominate the formation of the resonance lines, then $N_{ion} < N^*_{ion}$ at depths where continua are transparent but resonance lines are still saturated (pure C-type), and $N_{ion} > N^*_{ion}$ at depths where $b_1 > 1$ (i.e. where the A-type effects dominate over the C-type).

The ionization shift is an important NLTE effect because it can *systematically* affect the formation even of weak lines. Its relative simplicity indicates the possibility, to be employed more fully in future calculations, of considering NLTE effects for a large number of lines without performing heavy calculations. Specifically, were we able to estimate quantitatively the ionization shifts for individual ions, we could calculate synthetic spectra and NLTE blanketed model atmospheres by methods analogous to those employed in the LTE calculations.

On the other hand, the present state of the matter does not allow such quantitative estimates to be made reliably. We are able, at least, to summarize the favorable conditions for the occurrence of the ionization shift (cf. Snijders 1977a):

1) The next ion is the dominant ionization degree of a given atom.
2) The next ion is predominantly in the ground state.
3) The given ion is predominantly in a few low-lying states.
4) One has the C-type continuum effect at least for one low-lying level.

Briefly, conditions 1 and 2 are necessary to obtain an actual ionization shift. If they are not satisfied, an increase of the ionization increases the ground-state population of the next ion substantially, thereby also increasing the recombination rate and thus the population of the lower ion. If condition 3 is not satisfied, the C-type effects may be overwhelmed by coupling (radiative and collisional) with other lines and continua; condition 4 is obvious.

The following ions have already been proved to fulfill all these requirements for the conditions found in the atmospheres of A and late B stars: C I, N I, Si I, S I, Mg II, Ca II, Sr II, Ba II. Taking into account the atomic level structure and the relevant ionization potentials, we may reasonably expect ionization shifts for the following ions: P I, Cl I, Mn I, Fe I, Co I, Ni I, Cu I, Zn I, Y II, Zr II, Hg I, etc., and, importantly, for once ionized rare earths.

It is interesting to note that Cowley (1980) suggested, on purely observational grounds that the LTE ionization equilibrium for the rare earth elements is probably violated. Our estimates show that the NLTE ionization shifts act in the right sense towards removing the observational discrepancies, but, clearly, this question deserves further attention.

TABLE 2: NLTE Transfer Solutions for Individual Atoms

Atom/ion	Reference	Purpose	Continuum effects	Line effects
Be II	Boesgaard et al. (1982)	LWA	-	type A res.lines
Be I,II	Borsenberger et al (1984)	D		-
B II	Praderie et al. (1977)	LA(N)	Type C	$b_2 > b_1$
B II,III	Borsenberger et al. (1979)	DNWL	Type C	(weak res. line)
C I	Snijders (1977a,b)	N,C	type C	-
	*Hubený (1981)	N(C)	type C	-
C II	Freire (1979)	CLAN	-	type A res.lines
N I	*Hubený (1981)	N	type C	-
O I	Baschek et al. (1977)	NLWA	-	type A res. lines
Mg I,II	Borsenberger et al (1984)	DW		
Mg II	Snijders and Lamers (1975)	NLW	type C	type A res. lines
	Freire Ferrero et al. (1983)	CLAN		
Si I	Leckrone and Snijders (1979)	N	type C	-
	Vauclair et al. (1979)	D	type C	
Si II	Freire (1979)	CLAN	-	type A res.lines
S I	*Hubený (1981)	N	type C	-
Ca II	Snijders (1975)	N		type A res.lines
	Freire et al. (1978)	CLAN	type C	(competition
	Borsenberger et al. (1981)	DWLN		with C)
Mn III	Alecian and Michaud (1981)	D	-	-
Sr II	Praderie (1976)	NLW(D)	type C	weak type A
	Borsenberger et al. (1981)	DWN	type C	(competition with C)
Ba II	Borsenberger et al. (1984)	DW	type C	-

Purpose:
D – diffusion theory calculations
N – discussion of NLTE effects
A – abundance determination for individual stars
L – detailed line profile calculation
W – calculation of equivalent widths

C – search for chromospheres

Asterisk (*) denotes calculations performed self-consistently with the global atmospheric structure

4.3 Abundance Determinations

Experience gained in the aforementioned NLTE studies (see also Freire and Praderie 1974; Dumont et al. 1975) enables some conclusions to be drawn concerning the reliability of the various methods of abundance determinations. We shall try to summarize the main advantages and drawbacks of the individual methods in Table 3.

TABLE 3. Comparison of Various Methods of Abundance Determinations.

Means	Advantages	Drawbacks	Problems
Strong resonance lines (profiles)	–little influenced by blends –simple atomic models sufficient –smaller uncertainties in atomic parameters	NLTE obligatory (but)	–partial redistribution (interstellar absorption)
Intermediate lines (profiles eq. widths	–LTE source function probably satisfactory (but ionization shifts)	–uncertainty in using LTE –difficult NLTE (many levels) –influenced by blends!	–atomic data! –Zeeman broadening
Weak lines (eq. widths)	–LTE source function probably satisfactory (but ionization shifts)	–influenced by blends!	–atomic data –Zeeman broadening

It appears that the most advantageous method for *accurate* abundance determinations is the method of detailed comparison of the observed and predicted profiles of resonance lines (supplemented, if possible, by an analogous comparison for the lowest subordinate lines). It is, however, important to calculate the theoretical profiles in NLTE and to account, at least approximately, for the partial redistribution effects (cf. Hubený 1985, and references therein). As compared to classical methods, the present one requires more complicated calculations, yet this drawback is outweighed by an increased accuracy of the deduced abundances, which may easily be pushed up to 0.05 – 0.1 dex.

5. LINE BLANKETING

Excellent reviews of various aspects of line blanketing have been presented by Gustafsson (1981) and Carbon (1979, 1984). In this paper we avoid all the numerical aspects and restrict ourselves to the three following questions which appear to be the most important from the point of view of the spectroscopic diagnostics of CP stars:

1) Why is the LTE treatment of line blanketing so successful in explaining observations (i.e. both in construction model atmospheres and calculating detailed synthetic spectra)?
2) Yet, why and when is an NLTE treatment necessary and better?
3) What are the prospects of calculating blanketed model atmospheres and detailed synthetic spectra in an NLTE treatment?

We shall discuss these questions in turn.

5.1 Why is LTE Often Satisfactory?

The opacity in a given line i → j may be expressed (neglecting stimulated emission for convenience) as

$$k_l(\nu) = (\pi e^2/mc)\, f_{ij}\, b_i\, n_i\, H[a, \nu - \nu^0_{ij})/\Delta\nu_D]/(\sqrt{\pi}\Delta\nu_D)$$
$$= k^*_l(\nu)\, b_i$$

where k^*_l denotes the LTE opacity, b_i is the b-factor for the lower level; the other quantities have their standard meanings (see, e.g. Mihalas 1978). To the same degree of approximation, the line source function is given by

$$S_l = (b_j/b_i)\, B_\nu .$$

Now, for many weak lines, $b_i \simeq b_j$ while b_i may generally be different from unity (e.g. $b_i < 1$ for the ionization shift). Then $S_l \simeq B$, and $k_l \simeq k^*_l b_i$, i.e. the NLTE description only involves a single multiplicative factor in the opacity. However, b_i often deviates from unity by relatively small factors while the uncertainty in $g_i f_{ij}$ (the so-called gf-value), or the abundance adopted, may easily be comparable with or larger than b_i.

This is not to say that LTE is a perfect description. This simple example indicates that there are other sourced of uncertainties in treating line blanketing which may often be more important than NLTE itself. In this respect, an interesting conclusion follows from a study by Burger (1981): in computing a sample of a synthetic spectrum with the best available gf-values, and with all the gf-values set to −0.5, who found that all the observed gross spectral features were satisfactorily reproduced by both calculations. Since even this dramatic change in the oscillator strengths yields acceptable results, it is clear why NLTE, which would probably induce even smaller effects, can often be disregarded. Moreover, many spectral features, particularly in the UV, are seriously influenced by blends which may yield spurious asymmetries, shifts, and even appearances of spectral lines (Kurucz 1974); Hubený et al. 1985). Again, a fully LTE treatment seems to be better suited to such problems than the seemingly more exact NLTE study which neglects blends.

5.2. Why NLTE Blanketing:

The arguments in favor of the NLTE approach are precisely the same as those outlined in Section 3.2. However, very little work has been done so far that would demonstrate these methodological considerations quantitatively.

A preliminary analysis of this problem (Hubený 1981) suggests that the main error in the LTE blanketed model atmospheres is probably not caused by the LTE treatment of numerous weak lines, but rather by the LTE treatment of strong UV continua (C I, Si I, Etc.). Specifically, LTE largely overestimates the far UV opacity, thereby yielding too high backwarming in the continuum-forming regions.

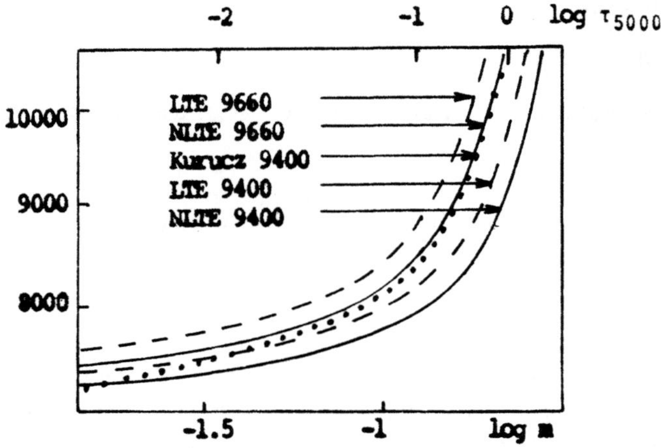

Figure 5. Temperature in the continuum-forming layers for various model atmospheres. Full lines represent NLTE-unblanketed models, broken lines LTE-unblanketed models for the indicated values of T_{eff}. The dots represent Kurucz's LTE blanketed model for T_{eff} = 9400K (after Hubený 1981).

This is illustrated in Figure 5. The LTE treatment, which ignores metal line blanketing, certainly underestimates the temperature in the continuum-forming regions (compare the curves Kurucz 9400 and the LTE 9400) but simultaneous overestimates the temperature as compared to the NLTE model (compare the curves LTE 9400 and NLTE 9400, or LTE 9660 and NLTE 9660). Yet, importantly enough, the difference T(blank., LTE) - T(unblank., LTE) is almost as large as the difference T(unblank., LTE) - T(unblank., NLTE) ! Since an analogous overestimate of the far UV opacity sources is necessarily present in Kurucz's models, the temperature in his models is probably systematically too high.

However, the above considerations do not take into account the influence of metal line blanketing on the ionization rates of carbon, etc., which would somewhat decrease the magnitude of the NLTE effects. Anyway, we may conclude that the effective temperature, deduced by means of Kurucz's models is uncertain by about 200K for early A-type stars. Studies (of normal stars), that do not require a higher precision, may employ the Kurucz grid, but studies that aim at a more precise determination of T_{eff} should wait until NLTE line blanketed models become available.

On the other hand, an almost identical behavior of temperature with depth for the LTE line blanketed model with T_{eff} = 9400 and the NLTE-unblanketed model with T_{eff} = 9660 (see Fig. 5) explains why detailed spectra, computed for both models, will nearly coincide under the same assumptions about line formation. Thus, an LTE spectrum synthesis using Kurucz's models may be quite reliable, provided that LTE is a good approximation for the formation of weak lines, even if the models themselves might be in error; the same applies to the NLTE-unblanketed models. In other words, this illustrates that it is, to a large extent, permissible to consider separately the effects of metal line blanketing on the atmospheric structure and on the synthetic spectra.

The problem of the uppermost layers, where NLTE is obligatory, has barely been tackled seriously so far. It is clear that the two often quoted results—large surface cooling in the LTE line blanketed models, and a temperature rise in the NLTE unblanketed models—represent crude oversimplifications of the problem.

5.3 Perspectives of NLTE Line Blanketing

The considerations presented above, together with other investigations (Phillips 1977; Anderson 1985), indicate that the problem of construction "reasonable" LTE line blanketed models is not hopeless. To treat millions of spectral lines in NLTE is certainly disastrous. However, it should be kept in mind that a large part of the total effect is due to a relatively small number of lines and continua. It is essential to treat some important lines and continua in NLTE; other lines may be treated in LTE or some approximate NLTE approach (as, for example, with the estimate of ionization shifts for relevant ions). This is perhaps more vital for the future research into the field of line blanketing than adding more and more lines in strict LTE.

The rationale for such an approach stems from the experience gained from numerical calculations: errors in the computed atmospheric structure due to an insufficient treatment of crucial opacity sources (e.g., C I ground-configuration continua in the case of early A stars - Hubený 1981; or C IV resonance lines in the case of O stars - Anderson 1985) are equivalent to, or may easily outweigh the combined effect of thousands and thousands of weak lines.

The most promising way forward seems to be a suitable extension of the approach developed by Anderson (1985). He uses a multifrequency/multigray algorithm which admits the inclusion of up to about 300 lines and any number of continua in full statistical equilibrium. Cleverly constructed artificial model atoms can extend the number of lines to 3000 or more, and opacity sampling techniques, similar to those employed in the LTE modeling, may account for still weaker lines. This procedure is thus ideally suited to meet the requirement specified above.

6. CONCLUSION

Recent work has amply demonstrated that the chemical peculiarities found in CP stars are not an artifact of an inadequate analysis. From this viewpoint, the NLTE physics is not an essential ingredient of our understanding of the CP phenomenon. However, whenever one looks at a stellar spectrum and tries to deduce from it the details of the physical state of the stellar atmosphere, NLTE physics *becomes* essential.

We have tried to demonstrate that, although many spectral features and/or many low-precision observed data may be satisfactorily explained by means of the traditional LTE modeling, there is still a wide class of observations which would require careful NLTE studies. Some work has already been carried out, but much more remains to be done. Although very laborious and computationally demanding, this effort will certainly be rewarding.

Acknowledgement: The author gratefully acknowledges the travel grant provided by the International Astronomical Union.

REFERENCES

Alecian, G., and Michaud, G. 1981, *Ap. J.*, **245**, 226.
Anderson, L. S. 1985, in *Progress in Stellar Spectral Line Formation Theory*, ed. J.E. Beckman and L. Crivellari, NATO ASI Series (Dordrecht: D. Reidel) p 225.
Artru, M. C., Jamar, C., Petrini, D., and Praderie, F. 1981, *Astron. Ap.*, **96**, 380.
Athay, R. 1972, *Radiative Transport in Spectral Lines*, (Dordrecht: D. Reidel)
Auer, L. H., and Mihalas, D. 1970, *Ap. J.*, **160**, 223.
Baschek, B., Scholz, M., and Sedlmayr, E. 1977, *Astron. Ap.*, **55**, 375.
Baschek, B., Heck, A., Jaschek, C., Jaschek, M., Köppen, J., Scholz, M., and Wehrse, R. 1984, *Astron. Ap.*, **131**, 378.
Bell, R. A., and Dreiling, L. A. 1981, *Ap. J.*, **248**, 1031.
Boesgaard, A. M., Heacox, W. D., Wolff, S. C., Borsenberger, J., and Praderie, F. 1982, *Ap. J.*, **259**, 723.
Borsenberger, J. 1979, *Thèse de 3ème cycle, Université Paris VII.*
Borsenberger, J., and Gros, M. 1978, *Astron. Ap. Suppl.*, **31**, 291.
Borsenberger, J., and Jamar, C. 1980, *Astron. Ap.*, **91**, 247.
Borsenberger, J., Michaud, G., and Praderie, F. 1979, *Astron. Ap.*, **76**, 287.
Borsenberger, J., Michaud, G., and Praderie, F. 1981, *Ap. J.*, **243**, 533.
Borsenberger, J., Michaud, G., and Praderie, F. 1984, *Astron. Ap.*, **139**, 147.
Bóme-Vitense, E. 1981, *Ap. J.*, **244**, 938.
Burger, M. 1981, *Astron. Ap.*, **94**, 199.
Buser, R., and Kurucz, R. L. 1978, *Astron. Ap.*, 70, 555
Carbon, D. F. 1979, *Ann. Rev. Astron. Ap.*, **17**, 513.
Carbon, D. F. 1984, in *Methods in Radiative Transfer*, ed. W. Kalkofen, (Cambridge: University Press), p. 395.
Carpenter, K. G. 1985, *Ap. J.*, **289**, 660.
Cowley, C. R. 1980, *Vistas Astron.*, **24**, 245.
Cowley, C. R. 1981, in *Upper Main Sequence Chemically Peculiar Stars*, 23rd Liège Astrophys. Coll., Université de Liège, p. 5.
Cowley, C. R., and Adelman, S. J. 1983, *Quart. J. Roy. Astron. Soc.*, **24**, 393.
Dreiling, L. A., and Bell, R. A. 1980, *Ap. J.*, **241**, 736.
Dumont, S., Heidemann, N., Jefferies, J. T., and Pecker, J.-C. 1975, *Astron. Ap.*, **40**, 127.
Frandsen, S. 1974, *Astron. Ap.*, **37**, 139.
Freire, R. 1979, *Astron. Ap.*, **78**, 148.
Freire, R., and Praderie, F. 1974, *Astron. Ap.*, **37**, 117.
Freire, R., Czarny, J., Felenbok, P., and Praderie, F. 1978, *Astron. Ap.*, **68**, 89.
Freire Ferrero, R., Gouttebroze, P., and Kondo, Y. 1983, *Astron. Ap.*, **121**, 59.
Gustafsson, B. 1981, in *Physical Processes in Red Giants*, ed. I. Iben, Jr., and A. Renzini (Dordrecht: D. Reidel), p. 25.
Hubený, I. 1980, *Astron. Ap.*, **86**, 225.
Hubený, I. 1981, *Astron. Ap.*, **98**, 96.
Hubený, I. 1985, in *Progress in Spectral Line Formation Theory*, ed. J. E. Beckman and L. Crivellari, NATO ASI Series (Dordrecht, D. Reidel), p. 27.
Hubený, I., Štefl, S., and Harmanec, P. 1985, *Bull. Astron. Inst. Czechosl.*, **36**, 214.
Jamar, C. 1980, *Astron. Ap.*, **89**, 22.
Jamar, C., Marcau-Hercot, D., and Praderie, F. 1978, *Astron. Ap.*, **63**, 155.
Kudritzki, R.-P. 1973, *Astron. Ap.*, **28**, 103.
Kurucz, R. L. 1974, *Ap. J.*, **188**, L 21.
Kurucz, R. L. 1979, *Ap. J. Suppl.*, **40**, 1.
Kurucz, R. L., and Furenlid, I. 1979, *Smithsonian Ap. Obs. Spec. Rep. No.* **387**.
Lane, M. C., and Lester, J. B. 1984, *Ap. J.*, **281**, 723.
Leckrone, D. S., Fowler, J. W., and Adelman, S. J. 1974, *Astron. Ap.*, **32**, 327.

Leckrone, D. S., and Snijders, M. A. J. 1979, *Ap. J. Suppl.*, **39**, 549.
Linsky, J. L. 1985, in *Progress in Spectral Line Formation Theory*, ed. J. E. Beckman and L. Crivellari, NATO ASI Series (Dordrecht, D. Reidel), p. 1.
Madej, J. 1983, *Acta Astron.*, **33**, 1.
Maitzen, H. M., and Muthsam, H. 1980, *Astron. Ap.*, **83**, 334.
Malagnini, M. L., Faraggiana, R., Morossi, C., and Crivellari, L. 1982, *Astron. Ap.*, **114**, 170.
Malagnini, M. L., Faraggiana, R., and Morossi, C. 1983, *Astron. Ap.*, **128**, 375.
Michaud, G. 1970, *Ap. J.*, **160**, 641.
Michaud, G. 1981, in *Upper Main Sequence Chemically Peculiar Stars*, 23rd Liège Astrophys. Coll., Université de Liège, p. 355.
Mihalas, D. 1972, *Non-LTE Model Atmospheres for B and O stars*, NCAR-TN/STR **76**.
Mihalas, D. 1978, *Stellar Atmospheres*, (2nd ed. San Francisco: Freeman).
Muthsam, H. 1978, *Astron Ap. Suppl.*, **35**, 107.
Muthsam, H. 1979, *Astron. Ap.*, **73**, 159.
Muthsam, H., and Cowley, C. R. 1984, *Astron. Ap.*, **130**, 348.
Muthsam, H., and Stepien, K. 1980, *Astron. Ap.*, **86**, 240.
Muthsam, H., and Weiss, W. W. 1978, *Astron. Ap.*, **69**, 155.
Peterson, D. M. 1970, *Ap. J.*, **161**, 685.
Peytremann, E. 1974, *Astron. Ap.*, **33**, 203.
Phillips, A. P. 1977, *Mon. Not. Roy. Astron. Soc.*, **181**, 777.
Praderie, F. 1976, in *Physics of Ap Stars,*, IAU Coll. 32, ed. W. W. Weiss, H. Jenkner, and H. J. Wood, Vienna: Universitätssternwarte.
Praderie, F. 1981, *Astron. Ap.*, **98**, 92.
Praderie, F. 1982, in *Ultraviolet Stellar Classification,*, ed. A. Heck and B. Battrick, ESA SP-182.
Praderie, F., Boesgaard, A. M., Milliard, and Pitois, M. L. 1977, *Ap. J.*, **214**, 130.
Praderie, F., Simonneau, E., and Snow, T. P. 1975, *Ap. Space Sci.*, **38**, 337.
Relyea, L. J., and Kurucz, R. L. 1978, *Ap. J. Suppl*, **37**, 45.
Snijders, M. A. J. 1975, *Bull. Amer. Astron. Soc.*, **7**, 469.
Snijders, M. A. J. 1977a, *Astron. Ap.*, **60**, 377.
Snijders, M. A. J. 1977b, *Ap. J.*, **214**, L35.
Snijders, M. A. J., and Lamers, H. J. G. L. M. 1975, *Astron. Ap.*, **41**, 245.
Stepien, K. 1978, *Astron. Ap.*, **70**, 509.
Stepien, K., and Muthsam, H. 1980, *Astron. Ap.*, **92**, 171.
Stepien, K., and Muthsam, H. 1981, *Astron. Ap.*, **100**, 159.
Strom, S. E., and Strom, K. M. 1969, *Ap. J.*, **155**, 17.
Thomas, R. N. 1983, *Stellar Atmospheric Structural Patterns*, NASA SP-471.
Underhill, A. B., Divan, L., Prevot-Burnichon, L., and Doazan, V. 1979, *Mon. Not. Roy. Astron. Soc.*, **189**, 601.
Underhill, A. B., and Doazan, V. 1982, *B Stars with and without Emission Lines*, NASA SP-456.
Vauclair, S., Hardorp, J., and Peterson, D. M. 1979, *Ap. J.*, **227**, 526.
Wolff, S. C. 1983, *The A Stars: Problems and Perspectives*, NASA SP-463.

DISCUSSION (Hubený)

MICHAUD: Is there a systematic effect of NLTE on line blanketing due to type C effects dominating for many ions?

HUBENÝ: Yes, such an effect may exist, and it will depend on T_{eff} and abundance. For example, in A0 stars, if Fe II is the dominant source of blanketing opacity, the type C effect is not large for Fe II, but it may be large for Fe I, which would be underpopulated. Thus, the overall blanketing in models is not affected very much for A0 stars. The situation is different for hotter stars, where the balance between Fe III and Fe II shifts towards the dominance of Fe III.

These statements should be verified by calculating NLTE line blanketed model atmospheres for A and B stars. A step in this direction has recently been taken by Anderson (Ap. J., in press), who has calculated NLTE models of hot stars ($T_{eff} \approx 35000$ K) taking into account blanketing effects. He has found a number of interesting phenomena.

There is hope that similar calculations will be performed for cooler stars soon. For example, in peculiar A stars, if rare earths are the dominant source of blanketing opacity, singly ionized rare earths surely are affected by type C. One needs always to bear in mind the type of star and the dominant source of blanketing.

ROMANOV: What is your opinion about the influence of the conditions of ionization on the determination of the abundance of elements in hot stars?

HUBENÝ: I think you are referring to the decrease of the ionization potential? Yes? I haven't made calculations, but it seems to me that those effects should be less than possible NLTE effects. The lowering of ionization potential is already included in the partition functions, even in LTE calculations, and NLTE calculations also take this into account. Comparing LTE calculations with and without lowering, and NLTE with LTE calculations taking lowering into account, which I think is the basis of your question, the answer depends on the ion. For those ions in which type C effects will occur, almost certainly NLTE effects will dominate. There may be some special situations in which the effect of lowering might be important, but to my knowledge this has not yet been systematically studied.

STĘPIEŃ: I think that it is generally accepted that if you have significant lowering of ionization potentials this is due to collisions playing a larger role, so the departures from LTE are generally smaller for small ionization potentials. It is the same for higher line excitation levels.

HUBENÝ: This is an important problem, because there are two approaches to the question: how do you approximate the higher lying levels? One way is to approximate the departure coefficient for the higher levels by the same coefficient used for the highest level explicitly included in the calculation. This approach has been adopted by Borsenberger, Michaud and Praderie in their diffusion calculations, and essentially says that coupling with the continuum is comparatively less than coupling with lower lying levels. On the contrary, one may very reasonably assume that high lying levels are strongly coupled with the continuum, so that departure coefficients would be more or less equal to

C. R. Cowley et al (eds.), Upper Main Sequence Stars with Anomalous Abundances, 77–79.

unity. To decide which approximation is more appropriate is very difficult, because the detailed form of the collision rates for those levels is, unfortunately, not very well known.

For example, in He I, the departure coefficients for levels with n = 5 are very similar to levels with n = 4 and nowhere near unity due to collisional coupling. So, this argument that very high lying levels must be near to LTE might not be quite correct in many situations.

STĘPIEŃ: Yes, I agree.

DROBYSHEVSKI: I wonder why you did not discuss effects of magnetic fields on NLTE. I mean not the atomic level splittings and other atomic effects, but so called gas discharge phenomena, with strong differences between electronic and ionic temperatures, particle beaming, laser effects, etc.

HUBENÝ: I have deliberately omitted these phenomena and concentrated on the "classical" stellar atmospheres problem (plane-parallel stratification, hydrostatic and radiative equilibrium). From the methodological point of view, I feel that it is necessary first to understand well the simpler situations before attempting to treat more complicated configurations. The primary question is: to what extent will a physically more consistent description of the interaction of radiation with matter affect the methodology of interpretation of observed stellar spectra? As I have shown, this problem is still far from being solved. Of course, there are large effects of magnetic fields on hydrostatic equilibrium, and so on, and these need to be studied.

STĘPIEŃ: So you are not against magnetic fields!

HUBENÝ: Not at all!

DWORETSKY: In some cases an observer can see three stages of ionization of an element in the spectrum of a normal or peculiar star simultaneously. Given that the atomic data are good, and that (usually) the lines seen are due to resonance transitions, then if an LTE analysis gives the same abundance for the element for all three stages, can we actually have confidence that the result is correct?

HUBENÝ: Well, this may be a strong argument in favour of the reliability of the result, but strictly speaking one can not definitely rule out the possibility that some unexpected physical phenomena produced a spurious interpretation of observations. In this respect, an interesting example is provided by an analysis of the C II resonance lines in Vega. The profiles synthesized assuming LTE and NLTE give the best fit for carbon abundances $3 \cdot 10^{-4}$ and $2 \cdot 10^{-4}$, respectively (Friere, Astr. Astrophys., 78, 148, 1979). The NLTE result was obtained without partial redistribution. My own result, for NLTE and partial redistribution, gives the value $3 \cdot 10^{-4}$, the same as the LTE calculation. This was purely accidental! Yet one can not conclude that the LTE approach is correct and NLTE incorrect. This does not answer your question, but shows that one should check very carefully whether the diagnostic method has not neglected some important physical phenomena which may spoil the interpretation. Generally, NLTE releases us from logical binds into which we may fall when trying to interpret observations solely on the basis of LTE.

STEPIEN: So, I think the conclusion is, Dr. Dworetsky, that your abundance may accidentally be correct! [laughter]

ALECIAN: One must also remember that abundance stratifications of some elements may affect strongly the radiative transfer in the lines. The abundances deduced for these elements by classical methods must be interpreted differently if these elements are strongly stratified (above $\tau_{5000} = 10^{-1}$).

HUBENÝ: I completely agree with you. Some situations require a detailed NLTE analysis.

STĘPIEŃ: When an element is pushed up in an atmosphere and is collected in a thin layer at the top, NLTE effects can be more pronounced than in the case of a uniform distribution. Would you expect to find any observable differences between these two cases and could you obtain a criterion to distinguish between a uniform and concentrated distribution of the element?

HUBENÝ: Differences between both cases do certainly exist, but they need not necessarily be easily observable. Each particular case requires careful study. Very few calculations of this sort have been made so far, and it is very difficult to make any quantitative (or qualitative) predictions.

INFLUENCE OF THE MAGNETIC FIELD ON THE POLARIZATION OF RADIATION SCATTERED BY ELECTRONS IN STELLAR ATMOSPHERES AND ENVELOPES

A. Z. Dolginov

A. F. Ioffe Institute of Physics and Technology,
194021 Leningrad, USSR

Yu. N. Gnedin, N. A. Silant'ev

Central Astronomical Observatory at Pulkovo,
196140 Leningrad, USSR

The radiation scattering by electrons play an important role for stars with the surface temperature $\geq 2.10^4$ ^oK. Many among these stars have magnetic fields of the order 1 - 100 G. We shall consider the role of the field for the radiative transfer in continuum spectra.

Electrons in the field form an axisymmetric medium similar to the single axis crystal. Hence, the medium will have dichroism and birefringency. If the gyrofrequency ω_B is much less than the radiation frequency ω then the dichroism can be neglected. The circular birefringency can be characterized by the angle of the Faraday rotation

C. R. Cowley et al (eds.), Upper Main Sequence Stars with Anomalous Abundances, 81–86.

$$\chi = 1/2\ \delta\ \tau_{Th} \cos\theta ,$$
$$\delta = (3/4\pi)(\omega_B/\omega)(\lambda/r_e) \simeq 0.8\lambda^2(\mu)B(G),$$
$$r_e = e^2/m_eC^2,\ \tau_{Th} = N_e\ \sigma_{Th}\ell,$$

σ_{Th} is the Thomson cross-section, N_e is the electron number density, ℓ is the path of the light, θ is the angle between the field B and the direction n of the light propagation. One can see that for the wavelength $\lambda \simeq 1\mu$ in the field B $\sim$ 1 G the Faraday rotation is large enough. For the field < 10^6G and λ < 1 mm the Stokes parameters satisfy the equations $\mu dI/d\tau = I - B_I$, $\mu dQ/d\tau = Q + U\ \delta \cos\theta - B_Q$.

$\mu dU/d\tau = U - Q\ \delta \cos\theta - B_u$, $\mu dV/d\tau = V - B_V$.

Here $B_m(\tau,n)$ are well known scattering terms (Chandrasekhar, 1950).

If $\chi \geq 1$ then the linear polarization is decreased for all directions except the one perpendicular to the field. Photons travel different geometrical paths and possess because of the Faraday rotation different directions of the polarization plane. In this case we can separate the equation for the intensity only:

$$\mu dI/d\tau = I - (3/16\pi) \int dn'(1 + (n.n')^2)I(n',\tau).$$

For the parameters Q and U we have:

$$-Q + iU = -\frac{F}{2\pi H_1} \cdot \frac{1-\mu^2}{1 + i\delta \cos \theta} \left[H\left(\frac{\mu}{1 + i\delta \cos \theta}\right) - \frac{3}{2} H_2 - \right.$$

$$\left. - \frac{3H_1\mu}{2(1 + i\delta \cos \theta)}\right] \left[3 - \frac{\mu^2}{(1+i\delta \cos \theta)^2}\right]^{-1}, \; H_n = \int_0^1 d\mu \mu^n H(\mu).$$

Here $H(\mu)$ is the well known Chandrasekhar's H -function. From this equation we can see that the maximum polarization for $\cos \theta = 0$ is 9.14% instead of the value 11.7% obtained in papers of Chandrasekhar (1950) and V. V. Sobolev (1956).

The large-scale magnetic field destroys the optical symmetry of the stellar atmosphere (and envelope) and leads to a non-zero integral linear polarization of the radiation from the spherical star. In the paper of Dolginov and Silant'ev (1979) this polarization has been calculated for the case of the dipole magnetic field. Its maximum value was about 0.2% for $\delta_e \simeq 2 \div 4$ for the stellar magnetic equator. The qualitative spectral dependence of polarization is presented in Figure 1.

The influence of the Faraday rotation is much more effective for the optically thin atmospheres and envelopes as was shown by Gnedin and Silant'ev (1980, 1984).

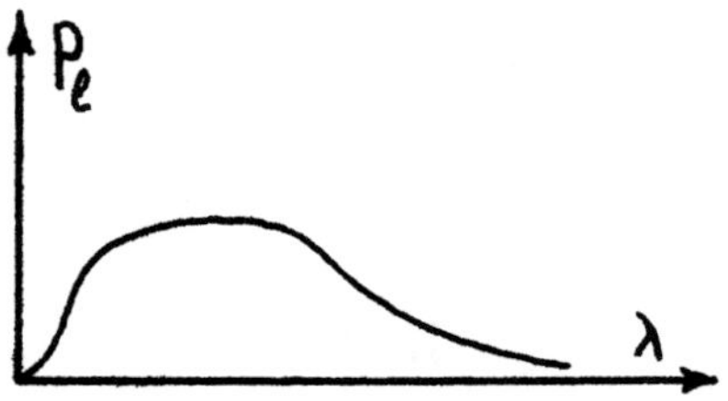

Fig. 1. Qualitative picture of the integral polarization spectrum of radiation emerging from spherical star with magnetic field.

The maximum of polarization may reach 10% if $\tau_{envelope} \sim 1$ for the case of a dipole field and with N_e = Const in the envelope. For small δ_e values the polarization degree P_ℓ increases as $\sim \lambda^4$ for any dependence of N_e the distance. For large Faraday rotation ($\delta_e >> 1$) the law of the polarization decrease depends strongly on the function $N_e(r)$. For a spherical envelope with N_e = Const one has $P_\ell \sim \lambda^{-2}$ and for the case $N_e \sim r^{-2}$ the polarization degree $P_\ell \sim 1/\sqrt{\lambda}$. If the magnetic dipole axis is precessing then the polarization varies with the period of the precession.

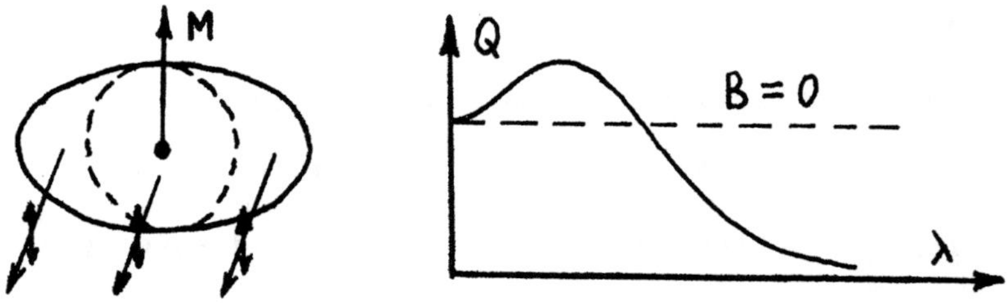

Figure 2. Polarization spectrum of radiation from ellipsoidal electron envelope when the axis of star's

magnetic dipole is perpendicular to the sight of telescope and coincides with the envelope rotation axis. The short arrows denote the direction of the electric vector oscillation of the light.

For a nonspherical envelope without a magnetic field the integral polarization exists and it is independent of the wavelength. If the magnetic field is present in this case then the Faraday rotation may as well enhance as diminish the polarization degree (see Fig. 2 an 3).

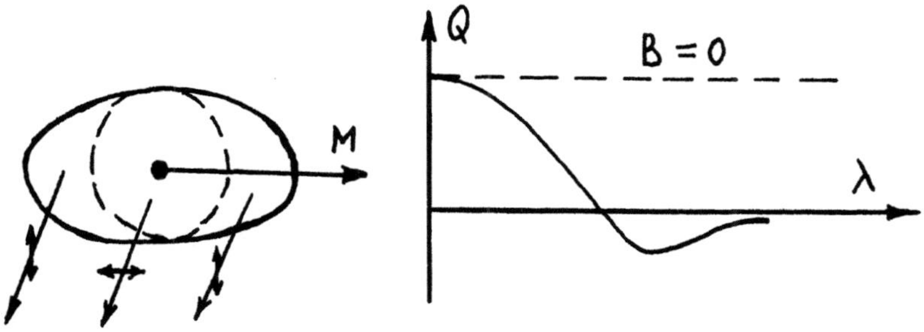

Figure 3. Polarization spectrum of radiation from an ellipsoidal electron envelope when the axis of star magnetic dipole is perpendicular to the sight of telescope and to the rotation axis.

The linear polarization spectra of hot stars can give information on the stellar magnetic field as well as on the electron number density distribution law in the envelope. This method may be useful also in those cases when the usual Zeeman-method is inefficient.

References

Chandrasekhar S., 1950, Radiative Transfer, Oxford.

Dolginov, A. Z., and Silant'ev, N. A., 1979, Pisma v Astron. Journal (in Russian), 5, 526.

Gnedin, Yu. N., and Silant'ev, N. A., 1980, Pisma v Astron. Journal (in Russian), 6, 344.

Gnedin, Yu. N., and Silant'ev, N. A., 1984, Astrophys. Space Sci., 102, 375.

Sobolev, V. V., 1956, Radiative Transfer in the Atmospheres of Stars and Planets (in Russian), Moscow.

A DETERMINATION OF MAGNETIC FIELDS OF T TAU AND Ae/Be HERBIG STARS USING THE PARAMETERS OF THEIR LINEAR POLARIZATION

Yu. N. Gnedin, M. A. Pogodin
Central Astronomical Observatory of
the USSR Academy of Sciences
Pulkovo, 196140 Leningrad
USSR

N. P. Red'kina
Institute of Astrophysics
Tadzhic Academy of Sciences
Dushanbe
USSR

ABSTRACT. Attempts of applying stellar polarization spectra to evaluation of magnetic fields of some types of stars are described.

Gnedin and Silant'ev (1980) proposed a method for determining stellar magnetic fields using linear polarization spectra of stars. The method was based on the effect of the Faraday rotation of the plane of polarization produced by the electron scattering in a magnetized circumstellar shell. The asymmetry of spatial distribution of the magnetic field should lead to linear polarization. Gnedin and Silant'ev gave spectral curves, calculated for spherical and ellipsoidal envelopes with different values of the dipole magnetic field B, optical thickness of the envelope τ_e and the inclination angle of the axis to the line-of-sight θ_m. The shape of these curves has characteristic maximum. Its position and polarization degree depends on the envelope form and the set of parameters B, τ_e, and θ_m.

Attempts of using the proposed method were made for determination of magnetic fields of T Tau and Ae/Be Herbig stars. These have gaseous envelopes and intrinsic polarization with the maximum in the visible spectral range. Moreover, the existence of magnetic fields of these stars predicted in a series of theoretical works (see, for example, Gnedin and Red'kina, 1984). One of the main difficulties in applying the method was the multicomponent nature of the observed linear polarization of these objects. Its parameters are influenced not only by the electron scattering in the gaseous envelope but also by a scattering in a) the circumstellar dust envelope b) a local nebula, associated with the star and c) interstellar medium with a highly inhomogeneous matter distribution at small galactic latitudes. The important problem was to correctly distinguish the component belonging to the gaseous envelope.

Firstly, this method was employed for the star T Tau (Gnedin and Red'kina, 1984). An assumption was used that the intrinsic polarization of T Tau is mainly caused by electron scattering, and the effect of the dust component could be ignored. This supposition was based on a discovery of a cool satellite of the star which can explain the IR excess in the T Tau

C. R. Cowley et al (eds.), Upper Main Sequence Stars with Anomalous Abundances, 87–90.

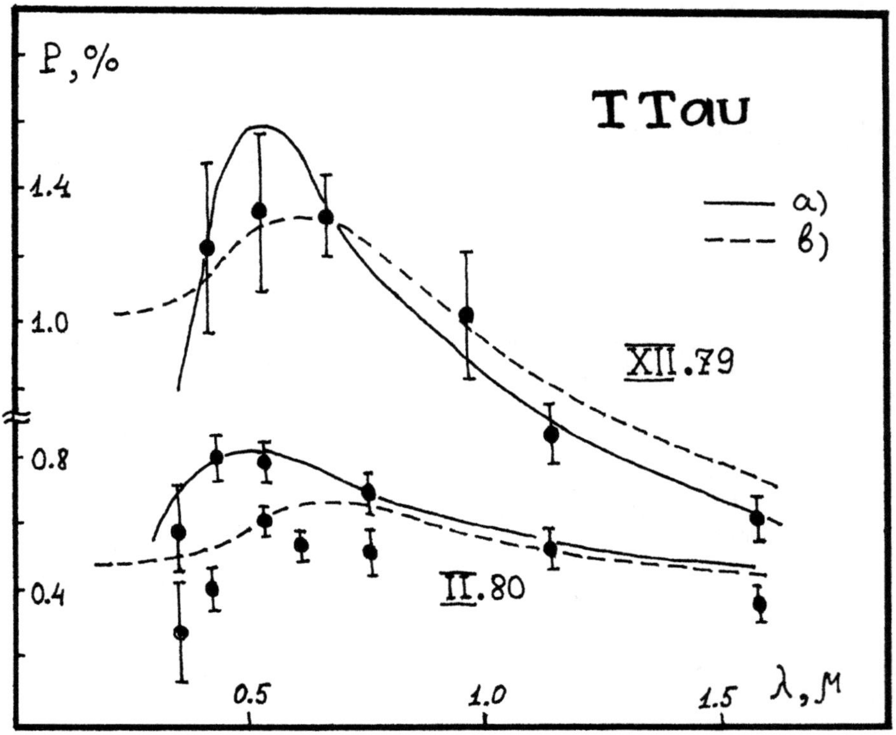

Figure 1. Comparison of the observed spectra of the intrinsic linear polarization of T Tau with theoretical curves. The values of the model parameters are:
December 1979:a)spherical model,B=250 Gauss,τ_e =0.20 b)ellipsoidal model with the ratio of principal axes 0.25,B=1500 Gauss, τ_e =0.14.
February 1980:a)spherical model,B=400 Gauss,τ_e =0.13 b)ellipsoidal model,B=1200 Gauss, τ_e =0.07.

spectrum (Dyck and Simon, 1982). Thus, this can justify the rejection of the hypothesis of the existence of a considerable dust envelope round T Tau. A comparison of the T Tau polarization spectrum free from interstellar component with calculated curves, gave a value of the magnetic field of the star B varying from 100 to 400 Gauss for the spherical model and from 300 to 1500 Gauss for the ellipsoidal model.

The evaluations of the fields of RY Tau, SU Aur, RW Aur and DI Cep were made similarly. A strong polarimetric variability of these objects (Figure 2) can be considered as an additional argument in favor of the mechanism of electron scattering producing the polarization. Their magnetic fields also attain the values of the order $10^2 - 10^3$ Gauss.

In the work by Gnedin and Pogodin (1985) devoted to a determination of magnetic fields of Ae/Be Herbig stars, an attempt was made to separate the gaseous and dust components of intrinsic linear polarization. Independent evaluations of τ_e obtained from the

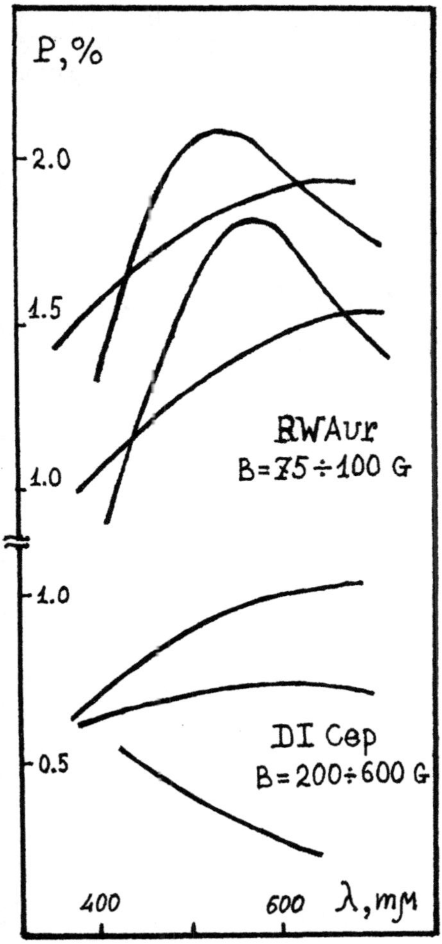

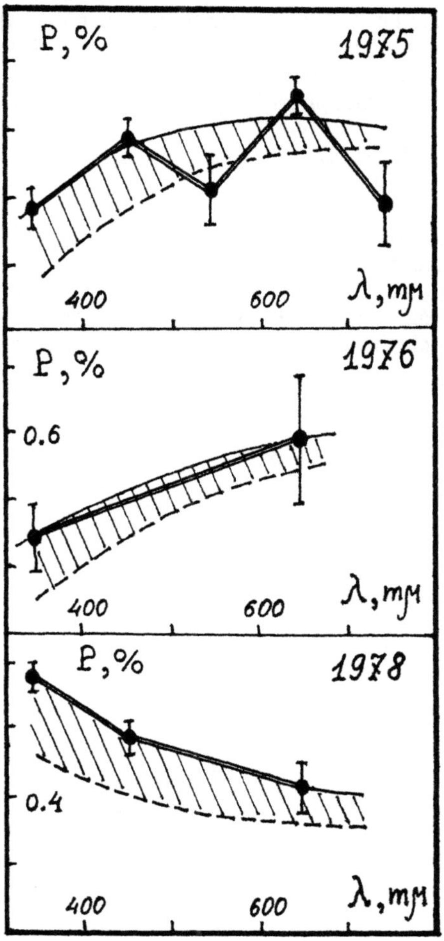

Figure 2. Variability of the spectrum of the intrinsic linear polarisation of some T Tau stars.

Figure 3. Separation of the gaseous and dust ($P(\lambda) \sim \lambda^{-1}$) components of the intrinsic linear polarisation of HD 200775. The dust component is shaded. The values of the model parameters are: τ_e =0.17, a)1975:B=110 Gauss, θ_m =70° b)1976:B=110 Gauss, θ_m =60° c)B=550 Gauss, θ_m =75°.

observed emission Balmer jump (see Garrison, 1978) were also employed. The results of a separation of the components in polarization spectrum of the star HD 200775 are given in Figure 3. Polarimetric variability can be explained in terms of variation of the magnetic field (B = 100 to 500 Gauss) and a precession for the magnetic axis of the star, the dust component being assumed invariable.

REFERENCES

Dyck, H. M., Simon, T., 1982, *Ap. J. Lett.*, **255**, 103.
Garrison, L. M., 1978, *Ap. J.*, **224**, 535.
Gnedin, Yu. N., Silant'ev, N. A., 1980, *Sov. Astron. J. Lett.*, **6**, 344.
Gnedin, Yu. N., Red'kina, N. P., 1984, *Sov. Astron. J. Lett.*, **10**, 613.
Gnedin, Yu. N., Pogodin, M. A., 1985, *Sov. Astron. J. Lett.*, **11**, 37.

PRACTICAL PROBLEMS IN THE ANALYSIS OF CP STARS

Charles R. Cowley
University of Michigan

Sveneric Johansson
Lunds Universitet

ABSTRACT. Saturation and curve-of-growth effects remain a dominant source of errors in abundance work. We suggest that the use of low noise material will enable future users to work more easily with weak lines and avoid these difficulties. The method of wavelength coincidence statistics is reviewed within the context of line identifications and abundance work. This technique is readily implemented, and should be a standard part of the analytical spectroscopy of stars. Analysis of the ultraviolet spectra of CP stars is severely handicapped by missing atomic data. Synthetic spectra are displayed showing the possible effects of hitherto unknown lines on spectra of IUE resolution.

1. INTRODUCTION

This paper will be concerned with selected, practical difficulties related to the analysis of stellar spectra.

The general methods of analytical stellar spectroscopy were detailed by Unsöld (1955) in the second edition of his famous textbook. Apart from the methods of non-LTE, whose successes have been limited to special cases, the improvements of the past 30 years have been primarily related to the speed of computation now possible. Especially in the analysis of CP stars, it can be said that current techniques hardly reflect the improvement in quality that one might have expected from an additional three decades of experience.

One of the reasons for this is that the stellar spectroscopists who have gained analytical experience have not remained in this field, although many are still active in astronomy. The lessons learned through experience have consequently not been retained, and abundance workers have repeated many of the mistakes of the past.

2. SATURATION PROBLEMS AND PALLIATIVE MEASURES

Of all of the errors that have affected the analysis of CP stars, the most important have probably been related to saturation, or curve-of-growth effects. These problems are not necessarily removed by spectral synthesis, although clearly, the use of information on line shapes as well as total strengths can be an advantage.

It is probably recognized by everyone that the use of weak lines avoids most saturation problems, but in spite of this fact, even very recent work has made use of strong lines. Why is this so? The most common reason for the use of stronger lines is that the spectral dispersion is insufficient to allow one to measure the weak lines accurately.

In typical abundance studies, the number of truly weak lines is often extremely small. Even in cases where hundreds of lines from a given spectrum may have been measured there may be less than half a dozen that are clearly uninfluenced by curve-of-growth effects.

The decision of a research worker to proceed with material of admittedly low quality (low dispersion, high noise) is often made for a variety of reasons, that are outside of the domain of scientific judgement. The question of importance is whether there is an obvious way

C. R. Cowley et al (eds.), Upper Main Sequence Stars with Anomalous Abundances, 91–98.

to relieve this situation, and we suggest that there is. The *Peculiar Newsletter* (see Hensberge and van Rensbergen 1985, henceforth, PN) publishes information on the availability of spectroscopic material at various observatories and from various research workers. Moreover, the PN often indicates that a worker is interested in sharing data. We suggest far too much analytical spectroscopy has been done with material inferior to that which might have been obtained from another astronomer or institution, simply by requesting it. In present times, it should be the simplest matter to borrow digitized intensity data on a magnetic tape.

Thus an early step in the analysis of a star should be a search of the literature for possible spectroscopic DATA, followed by an earnest attempt to obtain it. ONLY in the case where it is clear that new observations would be superior to those already made, or if the existing data is unavailable, need one apply for observing time to obtain new material.

It is now common to choose a microturbulence which yields no correlation of abundance with equivalent width. This technique has shortcomings. It must be recognized that the procedure in no way improves the abundances, which still depend on the weakest lines (and in rare instances on the very strong ones). Moreover, the accuracy with which a line can be measured degenerates rapidly as the strength of a line approaches the threshold of detectability. Thus, as we remarked many years ago, systematic errors that depend on the (magnitude of the) equivalent widths are to be expected (Cowley 1970 Section 4–13.1). They should be carefully investigated, but in practice, this may not be done. We suggest that it is more common to adjust the microturbulence in such a way that the systematic errors are not apparent.

All abundance workers are urged to make a careful consideration of possible systematic errors in their equivalent width measurements before arbitrarily adjusting the microturbulence parameter. Careful consideration should also be given the maximum line depths allowed by the models, *vis-a-vis* those observed in the star.

The systematic errors to be expected for weak lines will obviously be less serious, for a given line strength, the greater the signal-to-noise level. Every attempt should therefore be made to achieve signal-to-noise levels of at least 50. Of course, the signal-to-noise level that is necessary to measure a 20mA line with an uncertainty of, say, 10%, varies with the v.sin(i) of the star.

It is well to recommend the use of weak lines, but one cannot find critically evaluated gf-values for many of the weaker lines. Cowley and Adelman (1983) have stressed the importance of the "critical evaluations" by people at the US National Bureau of Standards, for astronomical applications. Usually, one can find gf-values from Kurucz and Peytremann (1975, henceforth, KP) for weak iron-group lines, but the accuracy of these values for non-LS-coupling allowed transitions, is known to be questionable. A critical problem of current interest is the upper limit for the iron abundance in some CP stars. Can it be as high as 9.0 (log[H] = 12), as suggested by the work of Muthsam and Cowley (1984) and others? Must the theoreticians take this seriously? It turns out to be quite difficult to evaluate this matter because the critically evaluated lines of Fe II, are not weak, at least not in those stars with the putative high iron abundances.

The problem is well illustrated in Table 2 of Muthsam and Cowley. The truly weak lines of Fe II are λλ4625.55 and 4666.77. The abundances from these lines are at the extreme upper limit of those listed for the Fe II lines. If uncertainties in the oscillator strengths or blends could be ignored, we would be forced to conclude that [Fe/H] must be AT LEAST as large as 9.0, and is quite possibly larger! At present, there does not appear to be a straightforward way for the astronomer to deal with this particular problem. We await more accurate calculations and measurements.

For many other atomic spectra, the critical evaluations of the NBS cover weak lines. Cowley and Adelman (1983) have suggested some additional sources of gf-values.

3. THE PROBLEM OF IDENTIFICATIONS

Authors of books on stellar atmospheres have apparently assumed the problem of line identification is so simple that no detailed discussion is needed. This is ironical. The problems can often be so difficult that no satisfactory solution can be found.

Line identification in late-type stars with solar abundances is straightforward in the traditional ($\lambda\lambda$3500 - 6500) region, for all but the weakest features. With CP stars, and even for "normal" stars in the satellite ultraviolet, one finds identification problems that can be vastly more difficult. The level of difficulty can vary enormously from one CP star to another, but in the satellite ultraviolet, line-by-line identification work is difficult or impossible, even for "normal" stars.

No stellar identification work is begun without some prior knowledge of what is present in the spectrum. Only a college freshman begins with the stellar wavelengths and the finding list from the Multiplet Tables. The identification worker has at his disposal — as a minimum — a knowledge of the strongest lines: the hydrogen lines, H and K of Ca II, etc. Usually, he has a list of identifications of wavelengths in some similar star to the one studied. Merrill's (1956) old book is still a useful compendium for work of this kind. The most useful sources for CP stars are unpublished studies by Bidelman (1966, 1968) who employed high dispersion spectra of stars with intrinsically sharp spectral lines. Of the published works, some of the more useful are those of Adelman (1974), Hiltner (1945), and the Jaschek's and González (1965). The researcher must keep in mind that considerable advances in atomic spectroscopy have been made since many of the older identification studies were done.

The identification of lines to use for abundances has often been made without measurements of wavelengths and we suggest that this can be a serious mistake. A measured wavelength that deviates by significantly more than the expected error is an immediate indication of a blend, and in rarer cases of isotope shifts (Hg II, Pt II). Fortran programs are available from a number of workers for the automatic measurement of digitized spectra, and for many CP stars digitized measurements are available on request (from CRC) for a large number of late B and A stars. The abundance worker who fails to take advantage of this material is remiss.

Identifications for a given element must start with a search for the strongest expected lines. The lines may be judged to be "present" or below some threshold for detection. In the latter case it is possible to set an upper limit to the abundance. This upper limit is often uninteresting, because a rather excessive overabundance of some exotic species could still be possible. Nevertheless, it is better to determine such an upper limit than to say the lines are "absent."

The traditional method of line identifications relies heavily on the relative (laboratory) intensities within multiplets. It is usually assumed that the relative intensities within a multiplet are unaffected by non-thermal excitations. This is a good working assumption, although one should always be aware of the possibility of selective excitations (Johansson and Jordan 1984, Cowley 1970 Table 2–6.1).

The method of multiplets works very well in many instances, but there are certainly cases where other techniques are preferable. First, we note that multiplets are traditionally defined on the assumption of LS coupling, which can be a very bad approximation. When this is the case, one must beware of theoretical LS-coupling intensities. Second, in the kinds of spectra that occur with the rare earths, there are a large number of lines with similar intensities that arise from levels not very different in excitation. Under these circumstances, the intensities of Meggers, Corliss, and Scribner (1975) are probably as good as the relative intensities in an LS-coupling approximation to a multiplet. These "natural" relative intensities should be useful even in the case of non-thermal excitations in the source. While the theoretical LS-coupling intensities may be perturbed by the mixing of (LS) states, laboratory intensities are not subject to this particular problem.

Neither the method of multiplets, nor any other method is of much use when the strongest expected lines of a given species are only weakly present. The argument that there is no other candidate for the identification of a given feature other than the strongest line of, say X, has little weight unless it is possible to securely identify virtually all of the lines in the spectrum apart from the one(s) in question.

In a heavily blended spectrum, the relative intensities can be so badly distorted, that even rather good laboratory data on line strengths can be of little use. If a line to be identified is in a multiplet, one must of course check to see if stronger lines in the multiplet are present with the proper intensities. Such a check should be made as a matter of course, but much of the power of the method is lost if all of the lines in question are rather weak, and the relative intensities vary only by a factor of two or three. When the candidate line is itself the strongest line in a multiplet, but it is very weak, it is rather meaningless to search for its congeners in a noisy spectrum.

C. Cowley and his collaborators have employed the technique of wavelength coincidence statistics (WCS). It is a global method, which gives no direct information on individual features, but provides an impersonal basis for rejection of the *null hypothesis, that lines of given ion are not present* in the stellar spectrum. Rejection of the null hypothesis at a high confidence level means (within a calculable risk) that an ion is present. Failure to reject the null hypothesis does not prove the absence of a species, which may be weakly present.

WCS has been applied with good success to high dispersion spectra (ca. 2A/mm) in the region λλ3000 – 6600 (see Cowley and Hensberge 1981). Limited applications have been made in the deeper UV (Hensberge *et al.* 1986, Bord and Davidson 1982, Davidson and Bord 1982).

Recent applications of WCS have incorporated information on intensities (Chjonacki, *et al.* 1984, Henseberge, *et al.* 1986). While the value of this information is more limited than one might think, important conclusions can be drawn from intercomparisons of the intensity levels at which sets of lines with different average intensities reach a given level of significance in stars with similar line widths (v sin(i)'s) and line densities. Likewise, one may intercompare the WCS parameters for the spectra of two elements in a given star; in such a comparison, one has the advantage that the line widths and densities are essentially constant. In the lanthanides, we usually employ three lists, a strong, an intermediate, and a weak set. In the spectrum of HR 7575 and β CrB, the WCS parameters for Nd II and Sm II indicate only marginal significances for all three lists, while for γ Equ, the WCS parameters are highly significant, for all three sets. Moreover, for La II, Ce II, and Gd II, the WCS parameters are highly significant in both stars. This result has enabled us to conclude that there is a "hole" in the lanthanide abundance distribution for the former stars, since we can think of no effect related to their spectra, such as line density, or broadening, that could account for such an apparently selective effect.

There are numerous ways of making line identifications. Each method has special advantages and disadvantages, many of which have been outlined above. The competent research worker selects those methods that are the most powerful for his special problems, and applies them with insight into their strengths and weaknesses. We recommend that WCS be employed at an early stage in any attempt at a line-by-line identification study, and that the worker look first for those species for which the WCS parameters indicate the presence beyond any reasonable doubt. Such elements will be typically from the iron-group, to be sure, but the clear presence of more exotic species may also be indicated.

For most of the iron-group elements, it would not be difficult to prepare a list of lines with intensity estimates made from gf-values, and to simply proceed with identifications down to a given threshold. Especially for the rare earths, the rough gf-values obtained from Monograph 145 (Cowley and Corliss 1983) should prove to be of value.

4. MISSING ATOMIC DATA, AND THE SATELLITE UV

It is generally well understood by astronomers that there is a pressing need for physical data on gf-values, collision cross sections, and line-broadening parameters. It is less often recognized that important work still needs to be done in the more basic domains of atomic structure—wavelengths and energy levels. This need has been demonstrated by various studies over the last decade, but the point still needs to be emphasized. The long path lengths and high temperatures of stellar atmospheres create powerful light sources that are difficult to simulate in the laboratory.

For ions, this problem is critical, because the densities are not high enough to allow absorption measurements. Absorption measurements have been made, of course, for neutral atoms. Thus in stellar photospheres, we "measure" the population of the lower level of a transition, while in the laboratory, one may see the corresponding line if one can manage to get sufficient population of the *upper* level. In practice, this means that for stars with high excitation temperatures, we will expect some unidentified lines, because the conditions for observing them in the laboratory are much less favorable than in the star.

We have referred to faint, high-excitation lines as "second generation" lines, because they do not appear in the classical lines lists such as the Multiplet or MIT Wavelength Tables. In fact, the Multiplet Tables do contain lines predicted on the basis of known (circa 1945) energy levels, mostly in Fe I. The second generation lines with which we are primarily concerned with here arise from levels that have been found in the laboratory within, say, the past decade.

Cowley and Arnold (1978) used WCS to demonstrate the presence of newly predicted lines in the region λλ3750 - 4650 of CP stars, and pointed out that chance coincidences with such lines have caused considerable confusion. The use of the tables of KP and their congeners (e.g. Kurucz 1981, henceforth, K-81) can eliminate many of the problems caused by such lines, but the tables that have been published as of this writing, were constructed before much of the recent work on the high level structure of iron-group spectra (see Johansson 1981). The KP Tables include only the Corliss-Bozman (1962) wavelengths for spectra of elements heavier than nickel, and therefore lack many of the important additions, especially in the rare earth spectra incorporated into in NBS Monograph 145, and in *Atomic Energy Levels—The Rare Earths* (Martin, Zalubas, and Hagan 1978).

The problem of second generation lines takes on an entirely new dimension when we consider the satellite ultraviolet, extensively studied now at high resolution for a decade with the help of the International Ultraviolet Explorer (IUE). In the near future, the more powerful instrumentation of the Space Telescope will be available, and the time is ripe for an examination of the question of how seriously handicapped we are by our present, incomplete knowledge of the high level structure, especially of iron-group elements.

We discuss here a preliminary study of a 5A region centered at λ1850, which was made to evaluate the importance of second-generation spectra on a study such as that of Leckrone (1984) of the Hg II line at λ1942. All calculations are based on a stellar model with T(eff) = 11000K, and log(g) = 3.5. The assumed abundances throughout are solar except for chromium, manganese iron, which are all enhanced by 1.0 dex. The enhancement for iron is greater than in a typical Hg-Mn star, but the difference should not be important in the present context, which is primarily illustrative. We also calculate strengths for a resonance line of a hypothetical element whose properties (ionization energies, partition functions, etc.) are those of mercury (Hg II λ1942) except that we assume a rest wavelength of λ1850.00. This wavelength was chosen before the lines within it were investigated. It turned out by chance that some important absorption by second-generation features fell very close to the rest wavelength assumed for the assumed line (hypotheticum: Hy II λ1850.00).

Figure 1 shows the region with only lines from the KP tables. In Figure 2, we give results with only second generation lines of Cr, Mn, and Fe. Details concerning these lines will be published elsewhere, but for the present, we note that these particular wavelengths all represent transitions between recently classified energy levels (Johansson 1986). Oscillator strengths were estimated, roughly, from the laboratory intensities. The theoretical spectra include neither instrumental effects nor rotational broadening, but a microturbulence of 2 km/sec was assumed.

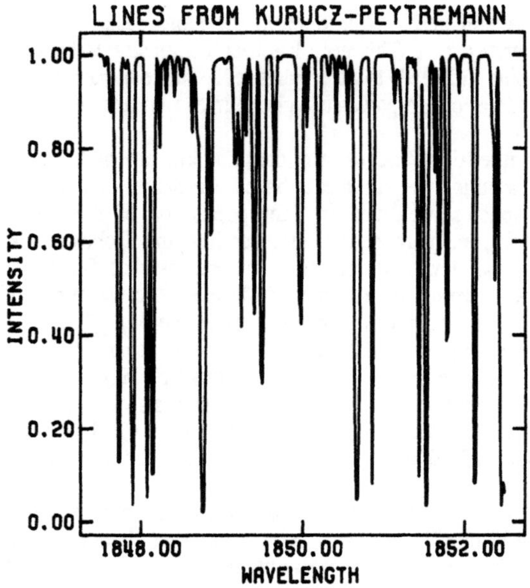

Figure 1. Theoretical spectrum KP lines: W_λ = 889mA.

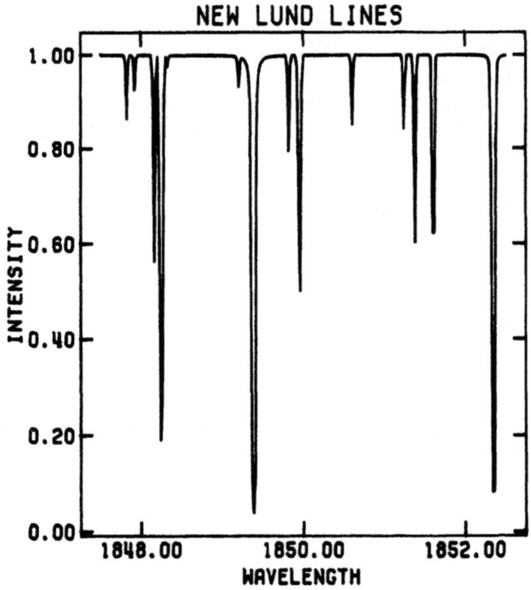

Figure 2. Theoretical spectrum including only new lines from Lund: W_λ = 228mA.

The second-generation lines provide a non-negligible absorption, although because of saturation effects, the combined absorption (988mA, not illustrated) is increased only by about 11%.

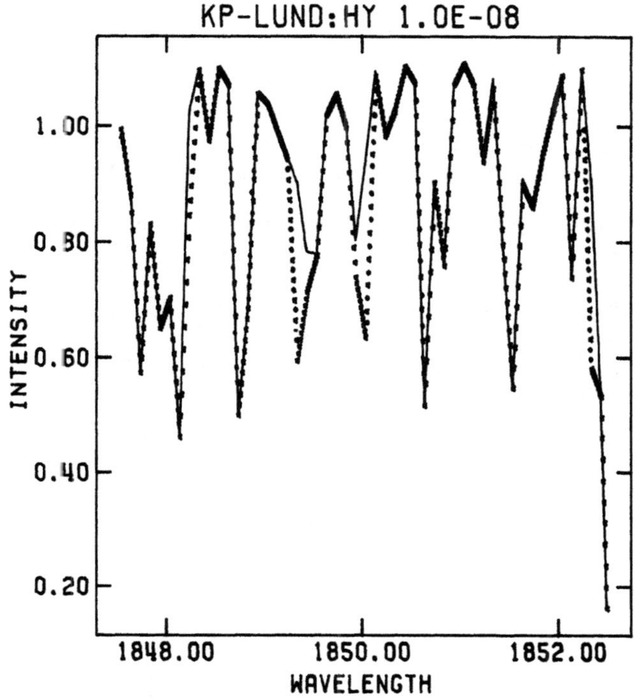

Figure 3. Simulated IUE spectra in the Hy II region. The Hy abundance fraction assumed is 10–8, corresponding to an excess of 100 (if Hy were Hg). The solid curve includes only lines from KP and Hy II. The points represent the additional contribution from second-generation lines.

Kurucz (K-81) updated the KP calculations for the Fe II spectrum only. His work rested primarily on analyses at Lund (see Johansson 1978), supplemented by purely theoretical calculations. The additional (non-KP) Fe II lines in λλ1847.5 – 1852.5 give a total absorption of 0.13A. They are comparable in importance to the absorption from the second-generation lines of our Figure 2, although somewhat smaller. It is important to note that the majority of the absorption from these K-81 lines comes from a set of lines near λ1848.5 for which the upper levels were not correctly classified in K-81. The correct amount of absorption from these particular levels is in reality much lower. This represents an unusual example where an astronomer might *overestimate* the contribution from Fe II.

We have made a number of sample calculations with various Hy abundances, including and excluding the second-generation lines. The calculated spectra were subjected to a 0.1A binning, with 15% noise added to simulate IUE spectra. We recall that Leckrone finds excesses of Hg of the order of 10^4. If we adopt a corresponding abundance for Hy, the resulting Hy II line is sufficiently strong that it is relatively unperturbed by the presence of the additional second-generation lines. However, for an Hy abundance roughly 10^2 that of (the solar value of Hg), the relative importance of the new lines is perceptible, as is shown in Figure 3.

It appears that Leckrone's conclusions regarding mercury excesses should be substantially correct, at least as far as the second-generation lines are concerned. When the Hg excesses are rather smaller than in the stars he investigated, more attention will have to be given to missing atomic data. We must keep in mind, of course, that in a different wavelength region, the second generation absorption could be substantially more, (or less).

5. REFERENCES

Adelman, S. J. 1974, *Ap. J. Suppl.*, **28**, 51.

Bidelman, W. P. 1966, in *Abundance Determinations in Stellar Spectra.*, IAU Symposium **26**, p. 229.

Bidelman, W. P. 1968, in *Nucleosynthesis*, ed. W. D. Arnett, C. J. Hansen, J. W. Truran, and A. G. W. Cameron (New York: Gordon and Breach), p. 63.

Bord, D. J., and Davidson, J. P. 1982, *Ap. J.*, **258**, 674.

Chjonacki, G. T., Cowley, C. R., and Bord, D. J. 1984, *Ap. J.*, **286**, 736–746.

Corliss, C. H. and Bozman, W. R. 1962, *N. B. S. Monograph*,

Cowley, C. R. 1970, *The Theory of Stellar Spectra*, (New York: Gordon and Breach).

Cowley, C. R., and Adelman, S. J. 1983, *Quart. Journ. Roy. Astron. Soc.*, **24**, 393.

Cowley, C. R., and Arnold, C. N. 1978, *Ap. J.*, **226**, 420.

Cowley, C. R., and Corliss, C. H. 1983, *Mon. Not. Roy. Astron. Soc.*, **203**, 651.

Cowley, C. R., and Hensberge, H. 1981, *Ap. J.*, **244**, 252.

Davidson, J. P., and Bord, D. J. 1982, *Astron. Ap.*, **111**, 362.

Henseberge, H., and van Rensbergen, W. 1985, *A Peculiar Newsletter*, Fourteenth issue, July 1.

Hensberge, H., Van Santvoort, J., van der Hucht, K. A., and Morgan, T. H., 1986, *Astron. Ap., in press.*

Hiltner, W. A. 1945, *Ap. J.*, **102**, 438.

Jaschek, M., Jaschek, C., and González, Z. 1965, *Zs. f. Ap.*, **62**, 21.

Johansson, S. 1978, *Physica Scr.*, **18**, 217.

Johansson, S. 1981, in *23rd Liège International Astrophysical Colloquium, Upper Main Sequence Chemically Peculiar Stars*, (Liège: Institut d'Astrophysique), p. 229.

Johansson, S. 1986, *in preparation.*

Johansson, S., and Jordan, C. 1984, *Mon. Not. Roy. Astron. Soc.*, **210**, 239–256.

Kurucz, R. L. 1981, *Smith. Ap. Obs. Spec. Rept.*, **390**,(K-81).

Kurucz, R. L., and Peytremann, E. 1975, *Smith. Ap. Obs. Spec. Rept.*, **362**, (KP).

Leckrone, D. S. 1984, *Ap. J.*, **286**, 725.

Martin, W. C., Zalubas, R., and Hagen, L. 1978, *Atomic Energy Levels—The Rare Earths*, NBS-NSRDS, **60**, (AE IV).

Meggers, W. F., Corliss, C. H., and Scribner, B. F. 1975, *NBS Monograph*, **145**.

Merrill, P. W. 1956, *Lines of the Chemical Elements in Astronomical Spectra*, Carnegie Institution of

Muthsam, H., and Cowley, C. R. 1984, *Astron. and Ap.*, **130**, 348.

Unsöld, A. 1955, *Physik der Sternatmosphären, 2nd Ed.*, (Berlin: Springer-Verlag).

Discussion appears after the following paper.

MISSING LEVELS AND LINES OF ASTROPHYSICAL IMPORTANCE

SVENERIC JOHANSSON and CHARLES R. COWLEY
(Department of Physics, University of Lund, Lund, Sweden)
(University of Michigan and Dominion Astrophysical Observatory)

The purpose of this paper is to make a brief survey of the knowledge of iron group elements as concerns energy levels and transition arrays in neutral and singly ionized atoms and to give rough predictions of missing levels and lines. The amount of data that have been published for these elements after the edition of the Multiplet Tables are considerable. For detailed information about sources of new data and compilations the reader is referred to the Report from Commission 14 in the Transactions of IAU. In this paper we will focus on still missing energy levels and associated missing transitions, which have to be considered in the construction of synthetic spectra.

The structure of the iron group elements is complex not only because of the filling of the 3d shell but also because of the competition in binding energy between the 3d- and the 4s electron. For a given number of valence electrons, k, we have to consider three different low configurations in the complex $(3d + 4s)^k$ having about the same energy in neutral and singly ionized atoms. In neutral atoms the $3d^{k-2}4s^2$ configuration generally occupies the ground state, while the $3d^k$ configuration appears at rather high excitation energy. In the singly ionized atoms all three configurations have about the same energy, giving the most pronounced complexity.

The importance of these three configurations is reflected in the observed spectra of the iron group elements, since the lower level in all strong transitions generally belongs to one of them. For an analysis of a stellar spectrum in absorption the knowledge of more or less all levels in the complex $(3d + 4s)^k$ is therefore nessecary. Detailed tables of all spectroscopic terms - known as well as unknown - in these configurations of first and second spectra are displayed on posters. Here we give the data in a condensed form in Table 1, which is aimed to illustrate the present knowledge of these low configurations. The first two columns give the ionization potential (I.P.) and the total number of spectroscopic terms in the $(3d + 4s)^k$ complex. The third column gives the number of known terms as a fraction of the total number and in the last column the excitation potential (E.P.) of the lowest unknown term is expressed as a fraction of the I.P.

The by far most prominent lines in the spectra of iron group elements are assigned to 3d-4p or 4s-4p transitions. For the neutral atoms these lines fall well above 3000 Å while they in singly ionized atoms build up the heavy resonance region below 3000 Å. In order to get an idea of in which spectral region different transitions occur we have plotted the position of various configurations as a function of element in Fig 1. As a zero level we have chosen the lowest term of the $3d^k4s$ configuration. Different curves then represent $3d^k nl$ configurations in the first (faint lines and dots) and second (bold lines and crosses)

C. R. Cowley et al (eds.), Upper Main Sequence Stars with Anomalous Abundances, 99–102.

Table 1

Spectrum	I.P.(eV)	Number of terms Total	Known	E.P.
Sc I	6.56	16	1.00	-
Sc II	12.89	8	1.00	-
Ti I	6.83	37	0.57	0.50
Ti II	13.63	16	1.00	-
V I	6.74	48	0.42	0.39
V II	14.2	37	0.95	0.52
Cr I	6.76	64	0.44	0.62
Cr II	16.49	48	0.88	0.48
Mn I	7.43	48	0.35	0.65
Mn II	15.64	64	0.64	0.47
Fe I	7.90	48	0.65	0.52
Fe II	16.18	48	0.79	0.53
Co I	7.86	16	0.81	0.51
Co II	17.05	37	0.51	0.31
Ni I	7.63	7	0.86	0.80
Ni II	18.15	16	0.88	0.42

spectra as a function of k in an iso-ionic comparison of the excitation energies. A configuration is represented by that particular level which is built on the lowest parent term and has the highest J-value. As we see in Fig 1 we get smooth curves and it will be easy to predict the position of missing configurations by extrapolation (e.g. 5p in Ti I).

If we now take the energy difference between two curves in Fig 1, which represent configurations of different parity in the same spectrum, we get the wavenumber of the strongest transition in that particular transition array. These differences are plotted in Fig.2 and the curves give an idea of where the strongest transitions occur in different spectra. Most of these lines are known (circles for predicted lines) since they represent transitions between levels built on the lowest parent term and thus have the biggest chance to show up in the spectrum. However, from Table 1 we can understand that there are still missing lines between levels built on higher parent terms. Generally, we can expect these transitions to occur at roughly the same wavelength as the corresponding electronic transition in Fig 2, since the excitation energy of the high parent term shift the two curves in Fig 1 by the same value. Since there are more than one term associated to one particular parent term we can expect the wavelength region covered by an electronic transition to be extended by about 2-300 Å around the plotted wavelength.

In Fig 2 we can see that the 4p-5s and 4p-4d transitions in the second spectra fall in the same wavelength region as the 4s-4p lines, which of course complicates the analysis of IUE spectra. (From present studies of Fe II in this region it is quite clear that a great number of strong lines remain unidentified.) The next lowest groups of lines extend below 2000Å and refer to the 4p-5d and 4p-6s transition in second spectra. The probability for them to occur in stellar spectra can be discussed in terms of branching ratios for different members in

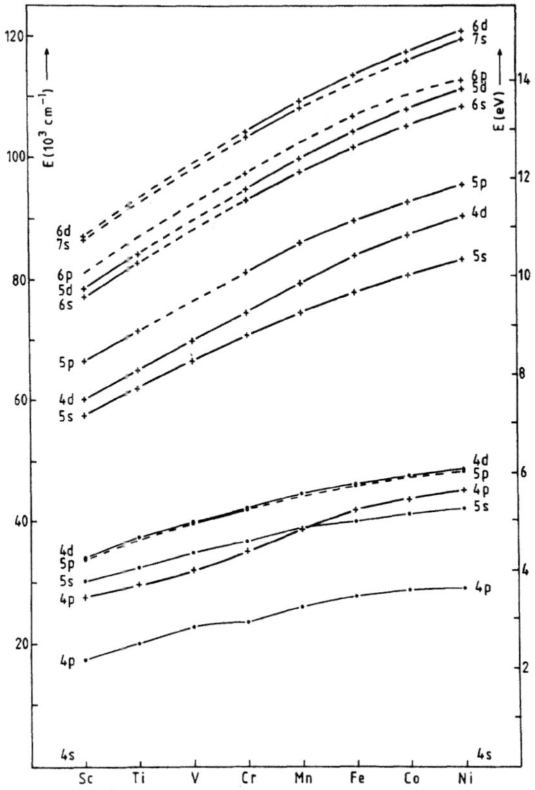

Fig. 1 Excitation energies for $3d^k nl$-configurations in first (dots) and second (crosses) spectra of the iron group elements

the s- and d-series. The f-values for the 4p-5s and 4p-4d lines are of the same order of magnitude (0.1 and 0.6 respectively) but roughly a factor of ten smaller for the 4p-6s and 4p-5d transitions.

In the wavelength region below 2500Å we also expect the 4s-np (n>5) transitions of the first spectra to occur. Laboratory measurements of the p-series have been done (in absorption) in a number of spectra. From oscillator strengths in Cr I (Huber et al 1975) we can get an idea of the scaling of the f-values in the p-series: 4s-4p f=0.11, 4s-5p f=0.014, 4s-6p f= 0.0022 i.e. roughly a factor of ten when going up in a Rydberg series.

In this brief review of missing lines we have only considered transitions within one of the three possible systems of configurations, namely the $3d^k nl$ system. As we pointed out above the $3d^k - 3d^{k-1}np$ transitions are very dominating in second spectra and give strong lines in the IUE region. They can not be represented in such a smooth curve in the diagram as the plotted transitions. In general we can

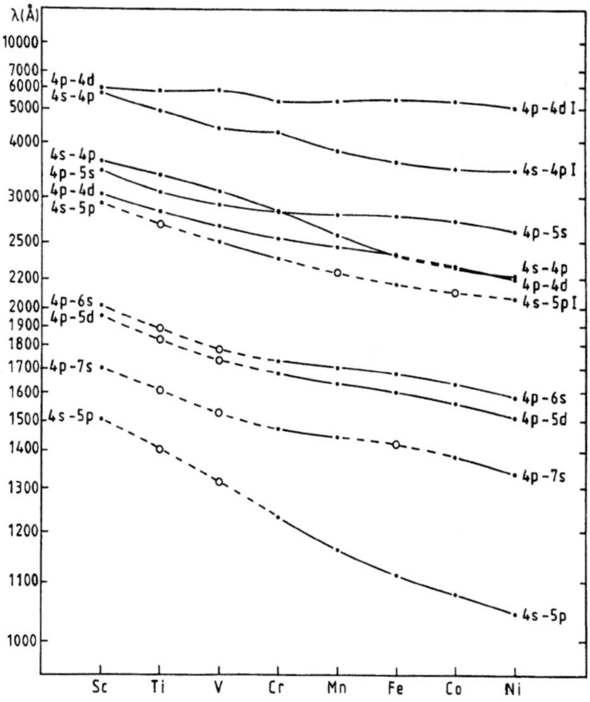

Fig 2. Representative wavelengths for some transitions within the $3d^k nl$ system in the first and second spectra of the iron group elements.

state that all possible LS transitions between $3d^k$ and $3d^{k-1}4p$ are strong. More or less all $3d^k$ levels are known in second spectra and since that is also the case for the $3d^{k-1}4p$ configuration nearly all prominent lines are known.

As was said in the introduction the $3d^{k-2}4s^2$ configurations play the key role in the neutral atoms and the associated system of configurations $3d^{k-2}4snl$ account for the major part of the known levels in these spectra. The lowest excited configuration is thus $3d^{k-2}4s4p$ which generally deexcites in the resonance lines. In order to get smooth curves for this type of configurations the levels must be plotted versus the grand parent terms of $3d^{k-2}$. The $3d^{k-2}nl$ configurations are fairly well known in neutral atoms but poorly known in singly ionized atoms.

Reference

Huber,M.C.E.,Sandeman,R.J.,and Tubbs,E.F.1975, Proc R Soc Lond A342,431.

DISCUSSION (Cowley and Johansson)

ADELMAN: Co-adding of spectra is an important technique for improving the signal-to-noise ratio for non-variable stars. Those who are interested should look at my paper on co-adding of Dominion Astrophysical Observatory spectrograms, which is a poster paper at this Colloquium. I want to make two comments about this technique. First, stars which do not have sharp lines do not have to be observed with high resolution. One can do better by observing them with lower resolution and co-adding the spectra. Second, it is a waste of valuable telescope time to obtain inferior data if better data can be obtained by requesting it from another observer.

Jeff Fuhr at the US National Bureau of Standards has given me some data on gf-values for Fe I, Fe II, Ti II, and Co II, and has authorized me to distribute it here. There are many NBS publications available which would interest everyone working on atomic spectra; Drs. W. Martin, J. Fuhr, and R. Zalubas are glad to give these free to anyone who requests them. They only ask that you write to them on your institution's official stationery when you request it. Their address is:

National Bureau of Standards
Physics Building - Room A267
Gaithersburg, MD 20899 USA

HUBENÝ: I would like to show you an example which illustrates the problem of misidentification in IUE spectra. Hubený, Štefl & Harmanec (Bull. Astr. Inst. Czech., 36, p. 214, 1985) show that blends of numerous Fe III lines produce two fictitious lines, closely simulating the strong resonance lines of C IV, for moderately and rapidly rotating B3 to B8 stars. Even at lower temperatures this can be a problem, because one gets what looks like C IV lines, but shifted about 100 km/s. This effect is seen in many Be stars! Here [shows slide] the fictitious blend is due to Fe II.

SEVERNY: You did not mention the possibility of astrophysical determination of gf-values using stars with known chemical composition. We have had problems analysing UV spectra obtained with Astron, because Kurucz and Peytremann gf-values are sometimes wrong by factors of 10^3 to 10^4. What is your opinion of this method?

COWLEY: This is certainly an important method. However, if there are additional unknown lines which are blends, we can make many errors with this method, as Dr. Hubený has just pointed out. Like all things, it has its good and bad parts. In general I prefer to use the NBS critically evaluated gf-values when they are available. Of course, there are many lines for which gf-values are not known, and we have to use the astrophysical method.

JOHANSSON: I'm glad to see that the data we are producing at Lund is of good help for your spectral analysis of stars. We have worked on Fe II since 1974, and this work is continuing. We are analysing the entire IUE region, in collaboration with Dr. Baschek in Heidelberg. We hope to produce a new multiplet table for Fe II. We are also working on other

C. R. Cowley et al (eds.), Upper Main Sequence Stars with Anomalous Abundances, 103–104.

iron-group elements. If you want some laboratory line lists, we can send these to you. In some wavelength regions there is a severe lack of atomic data for the iron-group.

COWLEY: I have one small additional comment on the Kurucz (1981) Fe II calculations. In the region discussed in my paper, the strongest contributions came from transitions to certain odd upper levels which had not been properly described quantum-mechanically in his work. These oscillator strengths are almost certainly greatly overestimated, and contribute roughly 10% to the general absorption there. Undoubtedly, that absorption belongs someplace, but not precisely at the wavelengths where it was calculated.

ARTRU: Even for Si II, which is considered to be a relatively well-known atomic spectrum, about 30% of the multiplets (many of them strong) have no known gf-values. About 25% are missing in the Kurucz & Peytremann list, because in Si II there are many doubly excited levels which have mixed configuration effects. So, even for a 'simple' ion like Si II there are problems.

DISCUSSION (Johansson and Cowley)

HUBENÝ: From the practical point of view, it would be nice to know that lines originating between newly found energy levels are generally weaker than previously known lines from well-established levels. Is there such a hope at present, or in the near future?

JOHANSSON: We know already that a great number of Fe II lines, which are not available in the literature, have log gf ≈ -1 and lower excitation potentials around 7 - 8 eV. They appear as strong absorption features in the IUE region. For Cr II, no lines below 1780 Å are available in existing compilations. We now know of about 1000 new lines in this region with the strongest lines around 1430 Å. The gf values for these lines are certainly not more than one order of magnitude less than the gf values for the resonance lines. The conclusion is that a lot of predicted lines will have intensities comparable to those of 'well-established' lines.

CHARACTERISTIC ABSORPTION FEATURES IN THE SPECTRA OF Ap-Si STARS BETWEEN 1250 AND 1850 Å.

Marie-Christine ARTRU

Département d'Astrophysique Fondamentale
Observatoire de Meudon
F-92195 MEUDON PRINCIPAL CEDEX, France

Thierry LANZ

Institut d'Astronomie de l'Université de Lausanne
and Observatoire de Genève,
CH 1290 CHAVANNES-DES-BOIS, Switzerland

Abstract : On the basis of IUE data, the specific absorption features in the spectra of Ap-Si Stars are displayed in the 1250-1850 Å range and identified when possible. The contribution of the Si II multiplets is calculated in LTE for a typical overabundance x 10^2 of silicon. It accounts for only a small part of the observed absorption. Common unidentified structures, smaller than the 1400 Å depression are pointed out.

INTRODUCTION

The broad absorption features observed at 1400 Å in the spectra of magnetic Ap stars is well correlated to the silicon overabundance (Jamar et al, 1978). It is interpretated as autoionization of Si II (Artru et al, 1981) but the precise identification cannot be established without new laboratory data on the highly excited levels of Si^+. Our purpose is to derive from IUE spectra a refined description of all the absorption features that appears specific of Ap-Si stars in the spectral range 1250-1850 Å.

More than 150 different stars, classified as Ap or Bp, have been observed by the IUE satellite. From these archives, we have selected a set of silicon stars with effective temperatures in the 10000-14000 K range and large absorption features at 1400 Å. We present here preliminary results obtained for the limited data sample described in table 1.

DESCRIPTION OF THE Ap-Si SPECIFIC ABSORPTION

To exhibit the specific absorptions in the silicon stars atmospheres, we display "relative" spectra by plotting the quantity log $(F_\lambda(p)/F_\lambda(n))$ versus the wavelength λ ; $F_\lambda(p)$ is the observed spectral flux of the peculiar star and $F_\lambda(n)$ the same for a normal comparison star of similar effective temperature. We use calibrated net spectra, either directly obtained from the IUE low-resolution images, or derived from high-resolution images after degrading the resolution.

Such "relative" spectra are given in fig. 1. The curves a, b and c correspond to the three silicon stars of table 1 and show the striking similarity of their absorption features. For comparison the curve d relates to each other the two normal stars (HD 87901 and HD 17081).

C. R. Cowley et al (eds.), Upper Main Sequence Stars with Anomalous Abundances, 105–108.

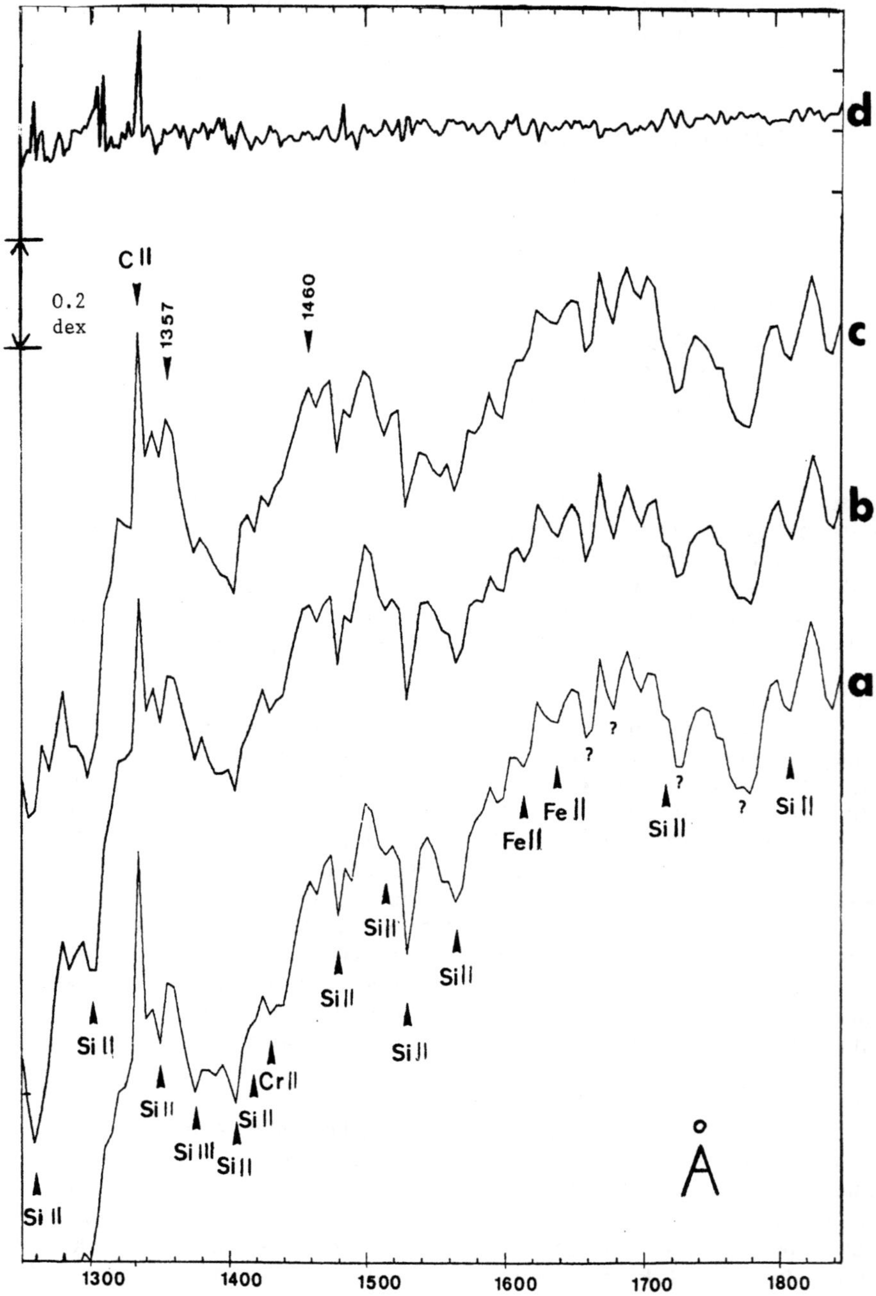

Fig.1 - Relative spectra of three Ap-Si stars (a, b, c) and one normal star (d). The logarithm of the flux ratio is plotted for HD27309 / HD17081 (a), HD25267 / HD17081 (b), HD34452 / HD17081 (c) and HD87901 / HD17081 (d).

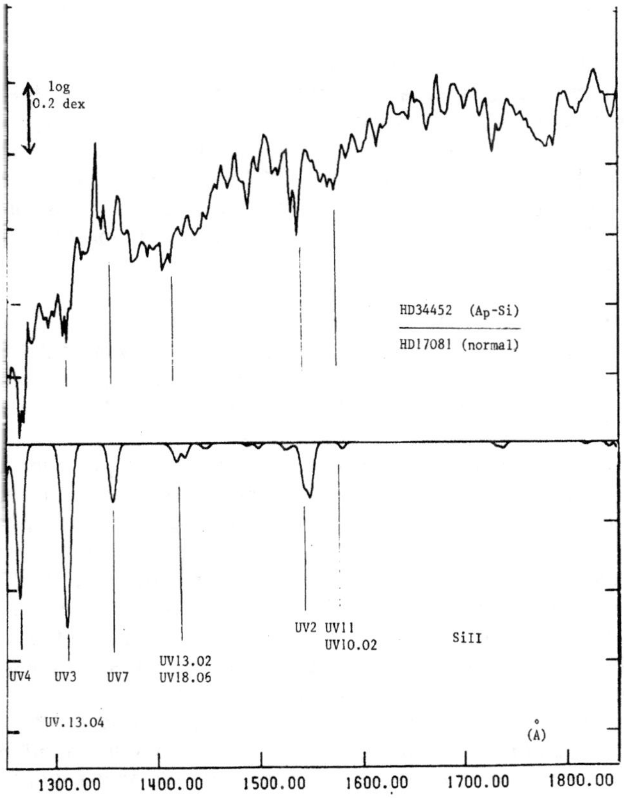

Fig. 2 - LTE synthetic spectrum of Si II multiplets for a silicon abundance of 100 x ⊙ (lower curve) and observed "relative" spectrum of HD34452 (upper curve) (the resolution is about 6 Å).

Table 1 - Sample of Ap-Si stars and IUE observations

Star identification	T_{eff} a)	δ_{1400} b)	IUE image number
HD 34452 Si	13400	1.14	H3 * 7542, L3 = 22167
HD 25267 τ^9Eri, Si	11700	0.97	H3 * 3071
HD 27309 56Tau, Si	11600	1.35	H3 * 7541, L3 = 7565
HD 17081 ΠCeti, normal	13500	--	H3 * 16256,L3 = 16255
HD 87901 αLeo, normal	12200	--	H3 * 8649, L3 = 8647

a) The effective temperature is estimated from photometric indexes (Lanz, 1985).

b) The δ_{1400} index measures the enhancement of the 1400 Å depression (Jamar et al, 1978).

Many of the sharper absorption lines, in the "relative" Ap -Si star spectra (fig. 1-a, b and c) are identified as strong multiplets of overabundant ions, mainly Si II and Fe II. The C II multiplet at 1335 Å appears as a peak, suggesting an underabundance of carbon. A similar weaker feature also appear for the comparison star HD 87901 (fig. 1-d).

The large 1400 Å feature has a similar shape for the three silicon stars and occurs between the same limits, 1357 and 1460 Å. Other smaller wide depressions are observed similarly for the three Si stars, around 1550 Å and 1750 Å. Their identification, possibly by autoionization of an overabundant ion is still an open question.

THE ABSORPTION SPECTRUM OF Si II

The strongest lines identified on the observed curves of fig. 1 are due to Si II multiplets (Moore, 1965) : for instance, UV4 at 1264 Å, UV13.02 and UV13.03 at 1405 – 1410 Å, UV12 and UV15.04 at 1485 Å and UV2 at 1530 Å.

In order to evaluate the contribution of the Si II spectrum to the opacity variations observed in fig. 1, we first calculated the total effect of the "normal" lines, as known from laboratory analysis and compiled by Moore (1965). A predicted LTE synthetic spectrum of Si II was computed using a blanketed model (13000 K, log g = 4) from Kurucz et al (1977). The silicium abundance was fixed at 100 times the solar one, which is the value generally given for HD 34452. The line list was limited to the experimental lines of Si II, with the gf-values taken from the compilation of Lanz and Artru (1985). The line spectrum of Si II obtained in this calculation is shown on fig. 2, with the same resolution and scale, as for the observed curves of fig. 1. Their comparison confirms the identification of the Si II multiplets, but shows that they contribute for only a small part, to the observed opacity. The "normal" spectra of other overabundant ions should also be calculated, but are likely not to fill the missing opacity. A systematic study of their autoionization spectra would be essential.

REFERENCES

- Artru, M.-C., Jamar, C., Petrini, D., Praderie, F., 1981, Astron. Astrophys. 96, 380.
- Jamar, C., Macau-Hercot, D., Praderie, F., 1978, Astron. Astrophys. 63, 155.
- Kurucz, R.L., Peytremann, E., Avrett, E.H., 1974, Blanketed Model Atmospheres for Early-type Stars (Washington, Smithsonian Institution Press).
- Lanz, T., 1985, Astron. Astrophys. 144, 191.
- Lanz, T., Artru, M.-C., 1985, Physica Scripta, in press.
- Moore, C.E., 1965, Selected Tables of Atomic Spectra, NSRDS-NBS3, National Bureau of Standards.

Discussion appears after the following paper.

SPACE OBSERVATIONS OF CHEMICALLY PECULIAR AND RELATED NORMAL STARS

David S. Leckrone
Laboratory for Astronomy and Solar Physics
Goddard Space Flight Center
Greenbelt, Maryland
USA

ABSTRACT. Progress in the spectroscopic study of CP stars and related sharp-lined normal stars from the IUE is briefly reviewed as a preamble to a discussion of the potential for research with the Hubble Space Telescope. The substantial gains in spectral resolution, signal-to-noise ratio and photometric accuracy that will be realized with the High Resolution Spectrograph on the HST will dramatically increase our ability to disentangle the complex ultraviolet spectra of these stars and to carry out accurate quantitative analyses.

1. INTRODUCTION

A decade ago at IAU Colloquium Number 32 in Vienna, Leckrone (1975) reviewed the properties of CP stars as observed from space. At that time the observational data were limited almost entirely to broad-band photometry and intermediate-band spectrophotometry in the wavelength interval 1100Å to 3000Å, which revealed the flux deficiencies of the CP stars in the ultraviolet, and which explained the complex photometric variations observed in the optical region as being due to variable backwarming. Since the time of that meeting a great deal of UV spectroscopic work has been carried out on CP stars, primarily with the International Ultraviolet Explorer (IUE). The extended lifetime of the IUE has allowed UV spectroscopy to advance from qualitative surveys to detailed quantitative abundance analyses. Results obtained by several groups have been reported in the literature for B, Al, Si, Cu, Ga, Sb, Pt, Hg, and Bi (Jacobs and Dworetsky 1981,1982; Dworetsky et al. 1984; Leckrone 1980,1981,1984; Sadakane et al. 1983,1985). In each case the information obtained from space has provided a test of the predictions of diffusion theory, has led to the identification of new anomalous species or has supported and strengthened previous ground-based work. We are only just beginning to reap the harvest of information from the IUE, as I hope to show in this discussion and we are on the threshold of a new era of quantitative accuracy, resolution and sensitivity which will be provided by the Hubble Space Telescope (HST).

C. R. Cowley et al (eds.), Upper Main Sequence Stars with Anomalous Abundances, 109–119.

2. COMPREHENSIVE ABUNDANCE ANALYSES WITH IUE AND OPTICAL DATA

Adelman and Leckrone have undertaken a long-term project to establish fundamental abundances for as many species as possible, beginning with the iron peak elements, in a selected group of CP and normal, sharp-lined B and A stars. The approach is to combine high quality, composited IUE spectra (Adelman and Leckrone 1985) with high signal-to-noise optical spectra in fully self-consistent abundance analyses. Since our data encompass the wavelength range from about 1250Å to about 6000Å, we can cover a larger fraction of the periodic table, a wider range of excitation and ionization states, and for many species a statistically more significant sample of lines than can be achieved from optical data alone. The ultraviolet spectra not only make available low excitation or resonance lines of many important species but also in some cases provide lines whose oscillator strengths and damping constants are more accurately determined than those of their counterparts at optical wavelengths. On the other hand all such analyses are currently limited by the high line density of ultraviolet spectra, the lack of accurate atomic data for most of the contaminate blending lines and the rather low resolution ($\lambda/\Delta\lambda \simeq 15000$) and signal-to-noise ratio (≤ 30) of the IUE data. We are still in the process of learning how best to ameliorate these problems in the IUE data. In the long term the observational problems will be greatly diminished by use of the High Resolution Spectrograph (HRS) on the HST.

TABLE 1

IUE ABUNDANCE PROGRAM STARS

Star	Type	Teff	V Sin i
θ Leo	A2 V	9250	20
o Peg	A1 V(Am?)	9625	12
ν Cap	B9.5 V	10250	17
134 Tau	B9 IV	10825	22
21 Aql	B8 III	13000	19
π Cet	B7 V	13150	18
ι CrB	HgMn	11380	3
κ Cnc	HgMn	13300	6
HD109995	A2(FHB)	8100	17

The stars currently being analysed in the Adelman and Leckrone abundance program are listed in Table 1. They were selected both for their low values of v sin i and also because a substantial amount of work on their optical spectra has already been completed. IUE data for all of these stars have been obtained and the data reduction, involving the co-addition of many spectra so as to reduce both random and fixed-pattern noise, is now complete. Our intent is to continue to update the abundances derived for these stars as improved observational or atomic data become available, thus establishing and maintaining them as reference standards.

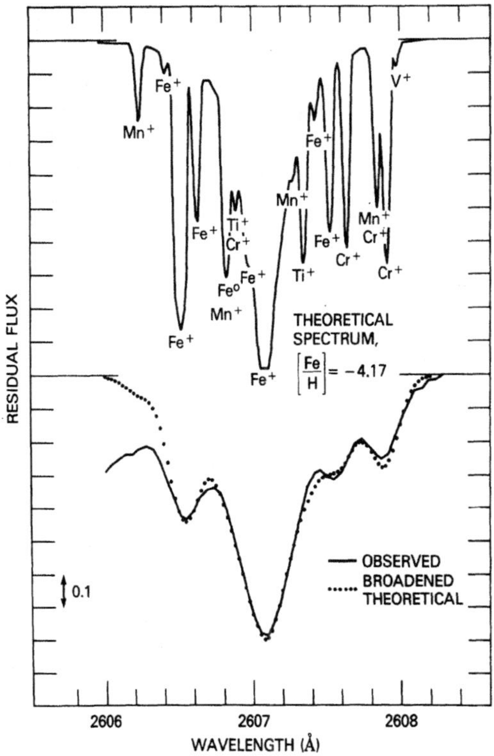

Figure 1. Spectral synthesis of the resonance line FeII λ2607 in o Peg. Top - theoretical spectrum with log[N(Fe)/N(H)] = -4.17. Bottom - IUE observation (solid curve) compared to theoretical spectrum, suitably broadened with rotation and instrument function.

The analyses have begun with iron, because its abundance is reasonably well established from the optical data alone. It thus provides a good opportunity to compare abundances obtained from both ultraviolet and optical regions for consistency. The top section of Figure 1 illustrates the computed synthetic spectrum of the resonance line, FeII(1) λ2607.086, in o Pegasi. Note the large number of blending lines. The line lists and log gf values of Kurucz and Peytremann (1975) and Kurucz (1981) were used initially for these blending lines. But it is well known that these data contain systematic errors, and it is necessary to artificially modify the semi-empirical log gf values of some of the blending lines in order to achieve a good fit to the observations. This procedure, which is discussed in detail in Leckrone (1981) for example, is bounded by the requirement that the adopted mix of log gf values for the blending lines must work for all of the stars we analyse. As long as the primary line being synthesized (in this case FeII λ2607) is the dominate member of the blend, rather large errors in the calculated strengths and positions of the blending contaminates can be incurred without seriously affecting the derived abundance. Errors will typically be less than 0.1 dex, even if we ignore a major blending contributor. In the computation illustrated in Figure 1, the log gf value used for the FeII resonance line is taken from a new critical compilation by Martin et al. (1986). Its uncertainty of approximately 10% is substantially lower than that of any currently available log gf values for FeII lines in the optical region. The lower part of Figure 1 shows this synthetic spectrum, convolved with the slit broadening function of the IUE spectrograph and rotated up to 12 km/s, compared to the line as observed with the IUE. The fit is reasonably good for an iron abundance log[N(Fe)/N(H)] = -4.17.

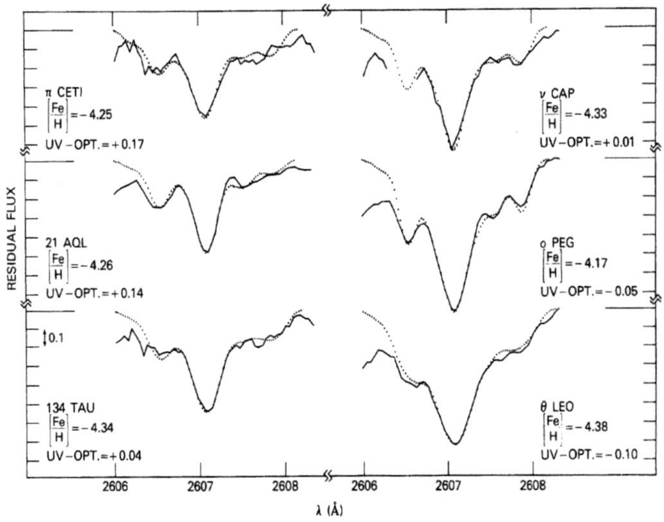

Figure 2. Spectral syntheses of FeII λ2607 in six normal stars. Solid curves are features observed with IUE; dotted curves are suitably broadened theoretical profiles.

Figure 2 illustrates the syntheses of FeII(1) λ2607 for the six normal stars included in Adelman and Leckrone's program. Using the procedures illustrated in detail in Figure 1, one obtains a good fit to all the observations. The preliminary abundances derived from this single ultraviolet resonance line of FeII are compared to the iron abundances derived from optical region observations of the same stars in Table 2. The generally good agreement between optical and ultraviolet values in the normal stars (top section of Table 2) is encouraging and gives us confidence to proceed with the analyses of numerous ultroviolet lines. There does appear to be a small systematic effect, which we do not yet understand, in that the optical abundances for the hotter stars tend to be lower than those derived from this single UV line. It will be of interest to see if this effect persists as we analyse additional Fe lines and as we continue to refine our data analysis techniques (for example, in the way we locate the continuum). We intend to complete similar analyses for about twelve additional ultraviolet FeII lines which have log gf values of quality B or B+ on the U.S. National Bureau of Standards scale of accuracy, i.e. with uncertainties around 10%. In addition we have good IUE data for about 25 other FeII lines, whose transition probabilities are less well known, for which astrophysical f-values could be derived. We are interested not only in accurate absolute abundances, but, for the normal stars, in the possible variations in abundances from star-to-star. After iron we will proceed to Mn, Co, Ni and other elements of the iron peak and ultimately to as much of the periodic table as possible.

TABLE 2

PRELIMINARY IRON ABUNDANCE RESULTS

Star	Optical	λ2607	UV - Optical
π Cet	-4.42	-4.25	+0.17
21 Aql	-4.40	-4.26	+0.14
134 Tau	-4.38	-4.34	+0.04
ν Cap	-4.34	-4.33	+0.01
ο Peg	-4.12	-4.17	-0.05
θ Leo	-4.28	-4.38	-0.10
Average	-4.32±0.11	-4.29±0.08	σ = 0.11
Sun	-4.37		
ι CrB	-4.20	-4.05	+0.15
κ Cnc	-4.47	-4.10	+0.37

3. PROSPECTS FOR RESEARCH WITH THE HUBBLE SPACE TELESCOPE

It is evident from the work done by numerous people on the IUE spectra of CP and related normal stars that a rich store of information about the nature and origin of chemical anomalies resides in the complex ultraviolet spectra of these stars. The best tool for accurate quantitative studies of this region will undoubtedly be the High Resolution Spectrograph (HRS) on the Hubble Space Telescope (HST), to be launched by NASA's Space Shuttle during the latter half of 1986. The HST will be maintained and repaired in orbit by Shuttle crew members over its lifetime of at least fifteen years. Its instruments can be replaced with even more sophisticated devices as time goes on. Recently several candidate second generation HST instruments were selected for study, including an echelle spectrograph with a two-dimensional, photon-counting imaging detector (the STIS or Space Telescope Imaging Spectrograph), which will allow simultaneous observations over a much wider wavelength interval than that covered by the HRS. Both the HRS and the STIS will have substantially wider dynamic ranges and lower levels of background noise than does the IUE.

In addition to spectroscopy the HST will provide other important capabilities for the study of CP stars. The basic characteristics of the scientific instruments which will be on board the HST at launch are summarized in Table 3. The High Speed Photometer (HSP), which is capable of accurate photometry in time bins as short as 10 μsec, might well be applied to the study of rapid variability throughout the ultraviolet and optical bands. The search for CP stars in clusters could be simplified by the high angular resolution (0.1 arcseconds) and the large selection of UV and optical filters in the two cameras. With regard to the HRS it is important to note that the highest resolving power ($\lambda/\Delta\lambda \simeq 95000$) is obtained with an echelle grating which introduces a significant amount of scattered light. The scattered light background can be measured and subtracted out but at the expense of longer exposure times. On the other hand the intermediate 25000 resolving power, obtained in first order, offers almost twice the resolution of the IUE and a superbly dark background. Thus one must exercise judgement as to whether the highest possible signal-to-noise ratio or the highest possible resolution is needed for a given observation. Note that the resolving power in the 25000 mode might be improved to about 50000 by deconvolving the instrumental broadening function from data with a sufficiently high signal-to-noise. The detectors in the HRS are linear diode arrays (Digicons). This limits the width of the spectral interval that can be measured in a single exposure to about 5 to 18 Å in the 95000 mode and to about 26 to 48 Å in the 25000 mode.

TABLE 3

HUBBLE SPACE TELESCOPE SCIENTIFIC INSTRUMENTS

	WAVELENGTH RANGE (A)	FIELD OF VIEW (ARCSEC)	ANGULAR RESOLUTION (ARCSEC)	SPECTRAL RESOLVING POWER	MINIMUM ACCUMULATION TIME (MSEC)	"LIMITING MAGNITUDE" (m_V)
WIDE FIELD AND PLANETARY CAMERA	1150-11500	70-160	0.1-0.2	FILTERS	100	28
FAINT OBJECT CAMERA	1150-7000	≤44	≤0.1	FILTERS	50	28
HIGH RESOLUTION SPECTROGRAPH	1100-3200	0.25-2.0	N/A	95000 25000 2500	200	12-16
FAINT OBJECT SPECTROGRAPH	1150-8000	0.1-4.0	N/A	1200 200	10	23-26
HIGH SPEED PHOTOMETER	1150-9000	0.4-1.0	N/A	FILTERS	0.01	24
FINE GUIDANCE SENSOR ASTROMETER	4670-7000	≅500	0.002	FILTERS	N/A	19

In the following discussion I use the example of the HgII resonance line at 1942 Å, which was the subject of my recent paper (Leckrone 1984), to illustrate the potential applications of the HRS to the study of the

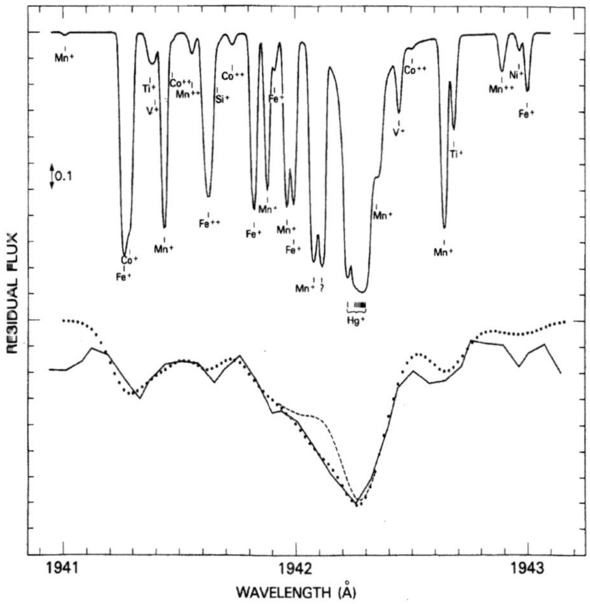

Figure 3. Spectral synthesis of the HgII λ1942 resonance line in ι CrB. Top - theoretical spectrum with log[N(Hg)/N(H)] = -6.1. Bottom - IUE observation (solid curve) compared to suitably broadened theoretical computation (dots) and to a theoretical computation from which an unidentified blending feature (question mark in upper figure) has been omitted (dashed curve).

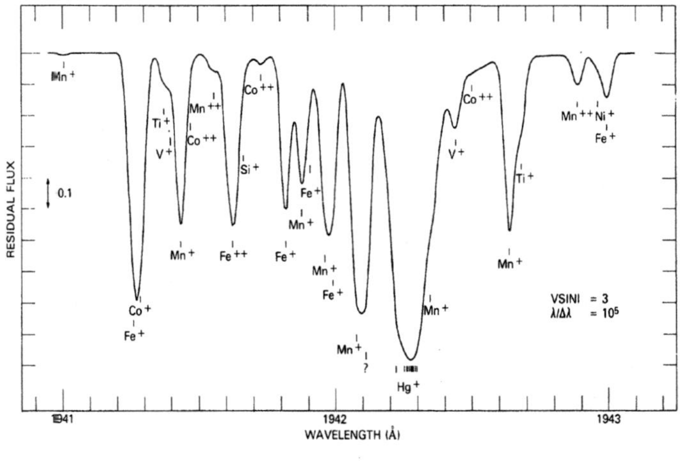

Figure 4. Theoretical spectrum from Figure 3 broadened with the expected instrumental function for the highest resolution mode of the High Resolution Spectrograph.

ultraviolet spectra of the sharpest-lined stars, many of which are included in the various classes of chemically peculiar stars. Why is one interested in the HgII resonance lines? Prior to the IUE the Hg anomaly rested entirely on observations of HgII λ3984, plus the very weak HgI λ4358 line seen in only a few stars. Figure 9 of Leckrone (1984) shows computed curves of growth for λ3984 as a function of isotopic mix parameter, q, effective temperature and microturbulent velocity, V_t. It is clear that the strength of λ3984 is extremely sensitive to the choice of q and V_t. Small errors in these parameters will produce large errors in the Hg abundance derived from this line. This is the cause of much of the large star-to-star scatter in Hg abundances found in the literature. In contrast, as is shown in Figure 3 of Leckrone (1984), the resonance line of HgII at 1942.275 Å is insensitive to all three parameters and is only slightly sensitive to the adopted value of the damping constant (it is dominated by radiation damping, which is easily calculated). Thus, this line should yield accurate abundances, provided the problems of line blending discussed previously can be overcome. The detailed synthesis of λ1942 in IUE observations of the HgMn star ι CrB is shown in Figure 3. The technique used is identical to that discussed earlier for the FeII lines. That is the Kurucz-Peytremann log gf values of the contaminate blending lines had to be adjusted to achieve an acceptable fit to the observations in all the stars considered. Also in this case an extra blending line of unknown identity (I assumed TiII) had to be added to the blend to explain the asymmetry of the line core on its short wavelength side. This is consistent with the findings of Cowley and Johannson (in this colloquium) that there are a significant number of missing lines in the available line lists. At the bottom of Figure 3 is this same synthetic spectrum, convolved with rotational and IUE instrumental broadening functions, and overlaid on the observed feature in ι CrB. The dashed line shows the theoretical spectrum without the addition of the unknown blending contaminate, for comparison. Figure 4 illustrates the synthetic spectrum from Figure 3 convolved with the 3 km/s rotational broadening function and with the instrumental broadening function of the HRS in the 95000 resolving power mode. Most of the blending lines, whose identities we could only guess in the IUE spectra are now resolved, and even some of the weak features show up as partially resolved bumps or shoulders on the stronger lines. In addition we are more likely to accurately locate the line-free continuum in such high resolution observations (this is a difficult but not insurmountable problem in the IUE spectra). The signal-to-noise ratio in HRS observations of this sort will be limited only by the length of time one chooses to devote to counting photons, but in a realistic situation would likely be at least a factor of two better than in the best co-added IUE spectra. The problem then becomes one of adequately supporting the atomic physicists who are interested in providing the accurate atomic data needed to do justice to such observations.

In Figure 5 I have focused on an approximately 1 Å spectral interval around λ1942.275 and have performed the computation of the synthetic spectrum for two extreme values of the isotopic mix parameter, q. The value q = -0.1 corresponds to a nearly terrestrial mix (as in the HgMn star μ Lep, for example), while the value q = +3.0 corresponds to an extreme isotopic anomaly (almost pure Hg^{204}) as is found in χ Lup. At this resolving power and with good signal-to-noise one will be able to discriminate to some degree the isotopic structure of this Hg line to compare it to previously reported results for λ3984.

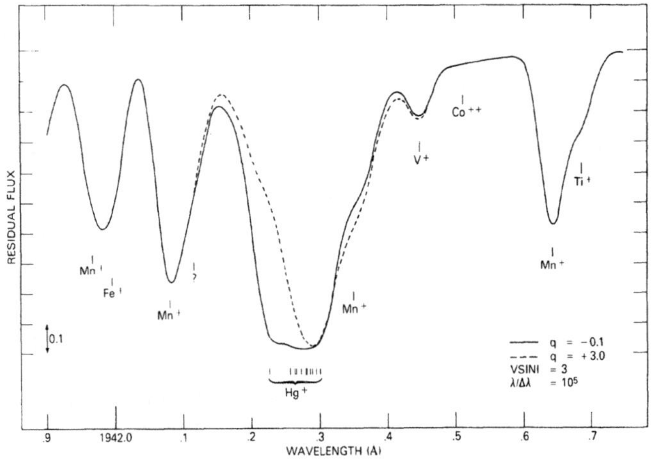

Figure 5. Theoretical computations of HgII λ1942 for two extreme values of the isotopic mix parameter, q, broadened to v sin i = 3 km/s and degraded to the highest resolution of the HRS.

4. EARLY OBSERVATIONS OF CP STARS WITH THE HST

The first observations of CP and related normal stars with the HST will be those I carry out as a member of the High Resolution Spectrograph Team. My program is made up of eight separate observational objectives, which focus on non-magnetic HgMn stars, a field horizontal branch star, and normal sharp-lined B and A main-sequence stars. The selection of objectives reflects my own judgement as to what scientific questions will be most amenable to immediate attack with the new observational capabilities of the HRS, as well as my own historical research interests. Certainly the magnetic CP stars, the Am stars, the stars with anomalous He lines, etc., will also provide fruitful research opportunities with the HRS and the other HST instruments. In the following discussion, I summarize my observational programs both to give an idea of the types of research one might undertake with the HRS and to point out the opportunity that is now being provided to astronomers from all over the world to submit proposals to observe with the HST. My programs are constrained to about 30 hours of observing time spread out over 2.5 years - sufficient only to scratch the surface of the important research to be done. The data obtained will be protected for one year after the completion of the observations, after which time they will become part of the archives of HST data accessible to the general community.

The eight major parts of my observing program are listed below. Table 4 lists the target stars associated with these objectives.

Part 1. - Survey BII λ1362 resonance line at $\lambda/\Delta\lambda \simeq 95000$ in HgMn and normal stars to establish B abundance trends.

Part 2. - Establish Hg abundance in normal stars based on observations of resonance lines at $\lambda/\Delta\lambda \simeq 95000$.

Part 3. - Study the systematic behavior of UV lines of HgI, HgII and HgIII and derive accurate Hg abundances in a wide range of Hg-rich stars at $\lambda/\Delta\lambda \simeq 95000$.

Part 4. - Explore odd-even abundance patterns in HgMn stars (program constrained by time allocation to search for low-excitation lines of Au in stars known to be rich in Pt and Hg) at $\lambda/\Delta\lambda \simeq 95000$.

Part 5. - Obtain complete UV spectra of selected HgMn stars with high signal-to-noise at $\lambda/\Delta\lambda \simeq 25000$ for line identification and abundance analyses.

Part 6. - Obtain complete UV spectra of selected normal B and A stars with high signal-to-noise at $\lambda/\Delta\lambda \simeq 25000$ to establish spectroscopic and abundance comparison standards.

Part 7. - Establish CNO abundances in a characteristic, bright field horizontal branch star using observations of selected low excitation lines at $\lambda/\Delta\lambda \simeq 25000$.

Part 8. - Obtain complete UV spectra of a bright sharp-lined normal A star and an extremely sharp-lined HgMn star at $\lambda/\Delta\lambda \simeq 95000$ (lower priority objective to be carried out if time permits).

TABLE 4

HST PROGRAM STARS

Star(HD)	Name	Objective	Star(HD)	Name	Objective
1909	HR 89	1,3	141556	χ Lup	1,3,4,8
17081	π Cet	2	143807	ι CrB	1,3,5
27295	53 Tau	1,3,4,5	145389	φ Her	3,5
33904	μ Lep	1,3	149121	28 Her	3,4
38899	134 Tau	2	174933	112 Her	1,3
48915	α CMa	8	182308	HR 7361	1,3,4
78316	κ Cnc	1,3,5	190229	HR 7664	1,3
89822	HR 4072	1,3,4	193432	ν Cap	1,2,4,6
97633	θ Leo	2	193452	HR 7775	1,3,4,5
110073	HR 4817	1,3	207857	HR 8349	1,3
129174	π^1 Boo	1,3	214994	ο Peg	2,3,6
130095		7	215573	ξ Oct	1,2,3,4,6

I strongly encourage my fellow CP-star aficionados to consider taking advantage of the new opportunities the HST will offer to help us resolve the mysteries of these enigmatic objects.

REFERENCES

Adelman, S.J. and Leckrone, D.S. 1985, NASA IUE Newsletter, No. 28, in press.
Dworetsky, M.,M., Storey, P.J. and Jacobs, J.M. 1984, Physica Scripta, T8, p.39.
Jacobs, J.M. and Dworetsky, M.M. 1981, in Upper Main Sequence Chemically Peculiar Stars, 23rd Liege International Astrophysical Colloquium, p.153.
Jacobs, J.M. and Dworetsky, M.M. 1982, Nature, 299, p.535.
Kurucz, R.L. 1981, Smithsonian Astrophysical Obs. Special Report, No. 390.
Kurucz, R.L. and Peytremann, E. 1975, Smithsonian Astrophysical Obs. Special Report, No. 362.
Leckrone, D.S. 1975, in Physics of Ap-Stars, IAU Coll. No. 32, (ed., W. Weiss, H. Jenkner and H. Wood), University of Vienna, p.465.
Leckrone, D.S. 1980, Highlights of Astr., 5, p.277.
Leckrone, D.S. 1981, Astrophys. J., 250, p.687.
Leckrone, D.S. 1984, Astrophys. J., 286, p.725.
Martin, G.A., Fuhr, J.R. and Wiese, W.L. 1986, in preparation.
Sadakane, K., Takada, M. and Jugaku, J. 1983, Astrophys. J., 274, p.261.
Sadakane, K., Jugaku, J. and Takada-Hidai, M. 1985, Astrophys. J., 297, p. 240.

DISCUSSION (Artru and Lanz)

MICHAUD: Are the theoretical calculations sufficiently accurate to predict if the absorption is strong enough to explain the strengths of the features, if not their positions?

ARTRU: Yes, the theoretical values of autoionization widths and oscillator strengths generally give the right order of magnitude for the width and strength, but we can not predict precisely the wavelength of the features.

MÉGESSIER: Did you try to compute synthetic spectra including all the elements expected to be overabundant, in order to estimate the amplitude of the depression which can not be explained by the overabundance of the element itself?

ARTRU: We only did the calculation for Si II, but we could introduce other species (Si III, Fe II, for example). These will not fill the gap. Contributions of unknown lines, especially autoionization lines, must be important.

HUBENÝ: Could you, for the average user of the data, please comment on the precision of the data from the paper by Artru, Jamar, Petrini & Praderie (Astron. Astrophys., 99, p. 401, 1981)? I am a little surprised that you have not included all of these results in the calculations here. To what extent can one trust this data, particularly for autoionization lines, when calculating synthetic spectra?

ARTRU: Yes, this paper looks like a step backward from the '81 paper, but in the earlier work we put in the synthesis all the data predicted in the calculation, including the autoionization widths. However, at that time we did not compare to IUE spectra, so we did not try to fit the predicted lines to the observed ones. Later, we tried to do this, and found that it is not trivial to assign each observed line to a predicted one, and vice versa. We think the predicted spectrum should be globally correct, especially the autoionization widths, but the wavelengths derived from differences of theoretical levels are not accurate enough for identification of individual lines.

COWLEY: A number of years ago, the Si star HD 34452 was studied by Tomley, Wallerstein & Wolff (Astron. Astrophys., 9, p. 380, 1970), and they found a result which was quite remarkable for Si stars, that the oxygen was apparently markedly overabundant. This interesting result should be confirmed and followed up, because oxygen is usually thought to be underabundant in magnetic Ap stars. My question is, do you think Si is capable of explaining all of the large absorption features, if the autoionization levels have the most favourable properties, or do you think that you must include absorption from other ions such as Fe, or if O is overabundant, possibly from that ion?

ARTRU: I think the Si II can explain the width we observe for, as an example, the 1400 Å depression. Unfortunately, we can not yet prove that the right line is located there. To do this, we need additional laboratory data.

ADELMAN: I have a comment on the previous question asked by Cowley. In normal B and A stars, one observes only a few moderate strength lines of O I in IUE spectra in the region 1300 - 1360 Å. Changing the oxygen abundance will not affect the total amount of line absorption very much.

C. R. Cowley et al (eds.), Upper Main Sequence Stars with Anomalous Abundances, 121–123.

There are no observed O II lines in the wavelengths covered by IUE spectra.

ARTRU: Also, chlorine is said to be overabundant in this star, but I did not see evidence of excess Cl lines in the range of wavelengths I studied.

DISCUSSION (Leckrone)

WEISS: Can you please comment on your experience with co-adding IUE spectra. In particular, how does the signal-to-noise ratio improve?

ADELMAN: Leckrone and I have studied the noise of high-resolution IUE spectra by comparing individual exposures with our co-additions. For properly exposed regions, the fixed-pattern noise is 6% of the total signal, while random noise is 4% of the total signal. The fixed-pattern noise, which is an instrumental effect variable over a time of order one year, can be shifted relative to the spectrum by ±25 km/s, by moving the star relative to the center of the large aperture. If one takes three exposures with the same exposure time, one with the star at the center, one at the left side, and the other at the right side, and then adds them with the stellar features in coincidence, then one can reduce the fixed-pattern noise by a factor of $\sqrt{3}$. A similar factor applies to the reduction in random noise. The detailed causes of the fixed-pattern noise are not understood very well.

MICHAUD: In the fit to the 1942 Å Hg II line, how many parameters were allowed to vary, and to what extent were the abundances known when the gf-values were allowed to vary?

ADELMAN: The abundances of blending species are known from the best available optical analyses, as are the stellar parameters: effective temperature, surface gravity, microturbulence, and v sin i. These are verified by examination of the ultraviolet data. The gf-values from the Kurucz and Peytremann list are used unless better ones are available. If necessary, and with some reluctance, the gf-values are adjusted to reproduce the observed region as well as possible in all the programme stars simultaneously. Essentially, we assume that we know gf times abundance, and we have to decouple them somehow.

DWORETSKY: The normal comparison star ξ Oct is probably an unfamiliar choice to most people at this colloquium. It is fairly bright and near the South Celestial Pole. Preliminary work on the IUE spectra by J. M. Jacobs at University College London showed that it appears to have nearly solar abundances of metals. Observations to obtain an accurate v sin i. are now being made at ESO (by Iye of Japan). It looks normal and very sharp-lined on my 15 Å/mm spectra. As we all know, whenever we find a sharp-lined late B star it usually turns out to be peculiar, but I think this will turn out to be a good comparison star with essentially solar abundances. Unfortunately it can not be observed from the Crimean Astrophysical Observatory!

HUBENÝ: I have two questions. First, how is the opacity in the additional fictitious lines treated? I mean, for example, how are the lower excitation potential, species, gf-value, etc chosen?

ADELMAN: The fictitious line was selected to fill in properly the missing opacity for all stars. Leckrone selected Ti II, as its abundance is fairly constant from star to star. The excitation potential chosen was a typical one for a line of this species.

HUBENÝ: So the choice is by trial and error? For many stars?

ADELMAN: Yes. For example, Leckrone can be sure that it isn't a Mn II line, because it does not vary in strength from star to star like those lines.

HUBENÝ: Second, why was α Lyrae (Vega) not chosen as a standard comparison star for Space Telescope?

ADELMAN: Vega is not as sharp-lined as the standards chosen by Leckrone. There are suggestions in the literature concerning its possible variability. Most of these, however, can not be taken seriously.

DWORETSKY: I will discuss Vega on Friday, in my review paper.

ADELMAN: The entire HRS team has chosen to observe Sirius rather than Vega. To observe the entire spectrum at the highest resolution will take 1.8 hours. If no one has reserved Vega, I am sure that there will be a group which will propose to observe it, perhaps in the R = 25,000 mode. The GTO programmes will be published in advance.

SEVERNY: Could you explain how you fixed the position of the continuum for the spectra?

ADELMAN: This is one of the most difficult and nasty problems one must handle. There are several approaches possible. In the 'heuristic' approach, I look at spectra of several stars at once and search for high points present in all the stars. I then assume that these define the continuum. This can be carried one step further. You compute synthetic spectra for these regions using the Kurucz and Peytremann line list, and study how abundance changes and rotation modify the appearance of these regions. This is the method I am actually trying to use, but I am not completely satisfied with it. Locating the continuum is one of the nastiest problems we have!

SCHÖNEICH: What do you know about proposed observations of magnetic CP stars on the Space Telescope?

ADELMAN: Dr. Leckrone chose to observe Hg-Mn stars and normal stars with his time as a Guaranteed Time Observer on the Hubble Space Telescope. I have no knowledge of planned observations of magnetic CP stars by any GTOs.

STUDY OF INHOMOGENEITIES ON THE SURFACE OF MAGNETIC CP STARS

V. L. Khokhlova
Astronomical Council of the USSR Academy of Sciences
48, Pyatnitskaya St, 109017 Moscow, USSR

ABSTRACT. The formulation and solution of the inverse problem to determine local Stokes parameters on the surface of a star , in particular magnetic chemically peculiar star, and hence to determine surface abundance distribution and magnetic field geometry is reviewed.

I will not speak here about those works in which the observed phase variations of the equivalent widths, mean radial velocities and the effective magnetic fields H_e are used to determine chemical inhomogeneities. Pioneer works by Deutch (1969) and Pyper (1969) are known to everybody. New formalism was also developed recently by Adelman et al., (1984).

Unfortunately, using these integrated over λ parameters leads to loss of a great deal of information, containing in spectra and therefore does not permit to get unique and detailed solution.

Using line profile variations Falk and Wehlaw (1974) have got more reliable result for Eu distribution over α^2CVn inspite of the fact that a crude simplified assumption of gaussian shape of local profiles was made. New method to determine surface chemical inhomogeneities and geometry of magnetic field is developed recently and described by Goncharsky et al., (1982) and partly had been reported at Liege in 1981. The outline of the essence of the new method is given below.

Spectral observations with the aid of circular and linear polarization analysers may give the information on Stokes parameters inside the spectral lines. The following expressions describe the observed Stokes parameters in a spectral line of a rotating star depending on and the phase of rotation ωt :

$$\tilde{R}_I(\lambda,\omega t)=\iint\limits_{\cos\theta>0} R_I(M)\,u(\theta)\,dM \qquad (1)$$

$$\tilde{R}_V(\lambda,\omega t)=\iint\limits_{\cos\theta>0} R_V(M)\,u(\theta)\,dM \qquad (2)$$

C. R. Cowley et al (eds.) Upper Main Sequence Stars with Anomalous Abundances, 125–132.

$$\tilde{R}_Q(\lambda,\omega t)=\iint\limits_{\cos\theta>0} R_Q(M)\,u(\theta)\,dM \tag{3}$$

$$R_U(\lambda,\omega t)=\iint\limits_{\cos\theta>0} R_U(M)\,u(\theta)\,dM \tag{4}$$

here $R_{I,V,Q,U}(M)$ are local Stokes parameters in the point M on the surface of a star, $u(\theta)$ - the limb darkening coefficient and dM - the elemental area around the point M.

The left sides of these equations are obtained from the observations. Solving the equations, one may determine $R_{I,V,Q,U}(M)$, that is to determine local abundances and surface magnetic field geometry.

Many physical parameters of a star are bound indirectly into these equations including even local model atmospheres because the local Stokes parameters result from radiation transfer in the atmosphere. Selfconsistent solution of the equations for all these unknowns is a very complicated problem and it is necessary therefore to introduce reasonable simplifications and try to solve it by parts. It is necessary first to clear up, which parameters may be neglected and when and what simplifications may be done.

Let us consider first the form of presentation of local profiles. To solve the equations (1 - 4) numerically, it is convinient to use analytical expressions obtained by Unno (1956) for Miln-Eddington approximation.

When there is no magnetic field (H=0) Unno's solution for the intensity profiles looks like as follows :

$$R_I(\lambda)=R(0)\cdot\eta(\lambda)/[1+\eta(\lambda)] \tag{5}$$

where R( ) is a central depth of a line,

$$\eta(\lambda)=k(\lambda)/\varkappa=k(0)\cdot f(\lambda)/\varkappa=\eta(0)f(\lambda)$$

The shape of the profile is assumed to be a Foigt one, so depending on Lorenz and Doppler widths γ_L and γ_D

This expression for $R_I($ $)$ resembles the known empirical Minnaerts formulae

$$R(\lambda)=R(0)\,\frac{\tau(\lambda)}{R(0)+\tau(\lambda)} \tag{6}$$

which nicely describes line profiles of different strengths in the Solar spectrum, if for R(0) the observed value is taken.

As the abundance changes over the surface of Ap-star, R(0) and η(0) are the functions of coordinates on stellar surface and so $R_I(M)$ contains two unknowns dependent on coordinates. But actually they are not completely independent. The dependance may be expressed by the analytical empirical relation as follows :

$$R(0)=R(0)_\infty\left(1-e^{-a\tau(0)}\right) \tag{7}$$

where $R(0)_\infty$ is the central depth of strongly saturated line when $\tau(0)\to\infty$, (Goncharsky et al., 1977, Pavlova and Khokhlova, 1983).

The parameter „a" as well as the parameters γ_L and γ_D in the Foight absorption coefficient profile $f(\lambda)$ may be considered in the first approximation as coordinate independent. Any "exact" profiles may be well represented by these analytical formula. The parameters of the representation may be easily determined in the way described below : the approximated profiles being computed using the formulae (5) for a grid of parameters $\tau(0)$, γ_L and γ_D, the parameter $R(0)$ being taken from the given theoretically computed "exact" profiles for a given model atmosphere for a set of abundances of an element. The r.m.s. deviation σ of an approximated $R(\lambda)$ from the exact one is computed along all the line and dependance of σ on $\lg\tau(0)$ is plotted for different γ_D (see Figure 1). One can see that σ has a minima at certain values of $\tau(0)$ and γ_D. On the field of parameters $\lg\tau(0)$ and γ_D one can easily determine the location of those values which provide a representation of all "exact" profiles by the approximation formula with the r.m.s. deviation σ =1%, 2%, 3% and so on (see Figure 2). When γ_D is fixed on Figure 2 (γ_D =0.04 for example), one obtains the set of values of $\tau(0)$ which provide σ being 1%, 2% and so on.

Let us now plot $R(0)$ as a function of $\tau(0)$ thus obtained, which must be described by formula (7). Parameter "a" may be found by fitting a computed dependance $R(0)$ on $\tau(0)$ using equation (6) to the values found above (Figure 3).

So introducing formula (7) we use in fact not a Miln-Eddington solution for local profiles but much better approximation to the "exact" theoretical profiles. The only problem is to choose properly the model atmosphere, but that is the problem not only for this case.

Inserting expressions (6) and (7) into the equations (1 -4) and solving the equations in some way, one should obtain local Stokes parameters and hence, the distribution of elements and magnetic field geometry.

When the magnetic field is weak ($H_p \simeq 0$) or lande-factor of a line under consideration is small ($z \simeq 0$), the only one equation (1) should be solved to obtain a map of a chemical element. The Tikhonov method of numerical solution of equation (1) for mapping chemical elements was used in the work by Goncharsky et al., (1977, 1982). The influence of the magnetic field on the line profiles was not taken into account that time.

Up to now five stars shown in the Table have been investigated by this method.

For the stars ε UMa and θ Aur high precision line profiles were obtained with a reticon detector by Rice and Wehlau and a very good agreement between the observed and computed line profiles has been achieved. The paper on θ Aur is presented at this Colloquium.

TABLE.

Star	Element		S/N	Reference
α^2CVn	Eu	0.02	Ph	Goncharsky et al., (1983)
	Ti, Cr, Fe	0.02	Ph	Khokhlova and Pavlova (1984)
CU Vir	Si	0.02	Ph	Goncharsky et al., (1983)
χ Ser	Sr	0.02	Ph	Goncharsky et al., (1983)
εU Ma	Cr, Fe	0.01	$\simeq$200	Wehlau et al., (1982)
θAur	Si, Cr, Fe	0.01	$\simeq$500	Wehlau et al., (1985)

In the process of solution the minimization of r.m.s. deviation σ of the computed profiles from the observed ones is performed. The values of σ obtained are shown in the 3-d column of the Table.

The assumption is made that the model atmosphere does not depend on the coordinates on the surface of a star. Principal limitations of the method should be noted : 1) the map may be obtained only for a strip + 45° along the "subsolar" line. 2) The longitude of a "spot" may be determined much better than the latitude.*/

Numerical modelling of profiles $\tilde{R}_{I,V,Q,U}(\lambda)$ made by Piskunov and Khokhlova (1983, 1984) showed the distortion of the intencity profiles by the magnetic field (see Figure 4).

Similar results were obtained earlier by Borra (1977), but he paied attention to the shift of a line as a whole and found it to be rather small. When rotational broadening becomes larger than the Zeeman splitting, line asymmetry does not depend on rotational velocity.

The distortion of profiles may be explained by the fact that the magnetic intensification of lines depending on the angle γ between the magnetic field vector H and the line of sight, reaches its maxima according to Unno (1956) at $\gamma \simeq 55^\circ$. When a star is rotating the regions of enhanced magnetic intensification move along stellar surface and that makes difference between them and spots of real enhanced abundances. This distortion of profiles appears as noises when solving the inverse problem for chemical elements distribution, preventing the convolution of iterative process.

*/

We tried to solve the inverse problem for a set of profiles presented on Figure 4 and got a homogenuous distribution of an element over the surface of a star with the r.m.s. deviation of profiles σ=0.04 (which is four times larger than that obtained for ε UMa and θ Aur). The influence of the magnetic field may be in principle considerably reduced when the lines with small Z are used for the analysis. It is also of no importance if the stars with weak magnetic field are investigated. For five stars mentioned in the Table only for α^2 CVn the effect may be noticeable, but seems to be also not very important.

The influence of chemical inhomogeneities on polarization profiles was studied by Piskunov and Khokhlova. One may see on Figure 5 that chemical spots considerably distort polarization profiles. It is clear that this effect should not be neglected when a magnetic field is being measured.

An algorithm of the solution of the inverse problem to determine the magnetic field geometry using circular polarization profiles, that is a solution of the equation (2), was developed by Piskunov (1985) for a particular case of shifted dipole field configuration. The distribution of an element is supposed to be known, that is determined independently. This code was used to determine the geometry of the magnetic field of α^2 CVn star on the base of circular polarization profiles of metallic lines observed photographically at the 6-m telescope (Glagolevsky et al., 1985). When inhomogenuous distribution of the elements determined by Khokhlova and Pavlova had been taken into account, centered dipole geometry was obtained from four lines of TiII and FeII and the uniqueness of the solution was confirmed by using different initial approximations and by the coinsidence of the results obtained from all the four lines of two elements.

It should be noted, that measurements of polarization in hydrogen lines cannot give the possibility to determine the geometry of a magnetic field by the inverse problem solution because the intrinsic widths of the lines are large as compared with the rotational broadening and so rotational shifts do not work.

The results of mapping the elements distribution over the surfaces of stars mentioned above in the Table show that the inhomogeneities exist on the surfaces of CP stars with strong magnetic field (α^2 CVn) as well as with very weak (if any) magnetic field (ε UMa, γ Ser). In some cases the distribution of elements is very complex (θ Aur) and cannot be connected with the magnetic field directly in a simple way.

Further investigations of the magnetic field feometry and distribution of elements using inverse problem solution method and precise observations of Stkes parameters in spectral lines are needed.

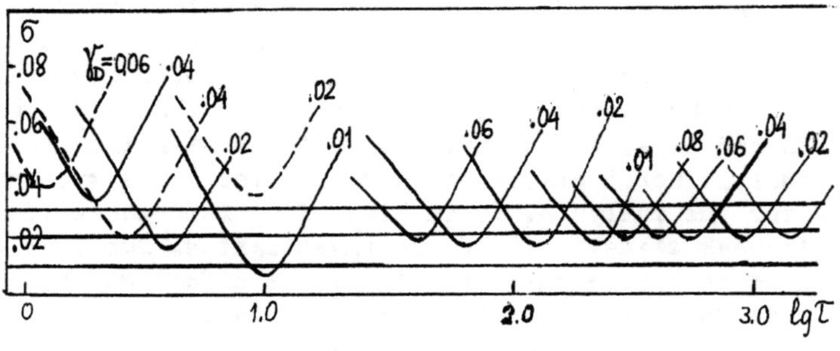

Figure 1. σ plotted versus $\lg\tau$ for different γ_D and abundances.

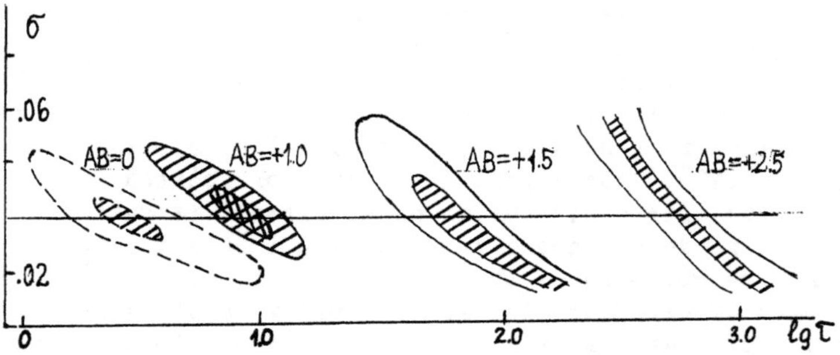

Figure 2. The area of given σ in the field of parameters γ_D-$\lg\tau$ for different abundances AB.

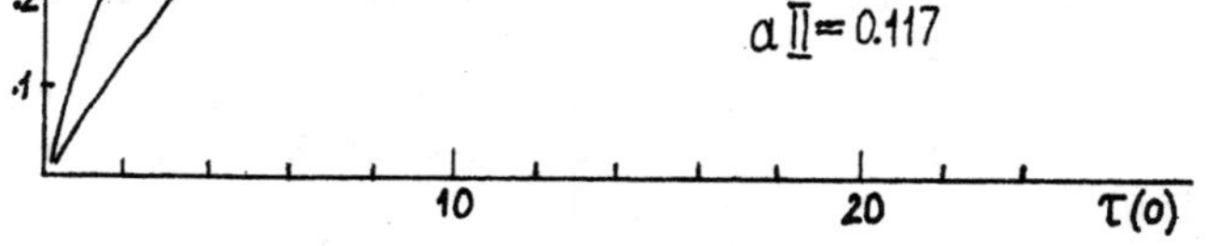

Figure 3. Fitting the analytical expression (7) (solid line) to the values found with the expressions (5), (6) and (7) (crosses and circles).

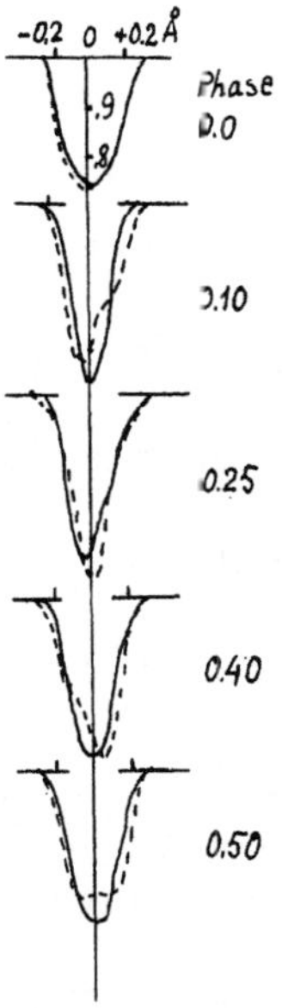

Figure 4.
Phase variations of the computed profiles of the line λ4520 FeII in the spectra of rotating magnetic star with Hp=10 KGs, V sin i=20 km/s, β=i=45° (solid line) and β=i=90° (broken line)

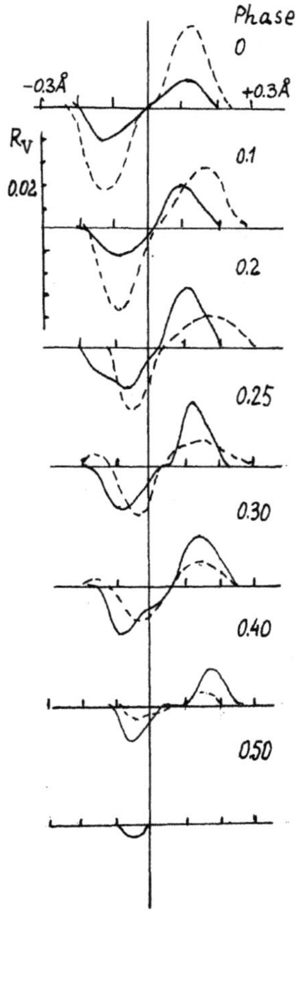

Figure 5.
Computed profiles of circular polarization of a centered dipole model Hp=10 KGs, β=i=45°, V_{eq}=10 km/s with a chemical spot situated on the magnetic pole (broken line) and a spot situated at the magnetic equator (solid line).

References

Adelman S., Hensberge H., van Rensbergen W., Astron. Astro-pys.Suppl.Ser. I984,v.57, N I, p.I2I
Borra E.F. and Landstreet J.D. Astrophys.J. I977, v."I";p.I42
Deutsch A.J. Astrophys.J. I970,v.I59,p.985
Glagolewsky Ju.V.,Piskunov N.E.,Khokhlova V.L. Soviet Astronomy Letters I985,(Russ.) v.II,N 5,p.37I
Goncharsky A.V.,Stepanov V.V.,Khokhlova V.L. and Yagola A.G. Soviet Astronomy Letters,I977,v.3,N 3,p.I47
Goncharsky A.V.,Stepanov V.V.,Khokhlova V.L.,Yagola A.G. Soviet Astronomy I982 (Russ.) v.59,p.II46
Goncharsky A.V.,Rjabchikova T.A.,Stepanov V.V.,Khokhlova V.L. and Yagola A.G. Soviet Astronomy I983,v.27(I),p.49
Falk A.F., Wehlau W.M. Astrophys.J.I974,v.I92,p.409
Khokhlova V.L.,Pavlova V.M.I984,Sovjet Astr.Lett.v.IO(3),p.I5
Khokhlova V.L. Soviet Scientific Rewiews,Section E,Astro-Astrophysics and Spase physics reviews ed.R.A.Syunyaev, I985,p.99
Pavlova V.M.,Khokhlova V.L. Nauchnye informatzii Astrofis.ser I980,N 43,p.65
Pyper D.M. Astrophys.J.Suppl.Ser.I969,v.I8,p.374
Piskunov N.E.,Khokhlova V.L.Soviet Astron.Lett.I984,v.IO(3), p.I87.
Piskunov N.E.,Khokhlova V.L.I983,Soviet Astron.Lett.v.9(6), p.346
Piskunov N.E. I985,Soviet Astron.Lett.v.II(i),p.346
Wehlau W.M.,Rice J.,Piskunov N.E.,Khokhlova V.L. I982, v.8(I) p.I5

DISCUSSION (Khokhlova)

STĘPIEŃ: For the work you do, it is very important to use correct model atmospheres. If one does not have a model of a particular Ap star one is studying, the question arises: which model approximation best fits an Ap star? Dr. Muthsam (Vienna) and I have produced a grid of models with different chemical compositions to compare with Osawa's star (HD 221568). When you look at T(τ) and the energy distribution in the ultraviolet and visible, you can see [shows some slides] that the model which best approximates HD 221568 is the model with solar composition increased by a factor of 30. This means that we should use a grid of models with abundances increased by a factor of this order to use it for different Ap stars. Even with this, there are differences between our best fit and the observed flux distribution for HD 221568, but it is much closer than the model with solar composition.

KHOKHLOVA: Certainly, to use the correct model would be very nice. But I did not use Kurucz models, I used the older Mihalas (1965) models to determine the parameter $R(0)_{\infty}$. Until one knows the distribution of elements, one can not use a proper blanketed model atmosphere. Remember that in our analysis we determine not the local abundance value directly, but local equivalent widths. These widths should be analysed with the aid of a proper model. There would be some sort of iteration process: as a first approximation we could get local abundances with the simplest model, then compute a new model and improve the abundance.

YAGOLA: I would like to supplement Dr. Khokhlova's report. The problem of mapping chemical elements on the surfaces of Ap stars is an example of an inverse problem in astrophysics. An astrophysicist often has to solve such problems when interpreting observational data. Unfortunately, many of these problems are incorrectly posed, so that a small change in experimental data can produce a large change in the solution. Since 1967, we have been using a stable method, a regularizing algorithm devised by Academician A. N. Tikhonov, and his scheme for solving inverse problems in astrophysics. To describe this algorithm now would be impossible, but I would like to recommend to you our books on this problem, published by Nauka Press. These are: Tikhonov, Goncharsky, Stepanov, and Yagola, A Regularizing Algorithm and A Priori Information, and two other books, The Numerical Method for Solution of Inverse Problems in Astrophysics (1978) and a new one, Incorrectly Posed Problems in Astrophysics, which should be published in September or October of this year.

MICHAUD: Have you tried to reproduce the profiles using rings, and if so, can you show how much worse the fits would be if you used rings? In other words, can you exclude rings?

KHOKHLOVA: We can not exclude rings, because in our method we can only get a map of the strip ±45° around the subsolar line.

MÉGESSIER: With your program, could you check whether rings can fit the observed profiles?

KHOKHLOVA: We do not fit any particular model. We solve for the distribution by inversion. The profiles are the data, and the maps are the result. [long pause] I can check any map you like!

POLOSUKHINA: How many lines were used to make the maps? Also, how

C. R. Cowley et al (eds.), Upper Main Sequence Stars with Anomalous Abundances, 133–134.

different were the maps from a number of different lines?

KHOKHLOVA: For θ Aurigae, for example, we used eight lines of iron. All maps obtained from lines of the same element are similar.

WEISS: Several years ago, it was a commonly expressed idea that elements are concentrated either at the magnetic equator or at the magnetic poles. From your maps, I gain the impression that the spots are more or less distributed at random on the surface, and not really correlated with the magnetic field. This might cause problems for diffusion. Can you comment?

KHOKHLOVA: For α^2 CVn, Eu is concentrated near the pole, and Cr near the equator, but Fe and Ti are more complicated. This is especially the case for θ Aur, which has a well determined rather weak magnetic field with a sinusoidal shape of the effective field variation, but a rather complicated chemical structure. Other stars, such as ε UMa, χ Ser and CU Vir also have prominent inhomogeneities but rather weak fields.

DWORETSKY: Dr. Khokhlova, is there some reason why you have not used magnetic null lines for element mapping? This would remove the complications due to the magnetic field. Is there any possibility that you may use them in further work?

KHOKHLOVA: Yes. There are not very many magnetic null lines, and many of them are blended. The first requirement for our analysis is the absence of blending. Certainly, we shall try to find such a line in the future.

HENSBERGE: You mentioned that the deviation from simple dipole magnetic structure found apparently in some stars might be caused by the spotted distribution of elements. The strongest indication for the deviation from simple dipole geometry is given by direct measurement of the surface magnetic field B_s in 53 Cam and HD 126515. In both stars, B_s varies approximately from 10 to 17 kG.

I guess that it would be very hard to explain the variation of a factor 1.7 by spots and a dipole structure; for a dipole, the field strength difference between equator and poles is only a factor of two.

KHOKHLOVA: For α^2 CVn the centered dipole is the case, but for other stars we do not know, but might guess, that taking into account chemical inhomogenieties may change the result of measurements of magnetic fields.

INVERSE PROBLEMS IN ASTROPHYSICS

A. M. Cherepaschuk, A. V. Goncharski,
and A. G. Yagola

Physics Department, Moscow University
Sternberg State Astronomical Institute
USSR

ABSTRACT. The applications of the regularizing algorithms to the solving of inverse problems in astrophysics are presented.

1. INTRODUCTION

The theory of incorrectly posed problems, created and advanced in the last years, has found applications for a wide range of inverse problems in optics, spectroscopy, electrodynamics, plasma diagnostics, geophysics, etc. (Tikhonov and Arsenin, 1977). Regularizing algorithms, when applied to the exerimental data processing, allow one to improve appreciably an accuracy of characteristics in studying physical phenomena. In some cases this is equivalent to increase of instrumental resolution. The increase of resolution is achieved because of computer employment in the data processing, rather than due to complication and development of the experimental installation itself, often connected with heavy expenses. The complex "device + computer" with the programming system of data processing, which involves regularising algorithms, has certainly great opportunities.

Astrophysics gives a large range of applications for solution of the "incorrectly posed problem" method, advanced in the last years. The matter is that presently astrophysics is basically an observational science. A researcher cannot exert an influence on processes taking place in distant stars and galaxies. He has to conclude about physical characteristics of such remote objects by their indirect manifestations visible from the Earth or near the Earth (from space stations). That is, an astrophysicist has to solve inverse problems while interpreting obtained observational data. Most of these problems are incorrectly posed. This means that a small change of experimental data can correspond with a however large change of a solution.

Since 1967 we have used stable methods (regularizing algorithms) advanced by the academician A. N. Tikhonov and his team (Tikhonov and Arsenin, 1977) for solving inverse (incorrectly posed) problems in astrophysics (Goncharsky, Cherepashchuk and Yagola, 1978). It needs to be said that investigation of specific astrophysical problems has stimulated development of some new mathematical methods. So, reconstructing the distribution of brightness over stellar disks in eclipsing binary systems, and using the natural physical information concerning monotonous decrease of the function to be determined, we have developed the effective numerical methods for solving incorrectly posed problems on sets of monotone functions (convex, monotone convex functions, and so on). The inverse problem for reconstruction of a strip-distribution of the radio-brightness over a source, which needs allowance for an error in representation of the radiotelescope directional pattern, has stimulated development of a theory and methods for solving incorrectly posed problems with an approximately specified operator. These methods have been realized as a complex of computer programs in FORTRAN (Tikhonov et al., 1983). Let us enumerate briefly astrophysical problems which have been solved by use of these and some other methods. The description of a problem statement and obtained results can be found in detail in our monograph *Incorrect Problems in Astrophysics* which will be published by "Nauka" press in 1985.

C. R. Cowley et al (eds.), Upper Main Sequence Stars with Anomalous Abundances, 135–136.

2. NATURE AND EVOLUTION OF WOLF-RAYET (WR) STARS IN CLOSE BINARY SYSTEMS

We formulated the problem of the light curve interpretation in eclipsing binaries (including WR-components) from a new point of view. On the base of modern methods of incorrectly posed problems regularization the effective numerical algorithms have been developed for solving this problem on a computer. Moreover, the observations of all known binaries with WR-components have been carried out. In the result of light curve processing the geometrical parameters of the eclipsing systems with WR-components, as well as distributions of brightness and absorption properties over disc of WR stars have been determined. The nature and evolutionary stage of these objects have been studied.

3. NEUTRON STARS AND BLACK HOLES IN BINARY SYSTEMS

The special inverse problem algorithms for interpreting optical light curves of the X-ray binary systems have been developed in terms of a parametric model. These algorithms enable us to determine not only the values of parameters, but besides their confidence intervals. X-ray binaries of various types were observed, including those in the southern sky. The analysis of these data by the new method yields the significant results. The lower limits for masses of relativistic objects in the systems Cygnus X-1 ($m_X > 7\ M_\odot$), SS 433 ($m_X > 6\ M_\odot$), LMC 3 ($m_X > 7\ M_\odot$) are estimated. Thus, the number of candidates for black holes has increased from one to three during the last decade. This gives a real opportunity to state a problem of investigation of black holes on the observational basis. For most other X-ray systems (for example, HD 153919, HZ Her, HD 77581) the analogous computations yield mass estimates $\leq 2\ M_\odot$ which is specific for a neutron stars. The optical parameters of accretion discs in X-rays binaries have been studied.

4. CONCLUSION

Moreover, the regularization method for solving incorrect problems has been used in following cases: a) for determining the observational structure and dynamical characteristics of accretion disks in nova and nova-like stars; b) for mapping a chemical element distributions over surface on Ap-stars (the problem was formulated by V. L. Khokhlova); c) for reconstruction of a strip-distribution of the brightness over a radio source.

Integration of the modern regularization methods for solving incorrect problems with observational practice in astronomy enables one to realize the effective regime of automated observations, when data processing and interpretations are carried out in real-time computations. An example of the problem which needs full automatization is the determination of a stellar angular diameters by the method of the lunar occultations of stars. At the High-Altitude expedition of SSAI (Sternberg State Astronomical Institute) the complex of equipment has been constructed which integrates a telescope with a computer and has a time resolution 10^{-3}s.

In conclusion let us notice that applications of the regularization methods for solving incorrectly posed problems have led to a number of prospective directions in astrophysics.

REFERENCES

Goncharsky, A. V., Cherepashchuk, A. M., Yagola, A. G., The *Numerical Methods for Solution of the Inverse Problems in Astrophysics*. – Moscow, "Nauka", 1978.

Tikhonov, A. N., Arsenin, V. Ya., *Solutions of Ill – Posed Problems*, Scripta Series in Mathematics, Wiley, N. J., 1977.

Tikhonov, A. N., Goncharsky, A. V., Stepanov, V. V., Yagola, A. G., *Regularizing Algorithms and a Priori Information*. – Moscow, "Nauka", 1983.

THE DISTRIBUTION OF FE, CR AND SI OVER THE SURFACE OF Θ AUR

V. L. Khokhlova
Astronomical Council, USSR Academy of Sciences, Moscow

J. B. Rice
Department of Physics, Brandon University, Brandon, Manitoba

W. H. Wehlau
Department of Astronomy, University of Western Ontario, London, Ontario

ABSTRACT. Very accurate spectroscopic line profiles have been obtained at various phases during the $3^d.618$ period of the magnetic variable Ap-star Θ Aur. These profiles were observed using a Reticon detector at the coudé focus of the 3.5 metre Canada-France-Hawaii telescope. The mapping of Si, Cr and Fe over the surface of the star was done by solving the Inverse Problem. Complex spotty structure has been revealed with the number of Fe spots found being as great as six. The distribution of Cr is found to be similar to Fe but with less detailed structure. Si is distributed quite differently from Fe and Cr. Discussion of the relationship of the magnetic field maximum phase and the light curve along with the maps of the distribution of elements suggests that the principal spots of Fe and Cr are in phase with the light variability but they are 90° out of phase with Si and the magnetic field variation. We are surprised by the Si variability seeming to be 90° out of phase with the light variability.

INTRODUCTION

Θ Aur (HD40132 = HR2095) is a magnetic Ap star (B9p) with a very well defined magnetic variability. Borra and Landstreet (1980) have measured an effective magnetic field that varies almost sinusoidally from about + 350 Gauss to − 250 Gauss. Rensbergen et al. (1980 a & b) made a detailed study of Θ Aur including an attempt to map the surface distribution of elements. They obtained the best ephemeris for the variability of Θ Aur by combining results from several papers as well as their own work. They give J. D. = 2442766.55 + $3^d.6190 \pm 0^d.0005$ for magnetic positive extremum and note that the light curve extrema do not coincide with the H_e extrema.

OBSERVATIONS

Observations of Θ Aur were made in 1981 and 1982 at the coudé spectrograph of the CFHT using a Reticon detector. The dispersion is 2.5 A/mm and the total instrumental profile is 90 mA FWHM. Pixel spacing is 35 mA so we smoothed the data from the Reticon over three pixels. The

C. R. Cowley et al (eds.), Upper Main Sequence Stars with Anomalous Abundances, 137–140.

signal-to-noise ratio for the observations is around 500 (see Fig. 1). Three regions of the spectrum were monitored: λλ 4890 - 4955, λλ 5010 - 5075 and λλ 5480 - 5545.

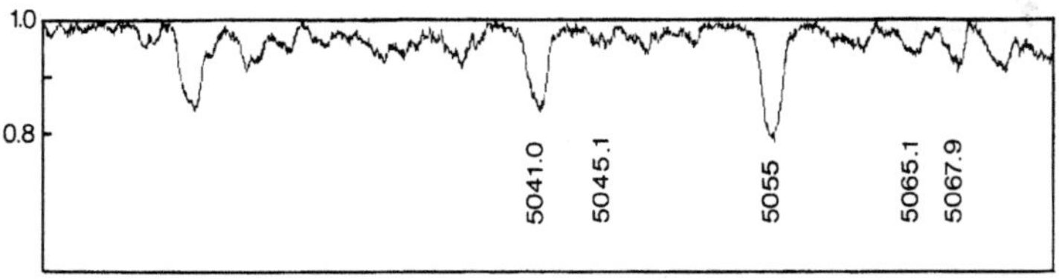

Fig. 1 An example of the spectrum of Θ Aur as observed using a Reticon detector. Wavelengths of some of the analysed lines are noted.

In order to map the star using the techniques of Goncharsky et al. (1982) we need to obtain V_r and V sin i. To eliminate the effects of the variable profiles in obtaining V_r and V sin i, the profiles for all phases were plotted on one diagram and the outer envelope used. From this it was determined that V sin i = 55 km/s and V_r = 31 km/s.

MAPPING THE SURFACE DISTRIBUTION OF LINE STRENGTH

Because of the width of the lines in Θ Aur, only a few lines were free enough of blends to be chosen for mapping using profiles. To use the approach of Goncharsky et al. for mapping, you must estimate i from the period, radius and V sin i of the star. Rensbergen et al. (1984 a & b) discuss the possible values of radius in Θ Aur and estimate it to be between 3.5 and 7.2 R_o. Consequently two intermediate values of radius were tried in this analysis, R = 4.0 R_o (with corresponding i = 80°) and R = 4.9 R_o (with corresponding i = 53°). The difference between the models for these different values of R (and i) was not great. Most of the maps presented here are for i = 53°.

Fig. 2 gives the maps for the Fe distribution. Fig. 2a is the only map that assumes i = 80° and it gives the result for the strong line Fe IIλ4923. We note that the contours are smooth and that one major spot at the longitude of light maximum dominates. Fig. 2b shows the more complex spot structure of the weaker lines; in this case the line is Fe IIλ5045. One might be inclined to dismiss the complex structure as noise except for the fact that the pattern repeats for the other weak Fe lines. Fig. 2c is a composite and shows that for the lines λ5045, λ5065, λ5067 and λ5510 there is remarkable repeatability in the location of the spots. For the purposes of this discussion the spots are labelled from I to VI. Spot IV coincides with the major spot of the map for λ4923. We note here that the longitude for these maps is arbitrary and that on this arbitrary scale the H_e positive extremum occurs when longitude 155° ± 25° is subsolar, light maximum when 83° is subsolar and light minimum when 255° is subsolar.

In Fig. 3 we see comparable maps for Cr and Si. Fig. 3a is for the CrII line λ5508 and Fig. 3b is for the SiII line λ5041. Note that the Cr pattern resembles Fe but the major peak for Si is displaced almost 90° earlier in phase so that it should correspond roughly to the longitude of the magnetic negative extremum.

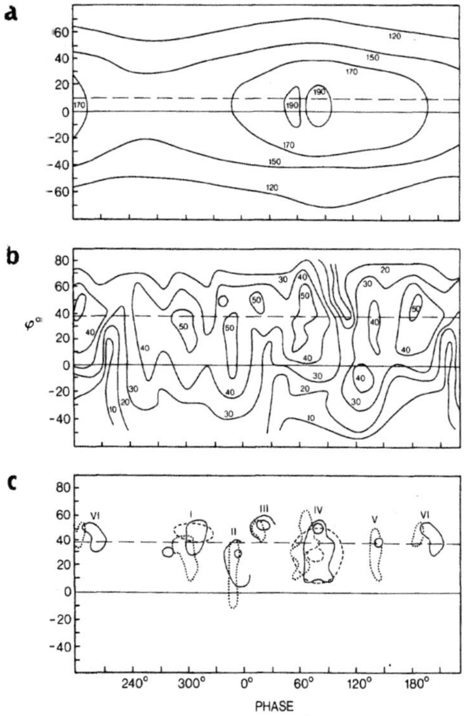

Fig. 2 Maps of the iron line strength distribution. a) Map of the strong line FeII λ4923.9 assuming i = 80°. Contours are marked in mA. b) Map for the weak line FeII λ5045.1 assuming i = 53°. c) Map showing peak values of local line strength for λλ5045, 4065, 5067 and 5510 of FeII assuming i = 53°.

Fig. 4 illustrates the closeness of fit between the observed profiles for three of the lines used for mapping and the profiles predicted by their corresponding maps. In all of the maps discussed above we suggest that the contours are only reliable in a belt within about 45° of the subsolar line. The subsolar line is indicated by a dashed line on the maps.

CONCLUSIONS

The first and simplest conclusion is that for θ Aur we have the Cr and Fe maximum in phase with the light variability but 90° out-of-phase with the Si variability and the variation in H_e. It is not unusual for Cr and Fe to be 90° out-of-phase with the magnetic field, that is what one expects if the Fe and Cr maxima are near the magnetic equator of Ap stars but it does seem unusual to have the light variation and Si variation 90° out-of-phase. The second point is that the maps for weak Fe lines show a fine structure of multiple separate spots that is fairly convincing and that seems to suggest we might not be dealing with a geometry that can be reduced to simple polar spots or a simple magnetic equatorial belt. We note also that Si seems to be associated with the magnetic poles in Θ Aur and with the negative pole especially.

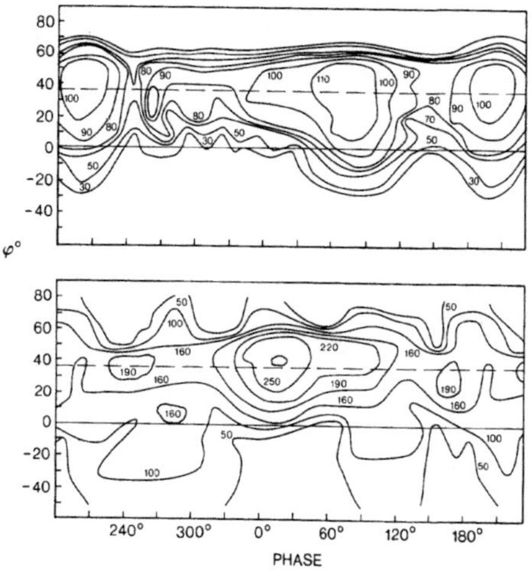

Fig. 3 Maps of the chromium and silicon distribution. a) Map of the CrII λ5508.6 line assuming i = 53°. b) Map for the SiII λ5041.026 line assuming i = 53°. Contours in mA.

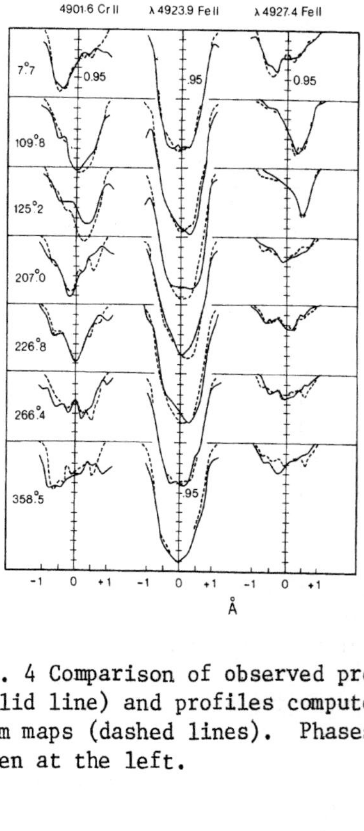

Fig. 4 Comparison of observed profiles (solid line) and profiles computed from maps (dashed lines). Phases are given at the left.

REFERENCES

Borra, E. F. and Landstreet, J. D., 1980, Ap. J. Suppl. 42, 421.

Goncharcky, A. V., Stepanov, V. V., Khokhlova, V. L. and Yagola, A. G., 1982, Soviet Astron. J. 59, 1148.

VanRensbergen, W., Hensberge, H., Adelman, S. J., 1984a, Astron. Astrophys. Suppl. 157, 121

VanRensbergen, W., Hensberge, H., Adelman, S. J., 1984b, Astron. Astrophys. 136, 31.

APPLICATION OF DEUTSCH'S METHOD FOR 53 CAM

I. Vincze
Gothard Astrophysical Observatory,
Eötvös University,
Vörösászló u. 112.
H-9707 Szombathely
Hungary

ABSTRACT. 53 Cam is a well-known magnetic, spectroscopic and photometric variable star of Ap-type. The line intensities of 53 Cam show a variation, which has a period of 8^h. The oblique-rotator model, developed mathematically by Deutsch /1970/, has been applied for the spectral and magnetic variations. In this paper, using Deutsch's method, the distribution of CrII, EuII, TiII, SrII on the stellar surface is analysed and the relation between the abundance of elements and the magnetic structure is studied. The Laplace and Fourier coefficients for each elements have been computed and also the distribution of the elements at the stellar surface has been drawn by a Sinclair ZX Spectrum personal computer.

1. INTRODUCTION

53 Cam is classified as an A2p star. Probably, this star has a red companion /V=12.06/. The magnetic field of the star is strong and variable. Borra and Landstreet /1976/ observed $-5400 \pm 350 < H_e < +4300 \pm 460$ gauss, with a period of $8^h.0267$. The star is a spectroscopic binary as well. The photoelectric magnetic curves of Borra and Landstreet are explained very satisfactorily by oblique-rotator models the observed variation in field strength it is possible to infer the distribution of magnetic flux over the stellar surface, although not always uniquely. Borra and Landstreet found, that one of the magnetic poles is stronger than the other. A field which has this property and easy to model can be obtained by displacing a dipole along its axis with a fraction of the stellar radius. Borra and Landstreet found that the observed magnetic curves of 53 Cam could be modelled this way, although an extremely large displacement /a=0.67/ was necessary to reproduce the $H_e(\Phi)$ curve of 53 Cam. They fitted a

C. R. Cowley et al (eds.), Upper Main Sequence Stars with Anomalous Abundances, 141–147.

decentred dipole model for the observed extremes of H_e and H_s to determine the values of i, β, H_p and a, where i is less than 90°, β is the angle between the observable stellar rotational pole and the positive magnetic pole, H_p is the field value at the stronger magnetic pole, and **a** is the fractional displacement in the direction of the positive pole. Preston /1970/ calculated the parameters $(i, \beta, H_p, a) = (65^\circ, 100^\circ, -28{,}000$ gauss, $-0.15)$ or $(80^\circ, 115^\circ, -28{,}000$ gauss, $-0.15)$. According to Huchra /1972/ $(i, \beta, H_p, a) = (50^\circ, 100^\circ, -28{,}400$ gauss, $-0.145)$. All of the model parameters vary about 10%.

2. APPLICATION OF THE OBLIQUE-ROTATOR MODEL

This paper applies Deutsch's /1970/ method and uses his notations. The variation in abundance is represented by the variation of the distribution function $\xi(\Psi, \nu)$ for a given spectral line. On the stellar surface the polar distance is Ψ, and the azimuth is ν. $\xi(\Psi, \nu)$ may be represented as the real part of a Laplace-series:

$$\Xi(\Psi, \nu) = \langle W \rangle \sum_{n=0}^{\infty} \sum_{m=-n}^{n} A_n^m \exp(im\varphi) P_n^{|m|}(\cos\Theta)$$

where $\langle W \rangle$ is the observed equivalent width, averaged over the cycle of the variation.

The Ψ, ν spherical coordinate system is related to the star, the Θ, φ system is related to the observer.

$$\Xi(\Psi, \nu) = \langle W \rangle \sum_{n=0}^{\infty} \sum_{m=-n}^{n} B_n^m \exp(im\varphi) P_n^{|m|}(\cos\Theta)$$

The variations of observable quantities such as $W/\langle W \rangle$ and V as functions of the phase can be expressed by Fourier series. The Fourier coefficients are functions of A_n^m.

$$W/\langle W \rangle = \sum_{n=m}^{\infty} D_{-m} \exp(-im\Phi)$$

$$(W/\langle W \rangle) \cdot V = (V_e \sin x) \sum_{m=0}^{\infty} E_{-m} \exp(-im\Phi)$$

$$D_{-m} = \sum_{n=m}^{\infty} P_n^m A_n^m \qquad E_{-m} = \sum_{n=m}^{\infty} Q_n^m A_n^m$$

We assume the classical limb darkening and line weakening laws:

$$\Lambda = 1 - \mu + \mu\cos\Theta \qquad J = 1 - \kappa + \kappa\cos\Theta$$

For further details of mathematics see Deutsch /1970/. It is found that the observations of $W/\langle W \rangle$ and $(W/\langle W \rangle) \cdot V$ can well be represented by Fourier series of second degree. The period can be observed. The parameters $\mu, \kappa, \Lambda_o, x, V_e$ of the star have to be determined and substituted into the following equations:

$$d_o = p_o a_o^o + p_1 a_1^o \cos x + \tfrac{1}{2} p_2 a_2^o (1 + 3\cos 2x) = 1$$

$$d_{-1} = p_1 a_1^1 \sin x + \tfrac{3}{2} p_2 a_2^1 \sin 2x$$

$$\delta_{-1} = p_1 \alpha_1^1 \sin x + \tfrac{3}{2} p_2 \alpha_2^1 \sin 2x$$

$$d_{-2}=3p_2a_2^2\sin^2\chi$$

$$\delta_{-2}=3p_2\alpha_2^2\sin^2\chi$$

$$e_o=0$$

$$e_{-1}(V_e\sin\chi)=V_e(-q_1\alpha_1^1\sin\chi-\tfrac{1}{2}q_2\alpha_2^1\sin2\chi)$$

$$\varepsilon_{-1}(V_e\sin\chi)=V_e(q_1a_1^1\sin\chi-\tfrac{1}{2}q_2a_2^1\sin2\chi)$$

$$e_{-2}(V_e\sin\chi)=V_e(-2q_2\ {}_2^2\sin^2\chi)$$

$$\varepsilon_{-2}(V_e\sin\chi)=V_e(-2q_2a_2^2\sin^2\chi)$$

This system of equations can be solved for a_n^m, α_n^m and we obtain the distribution function:

$$\xi=a_o^o+a_1^o\cos\psi+\tfrac{1}{2}a_2^o(3\cos^2\psi-1)+(a_1^1\cos\nu-\alpha_1^1\sin\nu)\sin\psi+$$
$$+3(a_2^1\cos\nu-\alpha_2^1\sin\nu)\cos\psi\sin\psi+3(a_2^2\cos2\nu-\alpha_2^2\sin2\nu)\sin^2\psi$$

This method was applied in many cases:

Pyper/1969/..................................α^2 CVn
Deutsch/1970/..............................HD 125248
Rice/1970/..................................HD 173650
Megessier/1975/...........................108 Aqr
Molnar, Mallama, Sorkey, Holm/1975/.......ι Cas
Floquet/1978/..............................HD 216533
Jasinski, Muciek, Woszczyk/1981/..........ι Cas

3. DETERMINATION OF STELLAR PARAMETERS

53 Cam /HR 3109=HD 65339/ is a spectroscopic and magnetic variable. Preston and Stepien /1968/ determined a period of $8^h.0278$ from photometric and magnetic observations of 53 Cam. A slightly shorter period, based on magnetic measurements, was determined by Borra and Landstreet/1977/:

JD/Positive crossover/=2435855.652+8.0267E

This paper uses the value $P=8^h.0267$.

The parameters μ and K could be determined from the model of the atmosphere of Ap star. For a given star, Θ_e, T_e and log g are to investigated. Selecting the $H\beta$, $H\gamma$ lines from the spectra of 53 Cam, and comparing these observed profiles with Mihalas' theoretical profiles the best fit has been obtained for log g=4 and $\Theta_{eff}=0.6$. An independent way of determining Θ_{eff} is given by the photometric indices from the Stömgren b-y through the Geneva B2-V1. Oke and Conti /1966/ found $0.606<\Theta_e<0.618$ from B-V. Hauck /1967/ obtained $\Theta_e=0.598$ from b-y and 0.608 from B2-V1. In this paper $\Theta_e=0.6$ is used, then $T_e=8400^oK$.

53 Cam is classified as an A2p star and from the model of

Mihalas /Crygar 1965/ μ=0.6.
Since

$$\kappa=1-\frac{1}{2-\mu}\left[\frac{2(3-\mu)}{1.05}-(4-\mu)\right],$$

from this equation κ=0.16. The corresponding value of is $1-\frac{1}{3}\mu$ /Chandrasekhar 1946/. We may now proceed to find the values for p_n, q_n. For 53 Cam:

p_o=0.9523809	q_o=0
p_1=0.6824829	q_1=0.2093537
p_2=0.3243197	q_2=0.3656341

To solve the system of equations we must know the angle x and the equatorial velocity V_e. There exist different procedures to obtain them.
One of them is:

$$V_e=\frac{50.613\ R}{P}$$

where V_e: equatorial rotation velocity in km/s
R : the stellar radius in solar units
P : the rotation period in days

Stift/1974/ studied the radii of Ap stars. The radii of 35 magnetic and other peculiar A and B-type stars were calculated with the help of luminosities and effective temperatures derived for the individual stars. The results indicate a range of the radii of Ap stars from 1.7$R_\odot$ to 4.4$R_\odot$. The radius of 53 Cam is 2.1$R_\odot$ according to Glagolevskij/1971/ and 2.0$R_\odot$ according to Stift/1974/. The corresponding equatorial velocity is V_e=12.61 km/s.
An alternative determination of V_e is

$$V_e=-\frac{3p_2(e_{-2}V_e\sin x)}{2q_{2-2}}=\frac{3p_2(\varepsilon_{-2}V_e\sin x)}{2q_2d_{-2}}$$

From this V_e=≈10 km/s
The parameter x was taken from the model of Borra and Landstreet/1976/.

4. RESULTS

53 Cam is a peculiar Sr-Cr-Eu star. For CrII, EuII, TiII and SrII I have computed the Laplace coefficients from the observations of Faraggiana /1973/. She measured the radial velocity and equivalent widths at different phases. From the measurements we deduce that TiII and EuII vary in antiphase, that is the minimum of EuII= the maximum of TiII at phase 0.35 and the maximum of EuII= the minimum of TiII at phase 0.7. TiII varies in phase with the magnetic field and EuII varies in antiphase with it. The Laplace coefficients for the elements are:

	CrII	EuII	TiII	SrII
a_0^0	2	1.5	0.4024709	1.1280839
a_1^0	-3	-1	-5	10.1
a_2^0	1	2	10	-10
a_1^1	-17.752941	-0.2160624	3.4075218	-23.927259
α_1^1	29.755943	0.500322	-5.3442993	39.728049
a_2^1	8.0887676	16.213051	-9.0918818	-0.9081441
α_2^1	-13.913373	-28.087616	14.635785	0.4707353
a_2^2	-0.0082621	0.0649432	0.7875502	0.8740808
α_2^2	-0.0110430	-0.0780557	0.4299339	1.2217495

5. DISCUSSION

In the photos of Fig 1, the distribution of the elements CrII, TiII, EuII, SrII on the stellar surface can be seen. From the point of view of the distribution of the elements, two groups can be distinguished: the first group consist of EuII and TiII, the second is of CrII and SrII. There are two patches of overabundance in the distribution of EuII and TiII. In the cases of CrII and SrII we have found one patch of overabundance on the stellar surface. From the observations we know, that EuII is in antiphase with TiII. The photos shows that, where EuII has patches of overabundance, there TiII has patches of underabundance. In the cases of SrII and CrII there is overabundance on one hemisphere and underabundance on the other hemisphere. In either cases the overabundance is in the same hemisphere. In the photos there is overabundance of TiII, CrII and SrII at the positive magnetic pole and overabundance of TiII at the negative magnetic pole. Thus overabundance of TiII at the magnetic pole and underabundance of TiII at the magnetic equator can be seen. The case of EuII is the opposite of TiII. Thus overabundance of EuII is at the magnetic equator, and underabundance is at the magnetic pole. Overabundance of CrII and SrII is found only at the positive magnetic pole, and underabundance at the negative magnetic pole. In the photos the light patches correspond to overabundance, the dark patches to underabundance of the elements. All photos show half of the stellar surface.

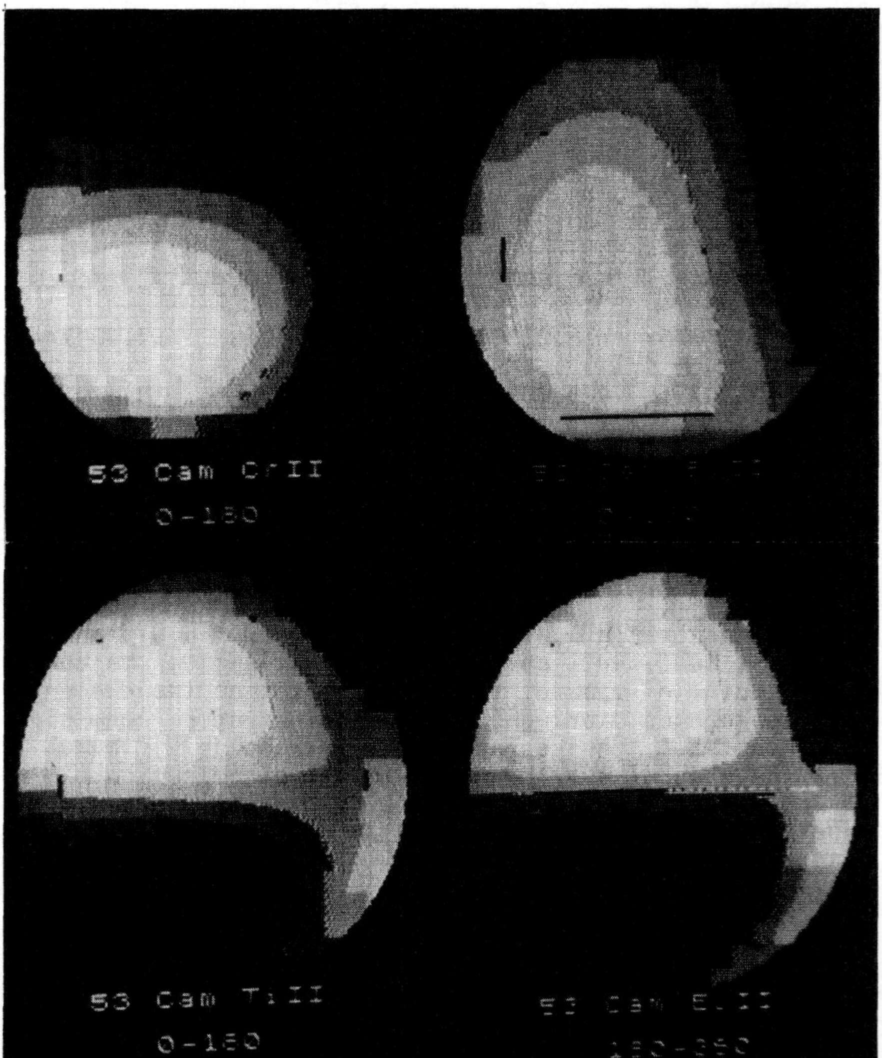

Fig. 1. The distribution of the elements CrII, SrII, TiII EuII on the surface of 53 Cam is shown in the photos.

REFERENCES

Borra, A.F. and Landstreet, J.D.:1977, Astrophys.J. 212, 141.
Crygar, J.:1965, Bull.Astron.Inst.Czech. 16,195.
Deutsch, A.J.:1970, Astrophys.J. 159,985.
Falk, A.E. and Wehlau, W.H.:1974, Astrophys.J. 192,409.
Faraggiana, R.:1973, Astron.Astrophys. 22,265.
Floquet, M.:1979, Astron.Astrophys. 77,263.

Huchra, J.:1972, Astrophys.J. 174,435.
Jasinski, M.,Muciek, M. and Woszczyk, A.:1981, Acta Astronomica. 31,321.
Megessier, C.:1975, Astron.Astrophys. 39,263.
Molnar, M.R. and et al.:1975, Physics of Ap-star. IAU Colloquium No.32.
Preston, G.W. and Stepien, K.:1968, Astrophys.J. 151,583.
Preston, G.W.:1971, P.A.S.P. 83,571.
Pyper, D.M.:1983, Astron.Astrophys.Suppl.Ser. 51,365.
Rice, J.B.:1970, Astron. Astrophys. 9,189.
Stift, M.J.:1974, Astron.Astrophys. 34,153.

THE NONHOMOGENEOUS DISTRIBUTION OF ABUNDANCES AND THE MAGNETIC FIELD MEASUREMENTS IN CP STARS.

F.A. CATALANO
Istituto di Astronomia dell Universita
Citta Universitaria
I 95125 CATANIA
ITALY

ABSTRACT. The influence of the non homogeneous distribution of the abundance of elements on the magnetic field measurements in chemically peculiar stars is discussed.

Magnetic fields in chemically peculiar stars of the upper main sequence are revealed by measuring the changes in the absorption line profiles and or the polarization in the lines induced by the Zeeman effect. Measurements of the circular polarization in spectral lines allow to determine the effective field (Beff) which is the mean value of the magnetic field component in the line of sight direction. In a very few stars, measurements of the spectral line splitting allow to compute the surface magnetic field (Bs).

Most part of magnetic data available in the literature are Beff values, but they are not a homogeneous set: to compare values from different observing sites some transformations are needed (Hensberge et al. 1979). The Beff values are also average values over different ions and elements and this causes loss of informations since stratifications of the ions of a given element may occur and patchy distributions of the abundances of elements are undoubtely ascertained. But the biggest problem to be faced in this context is given by the influence of this generally accepted nonuniform distribution of abundances on the magnetic field measurements. Some general relationship, such as the often observed coincidence of the extrema in the line strengths and magnetic fields, is generally accepted to exist but no quantitative investigation has yet been done. Really the problem is a very complicate one but we think some progress could be obtained if we first try to evaluate the polarization in a changing line profile for a very simplified case and then to look for applying the obtained results to a more realistic case. This is beeing done at the Astronomy Institute of Catania University.

Hensberge H., van Rensbergen W., Goossens M., Deridder G.: 1979 Astron. Astrophys. 75, 83.

C. R. Cowley et al (eds.), Upper Main Sequence Stars with Anomalous Abundances, 149.

SYSTEMATICS OF CP STARS

H. Hensberge, W. Van Rensbergen
Astrofysisch Instituut
Vrije Universiteit Brussel
Pleinlaan 2, B-1050 Brussel, België

ABSTRACT. This review is concerned with the following topics: 1. The occurrence of chemical peculiar stars in binaries; 2. Determination of masses of CP stars; 3. Determination of radii of CP stars; 4. CP stars in clusters. The paper emphasizes the literature from the 23rd Liège Colloquium in 1981 on, and does not refer systematically to earlier papers.

1. INTRODUCTION

This review discusses the empirical data on binarity, cluster membership and fundamental stellar quantities for CP stars, emphasizing the progress made in the last four years.

Binaries may give fundamental information about important physical quantities as stellar mass and radius. The distribution of orbital parameters, and the incidence of peculiarity among the companions of CP stars may give clues to the understanding of a either stellar structure-related or environment-related origin and development of the observed peculiarities. Aspects subject to evolution on time scales of the order 10^7-10^8 years may be studied preferentially using open cluster members, with known age.

2. CP STARS IN BINARIES

2.1. Occurrence of CP stars in binaries

Jaschek and Gomez (1970) analyzed a large sample of published radial velocities for normal stars and found a rather constant frequency of 47% ± 5% all along the main sequence from B0 to M.

The search for spectroscopic binaries among the brightest northern CP stars by Abt and Snowden (1973) revealed that duplicity among magnetic stars is significantly lower, while it is quite normal for HgMn stars. Many, but not all, Am stars are binaries (see Stickland,

C. R. Cowley et al (eds.), Upper Main Sequence Stars with Anomalous Abundances, 151–158.

1975 for references).

Studies of larger samples have confirmed the genuine difference in incidence of duplicity among magnetic and non-magnetic CP stars. However, there are indications that duplicity among magnetic stars might be less rare than believed previously (but nevertheless lower than along the main sequence, Floquet (1983), Gerbaldi et al. (1985)), while HgMn stars might have a larger than average incidence of duplicity (Schneider, this volume).

Recent analyses for HgMn stars (Schneider, this volume; Gerbaldi et al., 1985) are based on Schneider's (1981) catalogue. Depending on the adopted criteria, Schneider finds 57% to 72% binaries, included 7% of visual binaries. This agrees essentially with Gerbaldi et al. (1985) who find 54% of spectroscopic binaries from radial velocity data for stars in the BS catalogue; a percentage that has to be raised to nearly 70% if stars brighter than sixth magnitude are considered.

Gerbaldi et al. (1985) confirm the low binary frequency for magnetic stars. Radial velocity data from a sample of 31 He-w, 120 Si and 113 Sr-Cr-Eu stars show that the binary frequency of all but the coolest magnetic stars is not higher than 40%, in accordance with the 35% estimate of Floquet (1983) for magnetic stars. The coolest CP stars may occur as frequent in binaries as normal stars.

No data on He-rich were found, except for a qualitative remark by Bolton (1983) that their incidence in binaries is low.

2.2. Characteristics of CP binaries

Orbit determinations lack for most systems. Only 23 out of 97 HgMn binaries and 26 out of 233 classical CP2 studied by Gerbaldi et al. (1985) have known orbital parameters. Their sample is somewhat larger than the one studied by Zelwanowa et al. (1976), from the Osawa (1959) catalogue, and confirms the original results. There is a lack of short periods (HD15144 has P_{orb} = 3 days) and circular orbits. Double-lined spectroscopic binaries are rare, and only one marginally peculiar star is known to show eclipses (NGC2516- Cox nr.38, North, 1984). The secondaries may well have normal masses, since the statistical distribution of the mass function is not significantly different from systems with normal A primaries.

HgMn binaries share with the magnetic stars the lack of short periods (HD2019 has P_{orb} = 3.1 days) and the herewith connected deficiency of eclipsing systems. Eclipses are observed only in AR Aur = HD34364 (Nassau, 1935). But in other aspects, they are similar to binaries with normal late B primaries. More details are given elsewhere in this volume by Schneider.

Abt and Bidelman (1969) conclude that all stars in the range A4-

F1, IV-V, that are primaries of binaries with periods of approximately 2.5 to 100 days have metallic-line (Am) spectra. This does not exclude that some Am stars have shorter periods (δ Cap=HD207098 has P=1.2d; HD106112=HR4646 has P=1.27d).

3. MASSES OF CP STARS

Fundamental stellar masses are derived from the observations of eclipsing binaries. Popper (1980) lists 9 Am stars in five systems, having masses in the range 1.6 to 2.1 solar masses (and radii around 2 solar radii). This is what one expects for normal main sequence A stars. The mass of the B9 HgMn primary of the eclipsing binary AR Aur is 2.5 solar masses.

Similar mass determinations lack for magnetic stars, because the first eclipsing system has been detected fairly recently. Double-lined spectroscopic binaries give lower limits to the actual mass. Under favourable inclination conditions, this lower limit is of the same order of magnitude as the actual mass (e.g. HD98088, $M \sin^3 i = 1.7\ M_o$, Abt et al., 1968; HD141556, $M \sin^3 i = 2.3$ and 1.6 for primary Hg and secondary Am star resp., Dworetsky, 1972) and points again to roughly normal masses with respect to their MKK-classification. We like to call to your attention the recent mass determination for βCrB (Oetken and Orwert, 1985 and this volume). Knowledge of the visual orbit, the radial velocity variation of both components and the parallax - the latter introducing the largest uncertainty - leads to a mass of 1.8 solar masses for this cool Ap star.

We will not discuss here the body of circumstantial evidence, from positions in the HRD obtained generally after correcting for the influence of the peculiarities on the stellar flux distribution and interpreted with the use of standard evolutionary tracks. It places the CP stars in the mass range of 1.5 M_o for the coolest to over 5 M_o for the helium variables.

4. RADII OF CP STARS

The radius of a star is related through the Stefan-Boltzmann law to the total flux emitted by the star and its effective temperature,

$$M_{bol} = 42.31 - 5 \log R(R_o) - 10 \log T_{eff}$$

Both the stellar luminosity and T_{eff} are derived generally using the measured flux and its wavelength dependence in a restricted wavelength interval, and model atmosphere predictions outside that interval. A number of specific problems occur in the case of CP stars. Redistribution of ultraviolet flux produces a change in the slope of the Paschen continuum compared with normal stars (Leckrone et al., 1974), broad depressions in the continuum alter its shape, and differential line blocking influences the colours. All these factors are variable from star to star, depending on its peculiarity degree, and may be aspect dependent for a given star.

Ultraviolet astronomy has relieved part of the difficulties in obtaining the total stellar flux received at the earth, since at least for the cool and intermediate temperature CP stars the major part of the flux has become directly measurable. Information on the effective temperature may be obtained most accurately probably from the infrared, since our present knowledge indicates the absence of significant flux redistribution into this spectral region and since the IR line blocking is much less severe than at shorter wavelengths.

This is why the infrared flux method (IRFM) developed by Blackwell and Shallis (1977) is favourable. A monochromatic infrared flux $F_{E,\lambda}$ and the total stellar flux $\mathcal{F}_E$ at the earth are needed to solve the two basic equations

$$\sigma T^4_{eff} = (4/\Theta^2)\,\mathcal{F}_E$$

$$F_{E,\lambda} = (\Theta^2/4)\,\phi(T_{eff},g,\lambda)$$

for the angular diameter Θ and for T_{eff}. ϕ relates the dependence of surface flux on the stellar model atmosphere. Shallis et al. (1985) claim an accuracy on Θ of 8%. The conversion to linear radii is much more uncertain because of the less accurate parallaxes. They find in their sample of 9 stars $\langle R\rangle = 2.4 \pm 1.3$ (s.d.) R_o, and compare this to a mean value of 2.1 R_o obtained from normal stars of similar spectral types. The method is recently also applied by Lanz (1985).

Babu and Shylaja (1981) used observed fluxes in the wavelength interval 4000 -7800 Å and bolometric corrections related to T_{eff} for obtaining with a similar formalism radii for 23 Am and 69 CP2 stars. Their angular diameters for the 9 stars of Shallis et al. (1985) scatter around the IRFM results with a standard deviation of 17%. Their mean radii are 2.2 ± 0.6 (s.d.) R_o for Am stars and 2.9 ± 0.8 (s.d.) R_o for CP2 stars.

An indirect method for determining radii which has been applied with some success to diverse objects is based on the Barnes-Evans relation (Barnes et al., 1978). This relation has been established from a sample of stars having their radii estimated by more fundamental methods, mostly by lunar occultations. It is based on the correlation between the colour index determined from Paschen continuum measurements (reddening corrected V-R) and the surface brightness parameter, which in turn is related to the angular diameter of the star. In this way, the angular diameter of a star is found from photometric parameters alone.

Shore and Adelman (1979) conclude that the method may be useful to provide a first estimate for the radius, but cannot be applied with confidence indiscriminantly of the peculiar spectral characteristics of the CP stars. They find for individual stars significant discrepancies with radii determined from oblique rotator arguments (see further in this section), although the mean radius for their sample is in close agreement for both methods (2.7 $\pm$ 1.1 (s.d.) R_o on application of the Barnes-Evans relation, 2.5 $\pm$ 1.0 R_o from the oblique rotator model).

An alternative method for estimating radii of CP stars is based on the oblique rotator model. In particular the magnetic stars show light and spectrum variations which are modulated by the rotation period, P_{rot}. The rotation period is related to the radius R by

$$P_{rot}\, v_{eq} = 2\,\pi R,$$

so that R sin i is known when also the contribution of rotational broadening to the line widths is determined:

$$R \sin i = P_{rot}\, (v \sin i) \,/\, 50.613$$

when R is expressed in solar radii, P_{rot} in days and the apparent rotation velocity v sin i in km/s.

This relation has been used to derive a mean radius of samples of CP2 stars assuming a random orientation of the rotational axes relative to the line of sight. It may also be used for estimating radii of individual stars if sin i can be determined e.g. from magnetic field data when both H_e and H_s have been measured throughout the rotation cycle.

All mentioned methods agree statistically, in the sense that they add evidence to the conclusion that the radii of CP stars as a group are roughly equal or somewhat larger than those expected for main sequence stars of similar spectral type. More accurate individual radii may be expected after data of the Hipparkos project become available. Hipparkos satellite observations will extend the parallax horizon up to 75 parsec.

5. CP STARS IN CLUSTERS

The presence of CP stars in open clusters and associations has received an increasing attention in the last decade for both practical and theoretical reasons. From a practical point of view, an alternative to the time-consuming detection of CP stars by means of spectroscopic classification has been developed. The starting point of the photometric searches for CP stars is the recognition of the 5200 continuum depression as detection criterion for magnetic CP stars (Maitzen, 1976; see also Maitzen and Vogt, 1983). The detection probability is almost 100% for the silicon stars and the Sr-Cr-Eu stars hotter than spectral type A5, but diminishes significantly outside the corresponding temperature interval.

Photometric searches for CP stars are undertaken recently by Borra and coworkers (Joncas and Borra, 1981; Borra et al., 1982), by North and coworkers (North and Cramer, 1981; North, 1984; North and Waelkens, 1983; North, this volume), and by Maitzen and coworkers (Maitzen and Hensberge, 1981; Maitzen and Floquet, 1981; Maitzen, 1982; Maitzen and Wood, 1983; Maitzen and Schneider, 1984).

Borra and colleagues detected 3.5% of CP2 stars in the young Orion OB1 association, a percentage significantly lower than in the field. These stars have magnetic fields on the average a factor three stronger than the older stars (Borra, 1981). In this association as well as in Upper Scorpius, they find a relatively low peculiarity degree. They advanced the hypothesis that the frequency of Ap stars in the field would be reached ultimately in these associations when the helium-weak stars would develop classical Ap characteristics with time.

North and Cramer (1981) gave already a summary of their results in 36 clusters at the 23rd Liège Colloquium, using the Geneva Δ(V1-G) index as detection criterion. Since then, an important effort has been done to measure photometric periods for CP stars in clusters, to infer directly the importance of magnetic braking. They find no evidence for rotational braking on the main sequence, except for that expected from the conservation of angular momentum during the stellar evolution (North, this volume), a result that is recently confirmed by Borra et al. (1985).

Maitzen and colleagues concentrate on the detection of CP2 stars in a sample as large as possible, in order to obtain statistically significant results. Most of their work is presently unpublished. A summary of the meanwhile published results may be found in Hensberge et al. (1983). Additional clusters observed include IC 4725, NGC 5460, NGC 6087, NGC 1039, NGC 7092 (submitted papers) and NGC 225, 2232, 2244, 2264, 2301, 2323, 2343, 2437, 2447, 2451, 2547, 2548, 3114, 3532, 3766, 4103, 6231, 6871, 7160, IC 2391, 2395, 2602, Stock 2, Collinder 140, Trumpler 2 and 10, alpha Persei, Coma Berenices and the Plejades.

The most recent spectroscopic results are from Abt (1979) and Klochkova and Kopylov (1984 and this volume). Abt (1979) supported the occurrence of rotational braking on the main sequence, for magnetic stars, and has "the visual impression that the strengths of the peculiarity increases with time on the average". Klochkova and Kopylov, defining a quantitative spectroscopic peculiarity parameter and studying a larger sample, don't find any indication for a gradual change of peculiarity, rotational velocity or magnetic field strength during the main sequence life time.

The main theoretical reason for studying CP stars in clusters is the introduction of the parameter <u>time</u> in the empirical data set for these stars. Presently, simple, undisputable correlations with stellar age are not established. This may be due to several reasons. Firstly, it might be interpreted as evidence that the characteristic time scales of the involved processes are short i.e. the star reaches the main sequence as a full-grown peculiar star. In this case, by studying main sequence stars nothing can be learned about the order in which the different peculiarities develop (magnetic field decay or built-up, rotational braking, overabundances, spots...). Secondly, it might be interpreted as evidence that stellar individuality destroys any clear-cut correlation with age, although all physical processes involved are likely to be time-dependent. Among the factors that could be considered to produce this individuality are the star's environment and its chemical maturity which may be function of initial conditions and the way the interaction with the environment occurred. As a result, star-to-star, and thus also cluster-to-cluster differences can be expected. Such systematic difference on the level of clusters is claimed by Abt (1979) for the Am stars at a 85% confidence level.

With this in mind, Abt and Cardona (1983) considered an alternative approach. Instead of studying clusters, they determine the frequency of CP secondaries in binaries with O5-A1 main sequence primaries and secondaries in the absolute magnitude range of CP stars. The motive is that most field stars are escapees from the gradual desintegration of clusters, so that the condition of averaging out over many clusters is better fulfilled and the influence of stellar individuality is minimized. Although constrained by low number statistics, and by the fact that for the primaries only an upper limit of their age is known, the evidence is interpreted as support for an enhancement of the frequency of CP stars with time.

The conflicting conclusions of recent investigations clearly show that the situation is more complicated than perhaps hoped for originally. Much more empirical data and a more general approach, including other parameters than time, seem to be needed.

REFERENCES

Abt, H. A., Conti, P. S., Deutsch, A. J., Wallerstein G.: 1968, Astrophys. J. 153, 177.

Abt, H. A.: 1979, Astrophys. J. 230, 485.

Abt, H. A., Bidelman, W. P.: 1969, Astrophys. J. 158, 1091.

Abt, H. A., Cardona, O.: 1983, Astrophys. J. 272, 182.

Abt, H. A., Snowden, M. S.: 1973, Astrophys. J. Suppl. Ser. 25, 137.

Babu, G. S. D., Shylaja, B. S.: 1981, Astrophys. Space Sci. 79, 243.

Barnes, T. G., Evans, S. E., Moffett, T. J.: 1978, Mon. Not. R. Astron. Soc. 183, 285.

Blackwell, D. J., Shallis, M. J.: 1977, Mon. Not. R. Astron. Soc. 180, 177.

Bolton, C.: 1983, Hvar Obs. Bull. 7, no. 1, p. 241.

Borra, E. F.: 1981, Astrophys. J. 249, L39.

Borra, E. F., Joncas, G., Wizinowich, P.: 1982, Astron. Astrophys. 111, 117.

Borra, E. F., Beaulieu, A., Brousseau, D., Shelton, I.: 1985, preprint.
Dworetsky, M. M.: 1972, Publ. Astron. Soc. Pac. 84, 255.
Floquet, M.: 1983, in "Les journées de Strasbourg", Vème Réunion, ed. Observatoire de Strasbourg, p. 83.
Gerbaldi, M., Floquet, M., Hauck, B.: 1985, Astron. Astrophys. (in press).
Hensberge, H., Maitzen, H. M., Weiss, W. W.: 1983, The Messenger 34, 7
Jaschek, C., Gomez, A. E.: 1970, Publ. Astron. Soc. Pac. 82, 809.
Joncas, G., Borra, E. F.: 1981, Astron. Astrophys. 94, 134.
Klochkova, V. G., Kopylov, I. M.: 1984, in "Magnetic Stars", Proc. Riga Conference, ed. V. Khokhlova, D. Ptitsyn, O. Lielausis, p. 78.
Klochkova, V. G., Kopylov, I. M.: 1986, this volume
Lanz, T.: 1985, Astron. Astrophys. 144, 191.
Leckrone, D. S., Fowler, J. W., Adelman, S. J.: 1974, Astron. Astrophys. 32, 237.
Maitzen, H. M.: 1976, Astron. Astrophys. 51, 223.
Maitzen, H. M.: 1982, Astron. Astrophys. 115, 275.
Maitzen, H. M., Floquet, M.: 1981, Astron. Astrophys. 100, 3.
Maitzen, H. M., Hensberge, H.: 1981, Astron. Astrophys. 96, 151.
Maitzen, H. M., Schneider, H: 1984, Astron. Astrophys. 138, 189.
Maitzen, H. M., Vogt, N.: 1983, Astron. Astrophys. Suppl. Ser. 123, 48.
Maitzen, H. M., Wood, H. J.: 1983, Astron. Astrophys. 126, 80.
Nassau, J. J.: 1935, Astron. J. 45, 137.
North, P.: 1984, Astron. Astrophys. Suppl. Ser. 55, 259.
North, P.: 1986, this volume
North, P., Cramer, N.: 1981, in "Les étoiles de composition chimique anormale du début de la séquence principale", Proc. 23rd Liège Coll., p. 61.
North, P., Waelkens, C.: 1983, Inf. Bull. Var. Stars no. 2372.
Oetken, L., Orwert, R.: 1985, Astron. Nachr. 305, 319.
Osawa, K.: 1959, Astrophys. J. 130, 159.
Popper, D.: 1980, Ann. Rev. Astron. Astrophys. 18, 115.
Schneider, H.: 1981, Astron. Astrophys. Suppl. Ser. 44, 137.
Schneider, H.: 1986, this volume
Shallis, M. J., Baruch, J. E. F., Booth, A. J., Selby, M. J.: 1985, Mon. Not. R. Astron. Soc. 213, 307.
Shore, S. N., Adelman, S. J.: 1979, Astron. J. 84, 559.
Stickland, D. J.: 1975, in "Physics of Ap stars", Proc. IAU Coll. 32, ed. W.W. Weiss, H. Jenkner, H.J. Wood, p. 651.
Zelwanowa, E., Schöneich, W., Nikolova, S.: 1976, Astron. Nachr. 297, 229.

Discussion appears after the following paper.

SPECTROSCOPY OF CP STARS IN THE GROUPS OF DIFFERENT AGE

V. G. Klochkova, I. M. Kopylov
Special Astrophysical Observatory,
USSR Academy of Sciences, Nizhnii Arkhyz

ABSTRACT. The observational results for 108 stars, members of 10 galactic groups at the age range of lg t = = 6,4 - 8,7 (years), are briefly described. It is shown that the peculiarity index, rotational velocity and magnetic field strength of all the types of CP stars do not vary during their evolution across the main sequence stripe.

The most direct way for studiing the evolution of chemically peculiar (CP) stars is their observation in stellar groups for whose members one can suppose the close age and the similarity of the initial chemical composition.

Due to the wide variety of observational peculiarities in individual CP stars, a problem of the comparative analysis of the main parameters for these objects gets a statistical character (Hartoog, 1977; Abt, 1979; Wolff, 1981). It is necessary to study a rather large sample of CP stars of different peculiarity type in a large age range.

Observation of Bp, Ap stars, members of galactic clusters and associations of different age, there been carried out on the 6-meter telescope in 1978 - 1984 (Klochkova, Kopylov, 1985 a,b). The aim of this extensive program was to determine the quantitative characteristics of peculiar stars on the base of homogeneous spectral material of high resolution and in the framework of the common methods, to elicity the facts of possible dependence of spectral and physical parameters upon the ages of stars and to check the existing hypotheses on the origin and evolution of CP stars.

At the present time the set of parameters for 108 stars, members of 10 groups in the age range lg t = 6,4 - - 8,7 (t is given in years), have been determined. For each star 3 - 5 spectrograms with dispersion of 9 Å/mm (spectral resolution $\Delta\lambda$ = 0,25 Å, spectral range λ3900-

C. R. Cowley et al (eds.), Upper Main Sequence Stars with Anomalous Abundances, 159–162.

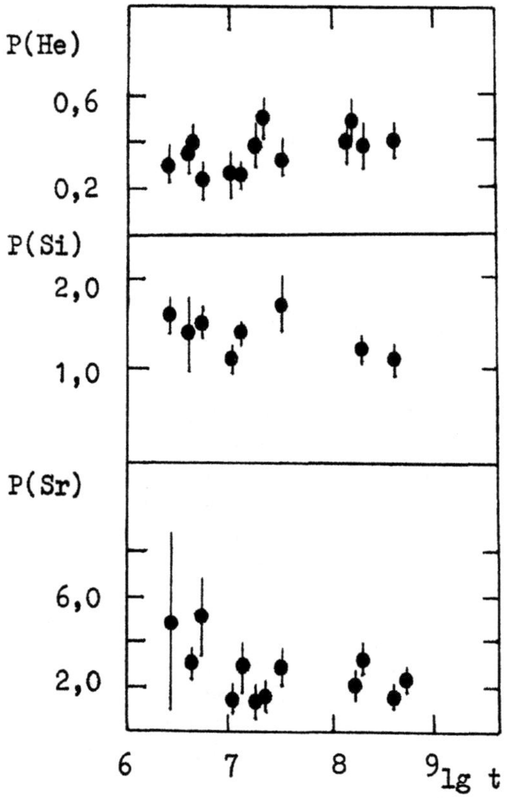

Fig 1. A dependence of mean quantitative index P(He), P(Si), and P(Sr) for CP stars in stellar groups on the age of a group.

The rotational velocities were considered separately for CP stars of four peculiarity types: 1) He-weak stars (n = 90), 2) SiII λ4200 and (or) SiII stars (n = = 49), 3) stars with strengthened SrII, CrII, EuII lines (n = 22), 4) Hg,Mn stars (n = 46). A great number of stars reveal several types of peculiarity simultaneously. The analysis of v sin i of CP stars has been carried out for the normalized rotational indices R according to the definition by Hartoog (1977):

$$R = v \sin i(CP) \,/\, \overline{v \sin i(norm)} ,$$

where v sin i(CP) is compared with averaged v sin i for normal stars in this cluster within the corresponding spectral type interval.

It was shown that for studied sample of clusters the $\bar{R}$ values do not correlate with age. We can draw a conclusion that there are no any significant losses of angular momentum by all the types of CP stars during their evolution within the MS stripe.

- 4900 Å) have been obtained. The minimum W_λ, measured with confidence, was near 30 mÅ.

We have determined spectral classes $Sp(B - V)_o$, Sp(Fe,Ti), $Mv(\beta)$, effective temperature Te, surface gravity lg g and rotational velocity v sin i. Index of peculiarity P is used for the quantitative characteristic of anomalous chemical composition of peculiar star atmospheres. Peculiarity index P is the ratio of W_λ for the line of some given element in the spectrum of studied CP star (i.e. He, Mn, Mg, Si, Sr, Cr, Eu) to the extremely possible (allowing for the errors) value of this W_λ in the spectra of normal stars with the same $Sp(B - V)_o$.

Age for each stellar group was found in conventional manner. The table presents the name of group, its age lg t, quota q % of CP stars among the stars of spectral classes B2 - A7 on the main sequence (MS), and the number n of the studied Bp, Ap stars.

Group	lg t	q	n	Group	lg t	q	n
Ori OB 1a	7,3	11	7	α Per	7,3	14	5
Ori OB 1b	6,6	8	6	Pleiades gr.	7,5	28	22
Ori OB 1c	6,4	8	11	Pleiades	8,1	6	4
Ori OB 1d	<6,0	0	2	M 39	8,2	10	6
Per OB 2	6,4	6	9	M 34	8,3	27	4
Upper Sco:				U Ma stream	8,6	14	9
nucleus	6,6	14	4	Coma	8,7	18	5
inner zone	6,7	33	13				
east zone	7,0	12	3				
west zone	7,1	22	10	Total			120

Remarks to the table: 1) for Upper Sco zones look for Klochkova et al., 1981; 2) 12 stars from ones of our primary program, at quantitative spectral classification we treated as the normal stars without any spectral anomalies.

The studied stars when plotted on Mv, $Sp(B - V)_o$ diagram, filling the whole band of the MS from zero age main sequence (ZAMS) up to the upper boudary of the MS in the mass range $1,0 \geqslant \lg \mathcal{M}/\mathcal{M}_\odot \geqslant 0,2$.

The correlations of peculiarity indices P of helium, silicon, strontium, etc with the age of the CP stars were analyzed. As the statistical analysis of the P values shows the peculiarity index of CP stars not depends upon their age (see figure). A dependence of the P indices upon the value lg g is not detected also. In this context we can consider lg g as some relative age characteristic of B2 - A7 stars inside the MS boudaries.

Taking into account the different evolution speed from ZAMS for stars with various masses we divided the program stars (independently on their peculiarity type) into three mass groups: 10 - 3,7 $\mathfrak{M}_\odot$; 3,6 - 2,6 $\mathfrak{M}_\odot$; and 2,5 - 1,6 $\mathfrak{M}_\odot$. There is no definite dependence of $\bar{R}$ value upon the age inside this mass groups; no systematical differences of $\bar{R}$ values between the groups are observed also. The dependence of $\bar{R}$ upon the temperature of stars is revealed for not a single type of peculiarity.

The root-mean-square values of magnetic fields $\langle Be \rangle$ for 52 stars were calculated using the published data, and our determination of $\langle Be \rangle$ for 19 cluster CP stars. As the comparison showed there is no any correlation of $\langle Be \rangle$ and lg t. The cross comparison does not reveal any correlation of $\langle Be \rangle$ and P(He), P(Si), ... ; v sin i and P(He), P(Si), ...

There is nothing but to suppose that:

a) CP stars become the slow rotators in comparison with the normal stars at the very early stages of stellar formation from the protostellar fragments. In the following evolution of a CP star near ZAMS and across the MS stripe its angular momentum does not suffer essential variations;

b) the magnetic CP stars have became the such ones yet before they came up to the ZAMS;

c) the chemical anomalies on the surface of CP stars begin to reveal themselves just before the ZAMS, when the future CP star approach the most quiet evolution phases - the stages of the ZAMS and the MS.

In the relatively stable atmosphere of slow rotating star rather effective diffusion processes can exist whose time scale is very short in comparison with duration of the MS stage. So the high peculiarity degree even in the youngest CP stars, members of Ori OB1, Per OB2, and innermost zone of Upper Scorpius associations ($t \approx 2 - 10$ millions years) can be explained.

References:

Abt H.A.,1979. Astrophys. J., 230, 485.

Hartoog M.R.,1977. Astrophys. J., 212, 723.

Klochkova V.G., Kopylov I.M., Kumajgorodskaja R.N.,1981. Pis'ma A. J., 7, 366.

Klochkova V.G., Kopylov I.M., 1985a. Astron. J.(Soviet), 62, (in press).

Klochkova V.G., Kopylov I.M., 1985b. Astron. J.(Soviet), 62, (in press).

Wolff S.C.,1981. Astrophys. J., 244, 221.

DISCUSSION (Hensberge and van Rensbergen)

KHOKHLOVA: It is difficult to distinguish between genuine spectroscopic binaries and stars with spotty distributions of metals, which show velocity variations as they rotate. What precautions were taken to minimize this problem, especially for Si and SrCrEu stars? A good example is the case of ε UMa, which for a long time was listed as a binary, until careful investigation revealed that the variations were due to the spotted distribution of elements.

HENSBERGE: If I recall correctly, variations from metal lines were rejected and only data from hydrogen Balmer lines, and other lines like Mg II 4481 Å (which tend not to be variable in these stars) were used. The percentage of binaries was determined by counting stars which had either a variable radial velocity or were mentioned as spectroscopic binaries, in a sample restricted to the Bright Star Catalogue. Suspected binaries or velocity variables were not taken into account. The numbers I quote may be in this sense upper limits.

KHOKHLOVA: How large was the sample for these statistics?

HENSBERGE: The sample contained 31 He-weak stars, 120 Si stars, 97 Hg-Mn stars and 113 cool CP2 stars.

KHOKHLOVA: Good, that is sufficient. Recently, Kopylov and Klochkova (Special Astrophysical Observatory) made an interesting investigation of stars in clusters.

HENSBERGE: I found a reference but I did not have time to trace the paper.

DROBYSHEVSKI: I wish to present some additional information concerning the duplicity of Am stars. This was done several years ago (E. M. Drobyshevski, Genesis and Classification of the Magnetic Stars II. Am Stars and Binarity of Early Stars, Preprint PhTI-445, Leningrad, 1973) but was not published for a wide distribution. I repeated G. P. Kuiper's analysis which was made for only 26 SB1 MS systems with orbits known at the time (Publs. Astron. Soc. Pacific, 47, p. 15, 1935). For random orbit orientations, he obtained a relation between the observed distribution $\phi(\nu)$ and the true distribution $\psi(\nu)$

$$\phi(\nu) = \nu \int_{\nu}^{\nu_{max}} \psi(\mu)(\mu^2-\nu^2)^{-1/2} \mu^{-1} d\mu$$

where $\mu = M_1/(M_1+M_2)$ and $\nu = \mu \sin i = [f(M)(M_1+M_2)^{-1}]^{1/3}$; f(M) is the mass function.

As data I used the second catalogue of SB systems (A. H. Batten, Publs. Domin. Astr. Obs., 13, p. 119, 1967) which contains data on 39 SB1 and SB2 systems with Am-type components and on 27 systems with normal A components. Only bright systems with 2 d $\leq P_{orb} \leq$ 200 d and quality a, b, c were used.

In Figs. 1a and 1b, distributions $\phi(\nu)$ for Am and A systems are presented. Also shown are distributions $\psi(\mu)$ and $\psi'(\mu)$ obtained by means of integration of the distribution. When obtaining $\psi'(\mu)$, the

C. R. Cowley et al (eds.), Upper Main Sequence Stars with Anomalous Abundances, 163–166.

observational selection was taken into account by assuming the SB detection probability to be proportional to ν^{-1}.

From these diagrams, one can see that Am SB systems separate into two groups: (1) $q = M_2/M_1 > 0.5$, and (2) $q < 0.4$. It is only natural to assume the first systems to be young, and the second group evolved, with invisible secondary components (white dwarf type?).

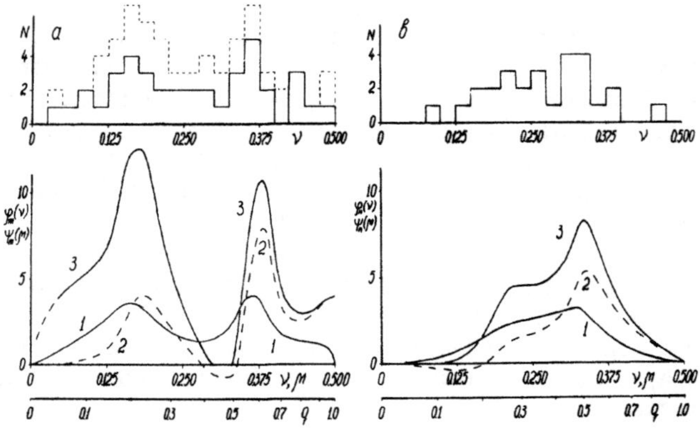

Fig 1.

Distribution for SB systems with Am (a) and normal A (b) components. The dashed line is a histogram for all the Am SB systems known (including faint systems, those of any period, and low quality orbits). Curve (1) is the $\phi(\nu)$, (2) is the $\psi(\mu)$, and (3) is the $\psi'(\mu)$ distribution, which takes into account observational selection.

The number of Am systems in group (2) is twice as great as the number in group (1). Such a result explains why Am stars are observed both in very young, and older, clusters.

To some extent, the distribution of systems with "normal" A star components fills the gap between the young and old Am systems. Nevertheless, for A stars as a whole, the bimodal distribution is conserved, as was found by M. Trimble, and by Kraicheva, Tutukov and Yungelson. In its turn, the number of Am systems is twice that of the normal A systems.

In 1973, when this work was done, only 223 bright Am stars were known, among them 92 SB (41%) including 62 systems with known orbits (7 eclipsing, 17 SB2 and 41 SB1). On the other hand, it follows from comparison of

$$\int_0^{0.5} \psi'_m(\mu)\, d\mu \qquad \text{with} \qquad \int_0^{0.5} \phi_m(\nu)\, d\nu$$

that only 40% of all SB stars with Am components are detected. From the coincidence of these figures (41% and 40%) one may conclude that all the

Am stars are binaries.

STĘPIEŃ: I wonder what meaning we can assign to the 'average' radius of an Ap star. These stars are found over a wide range of spectral types, from about B4 or B5, to around F2, which means that the range of radii would be expected to be a factor of, well, several. Would it not be more reasonable to restrict an analysis to a narrower spectral interval, say B8 to A2, and to compare an average from this interval with the radii of normal stars?

HENSBERGE: I agree with you. The best solution would be to derive for a well-defined sample of CP stars the distribution of the radii. With more photometric periods becoming available, and plausible assumptions about the distribution of the inclination angle of rotation axes with respect to the observer, this could be attempted using the period and line-width data.

However, only a few stars have been studied with the infrared flux method or by using the Barnes-Evans relation, perhaps only about ten stars. This means that definition of the subgroups is not possible. I quoted mean values here to show that systematic differences in results, from different methods and samples, do not amount to more than 20%. I emphasize with you that the use of an average radius has no physical meaning.

What you propose might be possible for the data assembled by Babu and Shylaja, because that sample contains about 80 to 90 CP stars, I think.

SCHÖNEICH: It is very important, for the theory of their origin, that the frequency of the magnetic CP stars in binaries is now higher. Previously, it was thought that this frequency was about 20-25%; now it is nearly normal. The former numbers could be interpreted as an absence of short periods and of circular orbits, but now we must assume that the orbits are changed, because the number of systems is the same as for normal A stars.

HENSBERGE: The comparison may be less straightforward than it seems, if one takes into account that the distribution of mass functions for CP and normal stars could be different. Indications for such a difference for Am stars was just described by Drobyshevski, who claims a bimodal distribution for f(m).

KOPYLOV: I would like to make two comments on the last part of this survey. First, it is necessary to take into account the individual properties of every cluster which contains CP stars; there are, for example, both young and old clusters which contain them. Second, we have examined CP stars in open clusters by quantitative spectroscopic methods (108 stars in 10 clusters). We find that there is no real change of v sin i, quantitative spectral peculiarity index, or $\langle B_e \rangle$ during the evolution of CP stars across the main sequence band.

DISCUSSION (Klochkova and Kopylov)

COWLEY: I'd like to make a comment on the weakness of the helium lines, which in the simplest interpretation reflects a lower He abundance. If the normal comparison stars have similar colours, then we all know that the He lines will be weaker in the peculiar stars. But, if you choose stars of the same effective temperature, we must consider other factors, because we know that the ground-based colours lead to a different effective temperature when the chemistry of the star is unusual. The effects of line blanketing suggest that we should use cooler normal stars for comparison purposes, and this should reduce the anomaly in the He line strengths.

KOPYLOV: Yes. This problem is well known to us. But, note that the abscissa in our plots (such as W_λ *vs*. Sp) is the effective temperature rather than the spectral type proper. For every star in our sample it was possible to find the effective temperature by the ionization equilibrium method and to compare this with the value obtained from space observations. For this reason, we think that the effective temperature scales for CP and normal stars do not differ significantly, and so the weakness of He is real, and not some mistake or error in the analysis.

DROBYSHEVSKI: From your data on the absence of a correlation between the peculiarity index and the age, it follows that the diffusion theory of chemically peculiar stars must be questioned. What can you say on this subject?

KOPYLOV: Our recent results are in some disagreement with the diffusion hypothesis. We are not surprised at this, because this theory is still being developed. Because many results for CP stars can be explained at present by this theory, we can express the hope that future modifications could lead to explanations for the main conclusions we have made in our statistical study. We see no real alternative to this theory for the explanation of the variety of individual properties of CP stars. This is a problem for the future.

MICHAUD: We could discuss that this afternoon, but I do not think there is any contradiction, because the time scales to establish the anomalies once the star arrives on the main sequence is very short.

ROTATION, VARIABILITY AND AGE OF Ap STARS

P. North
Institut d'Astronomie de l'Université de Lausanne
CH-1290 Chavannes-des-Bois
Switzerland

ABSTRACT. Photometrically determined gravities for field and cluster Ap stars, together with published periods, show that conservation of angular momentum alone is responsible for the observed increase of the periods with age. New photometric periods of a few cluster stars are presented which independently confirm this result. No outstanding change with age of the amplitude of lightcurves is noted.

1. RESULTS

1.1. Rotation

1.1.1. Ap stars in clusters. Table 1 lists a few periods of cluster stars recently obtained with the Geneva photometry. They confirm the already noticed lack of magnetic braking on the main sequence for CP2 and CP4 stars (Klochkova and Kopylov, 1984; North, 1984).

Table 1. New photometric periods of cluster Si stars

HD/HDE	CLUSTER	PERIOD	NO. OF MEASURES	LOG AGE	AGE REF.
16605	NGC 1039-37	NOT VARIABLE	19	8.3	North & Cramer 1981
45583	NGC 2232-9	1.177 D. OR 6.60 D.	44	7.35	Mermilliod 1980
66295	NGC 2516-26	2.454	37	8.1	North & Cramer 1981
66318	NGC 2516-24	NOT VARIABLE	38	8.1	North & Cramer 1981
68074	NGC 2547-5	1.1696 D.	21	7.65	North & Cramer 1981
145102	Upper Sco	1.42 D. (OR 3.33 D.?)	38	~ 7.0	Klochkova et al. 1981
304847	NGC 3114-234	2.3 D.? OR 4.6 D.?	25	8.1	North & Cramer 1981

C. R. Cowley et al. (eds.), Upper Main Sequence Stars with Anomalous Abundances, 167–170.

References to Table 1

Klochkova, V.G., Kopylov, I.M., Kumaigorodskaya, R.N.: 1981, Sov. Astron. Lett. 7(3), 203

Mermilliod, J.-C.: 1980, Thesis

North, P., Cramer, N.: 1981, in "Upper Main Sequence CP Stars", 23rd Liège Astrophys. Coll., p. 55

1.1.2 Field (and cluster) Ap stars with known gravity. The Geneva photometry gives a statistically reliable estimate of log g for most Ap stars, except for a few exceptionally blanketed ones (North & Cramer, 1984). Gravity is used as an age indicator in Fig. 1 and 2, where all stars with a known period (Catalano & Renson, 1984) and good Geneva colours are plotted. The gravities are probably less reliable in Fig. 2 because of increased blanketing by metallic lines (Teff being smaller) and because of the sensitivity of the Balmer jump to peculiarity (Gerbaldi et al., 1974). A linear least-squares fit through all 92 points of Fig. 1 yields a correlation at the 99% confidence level (r=0.387). Excluding the four anomalous points - whose initial period would be shorter than 0.5 d, while so short periods have never been observed - gives a better correlation (r=0.553) and exactly the same slope (b=-0.64) as that of the theoretical relation for the rather realistic case of rigid-body rotation (Endal & Sofia, 1979). The much greater dispersion of the 89 points of Fig. 2 prevents any fit from being done, but the lower envelope is compatible with the theoretical line.

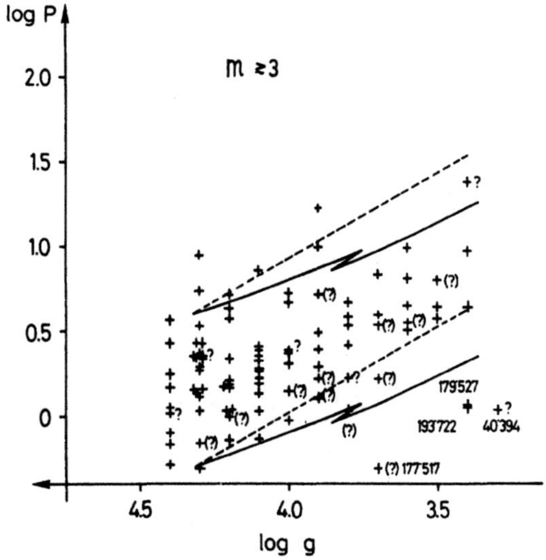

FIG. 1: PERIOD-GRAVITY RELATION FOR CP2 AND CP4 STARS MORE MASSIVE THAN 3 $M_{\odot}$. CONTINUOUS LINES: THEORETICAL RELATIONS FOR A ~4 $M_{\odot}$ STAR WITH COMPLETE RADIAL EXCHANGE OF ANGULAR MOMENTUM, WITH INITIAL PERIODS OF 0.5 AND 4.0 D. BROKEN LINES: THEORETICAL RELATIONS IN THE CASE OF CONSERVATION OF ANGULAR MOMENTUM IN INDEPENDENT SHELLS. QUESTION MARKS ARE FOR UNCERTAIN OR AMBIGUOUS PERIODS.

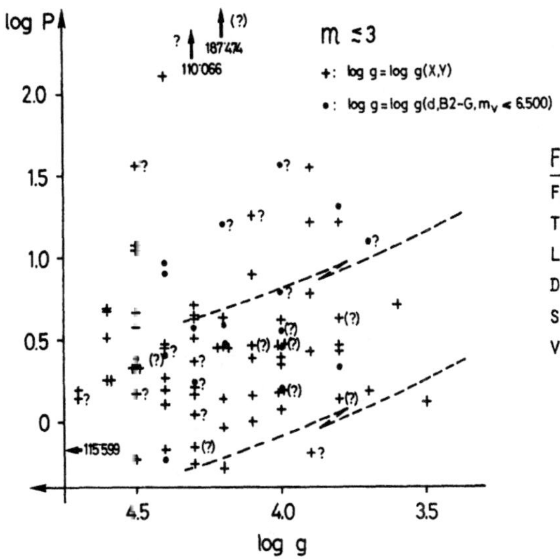

FIG. 2: SAME AS FIG. 1, FOR CP2 STARS LESS MASSIVE THAN 3 $M_{\odot}$. THE CONTINUOUS LINES OF FIG. 1 ARE REPRODUCED AS BROKEN LINES HERE SINCE THEY ARE NOT STRICTLY VALID FOR LOW MASS STARS.

1.1.3 Initial distribution of periods. Neglecting the influence of the initial rotational velocity on its subsequent evolution, one may estimate the period each star had on the ZAMS from its present period and evolutionary state. The distribution of initial periods is shown in Fig. 3 for stars with $M>3M_{\odot}$ and compared with the distribution for normal dwarfs. It is consistent with the observed periods of Ap stars in young associations.

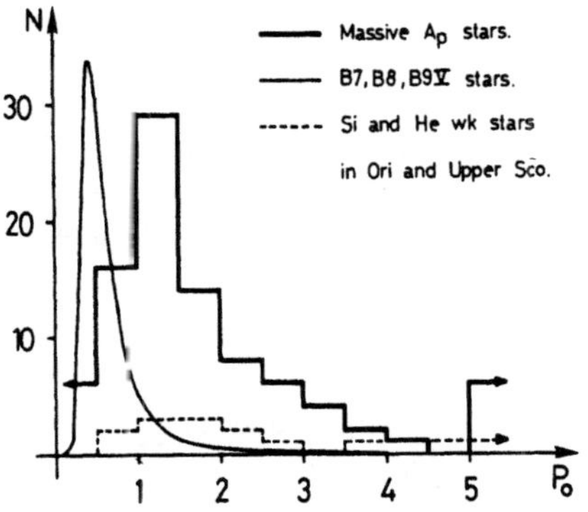

FIG. 3: DISTRIBUTION OF THE INITIAL PERIODS OF CP2 AND CP4 STARS WITH M > 3 $M_{\odot}$, DEDUCED FROM FIG. 1. THE DISTRIBUTION OF NORMAL CLASS V STARS IS SHOWN FOR COMPARISON.

1.2. Lightcurve amplitudes

Plots of the peak-to-peak amplitudes of an homogeneous sample of uvby lightcurves of 44 stars with any mass (Mathys & Manfroid, 1985) vs log g shows no significant trend, except perhaps in the u band, where less variation occurs at small gravities (Tables 2 to 5). More data are clearly needed.

Tables 2 to 5. Contingency tables for amplitudes in the u, v, b and y passbands. The χ^2 value and corresponding level of significance at which the null hypothesis can be rejected are indicated and show no significant correlation

TABLE 2 : U

A(U) \ LOG G	>3.9	≤3.9	TOTAL
> .06	15	3	18
≤ .06	15	11	26
TOTAL	30	14	44

χ^2 = 3.22 , ~ 92%

TABLE 3 : V

A(V) \ LOG G	>3.9	≤3.9	TOTAL
≥ .03	16	8	24
< .03	14	6	20
TOTAL	30	14	44

χ^2 = 0.06 , ~ 20%

TABLE 4 : B

A(B) \ LOG G	>3.9	≤3.9	TOTAL
≥.035	17	6	23
<.035	13	8	21
TOTAL	30	14	44

χ^2 = 0.73 , ~ 60%

TABLE 5 : Y

A(Y) \ LOG G	>3.9	≤3.9	TOTAL
≥ .03	16	5	21
< .03	14	9	23
TOTAL	30	14	44

χ^2 = 1.19 , ~ 70%

2. CONCLUSION

No significant braking (magnetic or other) occurs during the MS phase of field as well as cluster magnetic Ap stars in the range 2.5 to $6M_\odot$. Conservation of angular momentum alone accounts for the change of period on the MS. Slow rotation of Ap stars is thus related to stellar formation.

REFERENCES

Catalano, F.A., Renson, P.: 1984, Astron. Astrophys. Suppl. 55, 371
Endal, A.S., Sofia, S.: 1979, Astrophys. J. 232, 531
Gerbaldi, M., Hauck, B., Morguleff, N.: 1974, Astron. Astrophys. 30,105
Klochkova, V.G., Kopylov, I.M.: 1984, in "The Magnetic Stars", (eds. V.L. Khokhlova, D.A. Ptitsyn, O.A. Lielausis), Salaspils, p. 78
Mathys, G., Manfroid, J.: 1985, Astron. Astrophys. Suppl. 60, 17
North, P.: 1984, Astron. Astrophys. 141, 328
North, P., Cramer, N.: 1984, Astron. Astrophys. Suppl. 58, 387

Discussion appears after the Musielok paper.

Light curve analysis of rotating variable stars

A. Hempelmann and W. Schöneich
Central Institute for Astrophysics Potsdam, GDR

Abstract: A short description of a method for analysing light curves of rotating variable stars is given. This method is applied to light curves of HD 24712.

Introduction

According to the oblique rotator model, the typical brightness variations of magnetic CP stars are caused by an inhomogenous distribution of brightness over the stellar surface. Consequently, the light curves of spotted stars contain some information on large scale brightness distribution, which can be approximatively extracted from the light curves.

For this end we have to base the specific inverse problem (determination of some parameters from light curves) on a model of the type of the brightness distribution. The model has to be assumed in such a way that

i) the parameters to be determined should be interpretable by theory. For example, this model should allow a comparison with models of distributions of the magnetic field or of chemical elements.
ii) The model should not include a greater number of free parameters than can be derived from all the observed light curves of a given star.

The spot model

The model which serves as a basis is characterized by the following assumptions:

i) A very limited number of (more or less extended) spots on a spherical stellar surface may exist. (The light curves do not comprise any information on a possibly existing "small scale structure".)
ii) The fluxes derived from each spot and from the rest of the stellar surface should be described by a single parameter each of which depends on the wavelength λ.
iii) Each spot is to be of circular outline.

IV) We assume identical (linear) limb darkening laws inside and outside the spots. (Analogical to the analysis of light curves of eclipsing binaries we can expect that big errors in the limb darkening law - will result in only small errors of spot parameters - see Al Naimiy (1978)).

C. R. Cowley et al (eds.), Upper Main Sequence Stars with Anomalous Abundances, 171–174.

V) The values of the limb darkening coefficients u (λ) as well as the inclination angle i between the line of sight and the rotation axis have to be known.

The monochromatic flux L, integrated all over the visible hemisphere and going into the direction of the observer, will then be

$$L = L_o + \sum_{m=1}^{M} \frac{(q_m-1)\cdot L_o}{\pi\cdot R^2\cdot(1-u/3)} \iint_{G_m} (1-u+u\cdot\cos\gamma)\cdot\cos\gamma\cdot ds \quad (1)$$

$$\cos\gamma = \cos i\cdot\cos\beta + \sin i\cdot\sin\beta\cdot\cos(\vartheta-\varphi)$$

$$ds = R^2\sin\beta\, d\beta\, d\vartheta$$

where L_o - flux from the unspotted surface, q - ratio of flux densities inside and outside a spot, R - stellar radius, u - limb darkening coefficient, M - number of spots, G - area, occupied by a spot on the visible hemisphere, i - inclination angle, φ - rotation phase angle, β - polar distance and ϑ - stellar longitude.

To integrate numerically, we will devide the spot area into elements of selected extent ($D\vartheta$ x $D\beta$) and will test whether or not the given element is visible. Furthermore, we have to take into account that L and Lo are merely proportional to the observed values. In a simple way, we will replace L by the ratio of measured (and reduced) intensities of the variable and the comparison star, I/I_c. All the light curves of a given star (the observations I/I_c (φ, λ)) yield a system of equations, from which we have to estimate the following parameters:

I_o (λ) - the intensity level, which is observed in the case that the visible hemisphere is unspotted, and for each spot

q (λ) - the contrast between the spot and its surroundings,

ϑ_o - the longitude and

β_o - the angle to the rotation axis, of the spot center,

α - the spot radius.

With an approximative solution at α = DB/2, the spot radius will be expanded iteratively, and in each phase of iteration the other spot parameters will be calculated according to the method of differential correction of the parameters. In some special cases, β_o has also to be determined iteratively. As corresponding investigations have proved, this procedure secures a correct determination of even ambiguous solutions.

Light curve analysis of HD 24712

The rotational variability of this F0p star has been determined by Wolff and Morrison (1973) and also by Kurtz (1982). Kurtz determined the elements of the variability to be $JD_0 = 2440578.0 + 12.458\ E.$

Those light curves which have the greatest amplitudes were analysed, that is the light curve in v (Wolff and Morrison) as well as in B (Kurtz). Since each light curve is symmetrical to phase 0, a one spot model is capable to adapt the light curve completely in all colours. Our calculations have been based on an inclination angle of $i = 30^{o}$. (According to Kurtz (1982) a small i isto be expected). Limb darkening coefficients were taken from Al-Naimiy (1978).

The results of the light curve analysis for both independently investigated (and independently observed) light curves are completely adequate as to the solutions for the geometrical parameters (location of spot and spot radius). We obtained a double solution. The geometrical parameters are the following:

	solution 1	solution 2
longitude	$\lambda_{01} = 2^{o} \pm 1^{o}$	$\lambda_{02} = 2^{o} \pm 1^{o}$
polar distance	$\beta_{01} = 47^{o} \pm 1^{o}$	$\beta_{02} = 110^{o} \pm 40^{o}$
spot radius	$\alpha_1 = 25^{o} {}^{+20^{o}}_{-15^{o}}$	$\alpha_2 = 75^{o} {}^{+10^{o}}_{-15^{o}}$

In both cases the spot is darker than its environment. The strongly differing error ranges in the two solutions are the result of the special correlation of spot radius and polar distance. With increasing α, the polar distance remains almost constant in the range of solution 1, while β_o rapidly grows in solution range 2.

Kurtz (1982) developed an oblique pulsator model of HD 24712. Starting from the observation of a pulsational variability (Kurtz, 1982) and from the variation of the effective magnetic field B_{eff} (Preston, 1972), he defined an axis of symmetry for the pulsation as well as for the field distribution. According to his particulars and to the value we used for i, this axis completely agrees with the spot center in solution 1. From this we conclude that the field distribution (as well as the geometry of pulsation) and the surface brightness distribution are strongly correlated.

However we have to take into consideration the ambiguous interpretations of the light variability and of the variation of B_{eff}. Oetken (1977) developed a magnetic field model with equatorial symmetry for this star. The equatorial surface brightness distribution (according to solution two) is described by the following parameters:

$\vartheta_0 = 2^\circ$, $\beta_0 = 90^\circ$, $\alpha = 66^\circ$, q (B) = 0.88, I_0/I_c(B) = 0.796

The comparison between this model and the observations is shown in fig. 1.

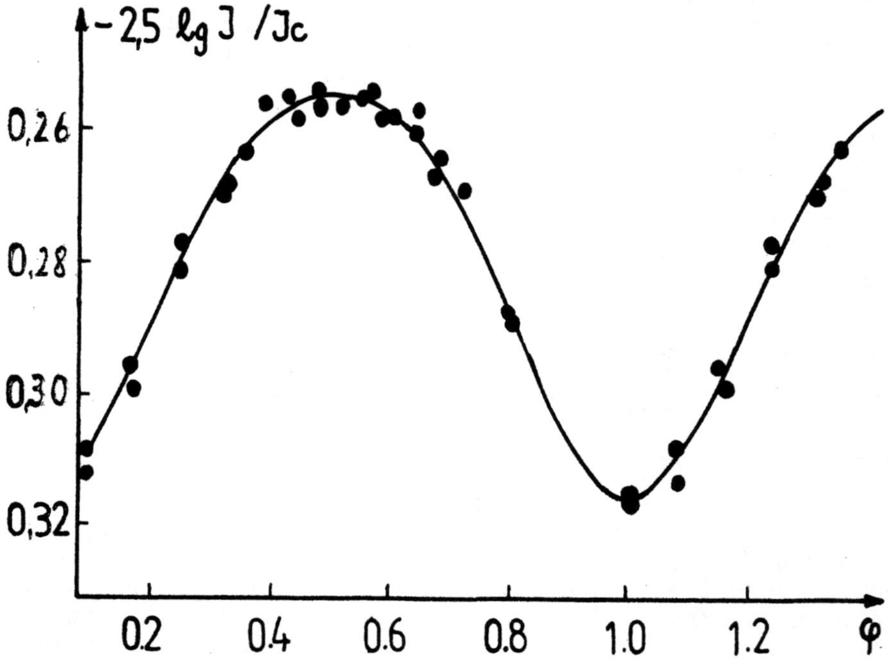

Fig. 1: Comparison between the observations (•) in B (Kurtz, 1982) and the one spot model of equatorial symmetry (—).

References:

Al-Naimiy, H. M., 1978, Astrophys. Space Sc. 53, 181
Kurtz, D. W., 1982, Month. Not. Roy, Astron. Soc. 200,807
Oetken, L., 1977, Astron. Nachr. 298, 197
Preston, G. W., 1972, Astrophys. J. 175, 465
Wolff, S. C., Morrison, N. D., 1973, Publ. Astron. Soc. Pacific 85, 141

A SEARCH FOR LONG-TERM PHOTOMETRIC VARIABILITY IN CP2-STARS

H. Hensberge
Astrofysisch Instituut, Vrije Universiteit Brussel
Pleinlaan 2, B-1050 Brussels, Belgium

ABSTRACT. Southern bright CP2 stars mentioned in the literature to be constant in light are observed to search for possible variations over months or years. Presently (up to end of April, 1985) four candidate long-period CP2 stars are detected, one of which had already been noticed in the literature as a slow spectroscopic variable.

1. OBSERVATIONS

In the frame of the programme of the group "Long-term monitoring of variable stars", a number of bright CP2 stars were observed regularly at the European Southern Observatory, La Silla, Chile. All observations were made relative to two comparison stars.

The stars were selected from the Bright Star Catalog according to the following criteria:

- they were published to be approximately constant over at least one week, and may have a low apparent rotational velocity v sin i, or
- evidence for long-term variations in spectral line strength or magnetic field has been found in the literature.

Part of the work is a continuation of a project started by the European ApWG since 1979 (Hensberge et al., 1984). Theoretical background and motives for this search are given in that paper.

2. RESULTS

The following stars are suspected to show long-term photometric variations:

HD 94660 = HR 4263	A0p Si	time base line = 5.5years
HD 116458 = HR 5049	Ap SrEu	time base line = 2 years
HD 151771 = HR 6244	B9p Si	time base line = 2 years
HD 187474 = HR 7552	Ap EuCrSi	time base line = 5 years

Except for the latter one, much more observations will be needed

C. R. Cowley et al (eds.), Upper Main Sequence Stars with Anomalous Abundances, 175–177.

to find the expected periodicity or even proof unambiguously the reality of the smaller variations. The present results for these stars are shown in Fig. 1.

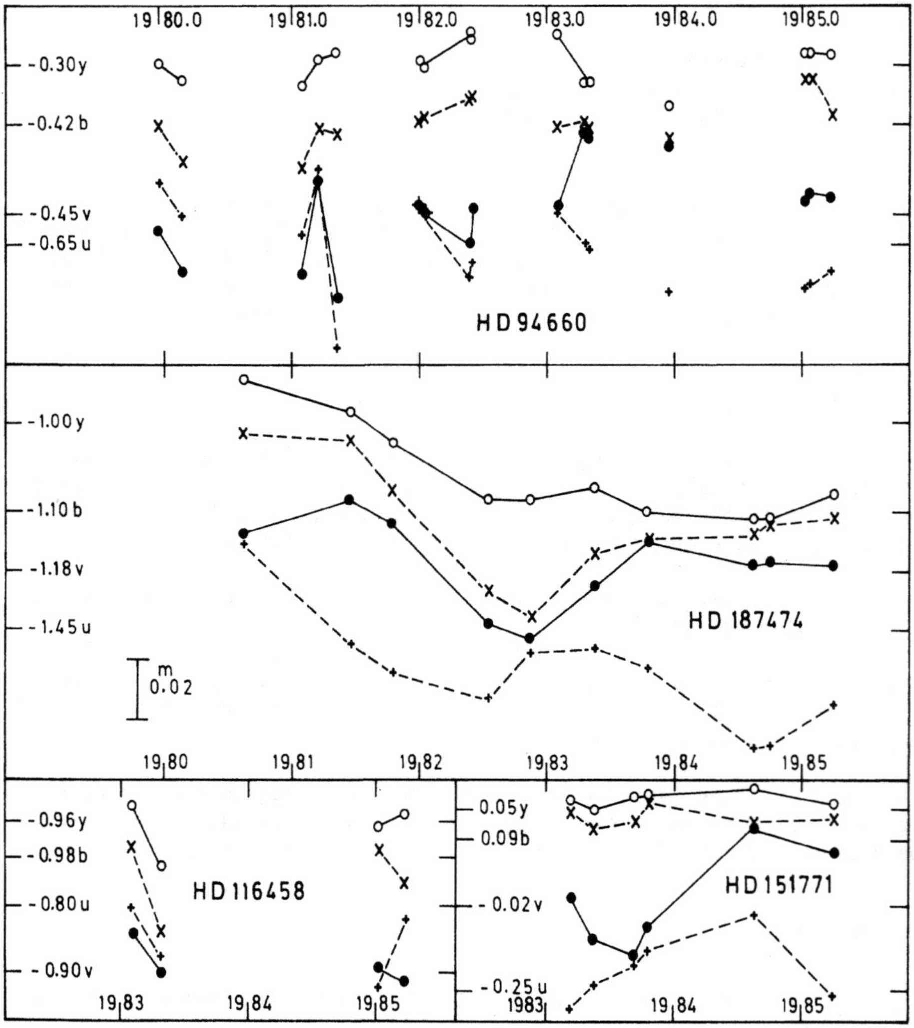

FIG. 1. Differential magnitudes for HD94660-HD94724, HD187474-HD189388, HD116458-HD116579 and HD151771-HD153072 (upper to lower panel resp.). The length of the bar in the left corner of the central panel corresponds to 0.02 mag in the vertical scale of all figures. Different symbols refer to u (+), v (•), b (x) and y (o).

The following stars have been observed without confirmed evidence for variations exceeding the 0.02 mag level:

HD 315	= HR 11	B8p Si	time base line = 450 days
HD 3326	= HR 151	A7p Sr	time base line = 500 days
HD 59256	= HR 2863	B9p Si	time base line = 2 years
HD 71066	= HR 3302	A0p Si	time base line = 5 years
HD 107696	= HR 4706	B9p	time base line = 2 years
HD 191984	= HR 7717	Ap SrCr	time base line = 550 days
HD 201601	= HR 8097	F0p SrEu	time base line = 450 days
HD 221760	= HR 8949	A2p SrCrEu	time base line = 2 years

3. CONCLUSION

These results, if interpreted in terms of the oblique rotator model, enhance the best estimate for the fraction of very slow rotators in the CP2 sample of the BS Catalog to 7% with an uncertainty of 2%. This relatively large fraction might be an indication for a large spread in efficiency of rotational braking among magnetic stars.

This work is based on observations by Catalano, Doom, Duerbeck, Floquet, Hageman, Hensberge, Maitzen, Mandel, Manfroid, Ott, Schneider, Schulte-Ladbeck, Stahl, Vander Linden, Weiss and Zickgraf. C. Sterken and J. Manfroid are gratefully acknowledged for the development of the infrastructure (administration and reduction facilities) for the "Long-term monitoring of variable stars"-group.

REFERENCES

Hensberge, H., Manfroid, J., Schneider, H., Maitzen, H. M., Catalano, F. A., Renson, P., Weiss, W. W., Floquet, M.: 1984, Astron. Astrophys. 132, 291.

THE NATURE OF BALMER LINE VARIABILITY IN CHEMICALLY PECULIAR STARS

B. Musielok
Astronomical Institute
of the Wrocław Univeristy
Kopernika 11
51-622 Wrocław
Poland

Photoelectric measurements of the β-index were made for six Ap-stars (56 Ari, 41 Tau, ϕ Dra, HD 188041, HD 215441, HD 221568) and one He-rich star (HD 184927). For all these stars the minimum of β occurs at a phase of maximum light at a wavelength longward of the null-wavelength region. Such coincidence can be explained by the blanketing mechanism proposed by Kodaira (1973) for the explanation of the Hγ line variations in HD 221568. According to Kurucz (1979), for stars hotter than 9000K an increase of metal abundance in the atmosphere causes an increase of the visible flux and an simultanous decrease of equivalent widths of Balmer lines. The same changes of the visible flux and equivalent widths of Balmer lines can be obtained using a model atmosphere with a suitable higher temperature. Using the Kurucz (1979) model atmospheres, temperature differences were calculated, which are necessary to obtain the observed decrease of the β-index and the increase of the flux in a given photometric band.

The relation between these temperature differences is given on Fig. 1, where the solid line is drawn assuming the "Balmer line" and the "photometric" temperature differences to be equal. Only the helium variable star HD 184927 evidently does not fulfill this relation. Applying the model atmospheres of Osmer and Peterson (1974) to the uvby photometry of Bond and Levato (1977) and spectroscopic measurements of Levato and Malaroda (1979), one can show, that within the observational errors the variations of the observed quantities (including the variations of hydrogen lines) can be explained as a consequence of variations of the relative helium to hydrogen content. This effect works also in the case of the Ap-star CU Vir. The measured variations of the equivalent width of the Hγ line (Riabchikova, 1972) and β-index (Weiss et al., 1976) are too large compared to the photometric variations (Weiss et al., 1976) --($\Delta T_{\text{Balmer lines}}$ = 645K; $\Delta T_{\text{photometry}}$ = 260K).

Khokhlova (1972) has determined the variations of the relative helium to hydrogen content for CU Vir. Applying to her data the model atmospheres of Klinglesmith (1971), one can calculate the variations of the Hγ line caused only by variations of the relative helium to hydrogen content and subtract these variations from the observed ones. The remaining variations of the hydrogen lines require an increase of the effective temperature of only 340K. This is marked by the arrow on Fig. 1.

CONCLUSIONS.

1) The Balmer line variations in Ap-stars can be explained by the blanketing mechanism.
2) In the case of helium variables, the Balmer lines are also affected by variations of the

C. R. Cowley et al (eds.), Upper Main Sequence Stars with Anomalous Abundances, 179–180.

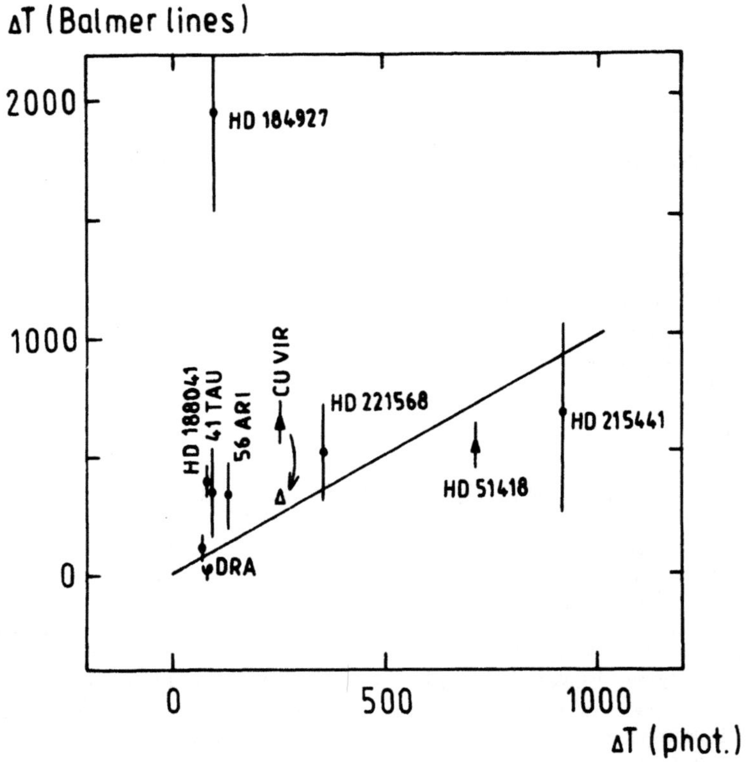

Figure 1. The relation between the "Balmer line" and "photometric" temperature differences. (• – present paper; ▲ – confirmed results of other investigations: HD 51418 – Gulliver and Winzer (1973), and Hardorp (1975); CU Vir – Riabchikova (1972) and Weiss et al. (1976).

relative helium to hydrogen content.

3) The quantitative dependence between the Balmer lines and the light variations gives strong support to the blanketing mechanism as the source of light variations in Ap-stars.

REFERENCES

Bond, H.E., and Levato, H. 1977, *Publ.Astron.Soc.Pacific*, **88**, 905.
Gulliver, A.F., and Winzer, J.E. 1973, *Astrophys.J.*, **183**, 701.
Hardorp, J. 1975, *Dudley Obs.Rep.*, **9**, 467.
Khokhlova, V.L. 1972, *Soviet Astron.J.*, **48**, 939.
Klinglesmith, D.A. 1971, *Hydrogen Line Blanketed Model Stellar Atmospheres*, NASA, Washington.D.C.
Kodaira, K. 1973, *Astron. Astrophys.*, **25**, 93.
Kurucz, R.L. 1979, *Astrophys. J. Suppl.*, **40**, 1.
Levato, H., and Malaroda, S. 1979, *Publ.Astron.Soc.Pacific*, **91**, 789.
Osmer, P.S., and Peterson, D.M. 1974, *Astrophys.J.*, **187**, 117.
Riabchikova, T.A. 1972, *Izv. Krym. Astrofiz. Obs.*, **45**, 146.
Weiss, W.W., Albrecht, R., and Wieder, R. 1976, *Astron. Astrophys.*, **47**, 423.

DISCUSSION (North)

COWLEY: Can you refresh my memory? Didn't Hartoog (Ap. J., 212, p. 723, 1977) suggest the same result as you, as far as the apparent braking of magnetic stars was concerned?
NORTH: Sorry, I did not understand the name of the person?
COWLEY: [attempting an accent] 'Ar-toog.
NORTH: 'Ar-toog, yes, of course! Indeed, he had already reached the same conclusion, and this started the controversy; I should have mentioned his result. The works by Abt (Ap. J., 230, p. 485, 1979) and by Wolff (Ap. J., 244, p. 221, 1981) came later, and were responsible for a consensus that magnetic braking did occur on the main sequence.
MÉGESSIER: The scatter in the relation log g(North) vs. log g (Mégessier) is not surprising, because in the method I used, the precision is no better than 0.25 dex.
NORTH: I also compared values of log g published by Klochkova and Kopylov for stars in the Sco-Cen moving cluster, and the agreement is significant, except that the slope of their relation is closer to 0.5 than 1.0. I do not know the reason.
SCHÖNEICH: In your log P vs. log g diagram there are four points which do not follow the general trend. We can help you to eliminate one of them: the period of HD 193722 is not ≈ 1.2 days as obtained earlier by ourselves and Winzer. The right period is about 8 - 9 days (Hildebrandt et al., Publs. Astrophys. Obs. Potsdam, Bd. 32, Heft 5, No. 112).
NORTH: Oh, that's fine! So the situation is improved. Thank you. [applause]

DISCUSSION (Musielok)

STEPIEŃ: The reason we observed Hβ in Ap stars was that we expected an influence of magnetic Lorentz forces on the formation of Balmer lines in the atmospheres of magnetic stars. We hoped to use the Hβ variations for diagnosis of the magnetic field. However, it turned out that the observed variations can be explained by the line blanketing alone.
KROLL: What is the expected variation in the equivalent width of Hβ?
MUSIELOK: The largest variations of the equivalent width are 25%, observed for the He-rich star HD 184927. For the other stars, the largest amplitudes were 20% for Osawa's star (HD 221568) and 10% for Babcock's star (HD 215441).
RYABCHIKOVA: I have a comment concerning CU Vir. In this case, there is a phase shift between variations of He and H lines, so it is difficult to explain the Balmer line variations by the blanketing of He in this star.
MUSIELOK: I am convinced that I took the effect of He variations in CU Vir into account correctly, but after your remark I will check this again.

C. R. Cowley et al (eds.), Upper Main Sequence Stars with Anomalous Abundances, 181.

THE VARIABILITY OF THE λ5200 FEATURE IN CP2 STARS

H.M. MAITZEN
Institut fur Astronomie
Universitat Wien
Turkenschanzstrasse 17
A-1180 Vienna, Austria

H. HENSBERGE
Astrofysisch Instituut
Vrije Universiteit Brussel
Pleinlaan 2
B-1050 Brussels, Belgium

ABSTRACT. 32 stars have been checked for Δa variability, with the purpose to learn which factor(s) influence the strength of the λ 5200 continuum depression. One third of the stars appear to be constant within a range of 0.005 mag. The results on the remaining ones indicate that maximum strength is related to at least one of the magnetic poles, but spectral inhomogeneities are almost certainly responsible for the largest amplitudes.

1. OBSERVATIONAL DATA

Maitzen (1976) discussed the detectability of Ap stars by Δa photometry. 32 Ap stars in his sample were observed on at least seven epochs, and are studied now for variability. Table 1 summarizes the results. We refer to Maitzen (1976) for details on the observing runs. The external one-sigma error for observations within each particular run has been estimated from standard stars, Ap stars with apparently constant Δa, and from comparison of observations at nearby phases: $\sigma \leq 0.003$ mag (see also column 8 of Table 1).

2. OCCURRENCE OF VARIATIONS

Only two stars in our sample have a total range in Δa exceeding 0.02 mag: EP Vir and CU Vir. Three other stars, not in our original sample, fall in this category: FF Vir (Hensberge et al., 1985), HD187474 (Hensberge et al., 1984) and HD133880 (Maitzen, unpubl.). They comprise the whole interval of photometric periods, from 0.5 days to over 6 years, but none of them is a cool Ap star.

The majority of the stars vary over a range of about 0.01 mag. (Fig. 2).

About one third of the stars (12/32) is constant within 0.005 mag.

C. R. Cowley et al (eds.), Upper Main Sequence Stars with Anomalous Abundances, 183–187.

TABLE 1. CP2 stars discussed in this paper. Columns give star identification (HD; name or HR); (rotation) period in days; number of Δa observations available; mean value of Δa and total range of variability (both as computed from sine-fit or, if necessary, from fit up to sin(2)). "<" means that the total range of variation is 6 or lower; shape of the Δa phase diagram; significance of the result: the first number given is the one-sigma of individual measurements around their mean value, the second one the one-sigma scatter of individual measurements around the best-fit curve. Unit = mmag. No fit was attempted when the original scatter did not exceed 3 mmag.

notation used for phase relations:

phase B_e, antiphase B_e indicates relation with absolute value of B_e when no (or very weak) polarity reversal is seen;

phase: weak (strong) extr. B_e indicates whether the weaker or stronger maximum of B_e corresponds to maximum Δa;

(prob) indicates an ill-defined phase dependence for one or some of the involved parameters.

HD	name	P(d)	N_{obs}	< a>	range	shape	signif.	phase relations; remarks
25823	41 Tau	7.23	10	30	9	narrow max.	4.0/2.2	phase: B_e (prob.), SrII, TiII antiphase: SiII
124224	CU Vir	0.52	9	20	22	see fig.	9.1/2.7	phase: strong extr. B_e, SiII antiphase: HeI
223640	ET Aqr	3.73	13	42	10	broad max.	4.5/2.1	P not accurate enough. SrII, TiII variable
37808	HR1957	1.10	10	24	8	sinusoidal	3.6/1.8	no data on magnetic field or spectrum variability
168733	HR6870		8	32	<		3.2/2.5	
74521	49 Cnc	4.24	8	73	<		2.7	
43819	HR2258	1.08	8	36	10:	sinusoidal	5.6/3.5	P suspect? R.sin(i)=0.26 seems to rule out variability
112413	α^2 CVn	5.47	8	42	11	sinusoidal	4.6/1.9	phase: weak extr. B_e, FeII, CrII antiphase: EuII, TiII
11503	γ_2 Ari	1.61 ?	7	39	<		2.8	
203585	θ^2 Mic	?	12	21	<		2.8	constant in light
192913	MW Vul	16.85	9	31	11:	double-0.9? inc. phase coverage	4.2/2.3	
10783	UZ Psc	4.13	13	46	12:	narrow max	7.7/5.4!	phase: B_e no spectrum variability
111133	EP Vir	16.3	9	57	27	sinusoidal	11.2/2.0	phase: B_e, CrI, CrII, FeI, FeII, SrII
72968	3 Hya	11.3 ?	9	47	<		2.3	
125248	CS Vir	9.30	7	46	<10:	incomplete phase coverage	2.5	phase: weak extr. B_e (prob.), Cr I (prob.), CrII (prob.) antiphase: EuII (prob.)
151525	45 Her	1.31	7	15	10:	double-0.8	3.6/0.8	Cr variable in strength, phase relation unknown
126515	FF Vir	130.0	42*	54	35	broad max.		phase: B_s, TiII, SiII, EuII, FeII, CrII observ. ApWG (Catalano, Pavlovski, Schneider, Vogt, Weiss) added
135297		?	7	32	<		3.2/ ?	inconclusive (too bad phase coverage Δa) CaII, SrII variable,
49976	HR2534	2.98	10	44	>11	incomplete phase coverage	4.8/ ?	B_e polarity reversal
220825	κ_1 Psc	0.59	13	32	<		2.4	
203006	θ^1 Mic	2.12	12	40	10	double-0.63	3.7/2.0	SrII max and min at <Δa> antiphase: Eu II
153882	HR6326	6.01	8	47	16:	broad max.	7.0/4.0!	phase: Be posit. extr. (symm. reversal)
130559	μ Lib	?	8	29	<		2.6	
221760	ι Phe	?	13	20	<		2.4	constant in light and spectrum
71866	TZ Lyn	6.80	9	47	<		2.8	B_e nearly symm. reversal, EuII and GdII double wave variab.
148898	ω Oph	2.99?	8	17	<8	double if var.	3.1/0.9	P not unambiguously established
30466		2.78	8	50	13	double-0.8	4.8/2.2	
22374		10.61	10	18	12	broad max.	5.5/3.5	
15144	AB Cet	3.00?	11	30	7	narrow max.?	3.3/1.5	Δa does not vary in 15.88d period proposed for B_e by Bonsack
134793		2.78	8	40	7	narrow max.	3.7/2.0	
3980	ξ Phe	3.95	10	38	<		3.0	
98088	SV Crt	5.91	9	34	12	double-0.5	4.9/2.3	phase: SrII, EuII; weak extr. B_e corresponds to primary max. Δa
137909	β CrB	18.5	8	23	<10:	incomplete phase coverage	2.4	phase: weak extr. B_e (prob.), B_s (prob.) no spectrum variability

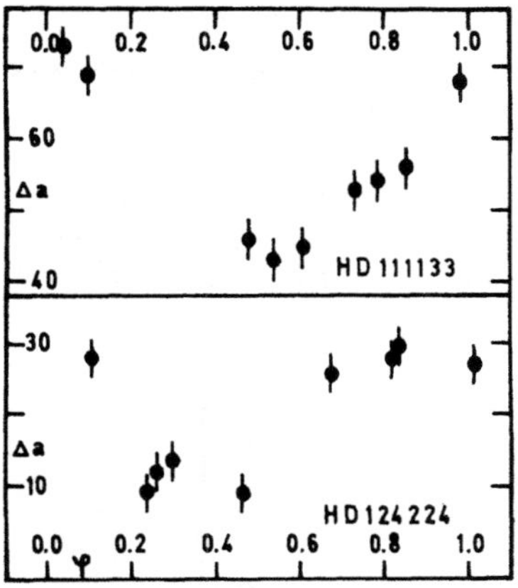

FIG. 1. Phase diagrams for EP Vir CU Vir. Phase zero corresponds to phase of maximum absolute value of effective field. The adopted periods are 16.304d and 0.520675d respectively.

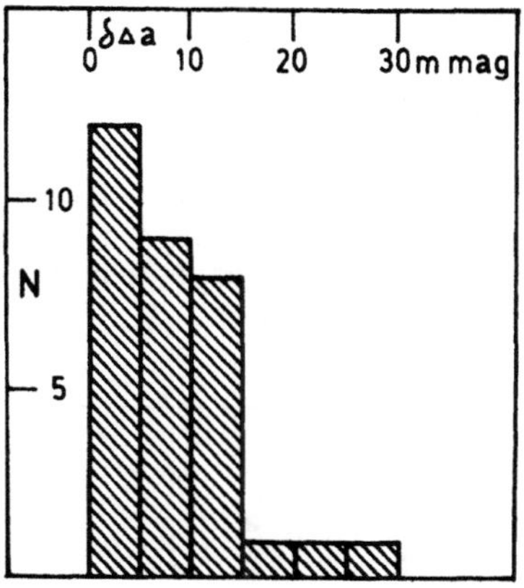

FIG. 2. Distribution of the range of Δa variability, δΔa, for our sample.

3. CORRELATION WITH SPECTRUM VARIATIONS

Δa variability correlates statistically with spectrum variability: 75% of the spectrum variables show a measurable variation, while this frequency is 50% for the other stars. This is most likely a moderate aspect effect.

Silicon shows in-phase variations in two stars with a very large Δa range (CU Vir, FF Vir), but the moderate variation of SiII in 41 Tau is definitely in antiphase. Chromium varies in-phase with Δa in the four stars where Cr variations were studied. Strontium varies in-phase, except for θ Mic where it varies 1/4 period out of phase with the double wave in Δa. Europium varies generally in antiphase with Δa, with the exception of SV Crt (Abt's star) and TZ Lyn. In the latter, the rare earts are variable, but Δa is constant, although the whole stellar surface is seen during the rotation cycle.

Although a statistical correlation between Si abundance and Δa, as found by Cowley (1981), cannot be ruled out, arguments against Si are: the antiphase variation in 41 Tau; Δa reaches only a secondary maximum at Si maximum in FF Vir; 3 Hya has very weak SiII (Babcock, 1958; Hensberge and De Loore, 1975), a rather weak magnetic field (B_s 2kG) but high Δa. It has very strong Cr and Sr. FF Vir, 3 Hya and the Cr spectrum variables in general lend support to a correlation Cr – Δa.

4. CORRELATION WITH MAGNETIC FIELD

Several papers have discussed the relation between B_s and the strength of the λ5200 feature (see e.g. North, 1980).

We computed estimates of $\langle B_s \rangle$ for as many programme stars as possible, using B_e data and information on oblique rotator geometry. Fig. 3 shows that the upper envelope of a (Δa,B2-G) graph is composed of stars with $B_s \geq 4$ kG.

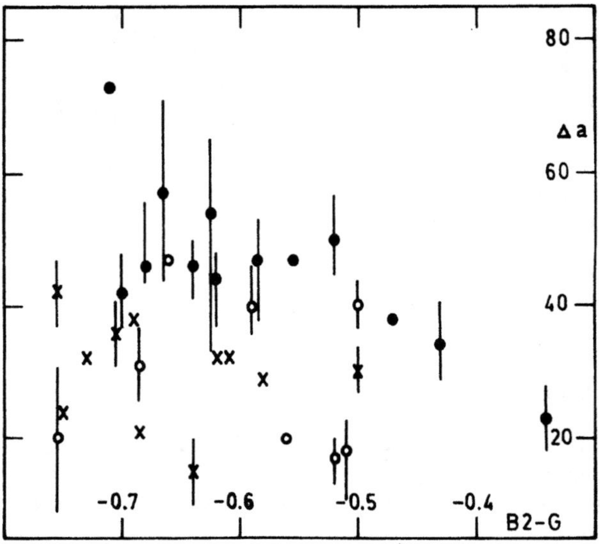

FIG. 3. Mean Δa values (time averaged over one cycle) and total variability range (=length of bars) against the temperature parameter B2-G (Geneva photometric system). Dots refer to stars with $B_s > 4$ kG, open circles to stars with $B_s < 4$ kG. Crosses are used when B_s could not be estimated.

Variability of Δa does not correlate with field strength. There is a strong tendency for Δa to show extrema when the field is longitudinal (Table 1, Fig. 3). If no clear polarity reversal occurs, then Δa varies in-phase with B_e . When both poles are clearly seen, high Δa may correspond as well to the stronger as to the weaker B_e extremum. This is not easily understood with a dipole model, but detailed calculations show that a relatively small quadrupole contribution (not larger than necessary to explain B_s in 53 Cam or FF Vir) suffices to create a "higher Δa = higher B_s" relation per star. In the case of α^2 CVn, the best-fit model of Borra and Landstreet (1977) predicts the measured antiphase relation.

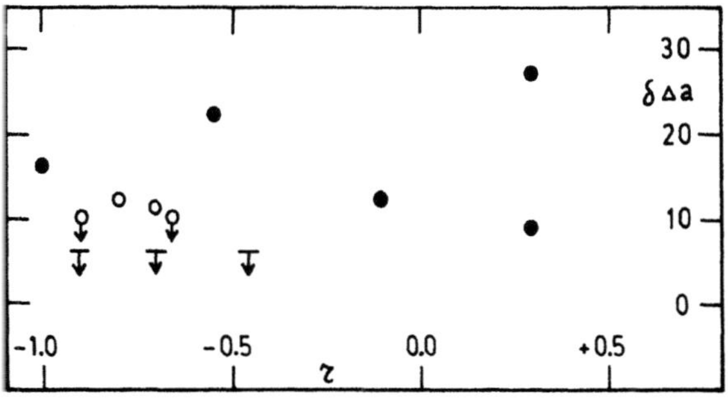

FIG. 4. Total variability range $\delta\Delta a$ against r, the ratio of the weak to the strong B_e extremum. Dots represent stars with Δa maximum coinciding approximately ($\Delta\phi<0.1$) with maximum $|B_e|$. Open circles represent stars with high Δa corresponding to the weaker B_e extremum. Arrows indicate upper limits for $\delta\Delta a$.

The Δa range for stars without polarity reversal is as large as when both poles are seen. Thus, the total Δa range seem to occur between regions with longitudinal and transversal fields.

It is very unlikely that the largest variations of Δa are dominated by magnetic line intensification effects. Variations in spectral line density are likely to dominate in the spectrum variables. Nevertheless, some stars without spectrum variability show a moderate Δa variability (UZ Psc).

REFERENCES

Babcock, H. W.: 1958, Astrophys.J. Suppl. 30, 141.
Borra, E. F., Landstreet, J. D.: 1977, Astrophys.J. 212, 141.
Cowley, C. R.: 1981, Astrophys.J. 246, 238.
Hensberge, H., De Loore, C.: 1975, Astron.Astrophys. Suppl. 20, 183.
Hensberge, H., Maitzen, H. M., Catalano, F.A., Schneider, H., Pavlovski, K., Weiss, W.W.: 1985, submitted to Astron.Astrophys.
Hensberge, H., Manfroid, J., Schneider, H., Maitzen, H. M., Catalano, F. A., Renson, P., Weiss, W. W., Floquet, M.: 1984, Astron.Astrophys. 132, 291.
Maitzen, H. M.: 1976, Astron. Astrophys. 51, 223.
North, P.: 1980, Astron.Astrophys. 82, 230.

Short time light variations of Ap-stars

G. Hildebrandt, W. Schöneich, D. Lange,
E. Želwanowa, A. Hempelmann
Zentralinstitut für Astrophysik der AdW der DDR
Sternwarte Babelsberg
DDR-1502 Potsdam-Babelsberg

From 1973 till 1981 sixteen Ap stars were investigated regarding to short time light variations with the twin telescope of the Zentralinstitut für Astrophysik stationed at the observatory Shemahka of the Academy of Science of Aserbaidshan. Only five of these stars show significant varyations with characteristical times in the region of 0.5 to 5 houres and amplitudes of about 0.01 mag. We will present here a short discussion of these five stars.

HD 173650

This object was observed 1979 and 1981. The frequence analysis of measurements from 1979 clearly show two periods of P1=0.06726 days (A1=0.0034 mag) and P2=0.06848 days (A2=0.0032 mag). The measurements from 1981 could not confirm these results. There are two possibilities for an explanation. Either the measurements from 1981 are placed in an unfavourable region of the beat period or the variations are not always present.

HD 184905

BRODSKAJA(1978) found variations with a period of about 70 minutes and an amplitude of 0.02 mag. PANOV(1981) observed one night only and found a cycle length of 25 to 30 minutes with an amplitude of 0.01 mag. We received from this star 32 measurement series which are longer than one hour. The frequence analysis as well as from the measurement series of different years and from all measurements together showed not any reference of periodical variations in the region of 1 to 4 houres. We found different cycle lengths from 70 to 188 minutes. Clearly variations are rare (about 20% of all nights) and they are not coupled with the rotation phases of this star.

HD 204411

RAKOSCH(1963) observed in one night variations in three spectral regions with different courses. We observed five measurement series from 1973 till 1974. All series show variations in cycle length and amplitude. We could not find any periodical variation.

C. R. Cowley et al (eds.), Upper Main Sequence Stars with Anomalous Abundances, 189–190.

HD 219749

This star was observed in 1977, 1979 and 1981 and we received 29 measurement series. We clearly found variations in amplitude and cycle length in many cases which points to a multiperiodical character. We found for:
1977 P1=0.09028 days (A1=0.0040 mag), P2=0.10694 days (A2=0.0028 mag); 1979 P1=0.08920 days (A1=0.0026 mag), P2=0.10601 days (A2=0.0034 mag), P3=0.11009 days (A3=0.0050 mag); 1981 P1=0.09022 days (A1=0.0054 mag), P2=0.10960 days (A2=0.0028 mag).
The measurements for all three years show nearly two identical pairs of periods. We found a third period for the observations of 1979 with dominating character. It is possible that the ratio $P2/P3 \approx 0.975$ points to a nonradial pulsation.

HD 224801

RAKOSCH(1963) found in four nights variations with a period of 125 minutes. JARZEBOWSKI(1981) observed in two nights unregulary fluctuations of light.
From 1973 till 1981 we received all together 18 measurement series. In different series we also found cycle lengths with a characteristical time of about 120 minutes. Other series compared to it show a constant course which points to a multiperiodical variation. Only for 1973 we could carry out a frequence analysis. We found for this year P1=0.09325 days (A1=0.0062 mag), P2=0.06737 days (A2=0.0028 mag). With these two periods we found a good fitting for the other measurements. Only the observations from 1979 show a deviation which is greater than the observational error. But this result stands not in opposition to the expection that the star shows a multiperiodical variation.

We can conclude that short time light variations with amplitudes from about 0.01 mag and characteristical times of houres are real but a rare phenomena for the Ap stars. In a few cases we observed irregulary variations. The reality of this effect one must confirm with more observations and higher accuracy. Our assessment agrees with that of WEISS(1983). He observed five Ap stars with low pulsation modes. We found two new Ap stars with these properties by our observations in addition to him. A detailed discussion of our investigations of short time light variations of Ap stars will be published in 'Publikationen des Astrophysikalischen Observatoriums Potsdam' 1985, Bd.32, Heft 5, Nr.112.

CP-STARS IN THE NEAR INFRARED : NORMAL !

R. Kroll
H. Schneider
H.H. Voigt
Universitäts-Sternwarte
Geismarlandstr. 11
D-3400 Göttingen
FRG

F.A. Catalano
Institute of Astronomy
of Catania University
Viale A.Doria 6
I-95125 Catania
Italy

ABSTRACT. 17 CP-stars have been measured in the IR filter bands J,H,K,L and M. No significant differences between CP- and normal main sequence stars can be found. Flux exzesses at 4.8 microns are not confirmed.

1. Introduction

Besides their function as cornerstones of the method of integrated fluxes (BLACKWELL & SHALLIS, 1979) to evaluate simultaneously effective temperature and apparent angular diameter of stars, IR-fluxes from stellar objects are the most convincing indicators for the presence of circumstellar material.
In 1983 GROOTE & KAUFMANN (hereafter referenced as 'GK') summarized their IR-measurements, obtained during three observing runs at the ESO 1m photometric telescope. From a sample of 105 CP-stars 60% showed excessive radiation in the M-band at 4.8 microns, exceeding the exspected flux from a KURUCZ model, found by the integrated flux method, by more than 20%. They proposed a model in which CP-stars are surrounded by dust disks with temperature of 300 to 650 K (GROOTE & KAUFMANN 1984).
Surely this picture would throw new light on the CP puzzle, but soon the GK results were doubted. BONSACK & DYCK (1983, 'BD' hereafter) compared IR-fluxes from CP stars with those of normal main sequence stars and could neither find any significant differences between both groups, nor evidences for excess flux at 4.8 microns. This survey was done with a 2.2m telescope at Mauna Kea, hence from the northern hemisphere. Therefore only a few stars are in common in both works, but in these cases magnitude differences of more than half a magnitude occured.
In this situation it seemed worthwile to try a decissive answer to this controversary by new comparative measurements.

2. Observations and Reduction

During our observing run from Nov. 8.-16., 1984 at the ESO 1m telescope CATALANO observed 17 CP-stars with an InSb detector in the five IR-filter bands J,H,K,L and M. Central wavelengths and widths of the filters are tabulated in Tab. 1. 16 of the stars can be found also in the GK compilation, whereas 7 are in common

C. R. Cowley et al (eds.), Upper Main Sequence Stars with Anomalous Abundances, 191–194.

with the BD survey.

The reduction was done in a straight forward manner, daily extinction values and zeropoints were obtained using

$$m_{inst} = m_{obs} - Ext. * Airm. + ZP$$

and a suitable set of standard stars. We selected 17 standards from the compilation of KOORNEEF (1983a, 1983b) that span approximately the same range in colors and magnitudes as our program stars.

Filter	λ	Δλ
J	1.24	0.32
H	1.63	0.28
K	2.19	0.39
L	3.79	0.68
M	4.64	0.63

Tab. 1 : Actual filter values; central wavelengths and widths as reported by ESO.

The transformation from instrumental to standard magnitudes may need a correction of the form

$$m_{st} = m_{inst} + corr(B-V, UT, m_{obs}, ...)$$

We checked such possible drift effects by plotting the differences between instrumental and standard star magnitudes versus J-L color, universal time and apparent brightness. Surprisingly we found a distinctive correlation with the brightness, which can be understood as a nonlinearity of the detector. The deviations could be satisfactorially fitted by a straight line that reversed its slope during most nights with increasing wavelengths, e.g. from J to M-Filter, as seen in Fig. 1. Because the corrections are in the order of tenths of a magnitude, hence fairly large, we recalculated extinction values (Ext'.) and zeropoints (ZP') with the corrections to the observed magnitudes of the form

$$m_{obs}' = m_{obs} + corr * m_{obs}$$

applied. After that no correlation of the deviation between standard and instrumental system could be found in respect to color, time or brightness that exceeded the errors of observations. Therefore the finally adopted equation for the transformation to the standard system is

$$m_{st} = m_{obs}' - Ext'. * Airm. + ZP'$$

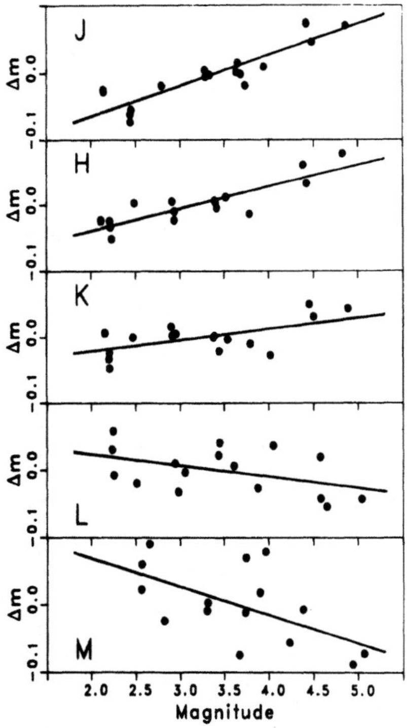

Fig. 1: Magnitude drift effect, difference of standard and instrumental magnitude versus apparent magnitude on Nov.8.,1984

3. Results

Tab. 2 shows the M-magnitudes of the program stars with the differences to the work of GK and BD, respectively. Positive sign indicates that our measurements imply fainter magnitude.

HD	M	GK	BD	n
3980	5.75	+.53		3
12447	3.80	+.03		4
22470	5.69	+.38	+.08	28
24712	5.37	+.29		6
25267	5.09	+.22	-.11	3
28843	6.30	+.56	-.05	12
37017	7.28	+.89	+.18	3
49333	6.70	+.52		3
54118	5.33			3
72968	5.75	+.34	-.28	9
74196	6.23	+.60		3
203006	4.72	.00	-.31	2
206088	3.08	+.04		2
220825	5.01	+.46		6
221006	6.15	+.58		6
221760	4.50	+.12		2
223640	5.68	+.24	+.01	4

Tab. 2 : M-Magnitude and differences to GK and BD surveys. Positive sign means our value is fainter. Last column gives number of measurements

	GK	BD
J	-.04	-.06
H	+.01	-.05
K	-.01	-.06
L	.00	(-.14)
M	+.36	-.07

Tab. 3 : Mean magnitude differences between this work and GK and BD surveys, respectively.

Tab. 3 shows the mean differences in all filters between this work and the two other author pairs. The agreement with GK is excellent in the J,H,K and L filters, whereas in the M-band a large discrepancy occurs. If we plot the M-band differences versus the apparent magnitude (in J-band), as done in Fig. 2, a strong correlation is seen in the sense that the differences get larger with decreasing brightness. We conclude that the GK work is affected by systematic errors. A reasonable explanation would be that they have been trapped by magnitude drift effects as discussed above, since they used primarily bright standard stars.

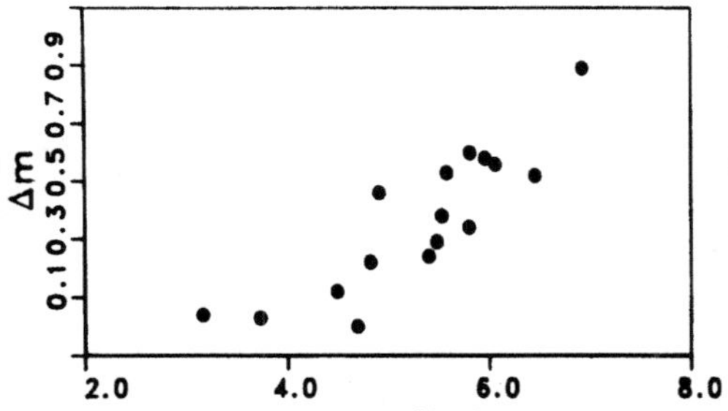

Fig. 2 : M-Magnitude differences between this work and GK versus apparent magnitude in J-band

Between our and the BD measurements there seems to exist a small zeropoint difference, since in all filters we measure notoriously about .06 magnitudes brighter than they do. The larger difference in the L-band is due to the fact that BD calibrated their filter in the L' standard system, shifted 0.2 microns redwards to the standard L-filter.

To answer the question if

there is still some of the claimed flux excess left, we apply the same procedure as done in the 1984 paper of GK. A double logarithmic plot of flux versus wavelength produces a straight line in the Rayleigh-Jeans approximation. For the flux calibration we used the values of KOORNEEF (1983b). Fig. 3 shows for each star straight lines computed with the first four filter values. All M-fluxes lay slightly below this line, thus the question now emerges as: are the CP-stars flux deficient in this range ? They are not. If we apply the same procedure to the standard stars, we also find flux deficiencies in the M-band of about 10% for stars of J-L color around zero. The reason for that behaviour are deviations from the Rayleigh Jeans approximation.
Now no more significant differences remain between CP- and normal stars.

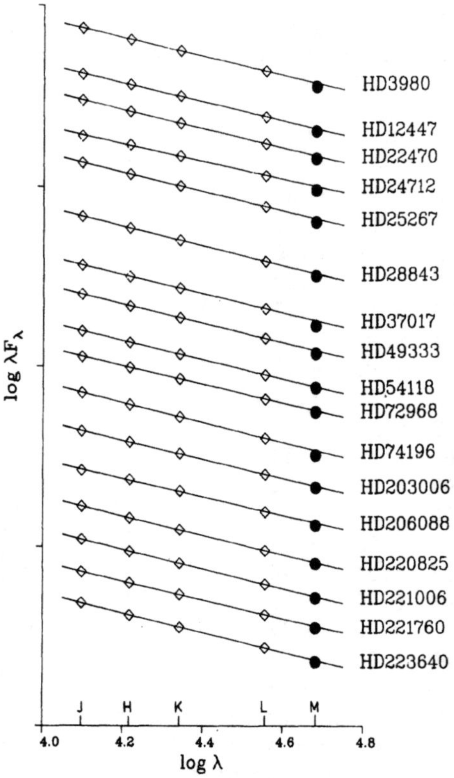

Fig. 3 : Exzess diagnostic diagram. M-values above line would indicate flux excess. For further explanation see text.

4. Conclusions

We have shown that IR fluxes from CP-stars in the range of one to five microns are not significantly different from those of normal stars. Therefore from such measurements we have no evidence for circumstellar matter around CPs, shining up at such wavelengths. HAVNES & GOERTZ (1984) have shown that a stellar magnetosphere, filled by mass loss from the central star could not account for the amount of IR-radiation needed to explain the GK results. We agree with their conclusion, that in order to detect circumstellar matter around CP-stars by the means of IR astronomy, observations at wavelengths longer than 5 microns are needed.

References :

Blackwell,D.E., Shallis,M.J. 1979 : MNRAS 188, 847
Bonsack,W.K., Dyck,H.M. : 1983, A & A 125, 29
Groote,D., Kaufmann,J.P. : 1983, A & A Suppl. 53, 91
Groote,D., Kaufmann,J.P. : 1984, A & A 130, 184
Havnes,O., Goertz,C.K., 1984: A&A 138, 421
Koorneef,J. 1983a : A & A Suppl. 51, 489
Koorneef,J. 1983b : A & A 128, 84
Kurucz,R.L. 1979, Ap. J. Suppl. 40, 1

Hα and OI PHOTOMETRY OF UPPER MAIN SEQUENCE STARS WITH ANOMALOUS ABUNDANCES

Eugenio E. Mendoza V.
Institute of Astronomy, University of México
P.O. Box 20-158, México 20, D.F.
01000 MEXICO

ABSTRACT. Additional α(16)Λ(9)-photometry of upper main sequence stars with anomalous abundances confirm our previous results, namely, Ap stars are neatly separated from normal main sequence stars. Furthermore, Ap stars are located on three different zones of the α(16)Λ(9)-plane, according to their abundance anomalies to form three photometric groups, (i) Si stars, (ii) Hg,Mn stars, and (iii) Cr,Eu,Sr stars.

1. INTRODUCTION

A preliminary report on the α(16)Λ(9)-photometric system of upper main sequence stars with anomalous abundances was given earlier (Mendoza 1977). The main result was that Ap stars are neatly separated from normal main sequence stars.

We have observed over 500 stars in the α(16)Λ(9)-photometric system mostly from the Bright Star Catalogue (Mendoza 1985a,b,c,). The observations include stars with spectral types from O to K, luminosity classes from I to V, and normal and abnormal stars. Figure 1 shows the α(16) Λ(9)-plane for the Bright Stars (Mendoza 1985a). The scatter in this figure is mainly due to luminosity effects and spectral peculiarities (Be, shell, Am, Ap, etc.). It is interesting to mention that stars with the Hα-line contaminated by emission have an α(16)-index <0.94, approximately; stars with the OI line (λ7774A) also contaminated by emission have an Λ(9)-index <0.26, approximately.

The photometric data indicate that there are probably two close sequences in the α(16)Λ(9)-diagram for normal main sequence stars, one for the O4-B9 V, and other for the A0-G2 V stars with a perceptible cosmic scatter around A0 V (Mendoza 1985c).

2. OBSERVATIONS

The new observations have been carried out with the 33-inch telescope of Observatorio Astronómico Nacional at San Pedro Mártir, in 1981-3. Table 1 contains the α(16)Λ(9)-photometry (in magnitudes) of 30 classical Ap stars. Jaschek and Egret (1982) classify them in four classes, for orien-

C. R. Cowley et al (eds.) Upper Main Sequence Stars with Anomalous Abundances, 195–198.

TABLE 1

H_α and OI-PHOTOMETRY of UPPER MAIN SEQUENCE STARS with ANOMALOUS ABUNDANCES

HD	Name	α(16)	Λ(9)	Sp	G
358	α And	1.233	0.325	B9	Hg,Mn
9996	GY And	1.368	0.294	B9	Cr,Eu,Sr
18296	21 Per	1.278	0.287	B9	Si:
19832	56 Ari	1.218	0.299	B9	Si
25823	41 Tau	1.192	0.289	B9	Si
28929		1.238	0.317	B9	Hg,Mn
33904	μ Lep	1.216	0.324	B9	Hg,Mn
40312	θ Aur	1.270	0.318	A0	Si
75333	14 Hya	1.221	0.319	B9	Hg,Mn
78316	κ Cnc	1.216	0.313	B8	Hg,Mn
106625	γ Crv	1.214	0.321	B8	Hg,Mn
108662	17 Com	1.338	0.285	B9	(Cr,Eu,Sr):
110066	AX CVn	1.398	0.291	A0	Cr,Eu,Sr
110073		1.209	0.323	B8	Hg,Mn
111133	EP Vir	1.352	0.296	A0	Cr,Eu,Sr
112185	ε UMa	1.359	0.287	A0	Cr,Eu,Sr
112413	α^2 CVn	1.285	0.289	A0	Si:
118022	78 Vir	1.390	0.271	A1	Cr,Eu,Sr
120198	84 UMa	1.328	0.284	B9	Cr,Eu,Sr
124224	CU Vir	1.242	0.286	B9	Si
130158	55 Hya	1.274	0.276	A0	Si
137909	β CrB	1.344	0.286	F0	Cr,Eu,Sr
148898	ω Oph	1.378	0.286	A7	Cr,Eu,Sr
151525	45 Her	1.314	0.294	B9	Cr,Eu,Sr
152107	52 Her	1.376	0.281	A2	Cr,Eu,Sr
174933	112 Her	1.245	0.318	B9	Hg,Mn
176232	10 Aql	1.322	0.275	F0	Cr,Eu,Sr
201601	γ Equ	1.330	0.288	F0	Cr,Eu,Sr
220825	κ Psc	1.372	0.277	A0	(Cr,Eu,Sr):
223640	108 Aqr	1.239	0.292	B9	Si

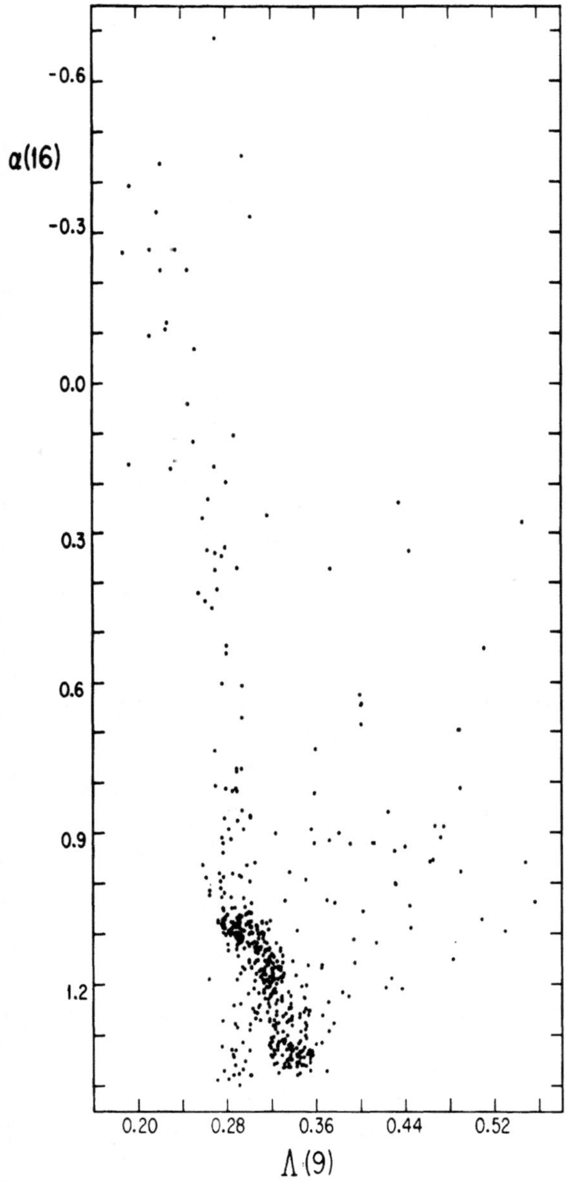

Fig. 1. The α(16)Λ(9)-plane for the Bright Stars (in magnitudes). CP-stars are located at the bottom and to the left side of this figure.

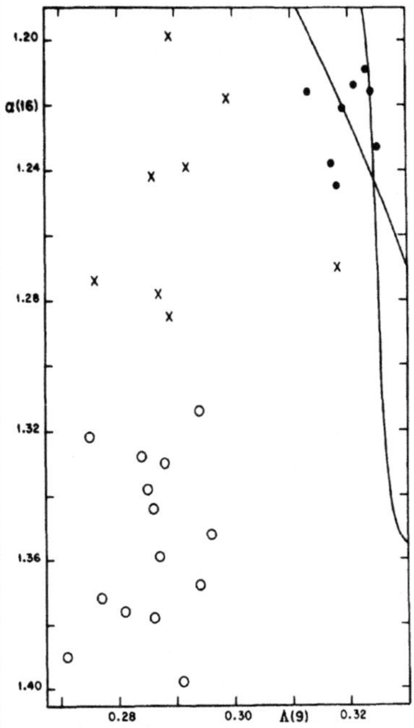

Fig. 2. The α(16)Λ(9)-array for the Ap stars listed in Table 1. Coding: solid line, normal main sequence, from O4 to G2; crosses, Si-stars; filled circles, Hg,Mn-stars; and open circles, Cr,Eu,Sr-stars (Am stars are not plotted; however, they do not mix-up with the Ap stars -loc. cit.).

tation purposes:

A: Si λ4200 and Si stars
B: Hg, Mn, and Hg-Mn stars
C: Si combined with Cr, Eu, Sr, etc.
D: Cr, Eu, Sr, and combinations

Last column of Table 1 lists these classes as:

A: Si
B: Hg, Mn
C: Si: or (Cr,Eu,Sr):
D: Cr,Eu, Sr

3. CONCLUSION

Figure 2 shows the α(16)Λ(9)-array for upper main sequence stars with anomalous abundances. Three groups are clearly seen, (i) the Si-stars; (ii) the Hg,Mn-stars, and (iii) the Cr,Eu,Sr-stars. Figure 2 also shows:

a) Si-stars and Hg,Mn-stars have approximately equal α(16)-index, but different Λ(9)-index, on the average.
b) Si-stars and Cr,Eu,Sr-stars have different α(16)-index and approximately equal Λ(9)-index, on the average.
c) Hg,Mn-stars and Cr,Eu,Sr-stars have both indices quite different.

The above can be interpreted as each class having slightly different physical parameters.

It is interesting to mention that other stars hotter than the Sun are also well separated in this photometric system (see Fig. 1 and Mendoza 1985c), such as the Am stars, which are part of this colloquium (we do not give herein more details because of lack of space -see Mendoza 1976 and 1985c). Thus, we conclude that the α(16)Λ(9)-photometric system is suitable to classify stars very accurately, especially the A-type stars.

REFERENCES

Jaschek, M., and Egret, D. 1982, Catalog of Stellar Groups. Centre de Donnees Stellaires.
Mendoza, E.E. 1976, Rev. Mexicana Astron. Astrof., 2, 29.
Mendoza, E.E. 1977, Rev. Mexicana Astron. Astrof., 2, 259.
Mendoza, E.E. 1985a, Rev. Mexicana Astron. Astrof., 12 (in press).
Mendoza, E.E. 1985b, Rev. Mexicana Astron. Astrof., 12 (in press).
Mendoza, E.E. 1985c, to be published.

A PHOTOMETRIC INVESTIGATION OF THE MAGNETIC STAR 53 CAMELOPARDALIS

M. Muciek[1], P. North[2], F. Rufener[3], J. Gertner[1]

[1]Institute of Astronomy, Nicolaus Copernicus University
ul. Chopina 12/18, PL 87100 Toruń, Poland

[2]Institut d'Astronomie de l'Université de Lausanne
CH-1290 Chavannes-des-Bois, Switzerland

[3]Observatoire de Genève, CH-1290 Sauverny, Switzerland

ABSTRACT. Geneva photometry of the important and relatively well known Ap star 53 Cam (HR 3109, HD 65339) is presented. These 27 data, which cover rather evenly the rotational phase interval, are compared with the photometric data in ten bandpasses published by Musielok et al. (1980). They allow to show that at least four "null-wavelength regions" occur between 3400 and 7800 Å.

A simple model of surface brightness distribution is proposed, which is axisymmetric about the magnetic axis. The magnetic geometry of 53 Cam, fortunately, is rather well known (Borra and Landstreet, 1977) and makes such a model possible. This model is applied to the fluxes measured through 15 different bandpasses and to the integrated flux. The local distribution of the peculiarity parameter Δ(V1-G) and of the equivalent width of the TiII and CaII K lines are modelled in the same way. It seems that the integrated flux increases from the negative magnetic pole to the positive pole, but the variation is very small ($0^m.01$) and should be confirmed by infrared and ultraviolet measurements.

The mean pseudocontinuum of the star has been drawn from ~3500 to ~6600 Å using 14 passbands. It is shown that it can be roughly fitted with a solar composition Kurucz model having T_{eff}=8200 K and log g=4.0.

All figures, tables and more detailed information will be published in the 3rd number of Acta Astronomica, 1985.

REFERENCES

Borra, E.F., Landstreet, J.D.: 1977, Astrophys. J. 212, 141
Musielok, B., Lange, D., Schöneich, W., Hildebrandt, G., Zhelvanova, E., Hempelmann, A., Salmanov, G.: 1980, Astr. Nachr. 301, 71

C. R. Cowley et al (eds.), Upper Main Sequence Stars with Anomalous Abundances, 199.

SPECTROPHOTOMETRY OF THE BROAD CONTINUUM FEATURES IN MAGNETIC AP STARS

Diane M. Pyper
Physics Department, University of Nevada, Las Vegas, Las Vegas, NV 89154 USA
Saul J. Adelman*
NRC-NASA Research Assoc., Laboratory for Astronomy and Solar Physics, NASA Goddard Space Flight Center, Greenbelt, MD 20771 USA

EFFECT OF THE λ5200 FEATURE ON PHOTOMETRIC COLORS

The strongest broad absorption feature in the peculiar energy distributions of the Ap stars is that centered at about 5200 Å, thus the Stromgren y band and the Geneva V1 band are most affected in stars in which this feature is strong. Fig. 1 shows bandpasses (full width at ½ intensity maximum) of three widely used photometric systems superimposed on two of our scans of Ap stars and two solar abundance line blanketed model atmospheres (Kurucz 1979). It is seen that both the y and V1 bands fall entirely within the λ5200 feature. The plot (Fig. 2a) of b-y vs. Tpc (the color temperature of the red end of the Paschen continuum), shows that the b-y colors for most of our sample of Ap stars are displaced to the blue of the b-y, Teff relationship of Relyea and Kurucz (1978). In Fig. 2b, Δ(b-y) = (model b-y) - (observed b-y), for a given temperature is plotted vs. $\Delta WS_2(5200)$, a spectrophotometric index measuring the equivalent width of the λ5200 feature. There is a strong correlation between Δ(b-y) and $\Delta WS_2(5200)$, indicating quantitatively the large effect of the λ5200 feature on the y band, previously discussed by Adelman (1979). The deviations in Ap star B2-G values from the normal star B2-G vs. T curve are much less than for b-y, as the Geneva G band is largely outside the λ5200 feature (Fig. 2c). Thus B2-G is a better temperature indicator for Ap stars than is b-y (also see Hauck and North 1982).

PROFILES OF THE λ5200 FEATURE

Our scans of the Ap stars enable us to look for trends in the shape of the λ5200 feature. We can detect structures found earlier at higher resolutions for smaller numbers of stars (Maitzen and Seggewiss 1980 and references therein), such as a deep core at λ5200. We have made a preliminary classification of Ap stars based on the profiles of the λ5200 feature. Fig. 3 shows the characteristic profiles for the five groups of stars in relation to the bands used for two of the λ5200 photometric indices, Δa and Δ(V1-G). In general, only the

*Permanent address: Physics Department, The Citadel, Charleston, SC USA.

C. R. Cowley et al (eds.), Upper Main Sequence Stars with Anomalous Abundances, 201–204.

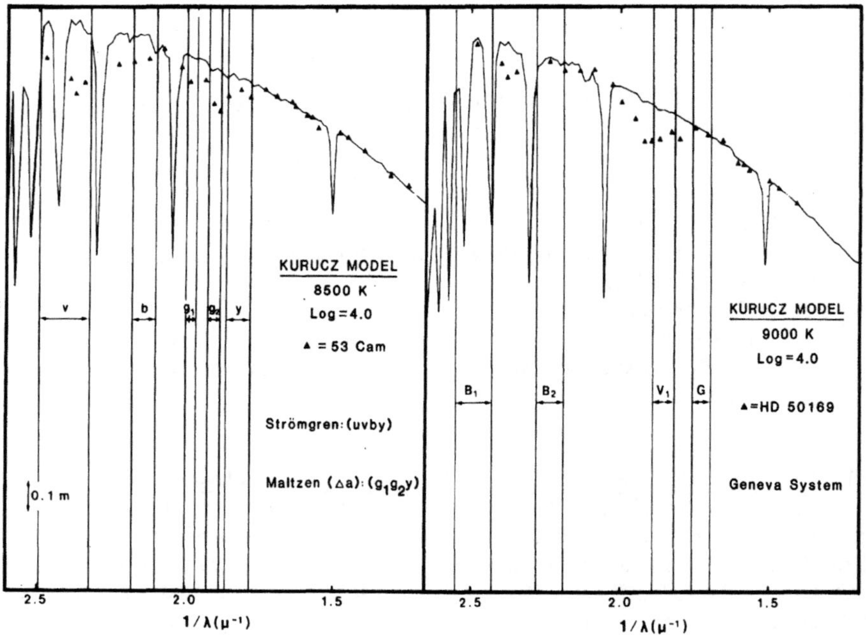

Figure 1. Photometric systems, normal star models and Ap star energy distributions in the visual spectral region.

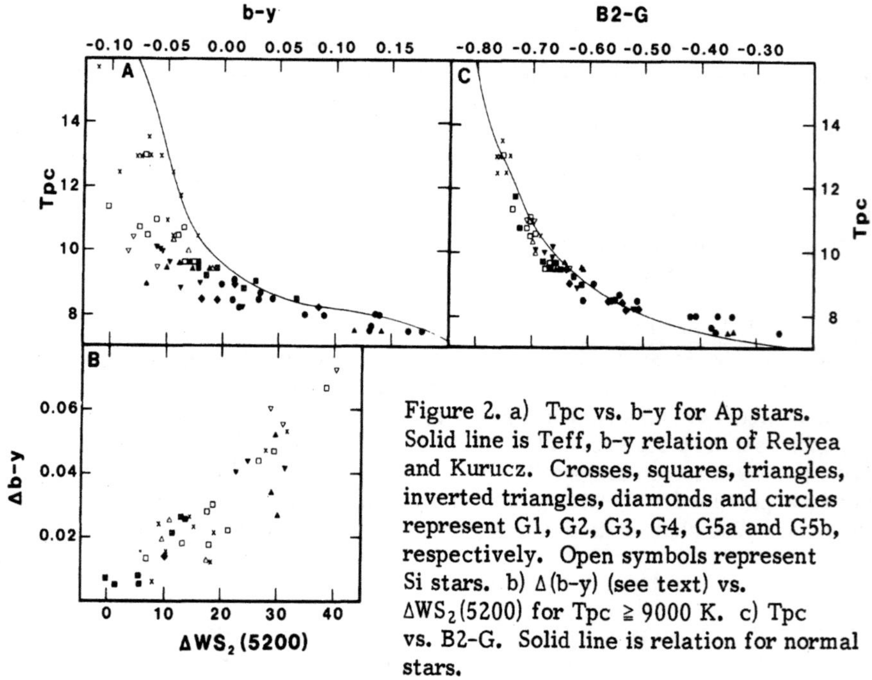

Figure 2. a) Tpc vs. b-y for Ap stars. Solid line is Teff, b-y relation of Relyea and Kurucz. Crosses, squares, triangles, inverted triangles, diamonds and circles represent G1, G2, G3, G4, G5a and G5b, respectively. Open symbols represent Si stars. b) Δ(b-y) (see text) vs. $\Delta WS_2(5200)$ for Tpc $\geq$ 9000 K. c) Tpc vs. B2-G. Solid line is relation for normal stars.

hotter Si stars of Group 1 (G1) show the very deep λ5200 core, while only the cooler SrCrEu stars of G5b show absorption minima at λ5000 and λ5556, with the central core shallower and shifted toward the red. In G3 and G4, which contain both Si and non-Si stars, the Si stars usually show more absorption at λ5360 and λ5470. In Fig. 4, (V1-G) is plotted vs. ΔWS_2(5200). The points for stars in G3, G4 and G5 in general fall below those in G1 and G2. This is because the V1 band measures only the red half of the feature, so in G3, G4 and G5, which have stronger redward absorption, Δ(V1-G) measures a stronger feature than in G1 and G2. In Fig. 5, our synthesized values of Δa (Adelman 1979 and references therein) are plotted vs. Δa', the photometric index that measures the absorption at λ5264 relative to points at λ4785 and λ5840, which are outside the feature. The plot shows the G1 stars to have the greatest slope and the G5 stars the smallest. This is due to the fact that the Δa index better detects the central core of the feature at λ5200, since the g_2 band covers most of it. However, neither the y nor the g_1 band is outside the broader shallow absorption, thus Δa values are stronger for stars in G1 and weakest for those in G5, where the central core is

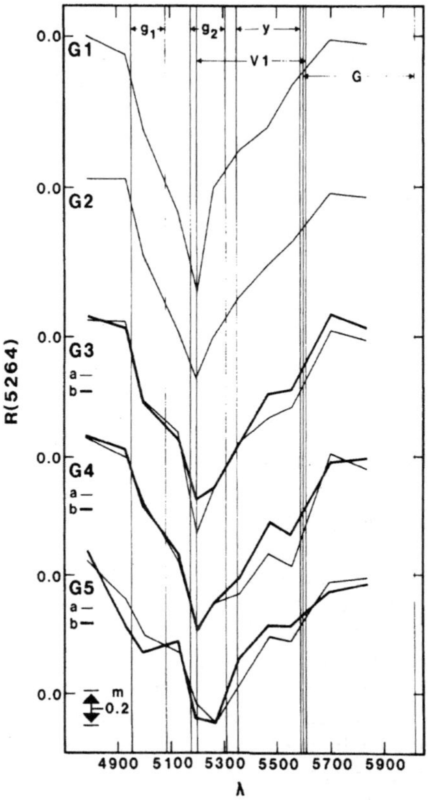

Figure 3. λ5200 profiles and Maitzen and Geneva photometric bands. R(5264) is the residual intensity normalized to λ5264.

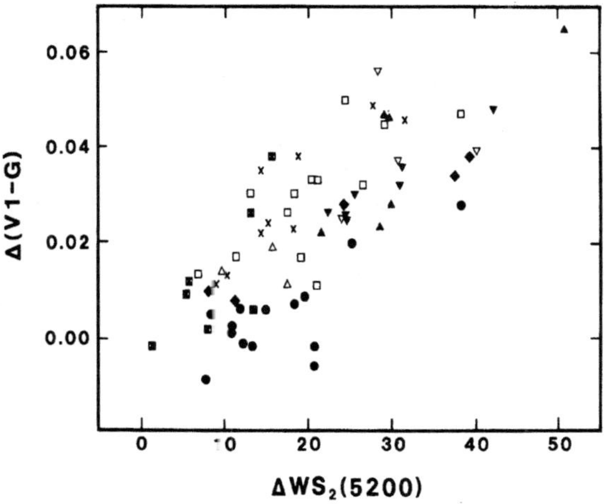

Figure 4. Δ(V1-G) vs. ΔWS_2(5200). Symbols are the same as in Fig. 2.

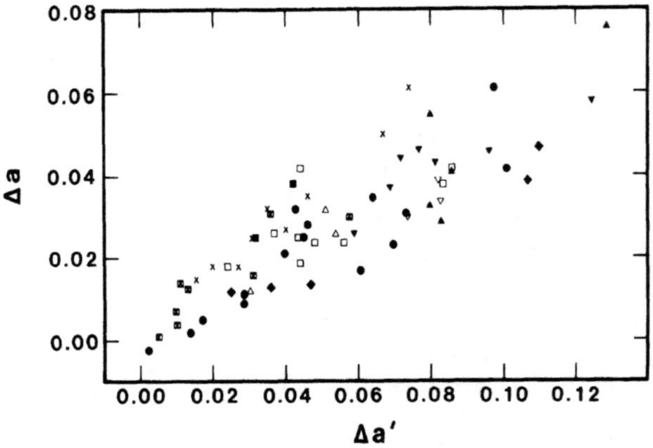

Figure 5. Synthesized Δa vs. Δa'. Symbols are the same as in Fig. 2.

weaker and absorption is stronger in both the g_1 and y bands.

CORRELATIONS

The spectrophotometric index $\Delta WS_2(5200)$ shows good correlations (r = 0.8) with the Geneva and Maitzen λ5200 indices, Δ(V1-G), Z and Δa. There is also a strong correlation (r = 0.8) between the spectrophotometric indices for the λ5200 and λ4200 broad absorption features, showing that both are strong in the same stars. A peculiarity index, PI, was calculated based on the presence of all four of the broad features, with PI = 8 if all four are definitely present. There is a good correlation between PI and $\Delta WS_2(5200)$, showing that stars peculiar in one feature tend to be peculiar in all. In addition, the cooler SrCrEu stars tend to have greater values of PI. We find that the average temperature decreases from G1 to G5, while the average PI value increases, however the correlation is not strong.

These investigations are part of a broader study which is in preparation.

REFERENCES

Adelman, S. J.: 1979, Astron. J., 84, 857
Adelman, S. J.: 1984, Astron. Astrophys. Suppl., 55, 479
Hauck, B., North, P.: 1982, Astron. Astrophys., 114, 23
Kurucz, R. L.: 1979, Astrophys. J. Suppl., 40, 1
Maitzen, H. M., Seggewiss, W.: 1980, Astron. Astrophys., 83, 328
Relyea, L. J., Kurucz, R. L.: 1978, Astrophys. J. Suppl., 37, 45

ON THE DUPLICITY OF CP3 STARS

H.Schneider
Universitäts-Sternwarte
Geismarlandstr.11
D-3400 Göttingen
F.R.G.

Abstract : Out of 111 CP3 stars 56.7 percent are binaries and further 15.3 percent are probable binaries. The distribution of the periods shows a maximum around 7 days and the deficit of periods <3 days is remarkable. Systems with periods <15 days seem to rotate synchronously, while wider binaries do not. Photometric observations confirm these, and, in addition, point to an occurance of slow pulsation.

I. Introduction

The catalogue of Hg-Mn (CP3) stars [1] contains 127 stars of which 102 are certain members of this group. The remaining stars are called 'suspected' (only one or different classification). During the last years it turned out that 16 stars from the latter subgroup are certainly CP3 stars, while 3 stars should be put back in the subgroup 'suspected', so I get 111 definite CP3 stars (literature updated until 1984), on which the following investigation will be concentrated.

II. Frequency of binaries

From the 111 stars 17 are SB2 systems (including the eclipsing binary AR Aur), 24 are SB1, and 14 have variable radial velocity or remarks about SB character in the literature. These 55 stars represent 49.5% of the sample. If I include the stars which occur in visual binary systems (7.2%) I get a value of 56.7%, which is slightly higher than the frequency of binaries among normal stars within the same spectral range.

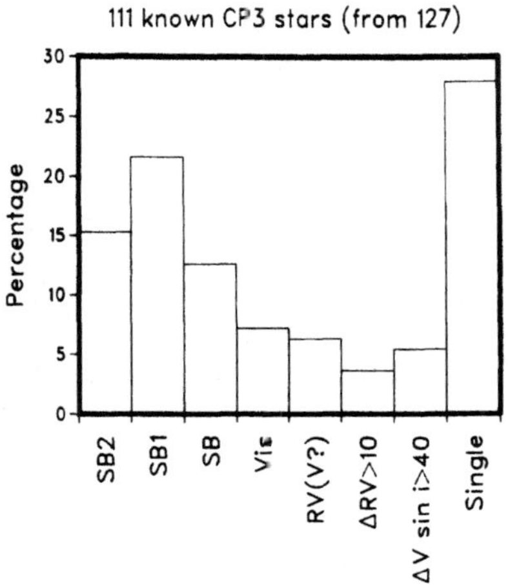

Fig. 1 : Distribution of CP3 stars of the groups as described in the text.

In general spectroscopic binaries will be detected by variable radial velocity (RV) or double-line patterns. If the double-line structure can not be separated, it results in variable projected rotational velocity (V sin i). During the compilation and updating of the catalogue many hints about probably vari-

C. R. Cowley et al (eds.), Upper Main Sequence Stars with Anomalous Abundances, 205–208.

able RV and differences in the published RV and V sin i data (more than 2 sources) were found for the single stars : 7 stars have remarks about probably variable RV, 4 show differences in the published RV data (>10Km/s), and 6 in the published V sin i data (>40Km/s). All together this yields 15.3%, and if one believes that all of these stars are binaries only 27.9% of the 111 CP3 stars remain to be single (many of these stars, especially cluster stars, do not have published RV and/or V sin i data). Figure 1 shows the distribution of the different groups.

From 8 stars the spectral type of both components are known : in 4 systems both show CP3 character, in one the other is a normal A5V star, while in the remaining systems the other component is classified as a metallic-line (CP1) star. This, the lack of a measurable magnetic field, and the high frequency of binaries points to a close relationship of the two groups.

III. Distribution of periods

Figure 2 presents the distribution of the periods. In comparison with the normal stars from Batten's [2] catalogue (B5-A0, III-V, without known CP and emission-line stars) the deficit of

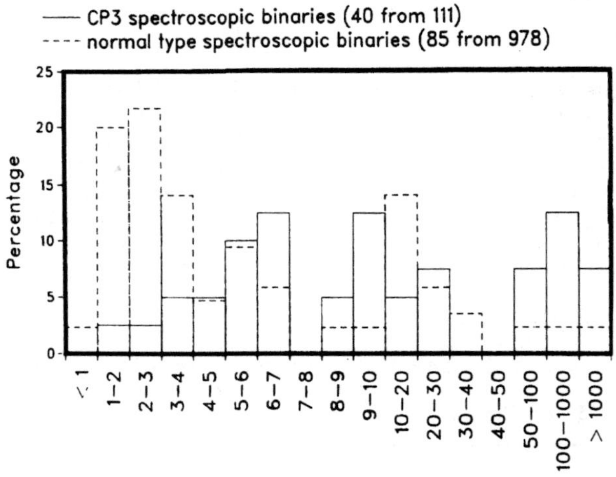

Fig. 2 : Distribution of periods [days] of the CP3 stars.

periods <3 days is remarkable (only 2 systems), while the enhancement of periods >50 days may be due to a selection effect.

If one assumes that in close systems the stars rotate synchronously and takes 90Km/s for the maximum rotational velocity (where diffusion can work), a calculation with $R=3.5R_{\odot}$ yields a period of approximately 2 days. This may explain the absence of systems with shorter periods.

IV. Rotational velocity

The distribution of the V sin i data of the systems shows a significant difference : the SB2 systems have values mainly <30Km/s while the data for the SB1 systems were more or less equally distributed. If I divide the binaries in systems with periods <15 and >15 days I find that the first group has a maximum for values <10Km/s, while the other group has one for 20-30Km/s

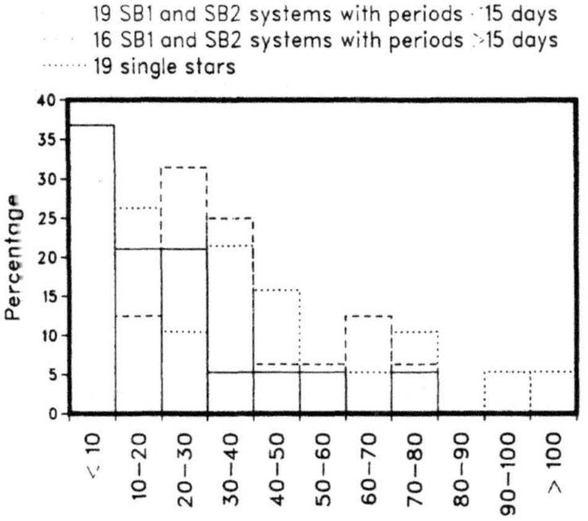

Fig. 3 : Distribution of V sin i [Km/s] of the three groups as described in the text.

(Fig. 3). Assuming random distribution for the axes I find average values for the rotational velocity of 30Km/s and 53Km/s, respectively, and for the single stars (no remarks about variable RV or V sin i) 61Km/s. Of course, one can argue that the axes are not randomly distributed, but more concentrated to smaller orbital inclinations. Checking the literature for published i data I find no evidence for the latter case : the values are more or less equally distributed with a slight peak around i=40°-50°. The low average value for the systems with periods <15 days stands in opposite to the assumption of random distribution : synchronous rotation for a star with a period of 7 days yields 37Km/s for V sin i, while from Figure 3 it is easily seen that most of the values are clearly smaller. This implies that the anomalies are more concentrated towards the pole, or that an unknown mechanism prevents synchronous rotation so that the stars really rotate more slowly.

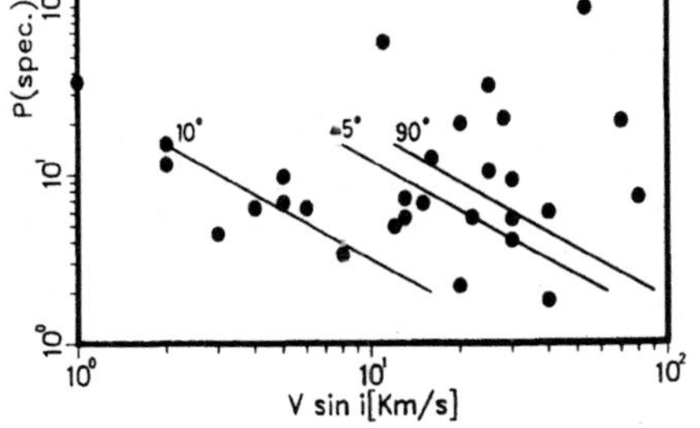

Fig.4 : Orbital periods vs. projected rotational velocity. The straight lines indicate synchronous rotation.

In Figure 4 the spectroscopic (orbital) periods are plotted versus V sin i. Assuming synchronous rotation for systems with periods <15 days (drawn in as straight lines in Figure 4, calculated with $R=3.5R_{\odot}$ and V=90Km/s) I find that the majority of these stars have orbital inclinations <45°, but some stars rotate with higher than synchronous velocity (see also chapter V).

The lack of long-period systems having smaller V sin i values may be due to a selection effect : the longer the periods the smaller the changes in RV and therewith the probability of discovery.

V. Photometric variability

About photometric variability of CP3 stars only little is known. Nevertheless, 17 stars are reported as variable, but because of small amplitudes (in the order of .01 mag.) the obtained periods are normally uncertain.

From 10 stars spectroscopic as well as photometric periods are known. Table I lists the periods.

Table I

HD	P(spec) [days]	P(phot) [days]	P(phot)/P(spec) =P(rot)/P(orb)
7374	~800	2.8	~211
145389	~560	7.8	~71
207857	~338	20.7	~16
143807	35.4	20-30(25)	~1.5
89822	11.6	11.5	~1
27295	4.4	4.4	1
129174	2.2	2.2	1
358	96.7	.96	~100
33647	21.4	.56	~38
27376	5.1	.51	~10

Comparing the spectroscopic (orbital) with the photometric (rotational) periods (Figure 5) I find the following : 4 systems with short orbital periods (<15 days) rotate synchronously, while 3 with long periods (>200 days) do not. Nevertheless the two groups do not show differences in rotational velocity. This means that duplicity can not be the reason for the low velocity.

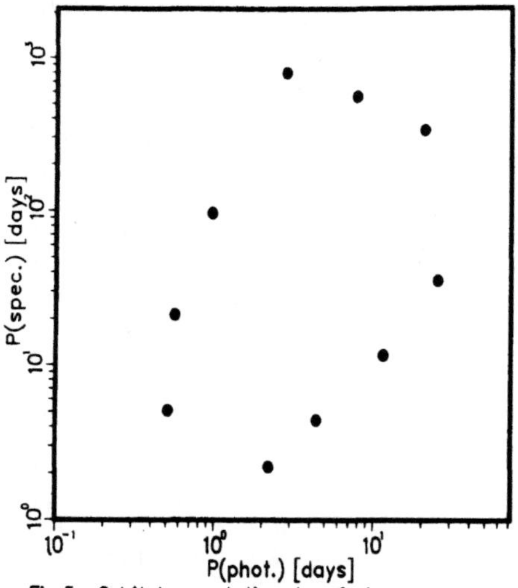

Fig.5 : Orbital vs. rotational periods.

Three systems do not fit in this picture. A straight forward estimation of their rotational velocities yields values over 150Km/s which stands in opposite to the diffusion theory. If one can trust the published periods the only explanation for this variations would be slow pulsation with periods between 12 and 24 hours decreasing with the orbital period.

References :

[1] Schneider,H. Astron.Astrophys. Suppl. **44**,137 (1981)

[2] Batten,A.H.,Fletcher,J.M.,Mann,P.J. Publ. DAO **15**,121 (1978)

INVESTIGATIONS OF CP AND NORMAL STARS WITH COADDED DOMINION ASTROPHYSICAL OBSERVATORY SPECTROGRAMS

Graham Hill
Dominion Astrophysical Observatory
National Research Council of Canada
Victoria, B. C. V8X 4M6 Canada

Saul J. Adelman[1,2,3]
Code 681, Laboratory for Astronomy and Solar Physics
NASA Goddard Space Flight Center
Greenbelt, MD 20771 USA

ABSTRACT. We describe the procedures developed at the Dominion Astrophysical Observatory to coadd high dispersion coude spectrograms and thus increase the signal-to-noise ratio in photographic data. These techniques can yield low noise spectra over a considerable wavelength range. Preliminary results of the analyses of such coadditions are shown to demonstrate the power of these procedures.

1. INTRODUCTION

Recently Cowley and Adelman (1983) examined the prospects for improved abundance analyses of CP stars and the normal B and A stars used as comparison standards. A major factor which will lead to such an improvement is the use of higher signal-to-noise data at high resolution. Much of this data is being obtained with solid state detectors, which usually are small and consequently can only obtain data over limited wavelength ranges. An alternative for non-variable stars is to add photographic spectrograms which will increase the signal-to-noise ratio by the square root of the number of plates provided there are no registration problems.

We are adding IIaO spectrograms obtained with the coude spectrograph of the 1.22-m (48-inch) telescope of the Dominion Astrophysical Observatory (DAO) (Richardson 1968). The elements of this horizontal spectrograph are mechanically isolated from one another. This extremely stable instrument has been maintained so that two spectrograms of the same non-variable star taken many years apart, exposed to the same amount of starlight, and developed using the same procedures will resemble two such spectrograms taken on successive nights.

---

[1] Visiting Observer at the Dominion Astrophysical Observatory
[2] NRC-NASA Research Associate
[3] On leave from the Department of Physics, The Citadel, Charleston, SC 29409 USA

C. R. Cowley et al (eds.), Upper Main Sequence Stars with Anomalous Abundances, 209–215.

2. DAO REDUCTION PROCEDURES

The DAO reduction programs are designed to make the reduction of photographic spectrograms as simple as possible without sacrificing accuracy and precision. The properties of each spectrograph are employed as early as possible in the reduction and scanning procedure. For a given camera and grating rotation except for minor differences, the location of clear plate, the comparison lines, the calibration steps, and the spectrum have the same relative positions on all exposures. Most of the time at the PDS microdensitometer is spent preparing to trace plates: aligning and focusing them and then entering commands for SCANN (Fisher, Morris, and Hoffman 1983), the DAO PDS spectrum scanning program. The use of lenses with a reasonably large depth of field allows one to usually obtain a good focus over the entire plate. One can attain with the reduction procedures positional accuracies comparable with those attained by standard measuring devices. Experimentally there are no systematic differences in the zero points between stellar and comparison spectrum positions.

The default values of various parameters needed to scan plates are contained on disk. Any small changes needed to properly scan a given plate are quickly made. The step calibrations are first scanned at two or more standard wavelengths. The PDS scans from standard initial reference positions to remove the difficulty of identifying the weakest step. Then cuts are made with respect to a known reference wavelength λ_o located near the center of the plate through the clear plate, through both the upper and lower comparison spectra, and through the stellar spectrum. The PDS Autolok system keeps the stage within one micron of the desired scan position at all times.

The output from SCANN is processed by the interactive spectrophotometric reduction program REDUCE (Hill, Fisher, and Poeckert 1982). Its uses VELMEAS (Hill, Ramsden, Fisher, and Morris 1982) to analyze arc spectra. A standard plate (X(mm) vs. λ(Å)) calculated with known spectrographic parameters and a list of arc wavelengths is used to predict the line positions. This method is very efficient as it uses differential measurements made with respect to known arc positions. Two cursor placements (line center and wing) for each line are made at a graphics terminal. Corrections are applied to predict the position of each new line. When the corrections have become sufficiently small, one can switch to an automatic mode. After the arc spectrum has been measured, a polynomial is fit through the data. One can accept the computed fit or adjust it manually.

The clear plate is scanned parallel to, but usually sampled less often than the stellar spectrum. One defines the number of points over which the clear values are to be averaged. The individual data and the averages are displayed on the screen. One can accept, alter, or delete any clear plate values. This clear plate file, converted to λ by the stored λ calibration from the comparison spectrum provides the zero point for calculating the densities.

The calibrations are converted to Baker densities Δ (see, e.g. Baker 1949, de Vaucouleurs 1968) which correlate almost linearly with

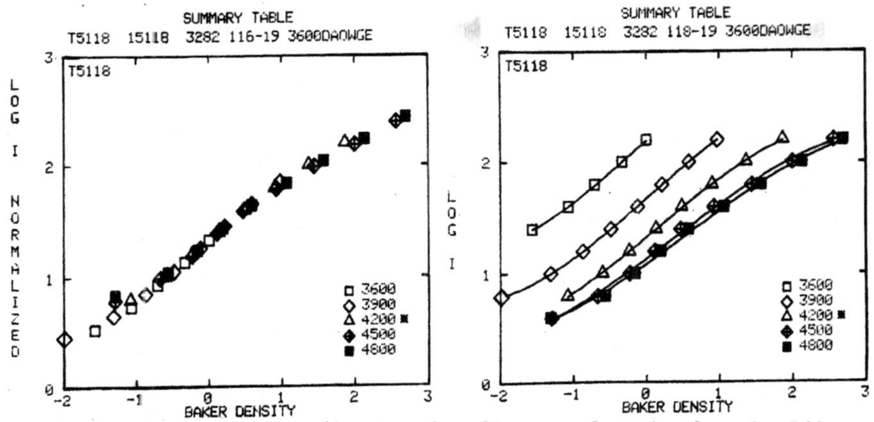

Figure 1. Log I vs. Baker density for five wavelengths for the DAO spectrogram 15118. a) shows the individual curves while b) the normalized curves.

log I. Baker density $\Delta = \log((c/t)^{\gamma}-1)$ where t and c are microphotometer measures of the transmitted light through the spectrum and clear plate, respectively, and γ which varies from emulsion to emulsion is approximately 1. The data for each calibration wavelength are displayed together with the mean values for clear plate and each calibration step (see Figure 1). Bad values can be edited as appropriate. Once a calibration has been verified the process is repeated for the next calibration. After the final calibration has been processed all calibrations are displayed together for comparison. The data is next normalized to a previously identified reference λ and displayed, a procedure which indicates the validity of adopting a single calibration for the entire wavelength region.

To manipulate stellar files, they are linearized in equal λ (or $\ln\lambda$) steps. One does not want to greatly exceed the limits provided by extreme arc positions as polynomial extrapolation can be dangerous. The PDS densities are converted to intensity by using the reduced clear plate and calibration files. Radial velocity corrections are applied for the Earth's orbital velocity. As the calibrations are possibly wavelength dependent, they are linearly interpolated, but fixed over a small λ range (often 10Å) to reduce the number of calibrations. Possible difficulties with the stellar spectrum falling below the clear plate are displayed and adjustments to the clear plate can be made before the final intensity conversion.

The stellar intensity files are next rectified so that the continuum is unity to aid the measurement of equivalent widths and the cross-correlations of individual spectra. A continuum is fitted automatically by a combining the stellar data with a file of continuum λ's and λ intervals over which the averages must be taken. Interpolation with the program INTEP (Hill 1982a), which uses Hermite splines, among the averaged data completes the rectification. INTEP draws smooth stable curves through the continuum points without the oscillations often accompanying polynomial interpolation. The adopted continuum may be verified by stepping through the spectrum and comparing the continuum with the data. This is very important for sharp-lined stars as slight radial velocity differences can shift continuum values into line cores. The continuum averaging is controlled by a factor which allows this average to be based on all or part of the data within a specified interval. There is a strong psychological aversion to placing a continuum through the middle of a noise spectrum. It should be placed near there in spectral regions with few lines and somewhat higher in richer spectra.

The program VCROSS (Hill 1982b) is used to perform cross-correlations to obtain the relative radial velocity of one stellar spectrum with respect to another selected as a standard for coaddition. Conversion of the rectified stellar files into the lnλ format is made to aid the contrast of the cross-correlation. Spectral regions containing Balmer lines are excluded from this analysis. The Fourier transform of the program star spectrum is calculated with FOURT (Brenner 1970), then the conjugate transform of the reference star data and the product of these transforms is evaluated, which when appropriately normalized (see Simkin 1976) is the cross-correlation function. The peak of the c.c.f. is measured on a graphics terminal by delimiting a region and fitting a parabolic, a Gaussian, or a Lorentzian function to the data. The advantage of cross-correlation over more conventional techniques to measure relative radial velocities is that most of the stellar spectrum is used rather than just a small part.

The program TSTACK shifts the individual stellar rectified files to a common radial velocity by correcting for their relative radial velocities, interpolates the values to a grid of wavelength spacings, and then adds them. Then REDUCE is used to extract the profiles of strong lines, in particular Hγ and the He I lines, for comparison with the predictions of model atmospheres. The program VLINE (Hill, Fisher, and Poeckert 1982), which is now also a subroutine of REDUCE, and a faster version NEWVLINE were used to fit Gaussian, Lorentzian, and rotational profiles, as appropriate, through the stellar data (Figure 2). These programs are based on the curve-fitting routine CURFIT (Bevington 1969). Up to 12 profiles and a linear continuum of edited pieces of spectrum can be fit simultaneously to yield the line positions and widths and equivalent widths of the coadded data. Examination of the FWHM of the fits can reveal possible line blending. The reality of very weak features can be judged by their FWHM and line depths.

The following approximate numbers provide some feeling for the speed at which reductions proceed. With practice one can scan between 8 and 12 spectra on 2" x 8" plates in an hour on the PDS. By reducing arcs, calibrations, and clear plates, once can produce between 4 and 6 linearized intensity spectra per hour. Rectification of each spectra takes between 10 minutes and an hour depending on the complexity of the spectra and the wavelength region covered. Cross-correlation of ten rectified spectra takes less than 15 minutes (and can be done automatically). The coaddition process takes about 2 minutes. In flat regions, one can measure of order 100 features per hour: equivalent widths, positions, and v sini. In regions of Balmer lines,

Gaus N 38

lam	err	EW(mA)	d	FWHM	vsini	Prof
4460.935	0.014	2.9	0.015	0.18	12	Gaus
4461.425	0.007	3.2	0.020	0.15	10	Gaus
4461.709	0.003	8.8	0.051	0.16	11	Gaus
4462.080	0.002	15.0	0.084	0.17	11	Gaus
4463.023	0.001	24.8	0.136	0.17	12	Gaus

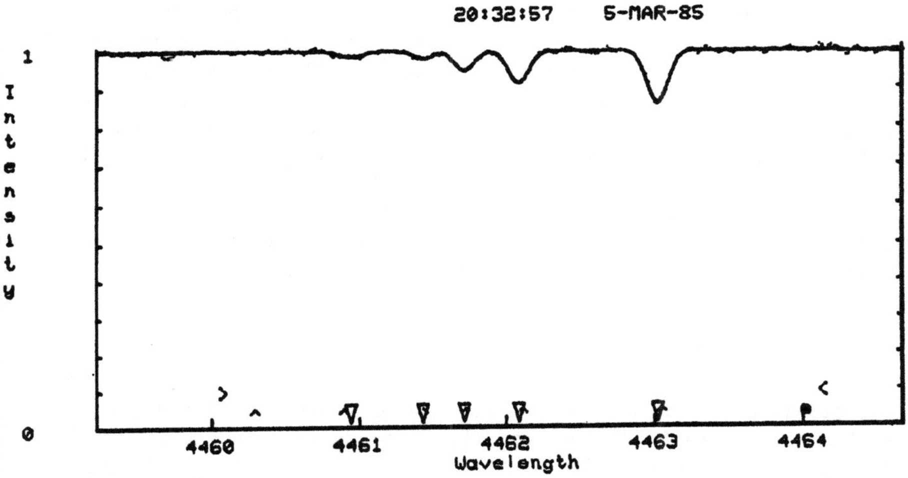

Figure 2. NEWVLINE measurements of the κ Cnc coadded spectra. The wavelengths, equivalent widths (in mÅ), the line depths (d), the FWHM (full width half maxima), and twice the v sini values are given.

the rate can be much slower. Obtaining values of the continua for the Hγ and He I line profiles takes 10 to 30 minutes. Many of these operations could be done faster if the graphics terminal worked faster (the current limitation at DAO is 9600 baud). To reduce the two halves of 2.4 Å/mm spectra of normal A stars takes about 3 to 4 days while for sharp-lined HgMn stars about twice as long is needed.

3. THE ANALYSIS PROGRAM

The combination of at least 6 similar spectrograms of the same star is an initial condition for our study. G. C. L. Aikman and C. R. Cowley have made available to us their DAO spectrograms of CP and normal stars. Our initial program stars, which are usually quite sharp-lined, for which we now have at least 6 2.4 IIaO Å/mm spectrograms include the normal stars τ Her, α Dra, and θ Leo and the HgMn stars μ Lep, ν Cnc, κ Cnc, ι CrB, υ Her, φ Her, and 28 Her. These stars include some previously analyzed by SJA using primarily Mt. Wilson Observatory 4.3 Å/mm IIaO spectrograms. This overlap will allow comparison of techniques and spectrographs as well as permit an initial comparison with 16 consistently analyzed sharp-lined B and A stars (Adelman 1985). As the DAO material for the stars in common consists of a larger number of spectrograms taken at higher dispersion and as the widths of the spectrograms are similar, we anticipate that the DAO material will have a higher signal-to-noise ratio than the Mt. Wilson material.

This material will be used to derive elemental abundances over a wide segment of the periodic table. Where possible resonance and/or low excitation line results will be compared with those of higher excitation lines. Searches will be made for evidence of atmospheric stratification and/or departures from LTE especially in the CP stars. Ultimately some of these derived abundances should provide critical tests of the radiative diffusion models for the production of the abundance anomalies. Ultraviolet analyses of the same stars should both complement and supplement this study (i.e. that of θ Leo, ι CrB, and κ Cnc by SJA and D. S. Leckrone with coadded IUE high dispersion images and planned Hubble Space Telescope observations by Leckrone). In addition, the relative radial velocities of the spectrograms, needed to coadd the spectrograms of a given star, will yield information on its binarity, furthering studies such as those of Aikman (1976). We also plan a stellar atlas as part of our line identification studies.

4. INITIAL RESULTS

Coadditions of 2.4 Å/mm IIaO spectrograms have been completed for the normal A star α Dra (6 plates) and the HgMn stars ι CrB and κ Cnc (10 plates each). The metal line profiles of α Dra are best fit in REDUCE with a rotational profile and those of ι CrB and κ Cnc by Gaussian profiles. Corrections for the instrumental profile (Fletcher et al. 1980) results in 12 km s^{-1} for α Dra, 3 km s^{-1} for ι CrB, and 5 km s^{-1} for κ Cnc. These values compare favorably with those in the literature: Slettebak et al. (1975) 15 km s^{-1} for α Dra, Wolff and Preston (1978) <3 and 6 km s^{-1} for ι CrB and κ Cnc, respectively, and Guthrie (1984) 6 km s^{-1} for κ Cnc.

Comparison of the Hγ profiles derived from the coadditions and spectrophotometry (Adelman 1978, Adelman and Pyper 1979) with the predictions of Kurucz's (1979) solar composition model atmospheres yield the following results:

α Dra T_{eff} = 10000 K, log g = 3.3
ι CrB T_{eff} = 10950 K, log g = 3.7
κ Cnc T_{eff} = 13250 K, log g = 3.5.

Corrections for a small amount of scattered light, 3-4%, (Fletcher, private communication) will improve the fit of the model predictions, which now match the line shoulders and wings.

For κ Cnc, the equivalent widths are systematically smaller than those of Aller (1970) in accord with both Heacox (1979) and Guthrie (1984). Compared with Aller (1970) who used similar dispersion spectrograms, many more lines, especially weak lines are seen, and new atomic species have been identified, e.g., S II, V II, and Y II.

We measured four lines on the ten rectified spectra of ι CrB. There is excellent agreement between the mean and coadded values. The formal 1σ (rms) values indicate errors of about 4 mÅ on single spectrograms, part of which is simply differences in the continuum placement.

λ	W_λ(mÅ) coadd	individual spectra
4029.395	18.0	18.0 ± 3.7
4030.077	7.9	8.2 ± 3.6
4030.203	6.7	7.1 ± 3.0
4030.472	25.5	25.4 ± 3.7

Differences in equivalent widths by making slightly different choices of continuum placement with REDUCE are typically about $1/2$ mÅ in flat spectral regions in the coadded spectra. This suggests that very weak lines can be measured with some confidence.

ACKNOWLEDGMENTS. We appreciate the assistance of Wesley A. Fisher for his support of this research. This research has been supported in part by a grant from The Citadel Development Foundation to SJA.

REFERENCES

Adelman, S. J. 1978, Astrophys. J. 222, 547
Adelman, S. J. 1985, in preparation
Adelman, S. J., and Pyper, D. M. 1979, Astr. J. 84, 1603
Aikman, G. C. L. 1976, Pub. Dom. Astrophys. Obs. 14, 379
Aller, M. F. 1970, Astr. Astrophys. 6, 67
Baker, A. E. 1949, Pub. R. Obs. Edinburgh 1, 15
Bevington, P. R. 1969, Data Reduction and Error Analysis for the Physical Sciences (McGraw-Hill: New York)
Brenner, N. M. 1970, "FOURT"; available COSMIC/SHARE, Univ. of Georgia, Barrow Hall, Athens, Georgia
Cowley, C. R., and Adelman, S. J. 1983, Quar. J. Roy. Astr. Soc. 24, 393
de Vaucouleurs, G. 1968, Appl. Optics 7, 1513
Fletcher, J. M., Harmer, C. F. W., and Harmer, D. L. 1980, Pub. Dom. Astrophys. Obs. 15, 405
Fisher, W., Morris, S., and Hoffman, W. 1983, Pub. Dom. Astrophys. Obs. 16, 131
Guthrie, B. N. G. 1984, Mon. Not. Roy. Astr. Soc. 206, 85
Heacock, W. D. 1979, Astrophys. J. Suppl. 41, 675
Hill, G. 1982a, Pub. Dom. Astrophys. Obs. 16, 67
Hill, G. 1982b, Pub. Dom. Astrophys. Obs. 16, 59
Hill, G., Fisher, W. A., and Poeckert, R. 1982, Pub. Dom. Astrophys. Obs. 16, 43
Hill, G., Ramsden, D., Fisher, W. A., and Morris, S. C. 1982, Pub. Dom. Astrophys. Obs. 16, 11
Kurucz, R. L. 1979, Astrophys. J. Suppl. 40, 1
Richardson, E. H. 1968, J. Roy. Astr. Soc. Canada 62, 313
Simkin, S. M. 1974, Astr. Astrophys. 31, 129
Slettebak, A., Collins, G. W., Boyce, P. B., White, N. M., and Parkinson, T. C. 1975, Astrophys. J. Suppl. 29, 137
Wolff, S. C., and Preston, G. W. 1978, Astrophys. J. Suppl. 37, 371

DISCUSSION (Hill and Adelman)

WEHLAU: Have you made any studies of systematic photographic errors which might persist from plate to plate, such as at the edges of sharp features?

ADELMAN: I have intercompared very limited regions of the spectra which I have coadded, to look for obvious problems. I intend to make additional checks for photographic problems which might lead to systematic errors. In particular, I plan to obtain parts of these spectra at the 1.2-m telescope at DAO with their Reticon to check the coaddition process. So far, I have not found any problems.

MÉGESSIER: I have a technical question concerning the PDS microdensitometer work. According to my experience, the most time-consuming problem is the focusing of the plate. I can not adjust and trace eight plates in one hour!

ADELMAN: The PDS microdensitometer at DAO has lenses with a relatively large depth of field. This, plus some tape strategically placed around the edges of the plate, usually allows one to obtain good focus over the entire plate. I have had problems similar to yours at Kitt Peak, and I was able to get good focus over only 4 inches [10 cm] of spectrogram.

DWORETSKY: I had the same problem (out-of-focus plates) when I used the PDS at the Royal Greenwich Observatory some years ago. The designers have not allowed for the fact that the photographic emulsion on a developed plate shrinks a bit, so that the plate is slightly curved. The PDS machine has no stage clips for holding the plate down. Our solution was to construct a weighted ring with clips (cut from a broken clock's main spring) to hold the plate down. This was an inexpensive way of solving the problem.

ADELMAN: It is true that plates do curl, and we even had a few plates where for some reason the emulsion shrinkage produced a twist, which made life very difficult.

C. R. Cowley et al (eds.), Upper Main Sequence Stars with Anomalous Abundances, 217.

RAPID OSCILLATIONS OF CP2-STARS

Werner W. Weiss
Institute for Astronomy
Tuerkenschanzstrasse 17
A - 1180 Wien
Austria

1. INTRODUCTION

With the announcement in 1978 (Kurtz, 1978) of short periodic photometric variations of HD 101065, the first member of a subgroup of CP2 stars was found. This gained attention due to the possibility of applying the methods of stellar seismology to late A to early F type stars. However, the extremely low photometric amplitude of these stars, which is typically few millimagnitudes in B, and the short periods, ranging from few minutes to about 15 minutes, make these stars difficult to discover and a considerable amount of telescope time is required to accumulate sufficient data for a reliable analysis of the frequency spectrum. As a consequence aliasing imposes serious problems. Synchronous observations from observatories well separated in longitude could overcome this dilemma, and have indeed proven to be sucessful (Kurtz and Seeman, 1983; Kurtz and Balona, 1984; Kurtz, Schneider and Weiss, 1985; Kurtz and Kreidl, 1985).

Presently, only speculations exist concerning the excitation mechanism for the oscillations. Insights in the stellar structure and in the magnetic field characteristics of the stellar interior are posssible from modelling rapidly oscillating CP2 stars.

Two remarks may be allowed in this context. First, as it is evident from the publications, it is possible even for small telescopes in the 0.5 m to 1 m class to contribute significantly to new astrophysical aspects. Second, as the history of the detection of rapidly oscillating CP2 stars demonstrates, one always should be open to the unexpected (and a word to the observers: do not trust

---

Based in part on observations collected at the European Southern Observatory, La Silla, Chile, with the financial support of the Austrian "Fonds zur Foerderung der wissenschaftlichen Forschung", project No. 4170.

C. R. Cowley et al (eds.), Upper Main Sequence Stars with Anomalous Abundances, 219–231.

theoretical results too much). In the case of HD 101065, indications for rapid photometric variations were already found in 1975 by Heck and Manfroid (Heck et al., 1976), but not followed up by the authors. For HD 24712, Weiss (1980) found some evidence for a variability in the order of 1 hour, but these observations were planned to detect periods expected for radial modes. These modes were the only ones considered possible for this group of stars by theoreticians, who, in any case, thought that main sequence stars in this temperature range would be dynamically stable. The spurious observations can be explained as interference of the pulsation and data taking frequency.

2. THE OBLIQUE PULSATOR MODEL

Not only did Don Kurtz discover the group of rapidly oscillating CP2 stars, he also provided the first and hitherto most successful model to explain the complex frequency pattern (Kurtz, 1982). According to this model, non-radial p-modes of low degree (i.e. a small l) are aligned to the magnetic field in such a way that the pulsation and magnetic field axes coincide. Generally, the magnetic field axis is inclined to the rotation axis, which gives rise to an amplitude-modulation of the pulsation light curve.

Some difficulties immediately arise for the theoretical understanding, among which are the questions why only very few modes (compared to the sun, e.g.) are excited and how the precession of the standing wave pattern about the axis of rotation can be suppressed. Dolez and Gough (1982) deal with this problem and they present arguments, based on a very simple model, that nonradial standing modes grow preferentially with a particular orientation to the magnetic field and decay after precession has destroyed the initial alignment. Their calculations give evidence that the required growth (decay) time of aligned (misaligned) modes is indeed short relative to the rate of precession, namely on a timescale of only few hours, depending only on the relative orientation to the magnetic field.

The most elaborated discussion of the oblique pulsator model is published by Dziembowski and Goode (1985). These authors criticize the model of Dolez and Gough (1982, op.cit.) because it does not account for the long-term coherence of the oscillation frequencies observed for most of the rapidly oscillating CP2 stars (exceptions might be HD 60435 and HD 201601). In their opinion, Dolez and Gough, as well as Mathys (1985, see later in this review), introduce unnecessary complications. Dziembowski and Goode (1985, op.cit.) treat the problem of a non-axisymmetric perturbation of the pulsation due to a rotating oblique magnetic field in a self-consistent way. They argue that magnetic field effects may dominate those of rotation.

$$\omega^{rot}/\,\omega_0 \cong m\cdot C(l,m)\cdot\Omega/\omega_0 \cong 10^{-2}\cdot\Omega/\omega_0 \cong 10^{-5} \div 10^{-6} \qquad (1)$$

with ω_0 being the unperturbed oscillation frequency, ω^{rot} the frequency splitting due to advection, C(l,m) the rotational splitting constant depending on the internal structure of the star and Ω its rotation frequency. Because these rotation effects will be quite small for oscillating CP2 stars the dominance of magnetic effects even can hold true for a fairly small ratio of magnetic pressure, p^{mag}, to the gas pressure, p, (Dziembowski and Goode, 1984) and thus for fairly weak magnetic fields.

$$\langle \omega^{mag}/\omega_0 \rangle \cong \langle p^{mag}/p \rangle \qquad (2)$$

Dziembowski and Goode's model reduces to Kurtz's simple oblique pulsator, if the effect of magnetism fully dominates that of rotation. An increasing ratio of rotational to magnetic frequency splitting results in an increasing contamination from other spherical harmonics and thus in a reduced degree of symmetry in this mode relative to the magnetic field axis and finally in an increasing deexcitation. The measure for the importance of rotation is the (observable) amplitude ratio of the prograde and retrograde frequencies, which have equal amplitudes in Kurtz's model. With the assumption of an approximately rigid rotation the authors even can estimate the strength of the magnetic field in the stellar interior.

Based on Kurtz's (1982) frequencies for HD 24712 an impressive numerical example is given by Dziembowski and Goode to illustrate the general properties of the model. However, the significance of the results largely depends on a correct identification of the pulsation modes.

As already mentioned, Mathys (1985, op.cit.) investigated in his so-called spotted pulsator model another possibility to explain some features which are typical of the frequency spectra observed for rapidly oscillating CP2 stars. He assumes aligned rotation and pulsation axes and explains the amplitude modulation of the oscillations due to inhomogeneities of the surface with a cylindrical symmetry about the magnetic-field axis. To account for the magnitude of the observed effects, Mathys also has to assume an inhomogeneous distribution of the ratio of the flux to radius variations and of the phase lag between these parameters. The frequency spacing in his model consequently is exactly equal to the stellar rotation frequency, which is indeed observed to a high degree of significance. The unequal amplitudes of the outer frequency peaks, observed for example in HD 83368, are a natural consequence of the model. One obviously can criticize that the number of introduced free parameters, most of which are presently unmeasurable, is too large to allow a conclusive test of the spotted pulsator model.

While Dziembowski and Goode have demonstrated that the heuristically established oblique pulsator model can be consistently derived and generalized, Shibahashi and Saio (1985) calculate frequencies for different realistic stellar models in analogy to the solar oscillations. They neglect, however, the effect of rotation and the magnetic field on the stellar equilibrium and atmospheric structure. In particular the authors study the influence of the T-τ relation, of an eventual helium depletion due to diffusion, and of the initial chemical composition on the eigenfrequencies for equilibrium models in the range of 1.3 to 2.4 solar masses.

It turns out that the ratio of the critical frequency above which no standing wave is stable, f(crit), and the characteristic frequency,

$$f_o \equiv 1/(2\cdot\int_0^R (1/c)dr) \cong (GM/R^3)^{1/2} \qquad (3)$$

is quite sensitive to T_{eff}/T_{surf} and thus to the particular peculiarities of CP2 stars. For frequencies higher than the critical frequency the surface becomes transparent and wavepatterns cannot be maintained stable. Shibahashi and Saio show that

- o $f(crit)/f_o$ is almost independent of the stellar mass and
- o only very weakly dependent on the stellar evolution.
- o $f(crit) - f_o$ keeps the homologous relation to the standard models over a large variety of chemically inhomogeneous envelopes and for models with different initial chemical composition.

With the relation $f_o/(GM/R^3)^{1/2} \cong 0.20$, derived by Gabriel et al. (1984) for main sequence models in the range of 1.5 to 2 solar masses, it is possible to estimate the radial order of the highest overtone, n(crit), to be about 30. It is worth mentioning that the ratio of 0.2 is very similar to the solar value of 0.216 which illustrates that many oscillation parameters scale indeed homologuous for a large range of different stellar models.

Shibahashi and Saio also discuss the problem of the rotational splitting of frequencies and calculate the coefficient C(n,l). This coefficient decreases with increasing n and is about 0.001 for l = 1 and n = 40, which is comparable with the upper limit determined by Kurtz and Seeman (1983, op.cit.) for HD 24712. A consequence of this small value is a lifetime of the wavepattern of more than 1000 days for a rotation period of the order of few days. This lifetime would be considerably longer than the growth and decay time for oscillations excited by the kappa-mechanism (Dolez and Gough, 1982, op.cit.) as well as by magnetic overstability. Shibahashi (1983) and Cox (1984) calculated the characteristic timescale for the latter case to be of the order of weeks or months.

3. THE RIDDLE OF THE EXCITATION MECHANISM

Overstable magnetic convection can be understood as convective instability in a thin superadiabatic zone close to the stellar surface in the presence of a moderate magnetic field. A rising mass-element will be hotter and less dense than the environment and thus tend to continue to move upward. Magnetic lines of force which are "frozen" in such a plasma tend to inhibit this upward motion and are responsible for the restoring force. The heat exchange in a superadiabatic layer results in a motion of the mass element which is faster on the way back to the origin than as it was for the upward direction. In an appropriate environment the oscillation amplitude will grow steadily, resulting in so-called overstable oscillations.

Another mechanism is related to the resonant excitation of large-scale non-radial oscillations by turbulent convection and is considered by Dolginov and Muslimov (1984), but not explicitly applied to CP2 stars.

Dolez and Gough (1982, op.cit.) investigate the effectiveness of the kappa-mechanism for driving non-radial oscillations. Their attempt to model a typical CP2 star is based on piecing together segments of two spherical symmetric stellar models which represent the magnetic poles and the magnetic equator, respectively. Growth and decay of modes is primarily determined by the outer stellar layers where the eigenfunctions can be taken in first order to be insensitive to l. Based on the argument that in these layers all low-degree p-modes with similar frequencies look similar, Gough and Dolez restrict their calculations to radial oscillations and assume that the calculated growth and decay times are also valid for the non-radial modes under investigation. Linear nonadiabatic radial pulsations were computed by the authors taking convective perturbation of the heat flux into account. They find a sharp maximum for the growth rate of n = 15, corresponding to a period of about 11 min, and conclude that some modes might be therefore self-exciting. The excitation of only very few modes in pulsating CP2 stars would find a simple explanation, if the most rapidly growing mode suppresses the other unstable modes.

Presently, nothing decisive can be said about the excitation mechanism working in real CP2 stars. The mechanism of magnetic overstability is related to magnetic inhibition of convection in sunspots (Bierman, 1941), however, it would mean accepting a new pulsation mechanism in an area of the HR-Diagram where already another mechnism (kappa-mechanism) is effective.

4. THE LIST OF RAPIDLY OSCILLATING CP2 STARS

Currently, eleven rapidly oscillating CP2 stars are known. In Table 1 a list of these stars is given together with references concerning the main parameters. It is obvious that for most of the

HD HR	m (V) (28,29)	H_{eff} in gauss	Puls. period (min) amplitude (mmag)	Rotation period (days)	References
6532	8.5	-	6.922 to 6.956 0.55 to 1.01	-	1
24712 1217	6.0	+300 to +1200 (18)	5.966 to 6.361 0.44 to 2.13	12.46	2, 3, 4, 17
60435	8.9	-	3.994 to 15.141 0.5 to 6	-	5
83368 3831	6.2	-700 to +700 (19)	5.819 to 11.705 0.18 to 2.14	2.85 (1.426)	4, 17
101065	8.0	-2200 (20)	6.070 to 12.140 0.26 to 5.40	-	6, 7, 8, 9
128898 5463	3.2	-300 (var) (21,22,23)	6.825 to 6.832 0.38 to 1.91	1.004 to 12	10, 16, 17
134214	7.5	-	5.650 3.23	-	14
137949	6.7	+1400 to +1800 (24)	8.272 1.39	7.194 to 23.26	4, 17
201601 8097	4.7	+500 to -800 (25,26,27)	12.448 0.86	312 to 72^{yr}	11, 15, 17
203932	8.8	-	5.942 0.66	-	12
217522	7.5	-	13.716 2	-	13

Table 1: List of rapidly pulsating CP2 stars. Whenever several pulsation periods were known, the smallest and largest period is given in minutes and the smallest and largest pulsation amplitude is given in units of 1/1000 of a magnitude (mmag). If more than one rotation period is given, several periods within the interval presented in Table 1 are discussed in the literature.

stars many parameters which are relevant for modeling the pulsation modes are unknown and still much has to be done by the observers.

The cross references for table 1 are the following: (1) Kurtz, Kreidl 1895; (2) Kurtz, Seeman 1983; (3) Kurtz, Schneider, Weiss 1985; (4) Kurtz 1982; (5) Matthews, Kurtz, Wehlau 1985; (6) Kurtz 1981; (7) Kurtz 1980; (8) Kurtz, Wegner 1979; (9) Weiss, Kreidl 1980; (10) Kurtz, Balona 1984; (11) Kurtz 1983a; (12) Kurtz 1984; (13) Kurtz 1983b; (14) Kreidl 1984a; (15) Weiss 1983; (16) Weiss, Schneider 1984; (17) Catalano, Renson 1984; (18) Preston 1972; (19) Thompson 1983; (20) Wolff, Hagen 1976; (21) Wood, Campusano 1975; (22) Borra, Landstreet 1975; (23) Borra, Landstreet 1980; (24) Wolff 1975; (25) Babcock 1958; (26) Bonsack, Pilachowski 1974; (27) Scholz 1979; (28) Vogt, Foundez 1979; (29) Blanco et al. 1970.

The references listed above contain to the best of our knowledge all publications related to observations of rapidly pulsating CP2 stars. In another review presented by D. Kurtz (1985) the observational data are nicely summarized and the reader is referred to it. In the following section we will try to pin down some observational and theoretical problems for which a solution seems to be crucial for a further improvement of our understanding of the phenomenon of rapidly pulsating CP2 stars.

5. SOME OF THE PROBLEMS

5.1 The photometric mode identification:

Following a recipe which seems to work quite successfully for Delta Scuti stars and Cepheids (Balona and Stobie, 1979), the first attempts of a mode identification were based on a discussion of the phase shift between light and color variations. Balona (1981) included gravity effects and has shown their importance for high frequency oscillations. Stamford and Watson (1981) included in addition pressure effects. Within the uncertainties in the assumptions, the phase lag between light and radial velocity variations yield reasonable mode identifications for HD 128898 (Weiss and Schneider, 1984), if neither gravity nor pressure effects were taken into account. However, including the gravity effect results in a phase shift of more than 100° for $l = 0$, 1 or 2, instead of 13° to 20° which were observed (Kurtz and Balona, 1984). On the other hand, the gravity effect is required to account for the observed phase shift of 63° for HD 83368 and for $l = 1$. In the case of HD 101065 with an observed phase shift of 39° (Kurtz, 1980) a value of 16° would be predicted for $l = 2$ including the gravity effect. This difference, however, is larger than the observational uncertainties in the phase shift.

Further evidence for a more complicated situation, hitherto unpublished, results from an observing run at ESO, La Silla. In December 1983, H. Schneider used the Walraven photometer attached to the Dutch 90cm telescope and observed HD 24712 and HD 83368. A full description of the instrument and reduction technique can be found in Weiss and Schneider (1984, op. cit.). We intended to use the wavelength dependence of the phase shifts relative to the Walraven-V channel to investigate whether or not certain frequencies originate from the same mode (Figure 1).

Within the errors for the phase determination, which are of the order of 10° to 15°, we find no systematic differences for the wavelength dependence of the phases for HD 24712. The frequency numbering corresponds to Kurtz (1982). A comparison with a similar figure obtained by Weiss and Schneider for HD 128898 yields a clearly different trend with wavelengths. The phase shifts for HD 83368 are again different. Basically, all five frequencies show a similar wavelength trend, except for the U-channel which contains the Balmer jump. For this wavelength region the low frequency triplet (f1 to f3) behaves differently to the high frequency triplet of which only f4 and f5 could be reasonably well detected in our data. This difference can be interpreted as a possible indication that the latter triplet is not just an overtone of the low frequency triplet.

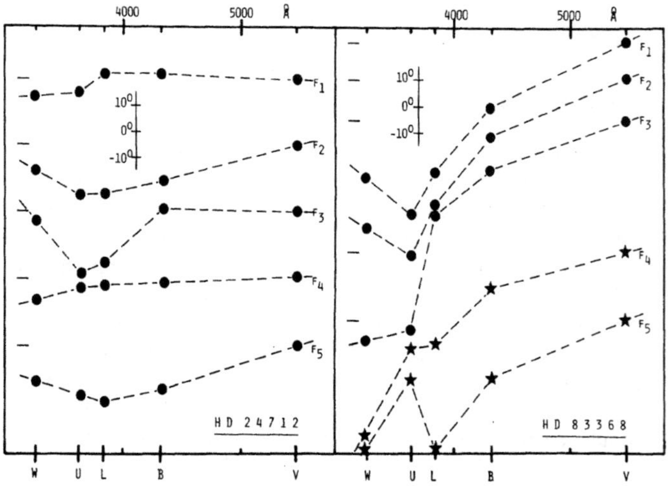

Figure 1: Phase shifts relative to Walraven - V.

Furthermore, we found that the wavelength characteristics determined for HD 128898 in May 1984 with the 4-channel Stroemgren photometer attached to the 50cm Danish telescope at La Silla, ESO, looked markedly different to what we have published earlier (Weiss and Schneider, 1984, op.cit.). Presently, we are unable to explain these observations conclusively and to use phase shifts for a mode identification based on the linearized theory.

5.2 The spectroscopic mode identification:

The negative results for a mode identification with photometric techniques motivated Schneider and Weiss (1986) to try spectral line profile variations for this purpose. They observed HD 128898 and HD 201601 with the ESO Coude-Echelle spectrometer at the 1.5m Coude-Auxiliary telescope and the Reticon detector in Chile. In Figure 2 an example is given of HD 128898 for the Ca I line profile at 6471.8 Å.

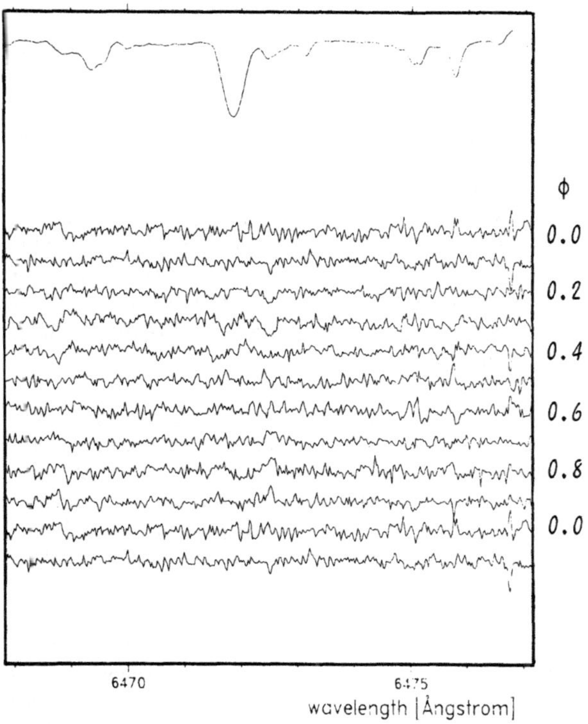

Figure 2: HD 128898 mean Ca I (6471.8 Å) lineprofile and residuals for pulsation phases 0.0 to 1.1 .

A total of 306 individual spectra were coadded according to the photometrically determined pulsation phase, binned in intervals of 0.1 in phase. The upper trace in Figure 2 shows the mean of all spectra, and below the residuals relative to this mean spectrum are plotted which were calculated for individual phase bins. The contemporaneously determined photometric amplitude was 3 mmag in Stromgren-u. No profile variations larger than 0.5% are present. A similar negative result can be reported for HD 201601. The latter star was also observed spectroscopically by Kreidl and Odell (1985) with the same negative result.

Synthetic line profiles for non radially pulsating CP2 stars are discussed by Odell and Kreidl (1984) and by Baade and Weiss (1986, abstract in this volume).

In any case, a correct mode identification is essential for a reliable discussion of the pulsation frequency spectrum and hence for a determination of several astrophysical parameters.

5.3 Temperature range for rapidly oscillating CP2 stars:

The hot and cool limits for rapidly oscillating CP2 stars are an important parameter for isolating a possible excitation mechanism. A coincidence of these limits with the instability strip known from Delta Scuti stars would lend some support to the kappa-mechanism. Superadiabaticity, on the other hand, decreases for the hotter CP2 stars and hence will be unfavorable for magnetic overstability (Shibahashi, 1983, op.cit.). In this context, the detection of a rapid variability in the slightly metal-deficient (!!!) F3Vp star HD 119288 (Matthews and Wehlau, 1985) is extremely interesting. Attempts to measure a possible magnetic field for this star are very much encouraged. The lack of any measureable field in this star would obscure things considerably.

Realizing the importance of observationally well defined temperature limits for rapidly oscillating CP2 stars, several surveys have regularly included also hotter stars. Such surveys were performed by Don Kurtz mainly at South Africa, by H. Schneider and W. Weiss mainly at ESO, La Silla, Chile, and at the Mauna Kea Observatory, Hawaii, by Tobias J. Kreidl at Lowell Observatoru, and by Jaymie M. Matthews and William H. Wehlau in Chile and Canada). However, none of these surveys verified short periodic variations reported by other observers (e.g. for 21 Com or HD 224801), nor did they detect such variations in hotter stars. These negative results do not neccessarily mean that hotter CP2 stars are pulsationally stable, as pulsation might be a transient phenomenon, and the amount of observing time might not have been sufficient. If hotter stars pulsate by some yet unknown reasons in systematically higher orders, then the already small photometric amplitude would be even smaller. These are only some of the most obvious explanations for the negative results.

5.4 $T - \tau$ relation and the frequency spectrum:

Shibahashi and Saio show that the critical frequency strongly depends on the $T - \tau$ relation. The most hitherto consistent and complete investigation of the atmosphere of CP stars is published by Muthsam (1979a,b). His models S76 and S77 give a ratio for $T_{eff}/T(\tau = 2.10^{-5})$ of 1.5 which is the value for the model series B in Shibahashi and Saio (1985, op.cit.). However, as these authors write, an even larger ratio of about 2.5 would be required to explain the observed frequencies in HD 24712 which are clearly higher than the critical frequency. Muthsam's model atmosphere, on the other hand, does not take magnetic pressure effects into account (nor do Shibahashi and Saio) and the model might be therefore still inadequate, despite its complexity. In addition, various theoretical models do not correctly reproduce the observed frequency-spacings in the cases of HD 24712 and HD 60435.

CP2 stars possibly are another example, like our sun, of contradicting results derived by astroseismology and by classical model calculations.

5.5 Long time stability of the frequency spectrum:

HD 60435 is characterized by the most complex frequency spectrum, including the highest and lowest frequencies yet observed for rapidly oscillating CP2 stars (Matthews et al, 1985, op.cit.). This star seems to provide a fairly strong case for transient oscillations, since well determined frequencies in one night can not be detected in another night. Considerably more data are necessary to prove this claim and for a reliable mode identification. As in HD 83368, resonant coupling of modes seems to be effective. For HD 60453 Matthews et al. speculate that an overtone of a very weak radial mode might excite the oscillations near 1.1 mHz. HD 201601 might be another example for transient oscillations. Although a rotation period of many years is discussed in the literature the pulsation amplitude changes much more rapidly. However, excellent photometric conditions are required to detect the oscillation which rarely exceeds the 1 mmag limit (Kurtz, 1983a; Weiss, 1983).

Dappen and Perdang (1985) investigate the theoretical implications for a non-linear, non-radial adiabatic mode coupling and raise the question for conditions under which chaotic oscillations become effective.

5.6 The group of low-harmonic pulsating CP2 stars:

It would be obvious to discuss in the context of this review the properties of the group of low-harmonic pulsating CP2 stars. Unfortunately, for most of the proposed members of this group contradicting observational evidence are found in the literature. A critical discussion of the relevant references clearly would be beyond the scope of this paper. We refer to a review presented at the Workshop on Rapid Variability of Early-Type Stars (Weiss, 1983) held at Hvar and to the references given herein. Furthermore, as discussed during this meeting, a survey program recently was finished by a group of astronomers headed by W. Schoeneich and the results are currently prepared for publication. Another group, coordinated by N. Polosukhina, investigated selected CP2 stars like 53 Cam (Burnashev et al., 1983), and they announced additional publications. T. Kreidl (1984b to 1984d) is still continuing his promising survey at Lowell Observatory. Polosukhina and Weiss currently are preparing a hopefully complete bibliography of publications related to low-harmonic pulsating CP2 stars.

Acknowledgements: I am grateful for financial support from the Max Kade Foundation, New York. Important input was received at the Institute for Astronomy, University of Hawaii, where I collaborated with Walter Bonsack and Jay Gaugain. It is also a pleasure to thank H. Schneider for cooperation over the past several years, as well as T. Kreidl, A. Odell, N. Polosukhina and C. Cowley for many interesting discussions and improvements of this review.

REFERENCES

Baade, D., Weiss, W.W.: 1986, Astron.Astrophys.Suppl.Ser. submitted
Babcock, H.W.: 1958, Astrophys.J.Suppl. 3, 141
Balona, L.: 1981, Mon.Not.Roy.Astr.Soc. 196, 159
Balona, L., Stobie, R.S.: 1979, Mon.Not.Roy.Astr.Soc. 189, 649
Bierman, L.: 1941, Vierteljahrsschr.Astron.Ges. 76, 194
Blanco, V.M., Demers, S., Douglass, G.G., Fitzgerald, M.P.: 1970, Publ. U.S. Naval Obs. 11,1
Bonsack, W.K., Pilachowski, C.A.: 1974, Astrophys.J. 190, 327
Borra, E.F., Landstreet, J.D.: 1975, Publ.Astron.Soc.Pacific 87, 961
Borra, E.F., Landstreet, J.D.: 1980, Astrophys.J.Suppl. 42, 421
Catalano, F.A., Renson, P.: 1984, Astron.Astrophys.Suppl.Ser. 55, 371
Cox, J.P.: 1984, Astrophys. J. 280, 220
Dappen, W., Perdang,J.: 1985, Astron.Astrophys. 151, 174
Dolez, N., Gough, D.O.: 1982, in "Pulsations in Classical and Cataclysmic Variables", J.P. Cox and C.J. Hansen eds., JILA, Boulder, p. 248
Dolginov, A.Z., Muslimov, A.G.: 1984, Astrophys.Space.Science 98, 15
Dziembowski, W., Goode, P.R.: 1984, Mem.Soc.Astron.Ital. 55, 185
Dziembowski, W., Goode, P.R.: 1985, preprint
Gabriel, M., Noels, A., Scuflaire, R., Mathys, G.: 1985, Astron. Astrophys. 143, 206
Heck, A., Manfroid, J., Renson, P.: 1976, Astron.Astrophys.Suppl. 25, 143
Kreidl, T.J.: 1984a, Inform.Bull.Var.Stars No. 2460
Kreidl, T.J.: 1984b, Inform.Bull.Var.Stars No. 2472
Kreidl, T.J.: 1984c, Inform.Bull.Var.Stars No. 2602
Kreidl, T.J.: 1984d, Inform.Bull.Var.Stars No. 2607
Kreidl, T.J.: 1985, Inform.Bull.Var.Stars No. 2739
Kreidl, T.J., Odell, A.P.: 1985, private communication
Kurtz, D.W.: 1978, Inf.Bull.Var.Stars No.1436
Kurtz, D.W.: 1980, Mon.Not.Roy.Astr.Soc. 191, 115
Kurtz, D.W.: 1981, Mon.Not.Roy.Astr.Soc. 196, 61
Kurtz, D.W.: 1982, Mon.Not.Roy.Astr.Soc. 200, 807

Kurtz, D.W.: 1983a, Mon.Not.Roy.Astr.Soc. 202, 1
Kurtz, D.W.: 1983b, Mon.Not.Roy.Astr.Soc. 205, 3
Kurtz, D.W.: 1984, Mon.Not.Roy.Astr.Soc. 209, 841
Kurtz, D.W.: 1985, in Proceedings NATO-Workshop "Seismology of the Sun and Distant Stars", D. Gough ed., Cambridge
Kurtz, D.W., Balona, L.A.: 1984, Mon.Not.Roy.Astr.Soc. 210, 779
Kurtz, D.W., Kreidl, T.J.: 1985, Mon.Not.Roy.Astr.Soc. in press
Kurtz, D.W., Schneider, H., Weiss, W.W.: 1985, Mon.Not.Roy.Astr.Soc. 215, 77
Kurtz, D.W., Seeman, J.: 1983, Mon.Not.Roy.Astr.Soc. 205, 11
Kurtz, D.W., Wegner, G.: 1979, Astrophys.J. 232, 510
Mathys, G.: 1985, Astron.Astrophys. 151, 315
Matthews, J., Kurtz, D.W., Wehlau, W.H.: 1985, Astrophys.J., in press
Matthews, J., Wehlau, W.H.: 1985, Inform.Bull.Var.Stars No. 2725
Muthsam, H.: 1979a, Astron.Astrophys. 73, 159
Muthsam, H.: 1979b, Astron.Astrophys.Suppl.Ser. 35, 107
Odell, A.P., Kreidl, T.J.: 1984, in Proceedings 25th Liege Internatl. Astrophys. Coll. "Theoretical Problems in Stellar Stability and Oscillations", pg. 148
Preston, G.W.: 1972, Astrophys.J. 175, 465
Schneider, H., Weiss, W.W.: 1986, in preparation
Shibahashi, H.: 1983, Astrophys.J.Lett. 275, L9
Shibahashi, H., Saio, H.: 1985, Publ.Astron.Soc. Japan, in press
Stamford, P.A., Watson, R.D.: 1981, Astrophys.Space Sci. 77, 131
Thompson, I.B.: 1983, Mon.Not.Roy.Astron.Soc. 205, 43p.
Vogt, N., Foundez, M.: 1979, Astron.Astrophys.Suppl.Ser. 36, 477
Weiss, W.W.: 1983a, Inform.Bull.Var.Stars No. 2384
Weiss, W.W.: 1983b, Hvar Obs. Bull. 7, 263
Weiss, W.W., Kreidl, T.J.: 1980, Astron.Astrophys. 81, 59
Weiss, W.W., Schneider, H.: 1984, Astron.Astrophys. 135, 148
Wolff, S.C.: 1975, Astrophys.J. 202, 127
Wolff, S.C., Hagen, W.: 1976, Publ.Astron.Soc.Pacific 88, 119
Wood, H.J., Campusano, L.B.: 1975, Astron.Astrophys. 45, 303

DISCUSSION (Weiss)

STĘPIEŃ: I would simply like to stress that there is a tremendous amount of information hidden in the observations of short-period oscillations, and we have to be aware of this. It is a real breakthrough in our knowledge of Ap stars. The frequencies, amplitudes and phases of the oscillations are sensitive to the internal structure of a star, so we can hope to gain some information about this structure. The influence of the magnetic field on these oscillations is proportional to

$$\int_0^R (1/c)(P_{mag}/P_{gas})dr$$

where c is the velocity of sound and the integration is over the stellar radius. Layers near the surface enter with a much larger weight. This means that we may also hope to learn something about the internal structure of the magnetic field.

WEISS: I agree completely with what you say. The complete frequency analysis is very important, and it helps to have the cooperation of at least two observatories separated widely in longitude so that continuous time intervals can be observed.

GERTH: Did you take into account the structure of the window function, which is connected with the temporal distribution of the observations?

WEISS: The spectral window was extremely clean since we were working with a continuous set of data taken at nearly constant time intervals. Of course, it is extremely important for frequency and power spectra discussions to include the spectral window to be sure there are no alias frequencies. In the case of α Circinus and the other stars this was done properly.

GERTH: Could you also describe the significance criterion used in order to decide whether or not a peak in the power spectrum was real?

WEISS: For the photometry, the criterion was the repeatability of the frequency peaks in different data sets, especially in data sets taken with different filters, by different observers, or with different telescopes. Kurtz assumes his line frequency analysis to be complete if the multi-frequency-fit residuals show only white noise.

C. R. Cowley et al (eds.), Upper Main Sequence Stars with Anomalous Abundances, 233.

COMPUTED SPECTRAL LINE VARIATIONS FOR OBLIQUE NONRADIAL PULSATORS

Dietrich Baade (1) and Werner W. Weiss (2)
(1) The Space Telescope European Coordinating Facility,
European Southern Observatory, Karl-Schwarzschildstr. 2,
D-8046 Garching, Fed. Rep. Germany
(2) Institute for Astronomy, Türkenschanzstrasse 17,
A-1180 Wien, Austria

ABSTRACT. Spectral line profiles are computed for nonradially pulsating CP2 stars. For a range which currently is thought to be typical for these stars, the influence of six parameters on the line profiles is considered: mode order ℓ and degree m, pulsation velocity amplitude, the angle between the rotation and pulsation axis, the angle between the rotation axis and the line-of-sight, and the phase angle of the rotation. In view of the expected low signal-to-noise ratio of observational data it is investigated to what extent easily measurable, simple quantities can still be useful in discriminating between different modes.

(The full text will be submitted for publication in Astronomy and Astrophysics, Supplement Series.)

C. R. Cowley et al (eds.), Upper Main Sequence Stars with Anomalous Abundances, 234.

Short-Periodic Radial Velocity Variations of the B9p Star ET And

E. Gerth

Zentralinstitut für Astrophysik
der Akademie der Wissenschaften der DDR,

DDR-1500 Potsdam
Telegrafenberg

ABSTRACT. Spectroscopic observations of the B9p star ET And secured at the Bulgarian National Observatory Roshen in the years 1981-1984, consisting of 97 plates, show clearly a radial velocity period of 0.198 d - with a ratio to the photometrical period of exactly 2 : 1. This behaviour would hint at a close binary system; but there arise difficulties in explaining it by this way because of the extreme short period, so that pulsation must be taken into account. However, besides of the confident period of 0.198 d, there is evidence for shorter periods in the region of 45 min, which may be attributed only to pulsational processes.

Since 1981 the peculiar B9 star ET And (HD 219749) was taken under cooperative investigation of the Bulgarian National Observatory Roshen and the Central Institute of Astrophysics of the GDR. The detection of a short-periodic photometric variation of about 0.1 d for this star by Panov /1/ suggested to carry out simultaneous photometric and spectroscopic observations. In the years 1981 to 1984 97 spectrograms with the reciprocal dispersion 9.2 Åmm^{-1} resp. 17.5 Åmm^{-1} were obtained, based on the observations of several consecutive nights each year. Thus the search for periods was restricted to time scales of the order of one day. Longer periods for this star are known, too. After Ouhrabka and Grygar /2/ ET And is a binary with an orbital period of 48.304 d and the eccentricity of $\varepsilon = 0.50$. Hildebrandt and Hempelmann /3/ determined from photometric observations a period of 1.61883 d attributing it to the rotation of the primary star. The period of 0.1 d found by Panov obviously has neither any rational relation to the orbital period nor to the adopted rotational period.

Using all measurable Balmer and Si II lines we determined from radial velocity measurements of the '81 plates a period of 0.198 d /4/, which corresponds exactly to the double of the photometric period, namely 0.099 d. Of course, this statement depends on the accuracy and the

C. R. Cowley et al (eds.), Upper Main Sequence Stars with Anomalous Abundances, 235–238.

significance of measurement and evaluation. Therefore, to strengthen the conclusions, some remarks on the period search programme should be given.

The search for periods was performed by a method according to that of Deeming /5/ but completed by the application of weights for all measuring values. A significance criterion is given by the power quotient, which means the ratio between the power of a wave of any frequency to the square standard deviation, i. d. σ_f/σ. Independently, in order to decide whether or not a conspicuous peak in the power spectrum may belong to a real period, the convolution-shift theorem was used, in consequence of which the special pattern of the spectral window function in the frequency space of the Fourier transform is shifted to every frequency contained in the unequally spaced data set. In the case of our data with observational windows in three consecutive nights the power spectrum exhibits a complicated structure, being ambiguous whether the highest peak corresponds to the true period (Fig. 1a).

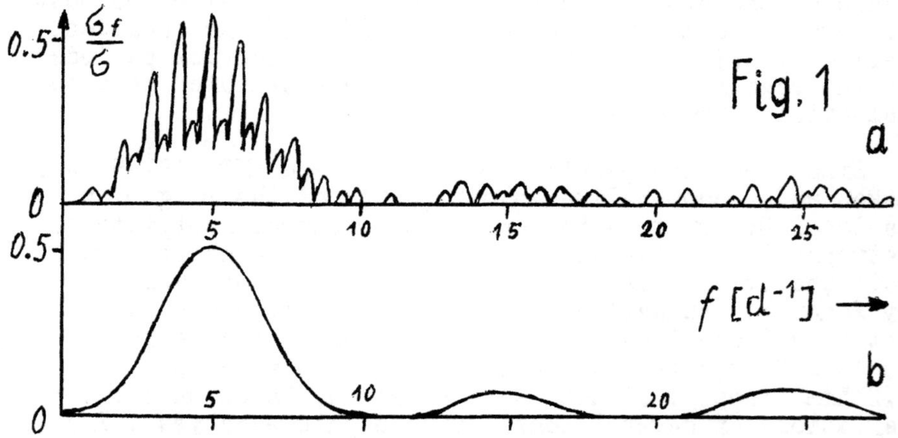

The ambiguity is eliminated by deriving the product of the power spectra of all nights. Such a procedure corresponds either to the multiplication rule of superposed probabilities or rather to the cross correlation of the data in the original space (Fig. 1b). The same procedure was performed for all results from 1981 to 1984, yielding a consistent power spectrum with a prominent peak at 0.198 d (Fig. 1c). The Fourier transform of the data of the 97 plates altogether gives the best fit of the phase curve for the period 0.198155 d (Fig. 2).

Hence we conclude that the radial velocity of ET And presumably continues to vary with a period of 0.198 d over the time of three years with an amplitude of 3.2 kms^{-1}. In Fig. 3a the double period 0.396 d is present, too, requiring further investigation. Besides of these

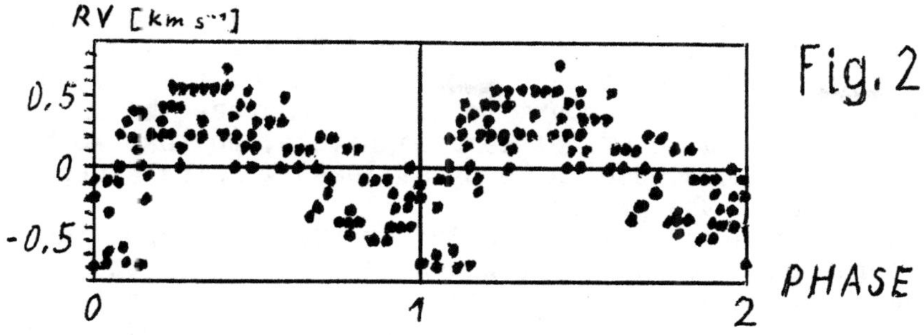

Fig. 2

periods another peak at the period 0.032 d and an amplitude of 1.9 kms⁻¹ is indicated. This feature is especially pronounced if we take into account the measuring results of the ´84 plates alone (Fig. 3a), but it does not contradict to all the other measurements. Subtracting the waves with these two periods, 0.198 d and 0.396 d, the significance criterion by the power quotient σ_f/σ for the period 0.032 d advances from 0.16 to 0.49 (Fig. 3b). Further support is given by comparing the power spectrum with the spectral window function (Fig. 3c) and regarding the site of the peak located before the limit set by the Nyquist frequency. Thus, based on the given data set, the existence of the period 0.032 d is secured from the mathematical point of view, but we cannot exclude completely the influence of measuring inaccuracy and artifacts.

Fig. 3

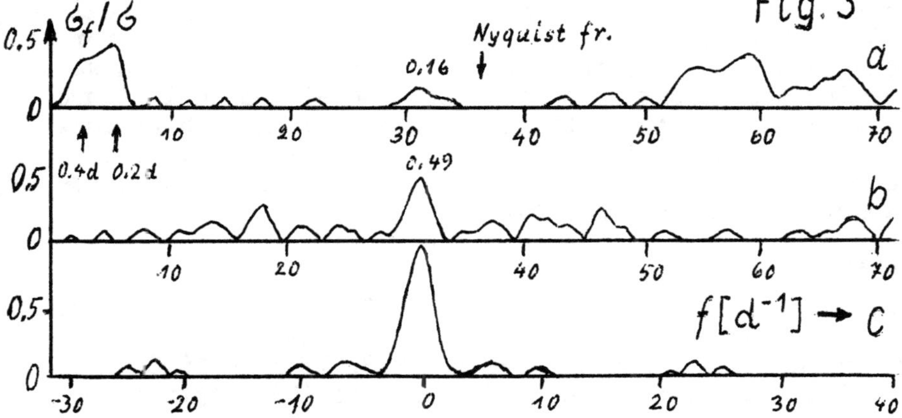

It was the intention of this report to present mainly observational results. Of ourse, we asked for the possible causes of the observed facts, too. The striking fact that the ratio of the periods of the radial velocity to that of the brightness is exactly 2 : 1 seems to hint at the existence of a close binary system, which was discussed already in /4, 6/. Even the phase shift between the waves of the radial velocity and the brightness secured by the simultaneous observations would not contradict. But there arise other serious difficulties. First of all the radius of a main sequence star with a adopted mass of 3.3 $M_\odot$ would amount to 2.2 $R_\odot$, whereas the orbital radius would be of the same magnitude, so that we had to conclude the companion being at least grazing along the surface of the primary.

Moreover, we have to assume a larger stellar radius than it corresponds to the ZAMS, which is asserted by the value of log g = 3.6 taken from the H_γ profile and further observations (line strengths etc.) indicating a location of ET And in the region of the luminosity classes II to III of the HR diagram (G. Scholz, paper in preparation). Therefore, approaching to the region of the instability strip of the HRD, this star can be expected to perform pulsations.

This view is supported by the minor period of 0.032 d resp. 46 min, too. Further we mention the photometric observations of Panov /7/ yielding periods of 9.3, 7.5, and 5.8 min.

The very short-periodic variations obviously are caused alone by the property of the primary as an evolved B9p star. But which kind of pulsation can we expect regarding the 0.2-day period? - The frequency ratio 2 : 1 suggests that nonradial pulsations are taking place which may be excited by external forces, for example by the changing distance of the secondary in the orbital motion producing surface oscillations at the eigenfrequency by the tidal action. The tangential component of the interaction forces induces the oscillations to run around the star as a wave forming the momentary surface into an ellipsoidal figure showing two times the bright side during one run around. - But such conclusions are based on a temporarily too poor observational material. Therefore, further observations of ET And are needed.

References:

1. Panov, K. P.: 1978, III Sci. Conf. on Magnetic Stars, Praha, Czech. Acad. Sci., Astron. Inst., p. 19
2. Ouhrabka, M., and J. Grygar: 1979, IBVS No 1600
3. Hildebrandt, G., and A. Hempelmann: 1981, Astron. Nachr. 302, 155
4. Gerth, E., G. Scholz, and K. P. Panov: 1984, Astron. Nachr. 305, 75
5. Deeming, T. J.: 1975, Astrophys. Space Sci. 36, 137
6. Gerth, E., G. Scholz, and K. P. Panov: 1984, VI Sci. Conf. on Magnetic Stars, Riga, Astr. Counc. SSSR, p.68
7. Panov, K. P.: 1984, VI Sci. Conf. on Magnetic Stars, Riga, Astr. Counc. SSSR, p. 77

FREQUENCY ANALYSIS OF THE RAPIDLY OSCILLATING Ap STAR HD 60435

Jaymie M. Matthews,+ Donald W. Kurtz,* and William H. Wehlau+

+Department of Astronomy, University of Western Ontario.
*Department of Astronomy, University of Cape Town.

ABSTRACT. The cool Ap star HD 60435 was monitored in a programme of rapid B photometry during 18 nights in January/February 1984, from two stations widely spaced in longitude (the University of Toronto 0.6-m telescope at the Carnegie Southern Observatory (CARSO) on Las Campanas, Chile, and the 0.5-m telescope of the South African Astronomical Observatory (SAAO)). On six of those nights, contiguous light curves from both sites were obtained.

Fourier analysis of these data confirms the rapid variability first reported by Kurtz (1984) and reveals several additional transient oscillations. HD 60435 exhibits persistent - but modulated - oscillations at a frequency near 1.4 mHz (period = 11.9 minutes), and short-lived oscillations at frequencies near 1.1 and 4.2 mHz (periods of 15.2 and 4.0 minutes, respectively). These latter two periods represent the longest and shortest yet observed in the class of rapidly oscillating Ap stars.

We have applied the oblique pulsator model (Kurtz 1982) to the fine-scale splittings detected in the frequency spectra of the 1.4 and 1.1 mHz oscillations. Also, the series of frequencies close to 1.4 mHz which fall into a pattern of roughly equal spacing is compared to such spacings predicted for overtones in pulsating main-sequence A stars (Shibahashi and Saio 1984). Both approaches suggest that HD 60435 is undergoing non-radial pulsations of odd *and* even degree (probably with $\ell \lesssim 3$).

The oblique pulsator interpretation of the splittings in the frequency spectrum and the amplitude modulation of the 1.4 mHz oscillations also predict a rotation period of approximately eight days for this star. Mean photometry of HD 60435, collected by the authors, supports a similar value of 7.7 days for the period.

Analysis of the oscillations is hampered by ambiguities due to daily aliases present in the data, and by the complicated structure and time-dependence of the frequency spectrum. Further observations of HD 60435 are essential if we are to fully understand its rapid variations.

REFERENCES.

Kurtz, D.W. 1982, M.N.R.A.S., 200, 807.
———— 1984, M.N.R.A.S., 209, 841.
Shibahashi, H. and Saio, H. 1984, submitted to *Publ.Astron.Soc.Japan*.

To be published in *The Astrophysical Journal*.

C. R. Cowley et al (eds.), Upper Main Sequence Stars with Anomalous Abundances, 239–242.

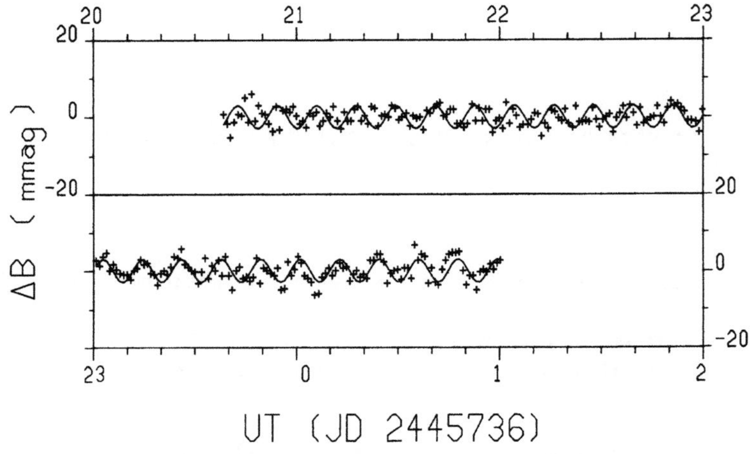

FIGURE 1. Light curve of rapid B photometry of HD 60435 from the night of JD 2445735. The crosses represent three-point averages of 20-second integrations. No comparison star was used. A sinusoid with a period of 11.8 minutes has been superimposed on the data.

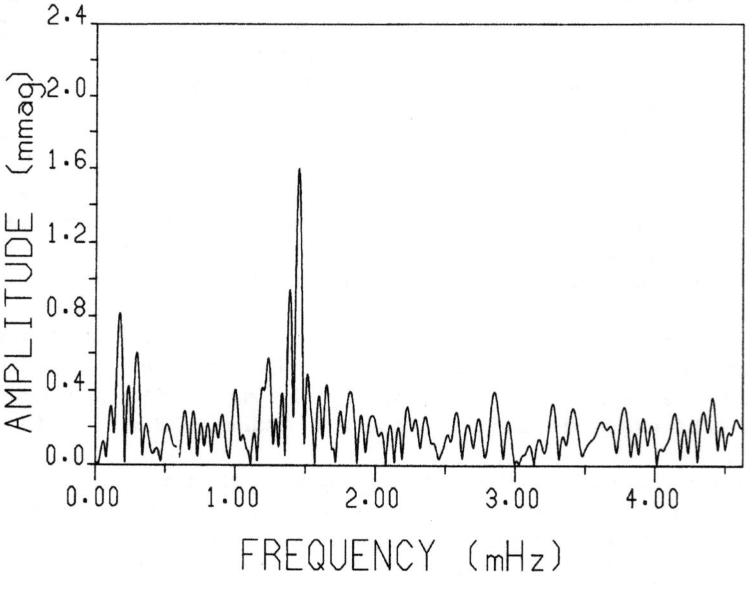

FIGURE 2. Amplitude spectrum of the light curve shown in Figure 1. The largest peak occurs at a frequency near 1.4 mHz. A second smaller peak is resolved just shortward in frequency. This hints at the more complicated frequency structure revealed by later detailed analysis.

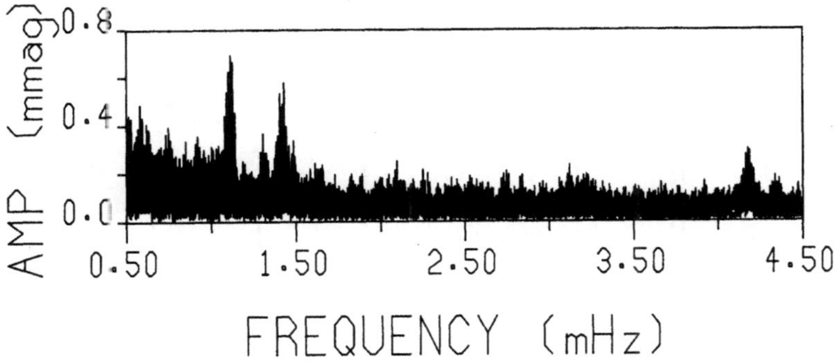

FIGURE 3. Amplitude spectrum of the entire data set, spanning JD 2445719-37, and including six nights of contiguous data from both observing sites. Additional oscillations are revealed at frequencies near 1.1 and 4.2 mHz (15.2- and 4.0-minute periods). An oscillation with a frequency near 2.8 mHz (period = 6 min) - not visible here - was observed on a few nights by DWK. We propose that this and the 4.2 mHz frequency may be resonances with the 1.2 mHz variation.

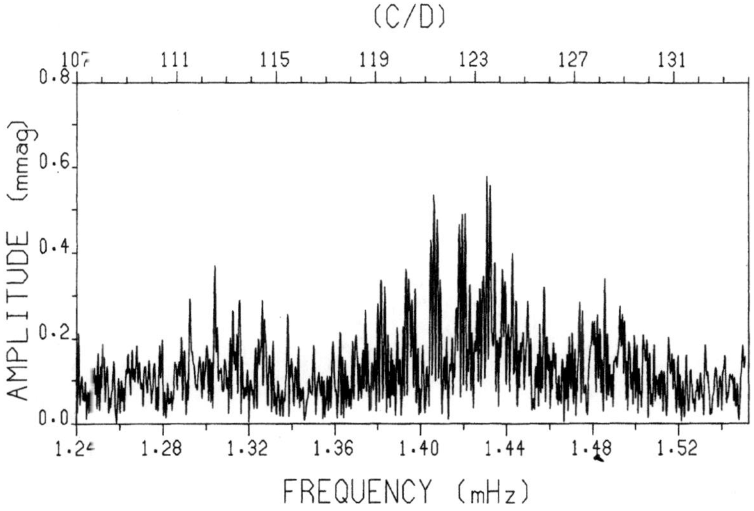

FIGURE 4. Amplitude spectrum in the frequency region around 1.4 mHz. Several peaks and their associated one-day aliases are apparent. The fine-splitting seen in the peaks between 1.40 and 1.44 mHz corresponds to a frequency spacing of about 1.4 ± 0.2 μHz (period = 8.3 ± 1.5 days). The oblique pulsator model predicts that an $\ell = 1$ ($\ell = 2$) oscillation will show up in the Fourier spectrum as a frequency triplet (quintuplet) with a spacing corresponding to the star's rotation frequency. This spectrum is therefore consistent with a rotation period for the star of approximately eight days.

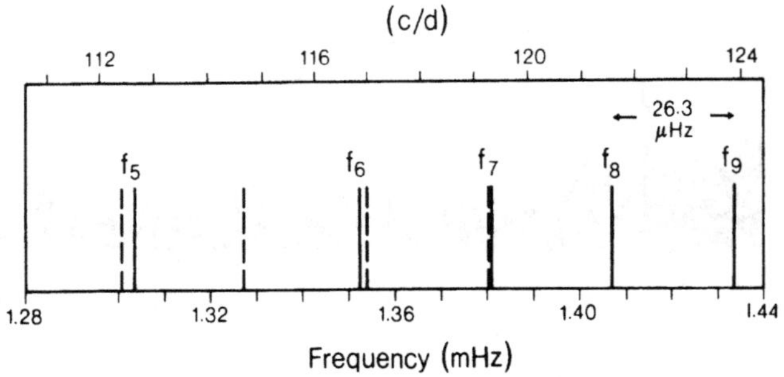

FIGURE 5. Schematic representation of the frequencies observed between 1.30 and 1.44 mHz in HD 60435. Frequencies f_7-f_9 were resolved in a single night of combined CARSO and SAAO data (in which one-day aliases are absent) from JD 2445728. These peaks appear in Figure 4, but the situation is confused by aliasing. Frequency f_5 occurs in Figure 4, and f_6 was detected by one of the authors (DWK) in a 1983 observing run. The dashed lines represent the expected position of frequencies if the spacing were a uniform 26.2 μHz.

By comparing this spacing with the predictions of Shibahashi and Saio (1984) for pulsating main-sequence A stars, we suggest that HD 60435 is undergoing non-radial pulsations of odd and even degree (probably ℓ = 1 and/or 3, and ℓ = 2.

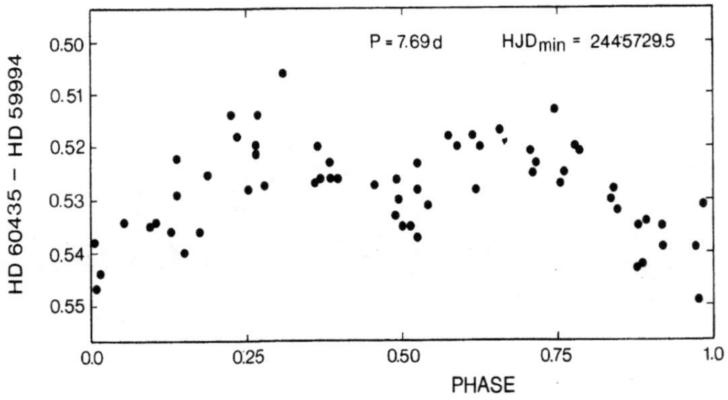

FIGURE 6. Phase diagram of mean photometry of HD 60435, using HD 59994AB as a comparison, collected during January/February 1985, plus 13 additional nights obtained by DWK in 1983. The data is plotted assuming a period of 7.69 days. If this is interpreted as the rotation period of HD 60435, it agrees with the values estimated from the splittings observed in the Fourier spectrum (see Fig.4) and the modulation of the rapid oscillation amplitude. (Note: The light curve possesses a secondary minimum, suggesting that the inclination and obliquity of HD 60435 are such that both magnetic poles are visible as it rotates).

INVESTIGATIONS OF THE MAGNETIC STAR 53 Cam VARIATIONS USING THE SPECTRA OF HIGH TIME RESOLUTION

N. S. Polosukhina and V. P. Malanushenko
Crimean Astrophysical Observatory
334413 p/o Nauchny, Crimea
USSR

I. Tuominen, H. Karttunen, H. Virtanen
Observatory and Astrophysics Laboratory
University of Helsinki
Finland

ABSTRACT. The results of automatic reduction of spectral observations of 53 Cam with high time resolution are presented. The statistical analysis of deviations of individual spectrograms of a star from the averaged one for one night permits one to allocate variable details of spectra and estimate the amplitudes of their variability.

1. INTRODUCTION

Using a complex of programs specially developed for statistical analysis of large series of spectroscopic data, about 1000 spectrograms of 53 Cam (with time resolution $\simeq$ 1 min) were reduced. This star was selected as a program star for cooperative observations in 1979 - 1982 (Polosukhina, 1979, 1980, 1981), since it is accessible for observations in a majority of the observatories. The aim of these observations was to search for the presence of spectral variations during a time significantly shorter than the period of star's rotation. 53 Cam is a typical Ap-star with the following peculiarities:

1) The presence of rather strong magnetic field ($H_{eff} = \pm 5$ Kgs), varying periodically with the period of star rotation (Borra et al., 1977).
2) Remarkable variations of the line intensities of some chemical elements, especially Ca II, Sc II, Si II and rare earth elements, whose behaviour correlates with the magnetic field variations. These investigations are discussed in detail by Faraggiana (1973) giving the values of the effective temperature T_{eff} = 8400K and log g = 4.0.
3) The presence of long-term variations of radial velocities that permitted one to draw conclusions about the binarity of this star (Scholz, 1978).

2. OBSERVATIONS

Here we discuss the results of spectroscopic observations of 53 Cam. The observations were made with the 2.6-m telescope using the spectrograph and image tube of the Crimean Astrophysical Observatory. The methods of the observations are described by Polosukhina et al. (1981). Spectral resolution was about 1 - 2 A at the inverse dispersion of about 40 A/mm; time resolution was about 1 - 2 min. in the spectral region 3800 - 4400A. The spectra were taken mainly on the photoemulsion "Izopanchrome type 17" having low photographic noise. Table 1 presents the observational material used in our paper.
The first column is the date of observations, the second is the phase in the middle of the observations according to the formulae: JD(pos. cross.) = 2435855.652 + 8.0267 x E (see

C. R. Cowley et al (eds.), Upper Main Sequence Stars with Anomalous Abundances, 243–252.

TABLE I

Data	Phase	N
20/21.12.75	0.12	10
18/19.01.76	0.71	20
22/23.11.77	0.70	12
23/24.12.77	0.55	47
16/17.01.78	0.50	10
24/25.01.78	0.54	49
17/18.03.78	0.00	81
18/19.03.78	0.13	85
15/16.04.78	0.62	66
31/01.04.80	0.82	121
26/27.11.80	0.74	106
26/27.12.80	0.48	17
15/16.01.81	0.96	49
15/16.03.81	0.30	82
16/17.03.81	0.43	104
12/13.04.81	0.79	41

Borra et al., 1977) and the third is the number of obtained spectrograms.

The main difficulty in our work was associated with the fact, that the effect of rapid variations sought is small, and to reveal it one needs, first, sufficiently long (no less than 2 hours) and homogeneous observations, and second - high enough accuracy to reduce the material in total.

3. METHOD OF REDUCTION

The method is based on the analysis of standard variations of individual spectrograms from the averaged one, as was used earlier (Polosukhina et al., 1981 and Bijaoui et al., 1979). At each point of the spectrum the dispersion of averaged spectrograms was calculated σ^2_{obs} and compared with the dispersion estimated by the analysis of instrumental noise σ^2_{exp}. They are compared according to statistical mathematics by F-Ficher criterion. When σ^2_{obs} at a given validation exceeds σ^2_{exp}, we may speak about the presence of variations at a given point λ_i of the spectrum and estimate the amplitude of variations.

1. The photometry of spectral material was carried out using the microdensitometer PDS (Elvius et al., 1978) of Lund observatory with a step of scanning 12 %mκ.

2. The characteristic curve was computed by the method described by Malanushenko et al., (1984). It is stored in the computer memory in a form of a table $I = F(D)$ with a step 0.01 D. This table permits one to convert "D" into "I" as quickly as required.

According to our method the instrumental noise is a sum of photoemulsion noise σ_{ph} and that of the image tube σ_{iT}. The influence of the image tube noise was estimated and it does not exceed 1%. The noise of photoemulsion was well studied (Palej, 1979) and thus it can be easily estimated. The calibration scale was converted into intensities and the value of noise was calculated with respect to the characteristic curve:

$$\sigma_{ph}(\bar{I}) = \sqrt{\Sigma(I_i-\bar{I})^2/(N-1)}/\bar{I}$$

for each level of density. The obtained error-curve is the dependence $\sigma_{ph}(I) = F(I)$ approximated by the polynomial and stored in the computer memory in a form of a table. Examples are shown in Figure 1.

3. The spectrograms were converted into intensities according to "DINT" program.

4. The continuum is calculated for each spectrogram according to "CONT" program which is functioning iteratively. At the zero approximation the polynomial of optimal degree (see Seber, 1977) is constructed by all points of the spectrum. Then, those points are eliminated which lie lower the polynomial within a half of the noise track. After that the polynomial is calculated according to the rest of points, the lowest are thrown away, and so on. The process of iteration is continued until the scatter of points on the spectrograms taken for continuum construction exceeds the noise track. Finally all spectrograms are normalized to individual continuous spectra. The inhomogeneities associated with the difference of treated spectra are eliminated by the correction program "CORR" (based on the comparison of individual spectra with the averaged one for a night). Figure 2 shows the examples of such a continuum construction.

5. The wavelength fit needs to involve all points of the spectrograms. The cross-correlation function is computed using the basic (first, as a rule) spectrum and the i-ths. The deviation corresponding to the maximum of cross-correlation function is the required step to fit the i-ths spectrum with the basic. The uncertainty of the fit by ± 0.5 step leads to so-called "slit" errors (Otnes et al., 1978). Its value is proportional to the gradient $\Delta I/\Delta\lambda$ (see Appendix).

6. Further reduction is based on the "STAT" program. Here the calculations of the mean spectrum and σ_{obs} are as usual. We should note however, that $\sigma_{obs,i}$ is a relative value (normalized to residual intensity of the spectrum at a given point I_i). While computing $\sigma_{exp,i}$ the noise of photoemulsion σ_{ph} is accounted upon as well as the influence of the "slit" error.

The allocation of variable points of the spectrum is realized according to F-Ficher criterion. If the inequality is valid:

$$\sigma^2_{obs,i}/\sigma^2_{exp,i} \geq F(\nu_1,\nu_2)_a \tag{1}$$

(where F is the value of distribution for ν_1 and ν_2 the degree of freedom for calculation of $\sigma_{obs,i}$ and $\sigma_{exp,i}$ and a is the confidence level) then $\sigma_{obs,i}$ exceeds $\sigma_{exp,i}$ by $a = 99\%$. Those points of the spectrum for which this requirement is fulfilled, are considered to be variable. Thus the estimation of amplitude variations will be as follows:

$$\sigma_{S,i} = \sqrt{\sigma^2_{obs,i} - \sigma^2_{exp,i}} \tag{2}$$

In case when the variability is not found (i.e. the requirement (1) is not fulfilled), one can estimate the upper limit of amplitude. Figure 3 shows some illustrations of individual spectrograms and Figure 4 – the results of statistical data treatment.

4. RESULTS OF THE OBSERVATIONS

Using the technique described above, we estimate the total effect of the spectrum variations for the time of observations during each given night. Figure 4 demonstrates our main results :

1. At the top of Figure 4, the mean spectrum of the star obtained during the whole observational time at a given night is presented.

2. The observed dispersion σ_{obs} and expected σ_{exp} are shown at the bottom of the same Figure. The confidence level is 99% for a given confidence interval.

3. The central part of Figure 4 shows the variable details of the spectrum derived from the comparison of both dispersions and the estimation of the variation amplitude σ_S for one night.

The allocated variability permits one first to estimate the total effect of variations during the time of observations for each night, and second to allocate the spectral details for which this effect is mostly noticeable. As a result of dispersograms analysis carried out for 16 nights of observations, we have come to the following conclusions :

1. The effect of short-term variations of the spectrum does exist and its value on an average does not exceed 5%.

2. The variable details in the stellar spectrum correspond to the lines of H, Ca, Sr, Si and rare earth elements.

3. The amplitudes of short-term variations vary with the phase of rotation from 3% to 10%, the most prominent variations correspond to Ca II K-line (see Figure 5).

4. Using the averaged spectrograms the variations of the spectrum with the phase of rotation were studied. Variations of the equivalent width of H-lines (with the amplitude 10%) were found, strong variations of the profile and equivalent width of Ca II K-line (with the amplitude 60%) were estimated as well as the variations of the equivalent widths of Si II, Sr II -lines (see Figure 6). A good agreement is observed with the data of Faraggiana (1973) for Ca II K-line (see Figure 7).

The fulfilled investigation permitted us to conclude, that there exist two types of variations of the spectrum of 53 Cam, having different intrinsic time and amplitudes. The most well-pronounced variations of the spectrum with the phase of star rotation have a known explanation in terms of an oblique rotator. The observed variations of the spectrum during one night is an overall effect due to first, the line intensity variations with the phase of star rotation in the time interval of several hours and seconds, probable influence of complex brightness variations of the star during one night, and finally, possible manifestations of chromospheric activity of a star with a strong magnetic field.

5. APPENDIX

Evidently, while shifting the i-th spectrogram in respect to the mean one by $\Delta\lambda_i$ there appears the difference of intensities :

$$\Delta I_{ji} = \text{grad } I_j \cdot \Delta\lambda_i \qquad (1)$$

where grad I_j is the intensity gradient at a point j of the spectrum. Summing up N spectrograms, we estimate the error :

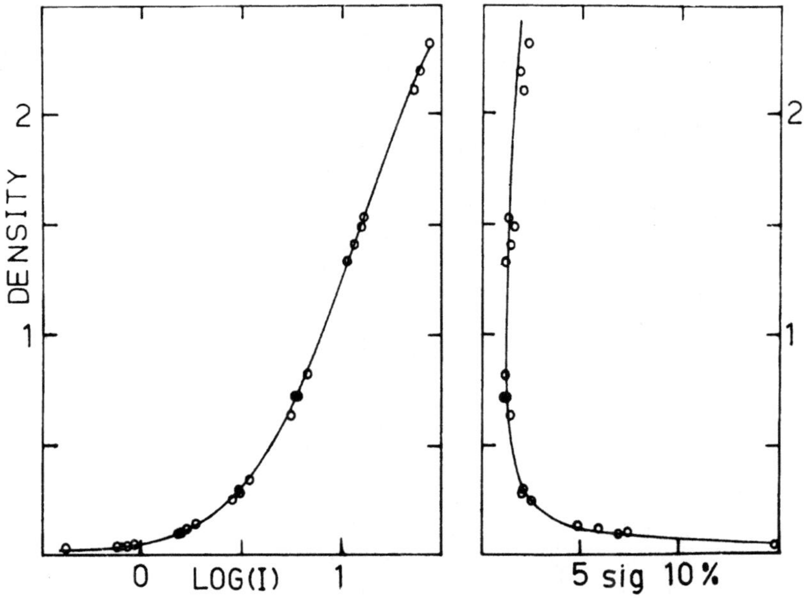

Figure 1. The measurements and approximation of the characteristic curve of errors. Log I and $\sigma_{ph}(\%)$ are shown along the abscissa for characteristic curve and curve of errors, correspondingly. On the ordinate – the densities.

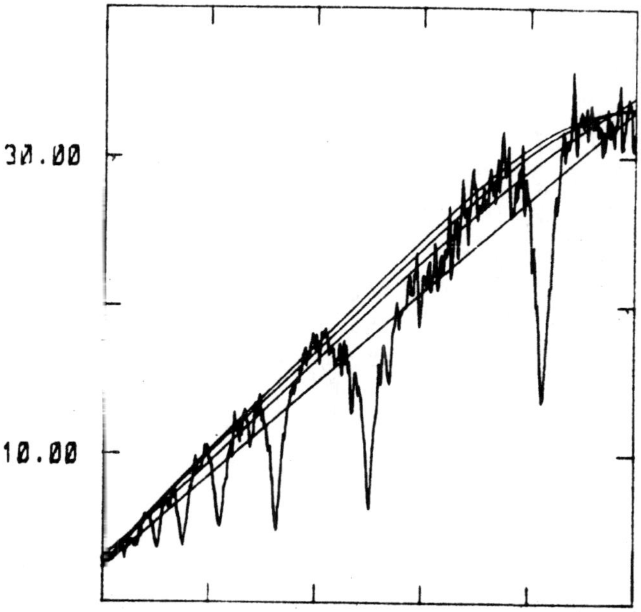

Figure 2. The examples of the continuum computations. Several lines show different approximations. The upper line is the resulting approximation.

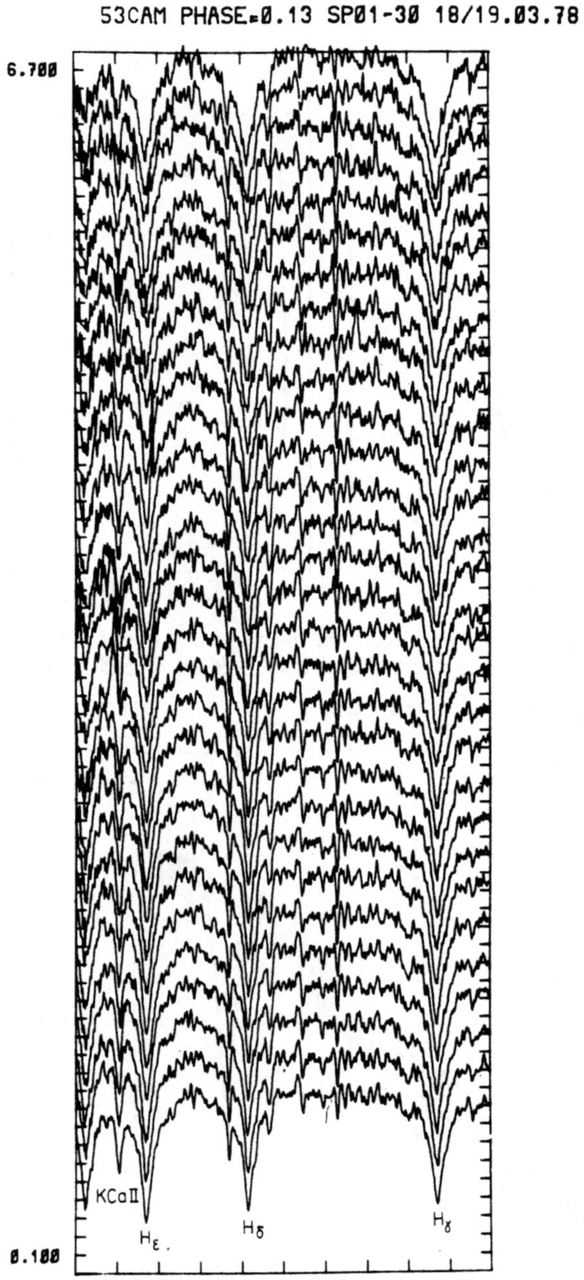

Figure 3. Individual spectrograms normalized to continuum.

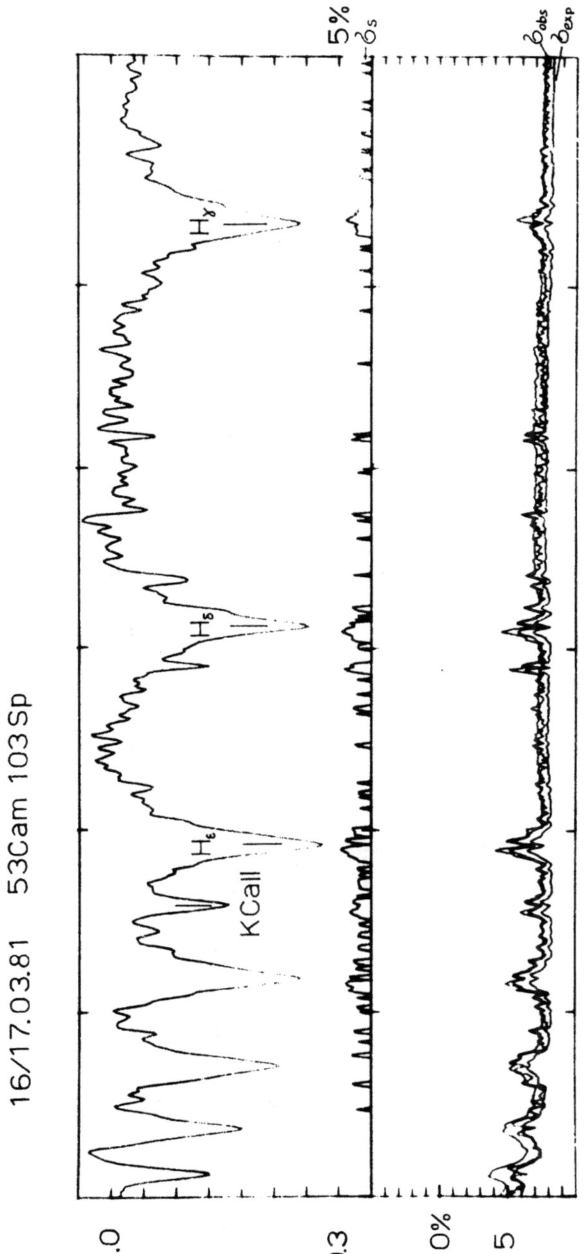

Figure 4. The results of data reduction. The abscissa corresponds to wavelength, ordinate: for the averaged spectrum – residual intensities in the units of continuum, for sigma – relative values in percents.

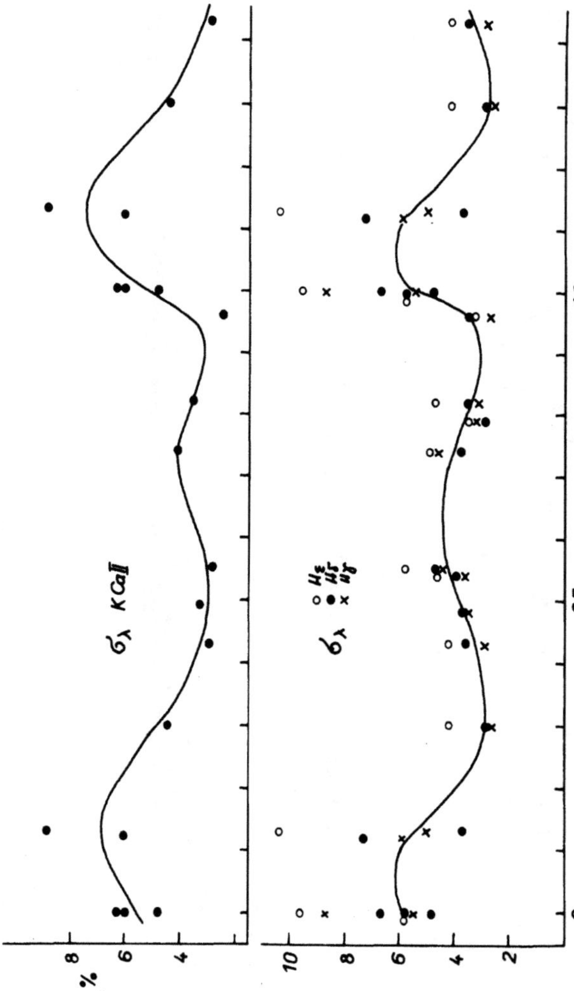

Figure 5. The fluctuations of rapid variations amplitude with the phase of rotation.

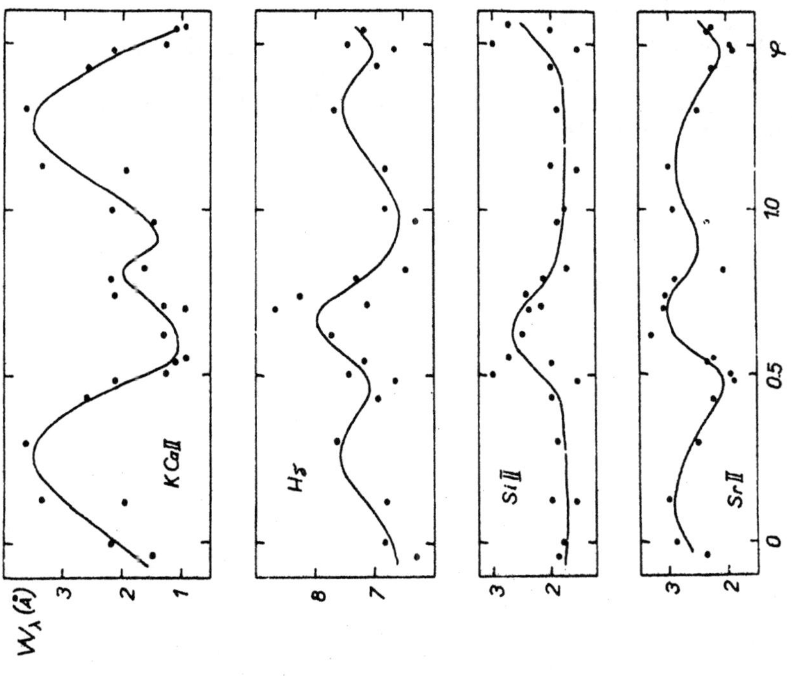

Figure 6. The variations of W_λ of the allocated details with the phase of rotation.

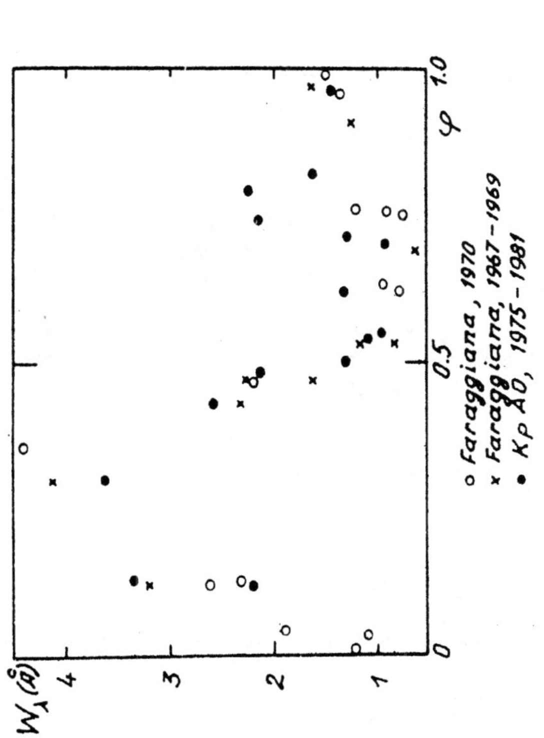

Figure 7. The comparison of W_λ measurements of Ca II K-line with the data of Faraggiana (1973).

$$\sigma_{slit,j} = \sqrt{\left[\frac{\Sigma \Delta I^2_{ji}}{(N-1)}\right]} = \text{grad} I_j \cdot \sqrt{\left[\Sigma \frac{\Delta\lambda^2_i}{(N-1)}\right]} \qquad (2)$$

at each i-ths point of the mean spectrum. The value ΔI is the difference between the discrete and precise fit of the i-ths spectrum with the mean one. It is calculated as a difference of places of discrete and precise maxima of cross-correlation function between the i-ths and mean spectrum.

In order to calculate the location of the precise maximum, the cross-correlation function is approximated by a cubic spline. The maximum is determined by the requirement that the first derivative of spline equals to zero. The value of gradient at each point of the mean spectrum is calculated by approximation :

$$\text{grad } I_j = (I_j - I_{j-1})/(\lambda_j - \lambda_{j-1})$$

REFERENCES

Bijaoui A., Doazan V. 1979, *Astron. Astrophys.*, 70, 285.
Borra E., Landstreet J. 1977, *Astrophys. J.*, **212**, 141.
Elvius T., Lindgren H., Lunga G., and Wihlborg N. 1978, *Reports from the obs. of Lund*, No. **14**.
Faraggiana R. 1973, *Astron. Astrophys.*, **22**, 265.
Malanushenko V., and Scherbakov A. 1984, *Izv. KrAO*, **72**.
Otnes R., and Enochson L. 1978, *Applied Time Series Analysis: 1. Basic techniques*, (New York: J. Wiley and Sons), ch. 3, 5.
Palej A. 1979, *Soobschenija GAISh*, **206**, 10.
Polosukhina H. 1979, *A Pec. Newslet.*, No. **2**., 1980. *Ibid.*, No. **4**, 1981., *Ibid*, No. **6**.
Polosukhina N., Chuvaev K., and Malanushenko V. 1981, *Izv. KrAO*, **64**, 37.
Seber J.A., 1977. *Linear Regression Analysis*, (New York: John Wiley and Sons), ch. 8.
Scholz G. 1978, *Astron. Nachr.*, **299**, 305.

HD 24975 : A NEW DELTA SCUTI STAR ?
(OR A MILD Ap STAR WITH SHORT PHOTOMETRIC VARIATIONS ?)

C.MEGESSIER
Observatoire de Paris-Meudon
F-92195 MEUDON, FRANCE

P.NORTH
Institut d'Astronomie de
l'Université de Lausanne
CH-1290 CHAVANNES-DES-BOIS
SWITZERLAND

M.BURNET
Observatoire de Genève
CH-1290 SAUVERNY
SWITZERLAND

INTRODUCTION

Looking in the literature for short period variations of Ap stars, we found the puzzling case of HD 24975. Used as a comparison star by Weiss (1978), it had been found then to present variations with a peak-to-peak amplitude of 0.01 in U, B and V with a period of about 45mn, but with no clear correlation between the three passbands.

HD 24975 has the same spectral type A2 as 21 Com which seems to present photometric variations with P ≃ 31mn and $\Delta V < 0{,}02$m (Percy, 1973, 1975). Both stars are near the blue edge of the δ Scuti instability strip in the HR diagram. The lack of photo metric data, however, prevented the precise location of HD 24975 from being found on the HR diagram.

OBSERVATIONS

We decided to observe this star in the Geneva photometric system in order to confirm its variability and have a better idea of its temperature and luminosity.

It has been monitored by one of us (MB) during two consecutive nights, on the 4-5th and 5-6th December 1984 at the 70cm Swiss telescope at La Silla. 72 measurements have been obtained each

C. R. Cowley et al (eds.), Upper Main Sequence Stars with Anomalous Abundances, 253–256.

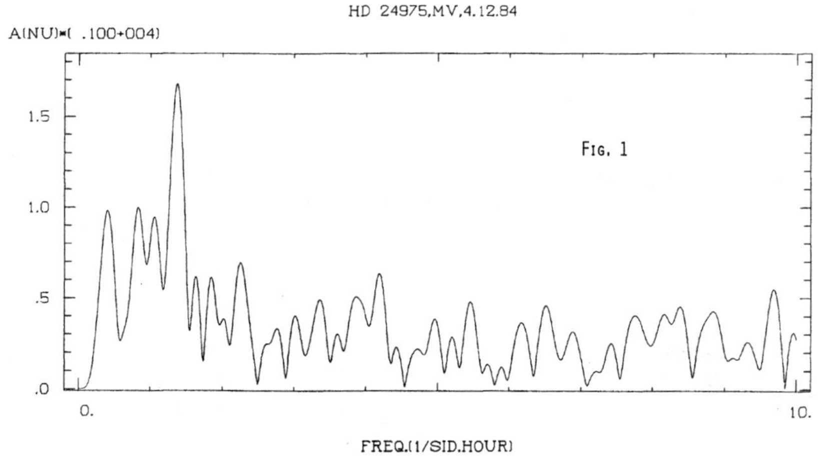

Fig. 1

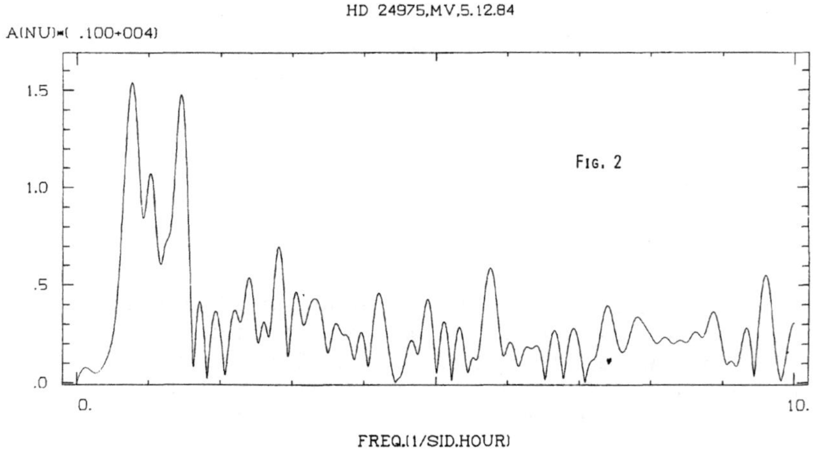

Fig. 2

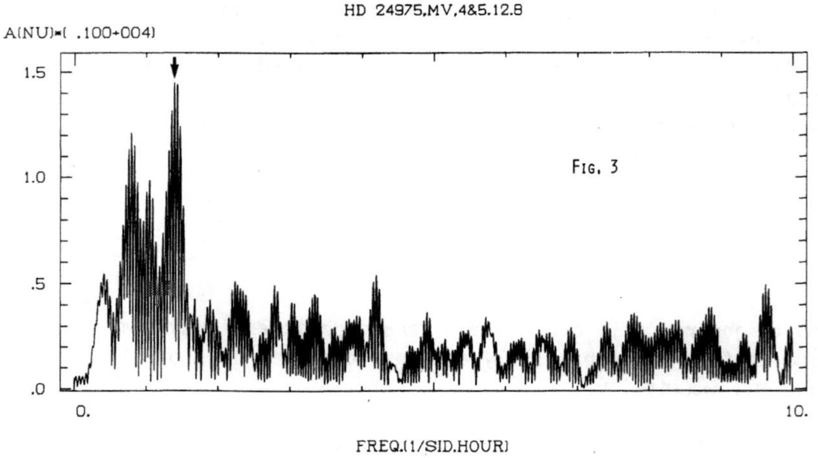

Fig. 3

HD 24975, 4-5.12.84

Fig. 4

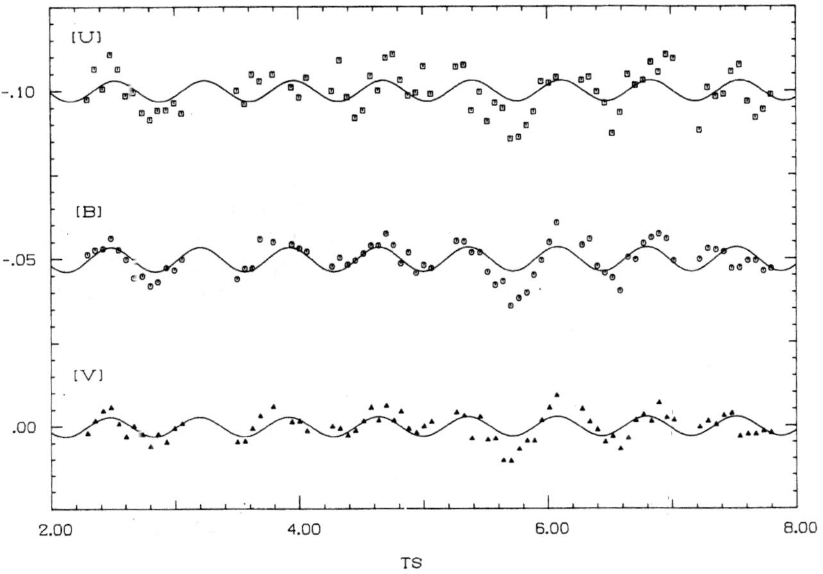

HD 24975, 5-6.12.84

Fig. 5

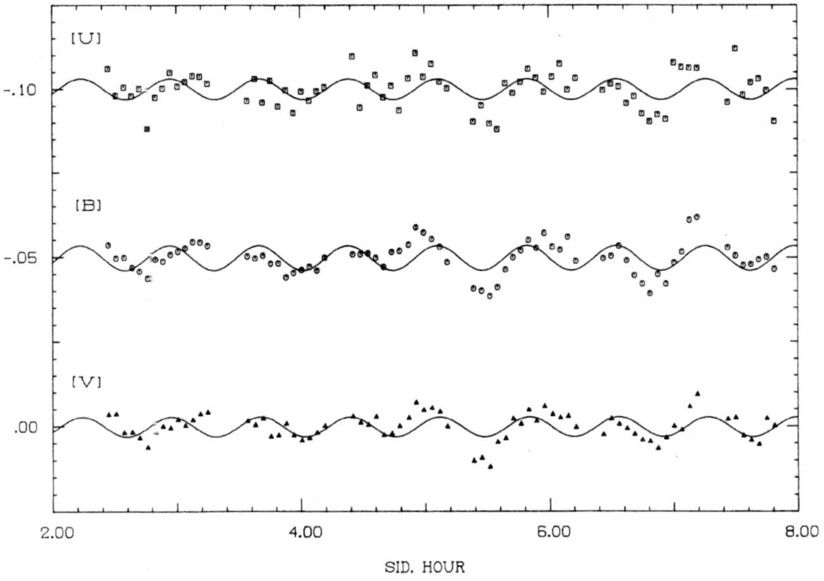

night. Standard stars have been observed at the same airmass about each hour, totalising 14 measures each night.

RESULTS

Discrete Fourier transform analysis (Deeming, 1975) performed with preliminary [U], [B] and [V] magnitudes in the normal system shows a significant peak at about $1.4h^{-1}$ on each night and in each colour. On the second night another peak appears near $0.8h^{-1}$, whose meaning is not yet clear. The periodograms of the [V] magnitude are shown for each night in fig.1 and 2. Analysis of both nights together (fig.3) yields two or three possible frequencies near 1.4, the most probable one being $1.394 \pm 0.002h^{-1}$. The most probable period is thus

$$P = 0.02989 \pm .00004 \text{ d.} = 0.717 \pm .001 \text{ h} = 43.05 \pm .06 \text{ mn}$$

The amplitudes are about the same (~ 0.003m) in all three passbands and no significant phase shift appears (fig.4 and 5).

Preliminary, approximate values of the main photometric parameters are the following : B2-V1 = 0.021, d = 1.22, m2 = -0.49, X = 1.38, Y = -0.16, Z = -0.02. These values imply the following physical characteristics (Hauck, 1985 and references therein) of HD 24975 : T_e = 7800 K, M_V = 2.1, F_e/H = -0.07.

CONCLUSION

The preliminary Geneva colours show that HD 24975 very probably has an A9V spectral type rather than A2. Thus it is probably a quite classical Scuti star, although it has one of the shortest periods known. A spectrum of this star would be interesting to confirm the spectral type and to look for any possible peculiarity.

REFERENCES

Breger, M. : 1979 Publ. Astron. Soc. Pacific 91, 5.
Deeming, T.J. : 1975, Astrophys. Space Sci. 36, 137.
Hauck, B. : 1985, Astron. Astrophys. (Submitted).
Percy, J.R. : 1973, Astron. Astrophys. 22, 381.
Percy, J.R. : 1975, Astron. J. 80, 698.

Hot Magnetic Stars

Abundances, Spectroscopy, Photometry, Systematics.

K. Hunger
Kiel

Systematics

The present discussion of hot magnetic CP stars is restricted to 17000 K $< T_{eff} <$ 30 000 K, the upper limit being the temperature up to which magnetic fields can be observed, the lower limit defining the boundary between He-rich and He-poor stars.

All of the hot magnetics belong to the class of intermediate helium stars which is defined by $n_H/n_{He} < 3$ (number ratio, Hunger 1975). Walborn (1983) lists 23 stars of this class, mostly with $n_H/n_{He} \approx 1$, and of spectral type B2 V. Magnetic fields, so far, have been discovered in 7 of them (Borra et al. 1983). All but one of these are spectrum variables, with periods $0.9 < P/\text{day} < 1.7$ (except one with P = 9.5 d), (Pedersen, 1979) and with rotational velocities of the order of $v \sin i \approx 150$ km/s which are compatible with the (P, v sin i)-relation of stars near the main sequence; i.e. these stars are oblique rotators.

As to the age and population: all evidence points to a rather young population as a large fraction is found in the Orion aggregate and the young cluster IC2944. The masses are well in excess of one solar mass, in contrast to the (old) extreme helium stars. Also the metal content of the subgroup seems to be solar (within a factor of 2) (Hunger, 1975).

Masses

The masses are crucial when one wants to know whether He-enrichment in the hot magnetics is due to diffusion and hence is confined to the surface, or whether the entire star is enriched, as is believed for the class of extreme helium stars. Since no binaries among this class are known, the only way to determine the mass is spectroscopical: from the distance d and the V-magnitude, the luminosity $L = 4\pi R^2 \sigma T^4_{eff}$ is derived. T_{eff} is obtained from spectral analysis, and hence the radius R, likewise the gravity $g = GM/R^2$, and hence the mass M.

The best way today to determine T_{eff} is from IUE fluxes, as the main flux of a B-star is carried in the IUE-band (Remie and Lamers, 1982). The fluxes are determined with a precision of $\pm$ 10%, and hence T_{eff} with $\pm$ 2.5%. As an example, the results for the best studied hot CP star, σ Ori E is given (Groote and Hunger, 1982): T_{eff} = 22500 K $\pm$ 600 K, independent of phase. In Fig. 1, the observed and theoretical fluxes are reproduced, which well agree, except for the IR-excess we come later to. While T_{eff} is determined largely independently of gravity g and of the helium number fraction ε_{He}, g and ε_{He} have to be determined

C. R. Cowley et al (eds.), Upper Main Sequence Stars with Anomalous Abundances, 257–269.

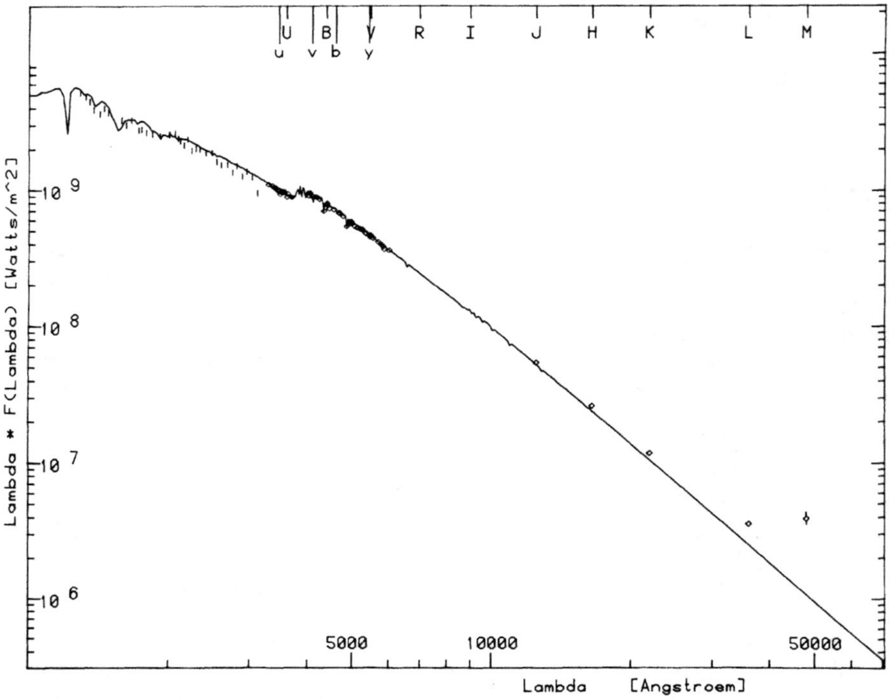

Fig. 1: The flux of σ Ori E, from 1100 Å to 50000 Å

simultaneously, from the profiles of the Balmer and helium lines, and of course phase dependent, as σ Ori E is a helium variable. The result is shown in Fig. 2: gravity is nearly constant, log g = 3.85, while ε_{He} reflects the familiar behaviour of the R-index variations (Pedersen, and Thomsen, 1977). With d = 400 pc (Heintz, 1974), the mass finally results: $M = 3\ M_{\odot}$.

Mass and gravity come out uncomfortably small. Both values consistently correspond to a star which evolved from a helium rich main sequence, with X = 0.3 (Fig. 3). However, a young star that begins its life as a fully mixed helium star is hardly conceivable, and hence the results have been questioned (Walborn, 1983). Either the distance is underestimated - which is rather unlikely - or the equivalent widths used are systematically too small. The ESO plate calibration has been carefully checked by intercomparison with other photographic systems and has been found reliable. The intercomparison, however, was purely photographical, and it may turn out that photoelectrically determined equivalent widths turn out to be systematically larger. A first indication for this comes from ESO CASPEC CCD spectrograms which for faint stars ($V = 10^m$) yield equivalent widths that are larger by

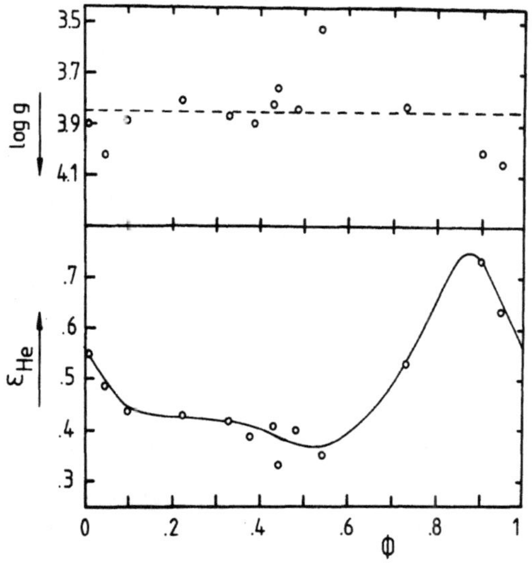

Fig. 2:
Gravity and helium number fraction of σ Ori E, as function of phase

40% in some cases. If this also holds for a bright star like σ Ori E, then gravity has to be corrected by a factor of more than 2 which means that the mass is 7 $M_\odot$ rather than 3 $M_\odot$, gravity and mass now well in agreement with the normal composition main sequence. Helium enrichment then is only confined to the surface.

The Oblique Rotator.

The variation of the helium content with phase can best be understood on the basis of an oblique rotator. The problem with the helium strong magnetic variables however, is that one does not observe variations of the radial velocities when the helium rich parts of the surface approach the observer or when they recede (Groote and Hunger, 1977). The missing radial velocities ($\Delta v_{Rad} < \pm$ 2 km/s) have some authors led to believe in helium-rich bands rather than helium-rich caps (Shore and Adelmann, 1981; Bolton, 1983). While this may be true in exceptional cases, it is certainly not true in the case of σ Ori E, for if one invents a surface structure that leads to small amplitudes in v_{Rad}, then one inevitably ends up also with small amplitudes in the equivalent widths. The answer to this problem is saturation. If the helium lines are core saturated both in the cap and in the disk, as is the case in our sample of stars, then the radial velocities seem to reflect rather the motion of the disk, because the disk lines are narrow, than the motion of the cap. This has been shown rigorously by Gruschinske (see Groote and Hunger, 1982) for the idealized case of a single helium cap that covers a full hemisphere of the star. The helium abundance inside and outside the cap is taken from Fig. 2, and v sin i = 150 km/s is assumed. The two phases, when the boundary is subsolar,

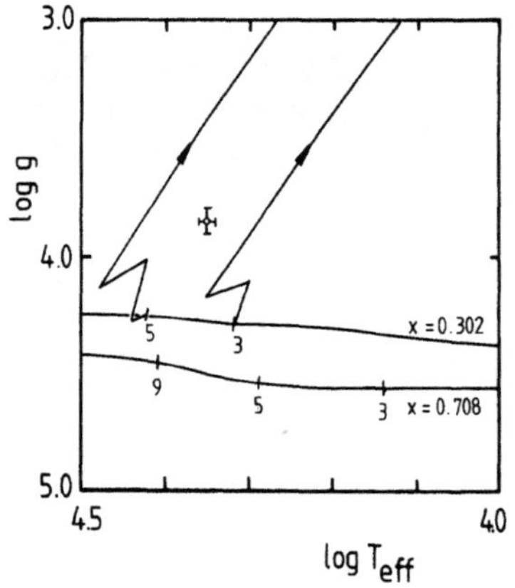

Fig. 3:

Two main sequences, with the hydrogen mass fractions X = 0.302 and 0.708 (normal composition), and two evolutionary tracks, of 3 and 5 solar mass stars, are shown in the (g, T_{eff})-diagram.

and when the cap is either approaching or receding is shown in Fig. 4. The example is the (symmetrized) profile of HeI 4471. While the wings (or equivalently the core of weak lines) indeed are shifted to the blue when the cap is approaching, the core is shifted inversely, by an amount of 40 km/s.

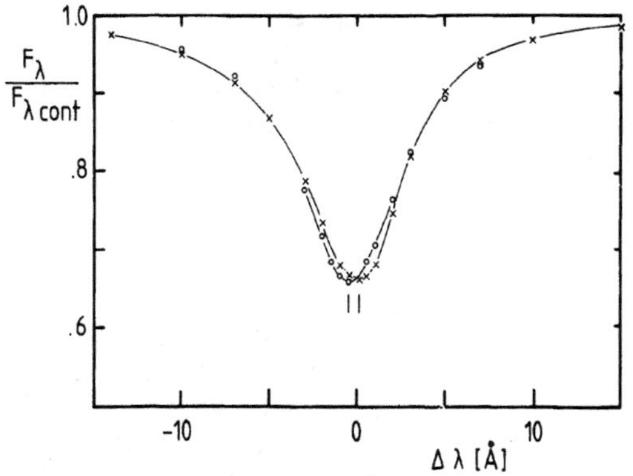

Fig. 4:

The symmetrized profile of HeI 4471, in two phases: ××× helium cap is approaching, ooo helium cap is receding. The core is shifted inversely, by 40 km/s.

This unexpected behaviour is indeed observed, without any exception, in all of the HeI lines, (Fig. 5). The (saturated) strong lines λλ 3820, 4026, 4471 have inversely shifted cores, the (unsaturated) weak lines 4121, 4438, 4713 are directly shifted, while the intermediate lines 4009, 4144, 4388 are unshifted. This explains why correlation techniques (Groote and Hunger, 1977) which use averages over many lines do not reveal R.V. amplitudes. Interesting in this context is that some He weak magnetic variables with their unsaturated lines do show R.V. variations.

An example for a banded structure represents HD 37776, where, for the first time, a magnetic quadrupole field with current loops has been discovered, Fig. 6 (Thompson and Landstreet, 1985; Groote and Kaufmann, 1981).

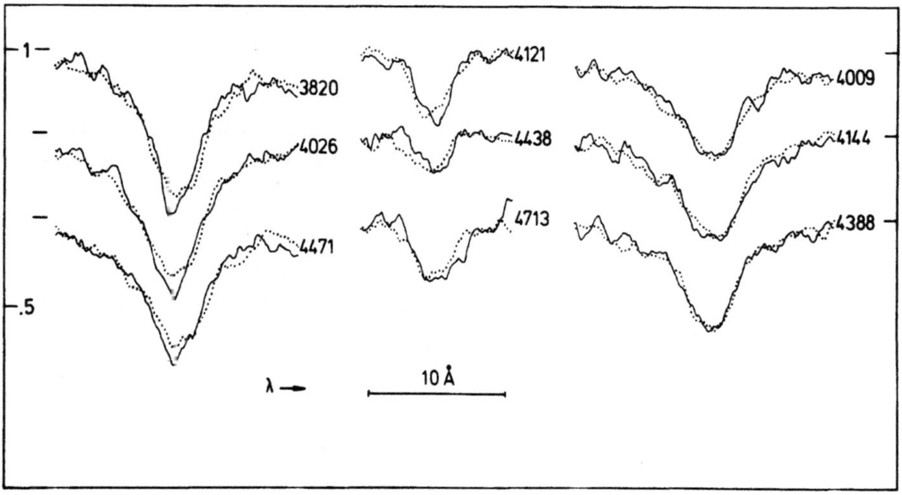

Fig. 5: Profiles of strong, medium strong and faint He lines of σ Ori E in two phases: ..... helium cap is approaching, —— helium cap is receding. Cores of strong lines are inversely shifted, while faint lines are directly shifted.

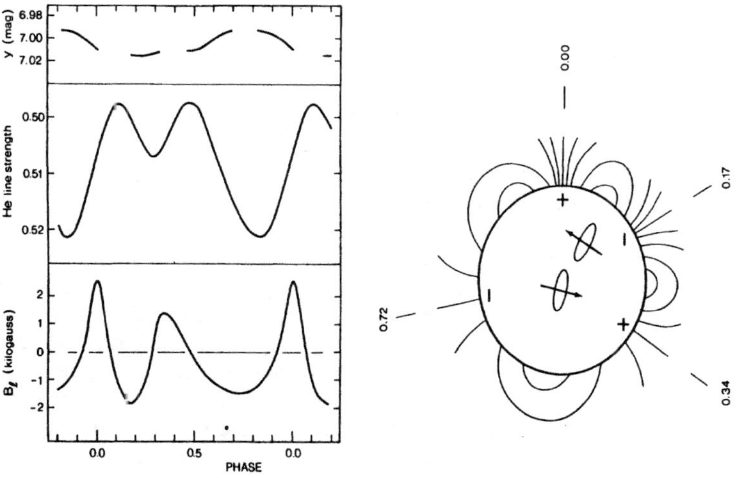

Fig. 6: Phase variation of y, He line strength and magnetic field of HD 37776. This variability is best explained by a banded helium distribution that is tied to a magnetic quadrupole field (Thompson and Landstreet, 1985).

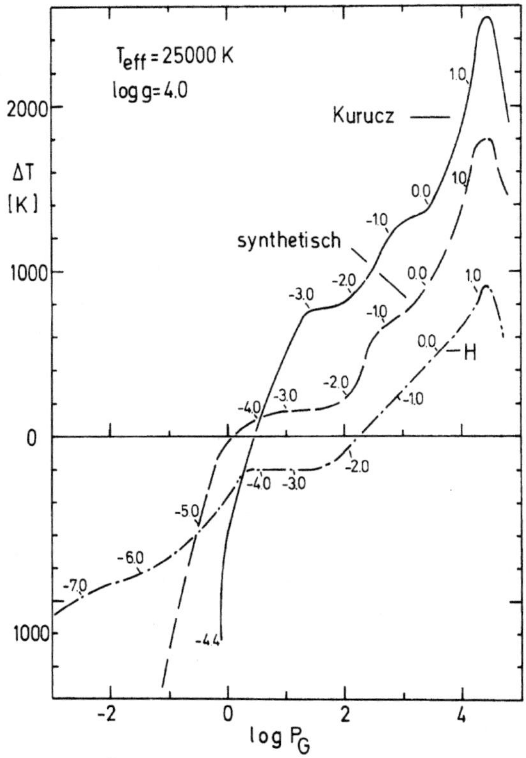

Fig. 7:
The temperature differences vs. an unblanketed model atmosphere, as function of depth (P_g): —·—·only H-line blanketing, - - - additional 500 metal lines, and —— full line blanketing. The temperature gradient is similar in all blanketed models.

Atmospheric Models

How realistic are the model atmospheres sofar in use? For, in the hot magnetic CP's we encounter the following problems:

1) metal line blanketing
2) horizontal abundance variations
3) vertical abundance stratifications
4) influence of magnetic fields on structure and spectrum

1) Fully blanketed models are available only for normal compositions (Kurucz, 1979). For He-enriched atmospheres, only a few models exist which moreover include "only" up to 500 metal lines (Heber, 1979). The effect of metal line blanketing on the temperature and hence on the emergent spectrum can best be judged from Fig. 7, where the temperature differences with respect to an unblanketed model are shown: for a Kurucz-model, for a Heber-model, and for a H-line blanketed model. For the same effective temperature, the temperature increases in the region of spectrum formation from the unblanketed model by 300 K (H-lines), by 700 K (500 metal lines), and lastly by 1300 K (full set of lines). The slopes of the 3 blanketed models, however, remain similar which means that the emergent fluxes, continua as well as lines, come out alike. Hence H-line blanketed models can safely be used, if one only corrects

T_{eff} by -1000 K. The latter was shown for normal composition atmospheres. It holds also for He-enriched atmospheres, as long as $\varepsilon_{He} < 0.5$.

2) Horizontal abundance variations, namely the variation of helium in spots (bands) and disk, do not pose any problems, as long as T_{eff} and g are constant over the surface: if one adds the line profiles of spots (bands) and disk, wheighted with the appropriate surface area fractions, then a profile results which is indistinguishable from a line profile with a helium content that is the area mean of the two helium fractions, of spots (bands) and disk.
It has been claimed that the sometimes observed antiphase variation of silicon does not reflect abundance variations, but is due to differences in excitation and ionization between disk to spot (see Bolton, 1983). This has been checked by Heber (1985), for the case of σ Ori E, on the basis of metal line blanketed atmospheres. The result is that the lines of SiII $\lambda\lambda$ 4128, 4131 increase by 6% in equivalent widths, when going from the disk (ε_{He} = 0.35) to the spot (ε_{He} = 0.90), while SiIII $\lambda\lambda$4553, 4568, 4575 decrease by 5%. Somewhat larger is the increase for SiIII λ 1417: 19%. This means that the silicon variations, if real, are due to abundance variations.

3) Vertical abundance stratifications are to be expected on account of diffusion. Other than in the case of compact stars like white dwarfs, in hot CP's helium floats atop of hydrogen. Whether the temperature structure is affected by such a "chemical profile" (which presumably is a diffusive "profile"), and whether the spectral lines will be affected depends not only on the strength of the discontinuity, but much more so on the depth of transition. If this depth is in the intermediate range of $\tau_{4000} = 10^{-2}...1$, then, from the calculations performed for white dwarfs (Jordan, 1985), models with temperature inversions and correspondingly strange line profiles occur. Analogous calculations for hot CP stars are still pending. But from the foregoing results one can extrapolate: as long as the line profiles are not anomalous, one can be fairly sure, that the transition depth is below $\tau_{4000} = 1$, and that the spectral analysis hence yields parameters and abundances of the chemically homogeneous part of the atmosphere, above the transition.

Anomalous profiles, i.e. profiles that cannot be reproduced by any conceivable model and any reasonable rotational velocity, are observed in the helium variable HD 49333. The intermediate temperature (T_{eff} = 15700 K) CP variable is depleted in helium which means that helium is diffused downwards, while hydrogen floats atop, similarly as in white dwarfs. Kaufmann and Theil (1985) have calculated profiles, assuming various abundance steps (strength of discontinuity), and various depths of transition. In Fig. 8a, the helium number fraction above the transition is taken as ε_{He} = 0.001 and 0.01 resp., while ε_{He} below the transition is 0.12. The depth of transition is kept at τ_{4000} = 0.8. (The model atmosphere, however, has been computed with the

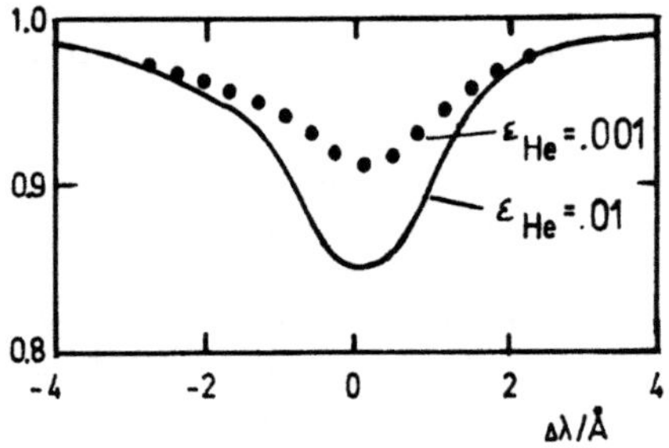

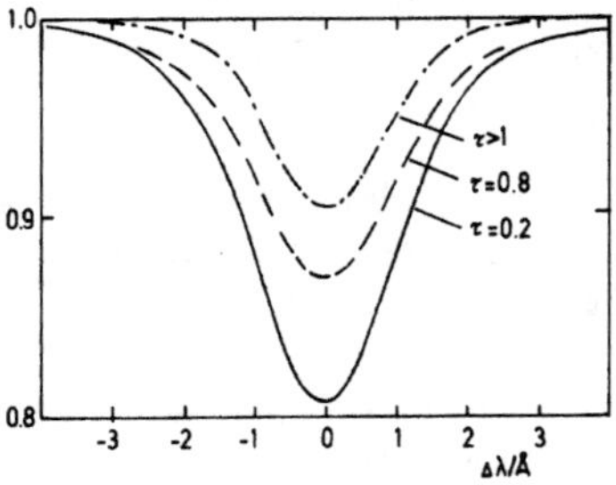

Fig. 8: How a vertical abundance stratification influences the profiles

a) at τ_{4000} = 0.8, the composition changes from ε_{He} = 0.001 (....) and 0.01 (——) resp. to ε_{He} = 0.12 of the bottom layer. (4471)

b) The composition above the transition depth varies from τ = 0.2 to 0.8 and > 1 (homogeneous model). (He I 4388)

assumption of an homogeneous atmosphere with ε_{He} = 0.12, as in this case the inhomogeneity has little effect on the model.) The reduction of helium in the upper layer reduces the core of HeI 4471, whereas the wings are not affected, i.e. the line appears shallow. In Fig. 8b, the effect of the depth of transition is shown: the transition from ε_{He} = 0.02 (above) to ε_{He} = 0.12 (below) takes place at τ_{4000} = 0.2 and 0.8 resp. HeI λ 4388 is greatly reduced when the transition depth moves downwards. In Fig. 8c, finally, the observed profile is compared with what is considered the best combination of depth of transition and strength of transition: ε_{He} = 0.001 (above) and 0.20 (below) with τ_{4000} = 0.4. (These results are provisional as they are based on noisy IIaO-plates. Furthermore, the surface flux integration effect - see above - is neglected.)

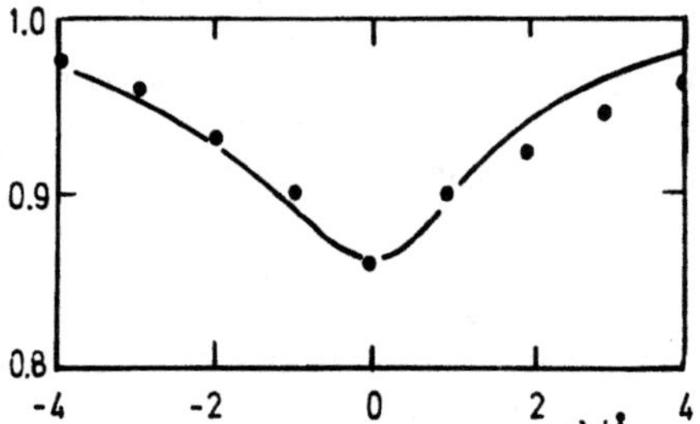

Fig. 8c:
The observed profile (. . .) of HD 49333 is best reproduced by a model with a transition depth at τ_{4000} = 0.4, and with ε_{He} = 0.001 (surface layer) and 0.20 (bottom layer) resp.

4) How the electromagnetic field influences the atmospheric structure and the emergent spectrum, via the opacity (Zeeman splitting of the contributing atomic lines) and via the magnetic forces (pressure stratification) has been studied in detail by Carpenter (1985) (see also Madej, 1981). The effects on the line profiles, however, are

small, and can only be detected under exceptional conditions (magnetic fields of the order of 20 KG, high quality spectrograms).

The Shell

Hot magnetic stars are believed to have corotating magnetospheres. In the example of σ Ori E it is shown that the magnetosphere is reduced to a truncated Van Allen belt (two corotating clouds) that contain a (predominantly) hydrogen plasma at a temperature of 15 000 K (Groote and Hunger, 1982). These clouds account for most of the variable features observed. (Even the eclipse like features of the light curve are due to the two corotating clouds. The conjecture that the brightness decline in the u-band is due to the helium caps, see Bolton, 1983, is erroneous as the helium caps are even (slightly) brigther than the disk, by $\Delta u = 0.^{m}1$, Kaufmann and Theil, 1985.)

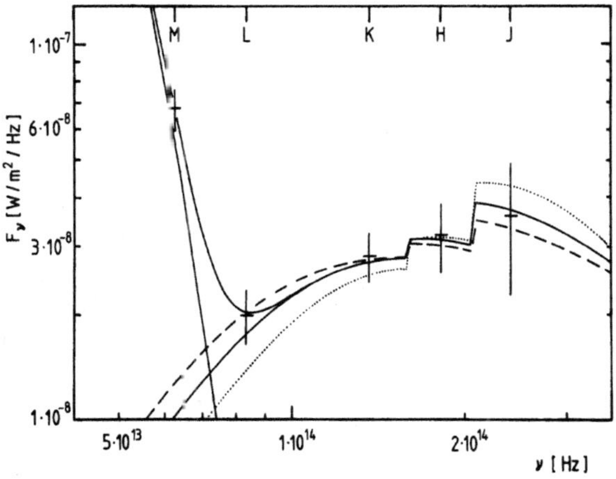

Fig. 9: The infrared excess of σ Ori E. Observations are marked by (long) crosses. The full drawn curve (lower part of the diagram) corresponds to free emission, with T = 15000 K.

The temperature of the clouds is derived from the IR excess up to the L-band. The additional excess at the M-band (Fig. 9) has often been questioned as the idea of corotating grains in the vicinity of a hot star is hardly conceivable. Also the excess could not be confirmed (Bonsack and Dyck, 1983; Odell and Lebofsky, 1984). If the excess were real, an alternative explanation would be synchrotron radiation due energetic particles (Havnes, 1981). In this case, radio emission should be observable. Attempts to detect emission with the 100 m Effelsberg telescope at 2 cm failed, because the spatial resolution was insufficient (see Groote and Hunger, 1982). At last σ Ori E, though, was detected at 2 and 6 cm, with the VLA - it was a serendipitous discovery (Drake et al., 1985). Also IRAS was able to see σ Ori E, the resolution however being too small to definitely ascribe it to the component E (Walker, 1985). These discoveries certainly open up the discussion of the M-band excess.

Magnetospheres and Winds

A major compagne has been started (Barker et al. 1985) to contemporaneously follow the observations in the optical, UV and Zeeman polarization, in order to study the interaction of winds with magnetic fields. A first example is the helium rich HD 184927. Fig. 10, left panel shows the variable features which have a common period of 9.5 days. The right panel reproduces the CIV resonance doublet at 5 different phases. The profiles vary with the same period: maximum (red shifted) emission occurs when the magnetic pole is subsolar, while maximum absorption occurs when the equator is subsolar. A detailed model is still lacking. It is evident, however, that the wind is strongly modulated by the magnetic field.

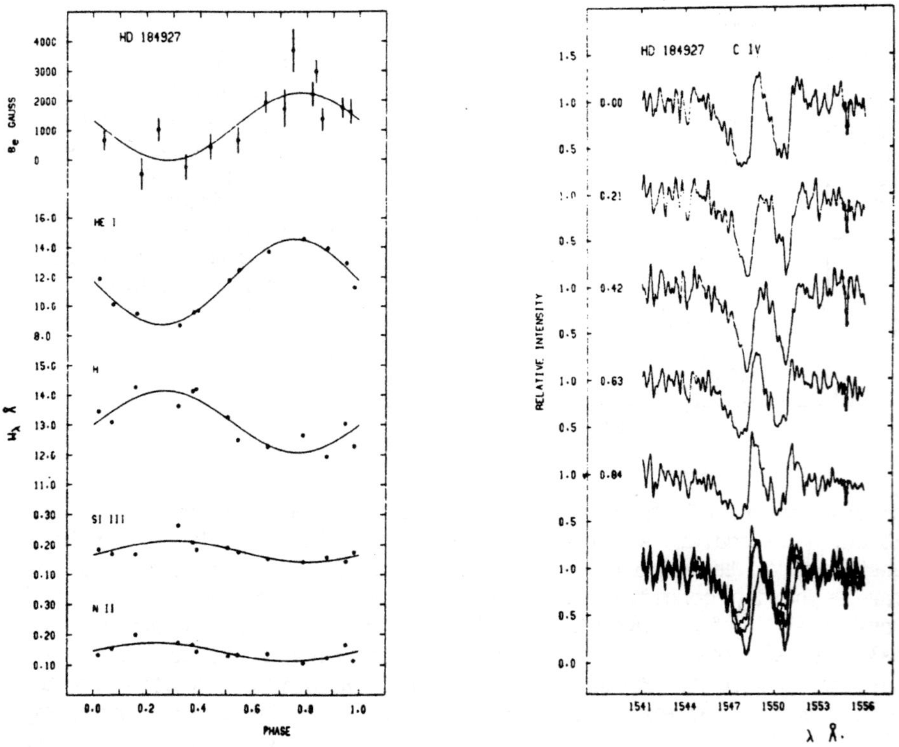

Fig. 10: The phase variation of HD 184927. Left panel: magnetic field, line strengths of He, H, Si III and N II (from top to bottom). Right panel: the resonance doublet of CIV, for various phases (left hand scale), and all phases superimposed, **at the bottom** (Barker et al. 1985).

Altogether, 8 helium strong stars have been checked for magnetically controlled winds, again CIV being the strongest indicator (Barker et al., 1985). A strange correlation has been found: the fast rotators do not show CIV in emission, but show H_α -emission instead, while for the slow rotators the opposite seems to be true (Fig. 11). More observations are needed before one can work out a definite model.

A similar campagne was started to observe helium weak variables, the first example being HD 21699 (Brown et al., 1985). Again, the CIV resonance doublet was found variable, with a period of 2.5 d, which is also the period of the magnetic field. In this case a model is forwarded. The approach however, is complicated, as the geometry in case of a strong magnetic field deviates from sphericity. In the case of HD21699, the wind occurs in form of a single jet or plume over one of the magnetic poles. It appears, though, that probably more work is

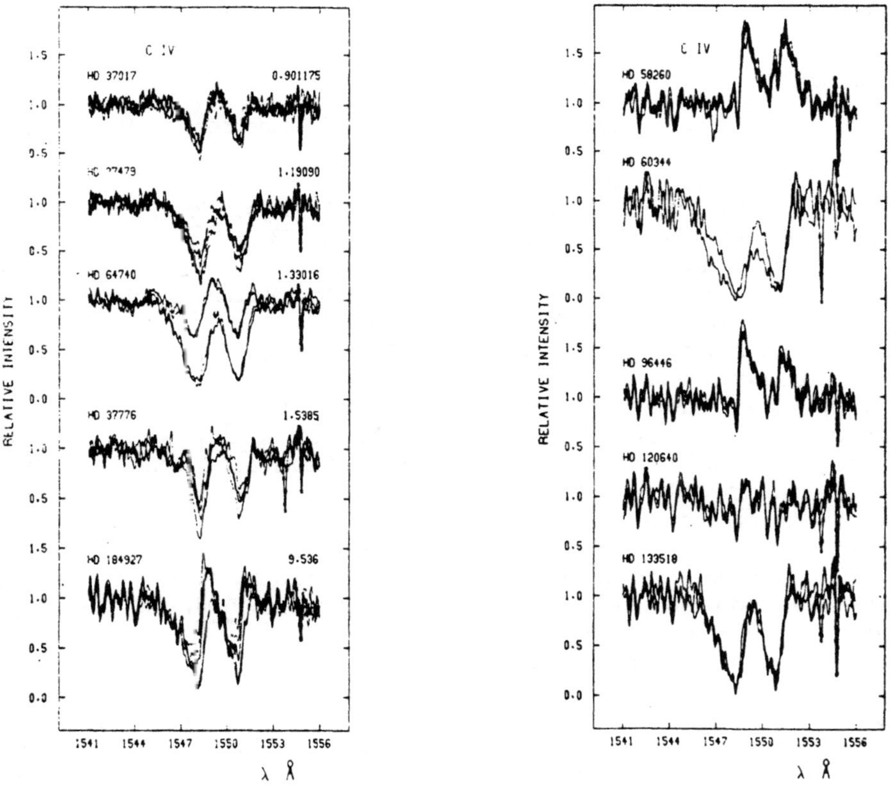

Fig. 11: Fast rotators (left panel) hardly show C IV in emission, while for the slow rotators (right panel) the opposite seems to be true (Barker et al. 1985).

needed, observational and theoretical, to corroborate this result. Of special importance in this context may be the large number of high resolution IUE spectrograms of σ Ori E which cover all crucial phases and which still wait unused in the databank of VILSPA.

Theory

This review cannot end without at least mentioning the important work done in the difficult field of modelling magnetospheric plasmas around early type stars. Some promising progress has been made by Havnes and Goertz (1984), but time does not permit discussion. When in the end, a selfconsistent picture of the structure of the magnetic field, of the magnetosphere and of the magnetically controlled wind around an early B type star emerges, it may be possible to make predictions as to magnetospheres of still hotter stars like O stars where magnetic fields cannot be observed, but which nevertheless may also have magnetically controlled winds.

References

Barker, P., Brown, D.N., Bolton, C.T. and Landstreet, J.D.: 1985, preprint

Bolton, C.T.: 1983, Harv. Obs. Bull. 7, 241

Bonsack, W.K. and Dyck, H.M.: 1983, Astron. Astrophys. 125, 29

Borra, E.F., Landstreet, J.D. and Thompson, I.: 1983, Astrophys. J. Suppl. 53, 151

Brown, D.N., Shore, S.N. and Sonneborn, G.: 1985, Astron. J., in press

Carpenter, K.G.: 1985, Astrophys. J. 289, 660

Drake, S.A., Abbot, D.C., Bieging, J.H., Churchwell, E. and Linsky, J.L.: 1985

Groote, D., and Hunger, K.: 1977, Astron. Astrophys. 56, 129

Groote, D., and Hunger, K.: 1982, Astron. Astrophys. 116, 64

Groote, D. and Kaufmann, J.P.: 1981, in The Chemically Peculiar Stars of the Upper Main Sequence, 23rd Liège Astrophysical Colloquium, p. 435

Havnes, O.: 1981, in The Chemically Peculiar Stars of the Upper Main Sequence, 23rd Liège Astrophysical Colloquium, p. 403

Havnes, O. and Goertz, C.K.: 1984, Astron. Astrophys. 138, 421

Heber, U.: 1979, Diploma Thesis

Heber, U.: 1985, private communication

Heintz, W.D.: 1974, Astrophys. J. 79, 397

Hunger, K.: 1975, in Problems in Stellar Atmosphere and Envelopes, Baschek, Kegel and Traving Editors, Springer-Verlag Berlin Heidelberg New York, p. 57

Jordan, S.: 1985, private communication

Kaufmann, J.P. and Theil, U.: 1985, private communication

Kurucz, R.L.: 1979, Astrophys. J. Suppl. 40, 1

Madej. J.: 1981, in The Chemical Peculiar Stars of the Upper Main Sequence, 23rd Liège Astrophysical Colloquium, p. 379
Odell, A.P. and Lebofsky, M.: 1984, Astron. Astrophys. 140, 468
Pedersen, H.: 1979, Astron. Astrophys. J. Suppl. 35, 313
Pedersen, H. and Thomsen, B.: 1977, Astron. Astrophys. Suppl. 30, 11
Remie, H. and Lamers, H.J.G.L.M.: 1982, Astron. Astrophys. 105, 85
Shore, S.N. and Adelman, S.: 1981, in The Chemically Peculiar Stars of the Upper Main Sequence, 23rd Liège Astrophysicsl Colloquium, p. 429
Thompson, I.B. and Landstreet, J.D.: 1985, Astrophys. J. in press
Walborn, N.: 1983, Astrophys. J. 268, 195
Walker, H.J.: 1985, preprint

ON THE ULTRAVIOLET PHOTOMETRIC VARIABILITY OF THE HELIUM-WEAK B STARS (FROM ANS DATA)

E. I. Želwanowa and W. Schöneich

Zentralinstitut für Astrophysik der AdW der DDR
15 Potsdam, Rosa-Luxemburg-Str. 17A, GDR

ABSTRACT: Photometric observations in the ultraviolet region obtained by the ANS satellite enabled us:
to confirm with a slight improvement the rotational period of HD 22470 (UV lightcurves of HD 22470 compared with u and magnetic curves are shown in Fig. 1);
to confirm the variability with very small amplitudes for HD 74196, a member of the open cluster IC 2391 (Fig. 2);
to compare the amplitude-wavelength relations of some other He-weak stars with known periods (HD 35298, HD 142884, HD 144334, HD 175362, and HD 109026).
A preliminary rotational period for the star HD 109026 using the 11 ANS observations was found to be about 1.5 days.

Up to now the ultraviolet lightcurves have been published only for one of the helium peculiar stars, HD 125823 (MOLNAR, 1974). We have investigated known helium-weak B stars for which three or more (up to 11) measurements are available from Astronomical Netherlands Satellite (ANS) observations. The observations were carried out during October 1974 - April 1976 with the ultraviolet photometer in five bands with central wavelengths and bandwidths (in Å): 1550(150), 1800(150), 2200 (200), 2500(150), 3300(100). For the instrumental characteristics and data reduction methods see WESSELIUS et al.(1982).

We have no place here to discuss all investigated stars in detail. We present only two stars: He-weak star with large amplitude of variation, HD 22470, and HD 74196 for which the ANS observations seem to confirm the variability with very small amplitudes.

C. R. Cowley et al (eds), Upper Main Sequence Stars with Anomalous Abundances, 271–274.

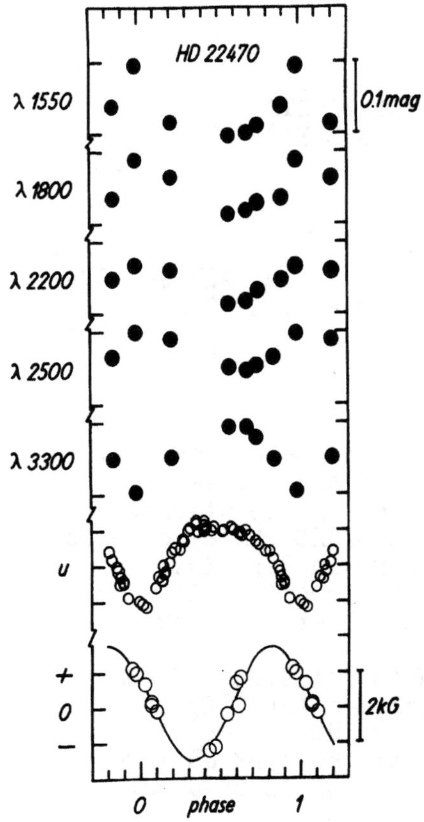

Fig.1
Ultraviolet lightcurves (ANS data) for HD 22470, plotted with the new ephemeris HJD (near uvby min.)= 2443456.539+1.92895 days (see text). Error bars are smaller than the size of the circles. Magnetic field and u curves are plotted for comparison.

HD 22470 = HR 1100 = EG Eri, He-weak, subclass Si

This star was found to be light variable (1.93 days) by RENSON and MANFROID (1981) and magnetic field variable by BORRA et al.(1983). Two periods (0.6785 days and 1.935 days) fit the set of 12 magnetic field observations almost equally well. A quantitative analysis of the EG Eri lightcurves shape in the Strömgren system is given by MATHYS and MANFROID (1984).The used measurements are nearly uniformly distributed over the period improved to 1.9387 days. All lightcurves are in phase. The maximum is broad and probably consists of two maxima merged together. The amplitudes are about 0.06 mag. in y,b and v bands and at least 0.10 mag. in u. Six UV ANS observations of HD 22470 are available (4 between JD 2442448.788 and ...450.008, and two at JD 2442634.292 and ...4.496). They permit to confirm and slightly improve the period given by MATHYS and MANFROID(1984). Using as zero point the real observation near the light minimum (the measurements were kindly placed at our disposal by Dr.Mathys) we thus adopt for HD 22470 the ephemeris:

HJD(near uvby min.) = 2443456.5389 + 1.92895 days·E

This period differs from that obtained by MATHYS and MANFROID(1984) only by 1·Δf (see table 2 in their paper). Ultraviolet lightcurves for HD 22470, compared with u and magnetic curves (MATHYS and MANFROID,1984; BORRA et al.,1983), are shown in Fig.1. This star is variable in y,b,v (not shown in Fig.1), u and λ3300 in phase. Maximum amplitude is in u and λ3300. The amplitude at λ1550 is comparable to that at u and λ3300,but is in antiphase!

The UV lightcurves in the phase range near the positive magnetic extremum is sufficiently well represented by observations. Unfortunately only one observation falls near the negative magnetic extremum, preventing the more detailed interpretation.

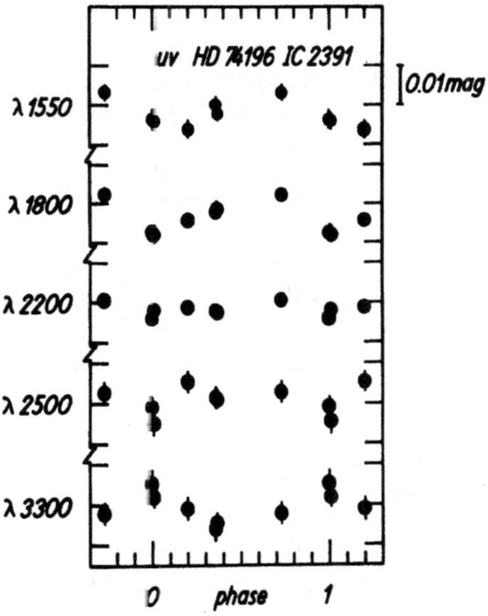

Fig.2
Ultraviolet lightcurves (ANS data) for HD 74196, plotted with the period 0.3880 days (JAKATE,1979).

HD 74196 = HR 3448, He-weak, member of IC2391

This star was found by JAKATE(1979) to be light variable with a period of only 0.3880 days and a small amplitude of 0.01 mag. in Strömgren b.

Six UV ANS observations in the time interval of 2.254 days are available for HD 74196. The ANS observations plotted in Fig.2 using the ephemeris:

JD(max λ3300) = 2442390.195 + 0.3880 days·E

suggest that the star is really variable. The amplitudes in λ3300 and λ1800 bands are about 0.01 mag. The star varies at λ1800 and λ1550 in antiphase to the λ3300 band.

Is HD 74196 indeed the He-weak star with the shortest known period? The high value of V•sini=235 km/s (UESUGI and FUKUDA , 1982) indicates a rapid rotation. Let us remember here that HD 74196 is a member of the open cluster IC 2391 and a spectroscopic binary (BUSCOMBE, 1965). Might it be that this star is also magnetic?

Conclusions

On the basis of the ultraviolet photometric ANS data we found that for the He-weak stars HD 22470 (Pleiades Group), HD 35298 (Ori OB1), HD 74196 (IC 2391,SB), HD 142884 (Sco-Cen, nonmagnetic, Si subclass), HD 144334 (Sco-Cen) and HD 175362 (Sco-Cen):

1) The variations in the λ1550 band is in antiphase to those in the λ3300 band and in the visible spectral region. This property agrees with the data for the stars HD 28843 (field), HD 125823 (= aCen, Sco-Cen) , and HD 175362, obtained for regions around λ2740 and λ1550 using TD-1 scans (SCHÖNEICH et al., 1983).
2) The amplitudes of the variations in the shortest (λ1550) and in the longest (λ3300) wavelengths bands are comparable to those in the visible light (u colour). This property of the He-weak stars is quite different from that of Ap stars, where the amplitudes at short UV wavelengths are in general considerably larger than in the visible.

If the period of the star HD 109026 (about 1.5 days, obtained from 11 ANS observations by SCHÖNEICH,1984) will be confirmed, this star would be an example of the He-weak stars, which vary in all spectral ANS bands in phase.

References

Borra,E.F., Landstreet,J.D., and Thompson,I.,1983, Astrophys. J.Suppl.Ser., 53, 151

Buscombe,W.,1965, Mon.Not.R.A.S., 129, 411

Jakate,S.M.,1979, Inf.Bull.Var.Stars, 1536

Mathys,G. and Manfroid,J.,1984, Preprint, submitted to Astron.Astrophys.Suppl.Ser.

Molnar,M.R.,1974, Astrophys.J., 187, 531

Renson,P., and Manfroid,J.,1981, Astron.Astrophys. Suppl. Ser., 44, 23

Schöneich,W., Želwanowa,E., and Jamar,C., 1983,unpublished

Schöneich,W., 1984,unpublished

Uesugi,A., and Fukuda,I., 1982,Revised Catalogue of stell. rotational velocities,Kyoto Univ.,Japan

Wesselius, P.R., van Duinen,R.J.,de Jonge,A.R.W., Aalders,J.W.G.,Luinge,W., and Wildeman,K.J.,1982, Astron.Astrophys.Suppl.Ser., 49, 427

Discussion appears after the following paper.

INTERMEDIATE PECULIAR STARS : THE Bp-Ap Si STARS.

C. Mégessier
Observatoire de Paris-Meudon
92195 Meudon Principal Cedex
France

ABSTRACT. This paper contains two parts :

1) A review of the most recent results on the silicon peculiar stars. Five topics will be reviewed : parameters of Bp-Ap Si stars - UV spectrum - spectrophotometry - spectrum variations and flux redistribution -magnetic braking.

2) The horizontal diffusion of the silicon in the presence of a magnetic field and, as a consequence, the evolution of the silicon abundance repartition in the stellar atmosphere. A first observational test is presented that supports the predicted scenario.

1. INTRODUCTION

The Bp-Ap Si stars span a wide effective temperature range between 10 000 °K and 16 000 °K. A magnetic field has been detected for each star of this group that was measured for this purpose. Among the new observational data that are continuously obtained, the spectrophotometry from the violet until the near IR and the high resolution UV spectra gives important and new information on the stellar atmospheres. Also the variations of the observed quantities as a function of the rotational period play an important part in the the progress of our understanding of the physical properties of the atmospheres.

Recently, controversial statements and observational tests have been published to precise the role of the magnetic field in the braking of the rotational velocities of the Bp-Ap stars.

Concerning the diffusion in the presence of a magnetic field, it has been predicted theoretically that during the main sequence stage of the star, the magnetic field lines guide the chemical elements in the stellar atmospheres into places where they cumulate. The case of silicon has been studied in details. The first results of an observational test are presented here.

C. R. Cowley et al (eds.), Upper Main Sequence Stars with Anomalous Abundances, 275–285.

2. REVIEW OF THE MOST RECENT RESULTS

2.1. Some statistical properties

2.1.1. Oscillator strengths.

A paper to appear soon will be quite useful to anyone interested in the study of the silicon in the stellar atmospheres : it is a compilation of the Si oscillator strengths for all the lines between 800 Å and 9500 Å by T. Lanz and M.C. Artru (1985).

2.1.2. Effective temperatures : determination and critical discussion of several photometric estimators.

T. Lanz (1985) determined the effective temperature of 13 normal B2 to A7V stars and for 11 Ap stars with the Blackwell-Shallis method (i.e. using the integrated flux through almost all the wavelengths). He showed that the $(B2-G)_0$ Geneva photometric index is a good photometric estimator for the Ap stars for which no UV flux measurement exists, although the derived T_{eff} are higher than those obtained by means of the integrated flux. The differences may be as large as 2000 °K. The $(R-I)_c$ index may also be good but few data are available. The others photometric estimators give overestimation of the T_{eff} up to 3000 °K. Using his $(B2-G)_0$ versus T_{eff} calibration, T. Lanz gives the T_{eff} of 28 Ap stars, among them 11 are Ap Si.

These results and the discussion are important since the choice of the effective temperature is quite critical for several purposes, among them the interpretation of the stellar spectra by means of atmospheric models.

2.1.3. Spectral line intensities and statistical derived properties.

P. Didelon (1985) made visible spectra line identification and intensity measurements for a sample of 20 Bp-Ap Si stars. He looked for correlations between the metallic line strengths and other properties of the stars such as $v_e \sin i$ or magnetic field strength. Among his conclusions we may notice the anticorrelation between λ 4233 Fe II line intensity and the projected rotational velocity. One can infer that a large rotational velocity induces hydrodynamical conditions that prevent overabundances, such as built by diffusion processes. Also "the intensity of the Si I lines seems to be correlated to the magnetic field strength, which can be explained by the radiative diffusion of Si in the presence of magnetic field".

2.2. UV Spectrum.

M.C. Artru and T. Lanz are studying the UV fluxes around the well-known absorption feature at λ 1400 Å. They compare the fluxes of some Ap Si stars of which T_{eff} are around 11 000 °K and 13 000 °K to the fluxes of normal stars. Their aim is to identify unknown absorption features and to look for their effective temperature dependence (see their communication).

M.C. Artru and R. Freire are identifying high resolution IUE spectra of 10 Ap Si stars and one normal star. They first focused their attention on the Ga II resonance line at λ 1414 Å (see their communication).

Takada-Hidai et al (this Colloquium) also identified GaII in UV spectra of CP stars.

2.3. Spectrophotometry.

In their extensive series of spectrophotometric observations, Pyper and Adelman (1985) measured recently the two Ap Si stars Cu Vir and HD 34452. The fluxes are well represented from 3300 Å up to 6000 Å by a single atmospheric model, which is not frequently the case among Ap stars which Balmer jump is generally smaller than those of normal stars. Cu Vir, which presents small λ 4200 and λ 5200 absorption features, is silicon variable, whereas HD 34452 which presents a strong λ 5200 absorption feature is not. However, Shylaja and Babu (1985) who observed also HD 34452 photoelectrically found variations centred around λλ 4200, 5200, 5700, and 6300 Å, as well as apparent variations of the continuum. So, for that star, the question is opened : are there variations or is the precision of the measurements questionable ? In any case the comparison between both stars is interesting.

2.4. Spectrum and flux variations.

Hempelman et al (1984) used TD1 spectrum of HD 170000 to test the influence on the UV spectrum of the presence of two spots on the stellar surface. They pointed out the effect of Fe II and Si II line intensities within the spots.

Iliev (1984) used visible and UV spectra to study the variability of HD 27309. Si, Fe, and Ti line intensities vary by about 40 % over the cycle of the star. He measured the blocking and showed that the spectrum variability causes the observed photometric variations. The continuum flux variation is crucial. The redistribution of the UV flux to the visible region is confirmed for that star.

2.5. Magnetic braking.

The large magnetic field of Bp-Ap stars is commonly involved to explain their low rotational velocities as compared to those of normal stars. The question arises to know whether the magnetic braking occurs before or during the main sequence stage of the star. Hartoog (1977), who made a statistic (over a sample of 25 Ap stars belonging to 9 open clusters), found no correlation between ages and projected rotational velocities. Abt (1979) distinguished the Ap Si stars from the cooler Sr Cr Eu Ap stars and he thus showed that for Ap Si stars v_e sin i decreases with the age. Wolff (1981), who increased the sample used by Abt, found a correlation between age and v_e sin i for Ap Si stars, but not between effective temperature and v_e sin i. She concludes that the magnetic braking should occur on the main sequence since the times spent before the main sequence are too much short to produce differential effect as a function of the time.

More recently, Klochkova and Kopylov (1984) and North (1984) infirmed this conclusion. The former did not get a correlation of the rotational velocities neither with the age nor with the stellar temperature. They conclude that the angular momentum of each star is obtained at its earliest stage or even during the stellar formation. North (1984) compared the period distributions as a function of the time, based on theoretical magnetic braking mechanisms, with the observed period distributions. His conclusion is that the magnetic braking mechanisms involved are too much efficient to explain the observed distributions, the hydrodynamical rotational braking proposed by Fleck (1980) being far too much efficient than the accretion theory proposed by Mestel (1975). North found that the Si stars do lose their angular momentum before the main sequence phase. The old Si stars have essentially the same periods as the young ones, though the shortest periods are absent among the oldest stars. The conservation of the angular momentum during the star evolution on the main sequence star is sufficient to account for the observed periods. North presents a new study in this colloquium which would confirm these conclusions.

In all these studies, the assumptions have to be discussed carefully. If it seems that the stars are slowed down during their main sequence stage, the question of the effectiveness of the magnetic braking, either before or during the main sequence stage, is not completely cleared up.

3. THE HORIZONTAL MIGRATION OF SILICON IN THE ATMOSPHERES OF Bp-Ap STARS. TIME DEPENDENT SILICON REPARTITION ON THE STELLAR SURFACE AND FIRST OBSERVATIONAL TEST.

We first remember the main principles of the diffusion in the presence of a magnetic field which give rise to ovserabundances tied to the magnetic field geometry. Then we give the main results obtained for the silicon. A first test of the predicted effects, based on field stars, is presented.

3.1. The theoretically expected scenario.

3.1.1. Diffusion in the presence of a magnetic field-general pattern.

The diffusion in the absence of a magnetic field gives rise to vertical displacements of the atoms in the stellar atmosphere (Michaud 1970). Using Chapman and Cowling (1970) results, Michaud, Mégessier and Charland (1981) showed that, in the magnetic stars, the ions are trapped by the magnetic field lines. Thus, everywhere these lines are inclined on the stellar surface, the ions diffusion velocity will have both a vertical and a horizontal component v_V and v_H, the latter being the smallest (see Fig. 1). The analytical expressions of v_H and v_V are given in Michaud et al (1981), Alécian and Vauclair (1981) and extensively in Mégessier (1984).

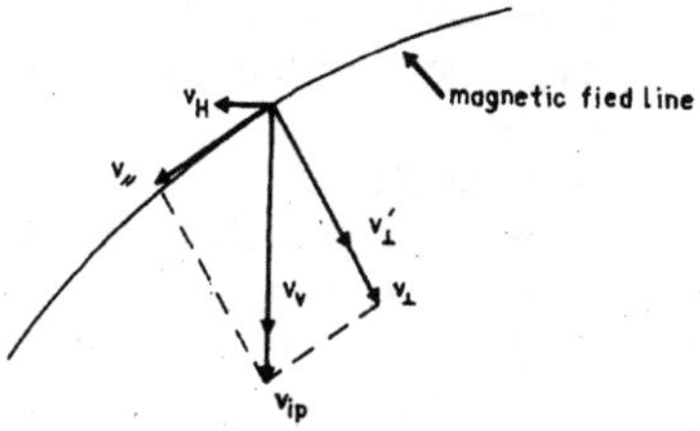

Fig. 1 : The vertical diffusion velocity v_{ip} splits into $v_{//}$ which is parallel to the magnetic field line and $v_\perp$ which is perpendicular to it. The perpendicular component $v_\perp$ is reduced, due to the magnetic line ($v_\perp'$). Finally, one gets v_H and v_V, respectively parallel and perpendicular to the stellar surface.

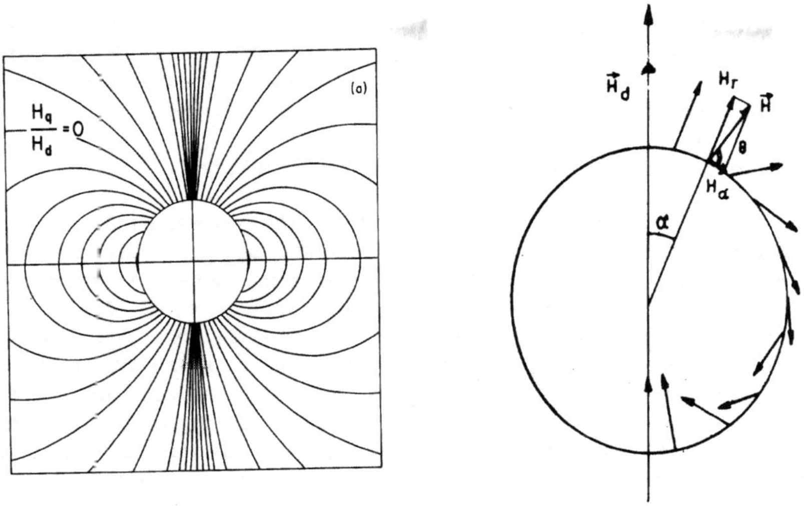

Fig. 2 : a) The magnetic field lines in the case of a dipolar field.
b) The corresponding local magnetic field on the stellar surface.

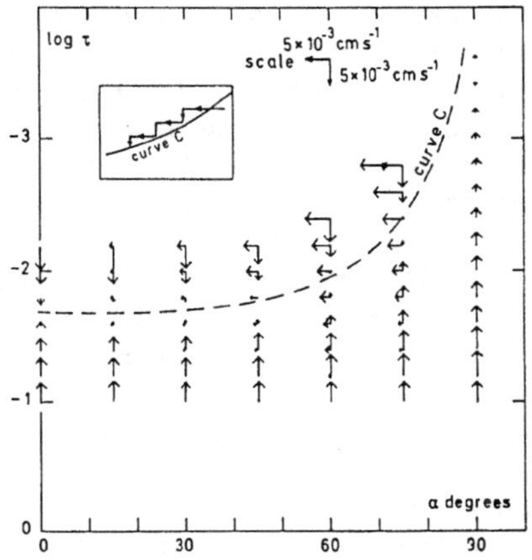

Fig. 3 : The local vertical and horizontal components of the diffusion velocity as a function of the optical depth τ and of α, where α refers to the position on the stellar surface (see Fig. 2a). At the magnetic pole $\alpha = 0°$ and at the magnetic equator $\alpha = 90°$. The dashed line C is the locus where $v_z = 0$. The insert shows the displacement of an ion along the curve C.

Due to the geometry of the magnetic field, these two components are a function of the position on the stellar surface and on the optical depth. Figure 2 gives both the shape of the magnetic lines for a dipolar field and the field strength as a function of the stellar latitude.

3.1.2. The case of silicon-time dependent diffusion.

Vauclair et al (1979) and Alécian and Vauclair (1981) computed the radiative forces on the silicon atoms. Due to the horizontal magnetic field lines, the silicon as a whole is prevented to sink and overabundances by factors up to 100 can be supported in the upper atmosphere. Michaud et al (1981) showed that the horizontal velocity is enough to allow the silicon atoms to migrate from the magnetic equator up to the magnetic poles in times shorter than the stellar life time, thus giving rise to different aspects of the star depending on its age. Mégessier (1984) studied in details the case of silicon. She computed the vertical and horizontal velocities of the silicon everywhere in the stellar atmosphere above the line forming region, from the magnetic equator up to the magnetic pole (see Fig. 3). The overabundances locally supported are depending on the inclination of the field lines on the horizontal. They are the largest at the magnetic equator and they decrease regularly up to the magnetic pole.

a) *Silicon fluxes in the atmosphere and predicted scenario.*
One may schematically summarize the situation as follows (see Mégessier, 1984). The atoms reach very quickly the optical depth where their vertical velocity cancels (less than 10^5 years) ; then they move slowly from the magnetic equator towards the magnetic poles under the optical depth where $v_z = 0$ (see curve c in Fig. 3). The horizontal displacements are such that the atoms moving up in the equatorial region can arrive near the magnetic poles in times shorter than the stellar life time.

The following scenario is expected. In the stars younger than 5 10^7 years, silicon is overabundant everywhere on the surface with a large maximum of abundance in a wide equatorial belt covering about one third of the stellar surface. After 5 10^7 or 10^8 years there exists no more equatorial maximum and the silicon is less overabundant everywhere else on the star. For the oldest stars, older than 5 10^8 or 10^9 years, small overabundance remains only in the two polar caps.

b) *Expected observational trends.*
From the scheme described above the silicon overabundance depends on three factors : the effective temperature, the magnetic field strength and the stellar age. Larger overabundances are supported higher in the upper atmosphere when the effective temperature is larger. Given the effective temperature, stronger the magnetic field, larger the silicon overabundance. The largest silicon overabundances are expected in the youngest stars.

An observational test of these effects is in progress, mainly using observations of Bp-Ap Si stars in open clusters. A first tentative test is presented here using field stars.

3.2 Observational test (Mégessier, 1985).

3.2.1 The data.

The data are collected in Table 1. We chose stars for which homogeneous values were available for the three parameters : T_{eff}, magnetic field, evolution estimator, and for the silicon abundances.

a) Effective temperatures and silicon abundances.
Homogeneous values of T_{eff} were obtained by means of the $(u-b)_0$ index calibrated as a function of T_{eff} (Matsushima 1969, Mégessier 1971). This index has been shown to be good enough for Ap stars when one needs to compare the stars in a sample (A discussion will be presented in a following paper). We used the silicon abundances determined by Mégessier (1971).

HD	$T_{eff\,(u-b)o}$	W(Hγ)	[Si/H]	$\lvert He_{max}\rvert$
173650	10 700	8.7	1.2	700
77350	10 375		0.3	470
133029	12 000	10.0	1.8	3 270
18296	11 900	8.6	1.2	1 350
219749	12 175	8.4	1.3	
112413	12 200		1.1	1 600
169952	12 600	8.3	1.8	
213871	12 600	8.3	1.5	
193722	12 600	7.5	1.3	
179527	12 600	6.8	1.2	
172761	12 800		0.0	590
27309	13 300	8.5	1.7	
224801	13 000	8.3	1.6	2 270
124224	13 700	8.1	1.3	1 260
21590	13 300	7.6	0.9	
224166	13 300	7.4	0.9	
25823	14 800		1.4	700
12767	14 800		0.75	290

Table 1

b) The magnetic field strengths.
The magnetic field is known only for 10 stars among the 17 stars of our sample. We considered the absolute value of the field maximum measured photographically by Bascock (1958) for 8 stars and we added two values measured photoelectrically by Borra and Landstreet (1980) and Landstreet et al (1975).

c) Age estimator.
The field stars exhibit a small but detectable evolution effect. Cramer and Maeder (1980) showed that the ages of the Ap stars of the main sequence are comprised between 1.5 10^7 and 3.2 10^8 years. From Mégessier (1984), at 1.5 10^7 years the silicon maximum centered on the magnetic equator is still present. After 3 10^8 years this maximum has disappeared and the overabundance elsewhere on the stellar surface is less strong than earlier.

For stars with spectral types earlier than A0, the Balmer lines equivalent widths may be used as a criterion of luminosity. $H\delta$ and $H\gamma$ equivalent widths were measured for most of the stars studied in Mégessier (1971). The values of $H\gamma$, available for 12 stars of our sample are given in Table 1.

3.2.2 Comparison between the observations and the predicted schenario.

Since the scheme we predicted depends on three parameters, we have to disentangle them.

a) Effective temperature and age effects.
The plot of the silicon abundances as a function of T_{eff} gives a cloudy pattern (Fig. 4). However, considering only the stars with the same $H\gamma$ equivalent width ($8.3 \leq W(H\gamma) \leq 8.7$ Å), i.e. with the same absolute magnitude, one gets a correlation between silicon abundances and T_{eff} (Fig. 4). The abundances increase with T_{eff}. This is in agreement with our prediction. Indeed in the HR diagram, along a line of constant M_V, the ages increase while T_{eff} decreases. Well the hottest and youngest stars are expected to show larger overabundances than the oldest and coldest ones.

b) Evolution effect.
To detect the age effect, we have to consider stars with same T_{eff}. It is why we arranged the data in Table 1 in sets of stars with about the same T_{eff}. For a given effective temperature, the silicon abundance decreases with the age (Fig. 5) as expected theoretically.

c) Magnetic field effect.
For a given effective temperature, the silicon abundance increases with the field strength, which is predicted theoretically (Fig. 6).

3.3 Conclusion.

Until now, the studies of the diffusion in the presence of a magnetic field do not take into account several effects which could modify the simplest scheme remembered here. Effects such as microturbulence and meridional circulation were mentioned by Michaud et al (1981). However, they would not strongly modify the main trends. The first observational test presented

Fig. 4 : The silicon abundance as a function of the effective temperature. The circled crosses are the representative points for the stars with : 8.3 ≤ W(Hγ) ≤8.7 Å, i.e. with about the same absolute magnitude.

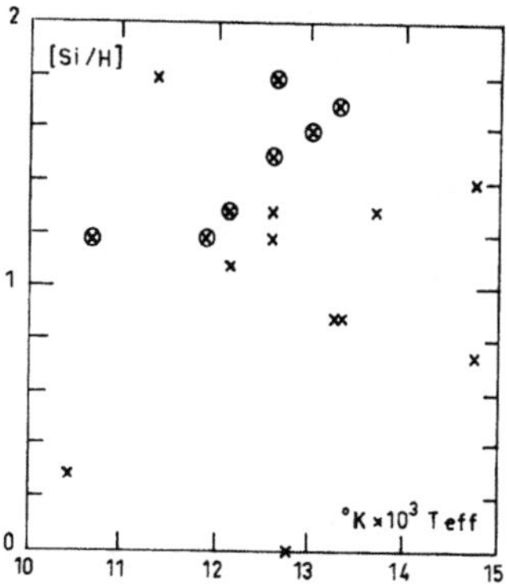

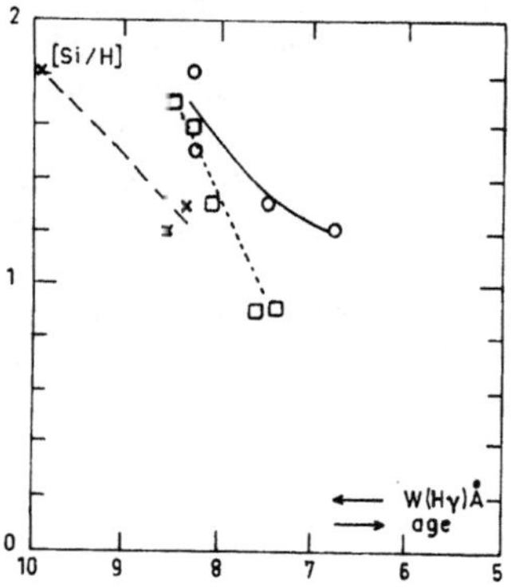

Fig. 5 : Silicon abundance as a function of the stellar age. Crosses : T_{eff} =11000, 12000°K ; circles : T_{eff} ≃ 12600°K ; squares : T_{eff} ≃13300°K. For a given effective temperature the silicon abundance decreases as a function of the stellar age.

Fig. 6 : Silicon abundance as a function of the maximum magnetic field strength. Symbols as in Fig. 5. Latin crosses : T_{eff} ≃ 10500°K ; stars : T_{eff} ≃ 14800°K. For a given effective temperature the silicon abundance increases with the magnetic field strength.

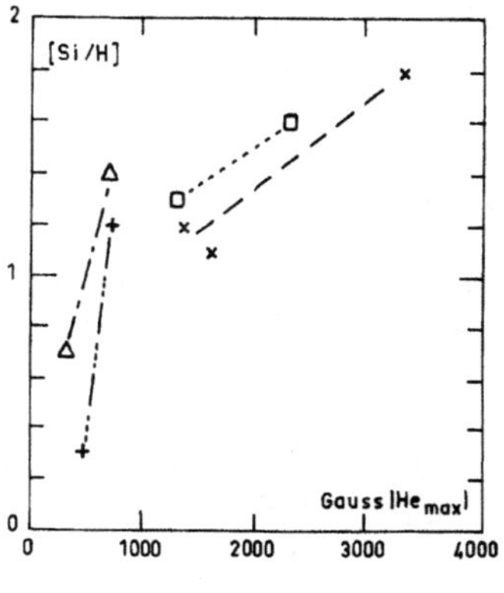

here clearly support the expected dependence of the silicon abundance on the three parameters : effective temperature, magnetic field strength and stellar age. If this first test is confirmed, it will be a strong support for the time dependent diffusion in the presence of a magnetic field, even if other phenomena have to be taken into account. Particularly, the dependence on the stellar age is a strong indication of the horizontal atoms migration along the magnetic field lines.

4. GENERAL CONCLUSION.

From the recent works, we can see that the increase of observational data confirms the general tendencies already known for Ap Si stars (and also for the other sub-types) : periodic photometric, line intensities and profiles variations, flux variations within and outside of the large spectral features (λ4200, 5200, 6300 Å). The correlation between these depressions and the other stellar properties has still to be improved ; it is not yet understood. This flux redistribution is confirmed by the study of some more individual stars. Progress are noticeable in the knowledge and the interpretation of the UV spectra although a great deal of work has still to be done in that field. Among other absorbents the role of silicon is stressed even if far to be completely explained. The role of the magnetic field to produce the surface inhomogeneities, mainly for the silicon, yields to a global description of the abundance in the atmosphere and to its stellar age dependence.

We feel that parts of the puzzle take place but the whole description of all phenomena acting in the Ap stellar atmospheres has not yet still arised and numerous fields of investigation are opened.

5. BIBLIOGRAPHY.

Abt, H.A. : 1979, Astrophys.J. *230*, 485.

Alécian, G., Vauclair, S. : 1981, Astron. Astrophys. *101*, 16.

Artru, M.C., Freire, R. : 1986, this colloquium.

Artru, M.C., Lanz, T. : 1986, this colloquium.

Babcock, H.W. : 1958, Astrophys.J. Suppl. *3*, 141.

Borra, E.F., Landstreet, J.D. : 1980, Astrophys.J. Suppl. *42*, 421.

Chapman, S., Cowling, T.G. : 1970, The mathematical Theory of Non-uniform Gases (Cambridge : Cambridge University Press).

Cramer, N., Maeder, A. : 1980, Astron. Astrophys. *88*, 135.

Didelon, P. : 1985, Revista Mexicana de Astronomia y Astrofisica, (submitted).

Fleck, R.C.Jr. : 1980, Astrophys.J. *240*, 218.

Hartoog, M.R. : 1977, Astrophys.J. *212*, 723.

Hemdelmann, A., Schoneich, W., Zelwanowa, E., Jamar, C. : 1984, in "Magnetic stars", 6th Scientific Conference of the Subcommission N°4 of the Multilateral Cooperation of the Acad. of Sc. of Socialist Countries", Riga, April 1984, ed.:V. Khokhlova, D. Ptitsyn, O. Lielausis, p. 46.

Iliev, I. Kh. : 1984, ibidem, p. 57.

Klochkova, V.G., Kopylov, J.M. : 1984, ibidem, p.80.
Landstreet, J.D., Borra, E.F., Angel, J.R.P., Illing, R.M.E. : 1975, Astrophys.J. *201*, 624.
Lanz, T. : 1985, Astron. Astrophys. *144*, 191.
Lanz, T., Artru, M.C. : 1985, Astron. Astrphys., in press.
Matsushima, S. : 1969, Astrophys. J. *158*, 1137.
Mégessier, C. : 1971, Astron. Astrophys. *10*, 332.
Mégessier, C. : 1984, Astron. Astrophys. *138*, 267.
Mégessier, C. : 1985, in preparation.
Mestel, L. : 1975, Mem. Soc. Roy. Sc. Liège, VIII, 79.
Michaud, G. : 1970, Astrophys. J. *160*, 641.
Michaud, G., Mégessier, C., Charland, Y. : 1981, Astron. Astrophys. 103, 244.
North, P. : 1984, Astron. Astrophys. *141*, 328.
Pyper, D.M., Adelman, S.J. : 1985, Astron. Astrophys. Suppl. *59*, 369.
Shylaja, B.S., Babu, G.S.D. : 1985, in press, abstract in "A peculiar Newsletter", N° 12, 1984, ed.:H. Hensberge and W. Von Rensberge.
Vauclair, S., Hardorp, P.J., Peterson, D.A. : 1979, Astrophys.J. *237*, 526.
Wolff, S.C. : 1981, Astrophys.J. *244*, 221.

DISCUSSION (Želwanowa and Schöneich)

KHOKHLOVA: Have any of these stars been investigated spectroscopically, and is there any information about the variations of lines of any elements?
ŽELWANOWA: Yes. For example, HD 175362 has well known variations of He I 4026 Å similar to those in a Cen. At the moment, I have no other information.
NORTH: I have a question about HD 74196. The period is really very short, and there are only six points, if I understood you correctly. To what extent can you guarantee the period?
ŽELWANOWA: [consults translator] We can't state it [the period] definitely for the moment, because we lack the observational data and we need to confirm them.

DISCUSSION (Mégessier)

STĘPIEŃ: Did you take into account centrifugal forces when calculating the motion of Si ions in the atmospheres of these stars?
MÉGESSIER: No, the model should be completed by taking into account some additional effects, such as the microturbulence, on the meridional circulation. As the rotational velocities of Ap stars are relatively small, the centrifugal forces may be considered to be second-order effects. A more precise model should take these into account.
STĘPIEŃ: In your diagram of Hγ equivalent width vs. Si overabundance, is the variation of Hγ with effective temperature taken into account?
MÉGESSIER: To study the evolutionary effects, I grouped the stars with the same effective temperature together, so that the only effect on the Balmer line is the gravity.
KHOKHLOVA: Could you please show the position of the Si stars HD 124224 and HD 34452, with large overabundances and weak magnetic fields, on your diagram?
MÉGESSIER: HD 124224 (CU Vir) is here [indicates]. I did not consider the other star because I did not have its Balmer line equivalent width. It seems to be in a consistent position on the [Si/H] vs. T_{eff} plot.
DOLGINOV: Magnetic traps are imperfect. It is well known from laboratory experiments with devices such as 'Tokamak', that the time for particles to escape from the trap is not governed by the normal diffusion in the field, but is due in some cases to Bohm diffusion which is faster, and is due to the various instabilities. The same is true for the trap in the terrestrial megnetosphere. Thus the Si particles can escape from the trap in a much shorter time than that obtained from the diffusion velocity. Have you taken into account these possibilities? Also, how long are the Si particles kept in the trap, and is this sufficient to get the observed overabundances?
MÉGESSIER: In the laboratory the conditions are not the same as in the stars. Here, the forces which maintain the Si ions trapped in the magnetic field lines are vertical, that is to say, normal to the field lines, which is not the case in the laboratory. Also, the equilibrium here is not static. There are permanent ascending and descending vertical Si flows in the atmosphere. In this way the descending atoms

C. R. Cowley et al (eds.), Upper Main Sequence Stars with Anomalous Abundances, 287–290.

are continuously replaced by the ascending ones.

MICHAUD: Mégessier's model is the simplest model that can be constructed using diffusion. In so far as it is confirmed by observations, no additional amplifications are needed. We realize that additional instabilities may well exist. However, they may not be important for the evolution of the abundances.

COWLEY: The lower effective temperatures for the Si stars which have been found as a result of the blanketing re-open the old debate between Leckrone and Preston concerning the He abundance. If the effective temperatures are really lower than was thought 10 - 15 years ago, then surely the underabundance factor of He must be smaller, that is to say, the He must be more nearly normal than we thought it was. Has this question been re-examined in the light of these new lower effective temperatures, and if so, what is the result?

MÉGESSIER: If the effective temperatures are in fact lower than those determined from visible wavelength colour indices, then perhaps He is not as underabundant as it was thought to be. However, one has to check the exact influence of those new temperatures on the stellar atmospheric structure and on the models.

COWLEY: The point is that the phenomenological spectral type would then be closer to the true type according to T_{eff}, because with low dispersion spectra we have classified these stars as later in type because the He lines are weaker. But, the He lines may be weaker because the T_{eff}'s really are lower than the colour temperatures. This would bring the spectral type more into line with the new effective temperatures, and they would be lower than the older colour temperatures.

MÉGESSIER: Yes, yes.

COWLEY: The plot of W_λ[Fe II 4233 Å] vs. v sin i is very interesting, but because this line is on the flat part of the curve-of-growth its strength can depend on factors other than the abundance of iron, such as microturbulence, etc.

MÉGESSIER: I think you are right. I mentioned this because it is a clearly observed effect, and Didelot, who made the statistical study, proposed that it is an abundance effect. However, he does not exclude any other effects; it's just a hypothesis.

ADELMAN: Heasley, Wolff & Timothy [Ap. J., 262, p. 663, 1982] have shown that one can obtain He/H values with errors of 10%, provided that one uses data with S/N ≥ 50. Such observations are needed for Si stars. Determinations of He/H with equivalent width values are notoriously unreliable.

I wish to make an additional brief comment about spectrophotometry. This is, to a large extent, a 'black art'. Most observers use mean extinction coefficients instead of properly determining it each night. A good check on the quality of the data is to synthesize intermediate band colours such as Strömgren or Geneva colours. Only if the agreement of the synthesized and directly observed values is good, should the spectrophotometry be trusted. Observers who do not take an adequate number of standard measurements (6 or more) can easily produce bad data. As referee, I recently rejected a paper based on only one or two standard star measurements per night.

DWORETSKY: You're not very nice, are you? [laughter]
ADELMAN: No! But they could not reproduce the standard colours!
ARTRU: The gallium overabundance, which was considered to be characteristic mainly of Hg-Mn stars, is also shown to be commonly overabundant in Si stars in the poster paper at this Colloquium presented by Takada, Sadakane & Jugaku. I would like to ask, are there definite sets of elements overabundant in each class of peculiar star?
MÉGESSIER: Each group of CP stars is characterized by the most prominent enhanced elements. From time to time, papers appear which claim the exclusion of overabundances of some specific species in the same stars, but so far nothing conclusive can be said. One may find several abundance anomalies in the same spectrum, with relative overabundances being different from star to star.
KHOKHLOVA: Two remarks concerning Cowley's question about helium. The first is that the meaning of "effective temperature" is not always explained. Is it the parameter in the Stefan-Boltzmann law, or is it the real temperature in the atmosphere, which determines the colour, line strengths, ionization, excitation, etc? Could you explain which meaning is used in your paper?
MÉGESSIER: The value of effective temperature was deduced from the use of the Blackwell-Shallis method, using the angular diameters, bolometric corrections, and infrared fluxes.
KHOKHLOVA: So it is the parameter in the Stefan-Boltzmann formula. So it doesn't reflect the real temperature in the atmosphere unless you take blanketing effects into account; this is what I meant.

I would like to mention another problem with He. Perhaps my paper on He NLTE excitation wasn't noticed, but there may be very strong NLTE effects for He excitation because everything depends on only one resonance line which is optically thick, so people assume that there is an equilibrium excitation. But, this resonance line is coupled with many Fe IV, Mn IV and other lines, and this could lead to very strong deviations from LTE.
MÉGESSIER: It is true that you have to make a careful study of the bolometric correction to get the right effective temperature, since it is realized that if you deduce it from the visual fluxes only, you get too high a value. If you use only the ultraviolet, you deduce too small a value. It is best to use data from all the wavelength ranges available, whatever the blanketing.
HUBENÝ: I would like to emphasize that effective temperature only makes sense as a measure of the total radiation flux coming from below through the boundary of an atmosphere. Its relation to the physical temperatures (e.g., electron temperature) as a function of depth has to be obtained from a model, and its relation to various 'temperatures' (colour temperature, excitation temperature, etc.) has to be obtained in each particular case by appropriate transfer solutions. This calculation can be done even in NLTE.

Concerning the He problem, I think that although the resonance lines could affect the formation of the singlet system, the triplet system should be affected to a much lesser extent, because these systems are not directly coupled, but are only connected by very weak intercombination lines. These are especially weak in A type stars.

Could you tell me if there is any other evidence for changes in effective temperature - in other words, the total flux - in Ap Si stars, besides those you mentioned, such as Muthsam and Stepien for α^2 CVn, and the recent work of Iliev?

MÉGESSIER: No, those are the first two.

HUBENÝ: I think it is very hazardous to speak of the concept of an averaged effective temperature, and from the methodological point of view, it is too easy to introduce new concepts which might explain everything.

MÉGESSIER: The point is that at two different phases of the same star, you can not use the same effective temperature model to compare what you see. But it does not mean that the effective temperature of the star is different. It just means that you can not invoke only line intensities to reproduce what you see. It is just a kind of fictitious parameter that you try to fit with some T_{eff} and some metal abundances. [barely audible, but apparently heated, background discussion] I think that Dr. Stępień is concerned!

STĘPIEŃ: I would like to correct one of your statements! The effective temperature is, of course, connected through the Stefan-Boltzmann formula with the total emission from a give element of surface, but it does not have to be directly connected with the total flux observed from a star. If you have a distribution of effective temperature over the surface, each element has its own value, which is a well-determined physical parameter, but the total flux integrated over the whole surface of the star is not related to any of these particular values of temperature. This means you need the concept of an average effective temperature. In other words, you may have the total flux obtained from the star integrated over all wavelengths. If the star is not spherically symmetric, or has spots, then you may still have the total flux from the star constant over all phases, but what you observe is something which represents the average temperature over the star at different phases. [interrupted by inaudible voice] No! The meaning of T_{eff} is that it is connected with emission from a given element, say one square centimetre. [interrupted again] But not from the whole star! Just from one square centimetre. Well, it is that way!

Is there really any observational evidence that the total integrated flux from a star changes with phase? I have seen a few papers, but I don't know whether the results are statistically above the errors. The paper given at this Colloquium by Polosukhina and Malanushenko on 53 Cam says that the flux may marginally change by 1-2%, but who knows if this is real or not? All I am saying is that some people have reported variations of the total integrated flux with phase.

ADELMAN: Much of the residual variability in the total flux in Leckrone's study of HD 215441, which is also seen in its effective temperature, is due to the lack of correction for the variability of the 5200 Å feature, which was unknown at the time the paper was written. Spectrophotometry shows that the strength of this feature varies by a factor of two. Correction for this substantially reduces the total flux variations.

BLANKETING HYPOTHESIS AND LIGHT VARIATIONS OF HD 27309

I. Kh. Iliev
National Astronomical Observatory
Bulgarian Academy of Sciences
Smolyan, BG-4700
Bulgaria

ABSTRACT. The influence of variable blanketing on the photometric variations of hot, silicon Ap star HD 27309 has been estimated. Spectrum variations in the visible are not the principal reason for the observed light variability, but flux redistribution from UV into visible fit well the observational data. An reference is made on the fulfillment of the variable blanketing assumptions at least for silicon Ap stars.

I. INTRODUCTION

As suggested by Peterson (1970) the photometric variations of the Ap stars could be produced by flux redistribution from shorter into longer wavelengths. Following Peterson's hypothesis the observed photometric variability of these stars is naturally caused by the spectrum variations. In the framework of the variable blanketing the light changes in a given band are connected with a total flux balance between energy redistributed into this band from shorter ones and energy blocked by the spectral lines in the same band. The observational data are in qualitative agreement with Peterson's assumptions, but any quantitative analysis gives ambiguous results (Schoeneich, 1981).

To check the variable blanketing hypothesis investigations of Ap stars with well established photometric and spectral behaviour are needed. One of these stars is the bright, Si-type star HD 27309.

C. R. Cowley et al (eds.), Upper Main Sequence Stars with Anomalous Abundances, 291–294.

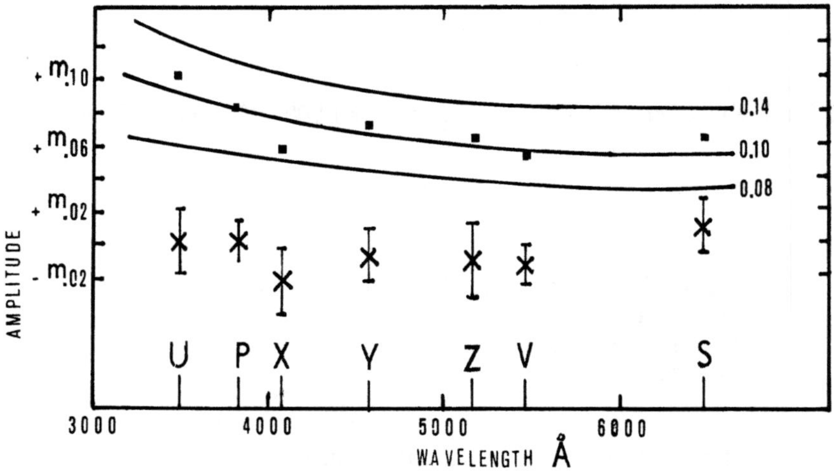

Fig.I. Observed and computed relations "amplitude - wavelength" for the Ap star HD 27309. Full squares represent the observed relation from Musielok et al. (1980); crosses with error bars - computed relation if only the spectrum variations in the visible are assumed; solid lines represent simulation results for the flux redistribution from the UV into the visible for three different values of the integral blocking coefficient.

2. OBSERVATIONAL DATA AND COMPUTATIONS

It is well known from Musielok et al. (1980) that HD 27309 exhibits a single-wave photometric variations with amplitudes up to 0.1 in the U-band of the Vilnius system. Jamar (1978) has found UV-flux variations in antiphase with the photometric ones. The amplitude of this UV-changes reaches 0.5. As shown recently by Iliev (1983a) HD 27309 is a moderate spectrum variable in the visible. The intensities of the Si, Fe and Ti lines vary by about 40 percent over the cycle of the star in phase with the photometric variations. Thus, light and spectrum changes are in qualitative agreement with the assumptions of Peterson's hypothesis.

During 1980-82 spectroscopic observations of HD 27309 were carried out at six-meter telescope with the main stellar spectrograph. Twenty spectrograms on IIaO and 103aF emulsions were obtained with dispersions of 7, 9 and 14 A/mm and spectral resolution of about 0.25 A in the wavelengths region 3300-6600 AA. The region of the Vilnius system is approximately the same.

After the usual reducing procedure line blocking coefficients in 25 A intervals were measured. Using the classical formula of Wildey et al. (1962) blanketing corrections for all bands of the Vilnius system are computed. The response curves are taken from Straizis and Zdanavicius (1970). The observed and computed relations "amplitude of light variations - wavelength" for HD 27309 are shown in figure 1. It should be pointed out that blanketing process means flux-redistribution or backwarming plus line blocking. This motivates the need to take into account not only the first or the second mechanism of flux variations.

3. DISCUSSION AND CONCLUSIONS

It becomes clear from figure 1 that the spectrum variations only in the visible region are not the cause for the observed light changes. The behaviour of the continuum is more crucial. The simulation results for the flux redistribution process from the UV-region for three different values of the integral blocking coefficient η are also shown in figure 1. The results for $\eta = 0.10$ fit well the observational data. Such value of the integral coefficient is in good agreement with Jamar's (1978) results. Thus, the photometric behaviour of HD 27309 is caused mainly by flux redistribution from UV into the visible.

It is important to mention that the results obtained for HD 27309 are very similar to these for HD 170000 (Musielok, 1981) and for HD 19832 and HD 184905 (Iliev, 1983b). This fact gives strong evidence in support of the variable blanketing hypothesis at least for hot, silicon Ap stars.

REFERENCES

Iliev, I.Kh.: 1983a, Soviet Astron. Zhurnal, 60, 597
Iliev, I.Kh.: 1983b, Soob. Sp. Astrofiz. Obs., 39, 5
Jamar, C.: 1978, Astron. Astrophys., 70, 379
Musielok, B.: 1981, Acta Astronomica, 31, 443
Musielok, B., Lange, D., Schoenaich, W., Zelwanowa, E., Hempelman, A., Salmanov, G.: 1980, Astron. Nachr., 301, 71
Peterson, D.M.: 1970, Astrophys. J., 161, 685
Schoeneich, W.: 1981, In Upper Main Sequence CP Stars, 23rd Liege Astrophys. Colloq. (Univ. Liege), p.235
Straizys, V., Zdanavicius, K.: 1970, Bull. Vilnius Astron. Obs., 29, 15
Wildey, R.L., Burbidge, E.M., Sandage, A.R., Burbidge, G.R.: 1962, Astrophys. J., 135, 94

A SPECTROSCOPIC STUDY OF THE Bp-STAR θ Aur

D. Kolev and E. Georgeva

National Astronomical Observatory
Bulgarian Academy of Sciences
P.O. Box 136
4700 Smolian
Bulgaria

ABSTRACT. The atmosphere parameters, spectral variability and the magnetic field geometry of the Bp-star θ Aur are evaluated from 30 coude-spectrograms. We adopt $T_e = 10\,800$ K and lg g = 3.40 (with lower reliability of lg g). The equivalent widths of the metalic lines show a relatively great scattering which troubles the period search. The best fit to our data gives the period $P = 3.620^d$, but the variability is clearly seen only for SiII lines. The uncertainties in the star parameters obtained bring to a great uncertainty in the magnetic dipole geometry : $i = 86^o \div 39^o$ and $\beta = 18^o \div 80^o$. We adopt the most plausible star parameters to be: $R = 3.8\,R_\odot$ $M = 3.5\,M_\odot$, $i = 80^o$, $\beta = 20^o$, Bp = 1 KGs .

1. STELLAR PARAMETERS

1.1. Observations

The spectrograms of θ Aur are obtained at the coude-focus of the 2-m telescope in BNAO mainly on IIaO plates with a dispersion 4 and 9 Å/mm. The aim of this work is to estimate the main stellar parameters and to study the spectral variations of θ Aur. We selected 30 "blue" plates of comparatively good quality and processed them by registrograms. We estimate (from sets of spectra obtained consecutively) the accuracy of W_λ to be about 10 ÷ 20% for lines with $W_\lambda \sim 300 \div 100$ mÅ. Theta Aurigae have a difficult for studying spectrum because of the broad lines - one can see the great scattering of W_λ noted in the very important papers by Adelman et al.(1984) and Van Rensbergen et al.(1984).

C. R. Cowley et al (eds.), Upper Main Sequence Stars with Anomalous Abundances, 295–298.

1.2. Hydrogen lines and stellar atmosphere parameters

We did not detect any significant variations of $H\gamma$ and $H\delta$ with the known periods of θ Aur. Therefore we averaged the profiles to improve the accuracy - $H\gamma$ coinsides exactly with the published ones by Gray and Evans(1973) and Van Rensbergen et al.(1984) from photoelectric scans.

The averaged profiles of $H\gamma$ and $H\delta$ were compared with the theoretical ones (Kurucz,1979). Our profiles fit the theoretical ones in the range: T_e=10 000 ÷ 12 000K and lg g = = 3.00 ÷ 4.00. The best agreement we have for the lowest T_e but then lg g must be embarrassly small. In addition $H\delta$ gives systematicaly lower (by 0.2 dex) lg g. Using the calibrations of $(B-V)_0$ vs. T_e given by Underhill(1982) we estimate T_e to be between 10 800 ÷ 11 000 K. Finally we adopt T_e=10 800 K and lg g =3.40 (with a lower reliability). These values are close to the published ones by Van Rensbergen et al.(1984) and differ rather significant from our first estimations (11500K;3.85), based only on $H\gamma$.

1.3. Radius and mass

We obtain V sin i = 53 km/s and must note the relatively great differences in the published values of V sin i - from 45 km/s(Babcock,1958) until 55km/s(Khokhlova et.al.,1985). The oblique rotator formula gives for our P and Vsin i a radius $R \geqslant 3.8\ R_\odot$. The same value we obtain from the calibrations of Underhill(1982) and Straizys and Kuriliene(1981) for the temperature adopted. The mean absolute visual magnitude from both calibrations is $M_V = -0.62^m$ and with the bolometric correction from Straizys and Kuriliene(1981) we obtain $M_{bol} = -1.07^m$. This allows us to estimate the mass to be 3.3÷ 3.6 solar masses. The well known relation between R , T_e and M_{bol} gives the same value for the radius. Assuming a mass of 3÷ 4 solar, for lg g=3.40 we obtain a radius of 5.6 ÷ 6.6 $R_\odot$, while $R \simeq 4 R_\odot$ needs lg g - 3.8. Because of the low reliability of lg g values we can conclude, that the actual value of the radius of θ Aur may to be in a wide range: 3.8 ÷ 6.6 solar radii .

2. SPECTRAL VARIATIONS

We used the W_λ obtained for SiII(5 lines),CrII(5 lines), FeII(3 lines), CaIIK and MgII 4481 for a period search in a range 1 ÷ 5 days. The intervals: 1 ÷ 1.5 ; 1.8 ÷ 2.1 and 3.6÷ 3.8 days are suspected,which covers the intervals obtained by Borra and Landstreet(1980) from magnetic field measurements. The best fit to our data gives $P = 3.620^d$, which practicaly coincides with $P = 3.618^d$ (Borra and

Landstreet, 1980) and $P = 3.6190^d$ (Adelman et al., 1984).

The averaged values $W_\lambda / \overline{W}_\lambda$ for the ions considered are plotted on the figure below. Nevertheless the great scattering at least for SiII it may be possible to suppose the presence of a "single wave" variations. The ephemeris are

$$JD(Si\ max) = 2\,444\,538.56 + 3.620\,E\ .$$

No attempt was made to map the plausible surface inhomogenities because of the low accuracy of our data.

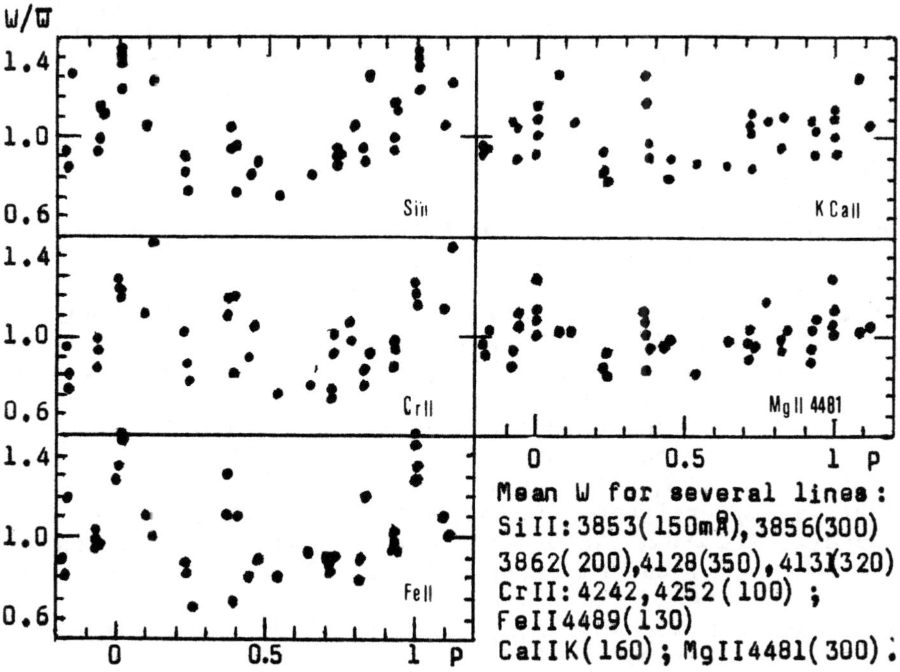

Mean W for several lines:
SiII: 3853(150mÅ), 3856(300)
3862(200), 4128(350), 4131(320)
CrII: 4242, 4252 (100);
FeII4489(130)
CaIIK(160); MgII4481(300).

Fig.1. The spectral variations of θ Aur.

3. MAGNETIC FIELD GEOMETRY

Following Borra and Landstreet's (1980) conclusions about a dipole geometry of the magnetic field of θ Aur and using theirs measurements, we made an attempt to re-obtain the possible values of the dipole parameters. The inclination of the oblique rotator "i" have a possible value between 86° and 39° for R of 3.8 and 6.0 $R_\odot$ respectively. The dipole-geometry formulae give values for the angle between rotational and dipole axes "β" in the range : 18° ÷ 80°! The polar field Bp maybe vary much more less – from ~1 to 2 KGs. In addition the uncertainties of V sin i influence the accuracy of the field geometry derived too. So, taken into account a possible error of $V\sin i$ about 10% for Aur we can estimate an additional uncertainty of "i" about 25° $\pm 10^{\circ}$ for different radii !

4. CONCLUSIONS

Some of our results are in good agreement with the results in the papers ouoted. There is no doubt that the actual period of θAur is about 3.6^{d}. The star is most probable a B9 III Si-variable. At present, we suppose, the greatest uncertainties occur in the field geometry. Obviously we need more precise data and, perhaps, a more extensive statistics to be able to make more definite conclusions.

REFERENCES

Adelman,S.J.,Hensberge,H.,Van Rensbergen,W.:1984, Astron. Astrophys.Suppl. 57, 121
Babcock,H.W.:1958, Astrophys.J.Suppl. 3, 141
Borra,E.F.,Landstreet,J.D.:1980,Astrophys.J.Suppl.42,421
Gray,D.F.,Evans,J.C.: 1973, Astrophys.J. 182, 147
Khokhlova,V.L.,Rice,I.B.,Wehlau,W.H.: 1985, poster:"The distribution of Fe,Cr and Si over the surface of the magnetic Ap-star Aur", IAU Coll.90, Crimean Astrophys.Obs., May, 1985
Kurucz,R.L.: 1979, Astrophys.J.Suppl. 40 ,1
Straizys,V.,Kuriliene,G.: 1981, Astrophys.Sp.Sci.,80,353
Underhill,A.: 1982, in:"B stars with and without emission lines",ed./aut. A.Underhill and V.Doazan , NASA- CNRS , p.60
Van Rensbergen,W.,Hensberge,H.,Adelman,S.J.: 1984, Astron. Astrophys. 136, 31

THE ULTRAVIOLET SPECTRAL ENERGY DISTRIBUTION OF THE MAGNETIC Ap STAR HD 170000 INSIDE AND OUTSIDE OF THE SPOTS

W. Schöneich, A. Hempelmann and E.I. Želwanowa

Zentralinstitut für Astrophysik der AdW der DDR
15 Potsdam, Rosa-Luxemburg-Str. 17A, GDR

ABSTRACT: Sixty five TD-1 scans of the magnetic silicon star HD 170000 have been investigated using the method for lightcurves analysis proposed by Hempelmann and Schöneich. The fluxes originating in two spots and in the rest surface of the star have been separated. The conclusion is that the whole stellar surface is strongly peculiar. Spot 1 is a region with the same peculiarity as the rest surface, but strengthened, whereas spot 2 shows additional peculiarities with respect to spot 1 and the rest surface.

Introduction. The magnetic CP stars are rotational variables. The properties of the variability of the peculiar spectral lines and of the brightness indicate that the peculiarity is correlated with peculiar regions on the star. But it is not clear until now, what are the properties of the stellar surface outside of such spots, which is not seen directly. The method for rotational lightcurve analysis proposed by Hempelmann and Schöneich (this conference) permits to separate the flux originating in the spots from that of the rest stellar surface. We will present here the results of the application of this method to the spectral flux variations of the magnetic Ap star HD 170000 in the ultraviolet.

Observational data. The used ephemeris of HD 170000:

$$JD(U_{max}) = 2442229.40 + 1.71646 \text{ days} \cdot E$$

was taken from MUSIELOK et al. (1980).

For the spectrophotometric investigation we used 65 scans obtained by the ultraviolet Sky-Survey -Telescope on TD-1A which were kindly extracted by Dr. Jamar from the Liege magnetic tapes. The observations are presented as lightcurves in Fig. 1.

In order to reduce the number of parameters to be obtained from TD-1 scans the geometric parameters of the model assumed as wavelength independent have been obtained

C. R. Cowley et al. (eds.), Upper Main Sequence Stars with Anomalous Abundances, 299–304.

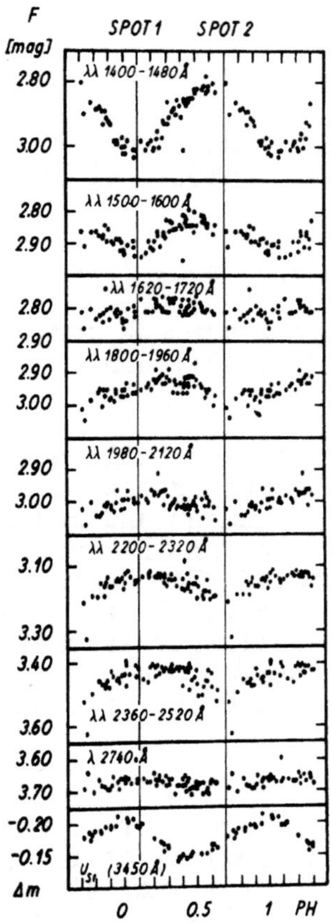

Fig. 1.
TD-1 lightcurves of
HD 170000

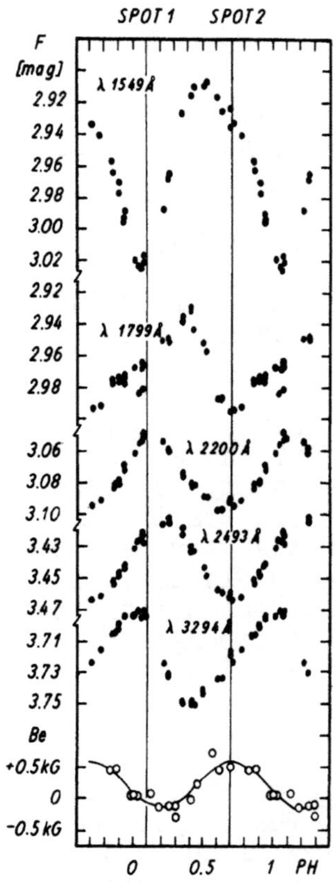

Fig. 2.
ANS lightcurves of
HD 170000

from 29 photometric observations from ANS (kindly placed at our disposal by Dr. Wesselius) and from the ten colour photometry (MUSIELOK et al.,1980). The U lightcurve (ten colour photometry) is shown in the bottom of Fig.1. Fig.2 shows the ANS lightcurves of HD 170000 and includes the magnetic field curve from LANDSTREET, BORRA (1977).
The transformation from fluxes to magnitudes was carried out using the relation:

$$m = -2.5 \operatorname{Log} F - 21.1 \quad .$$

Results of the model computation. The shape of the curves at different wavelengths suggests a two spots model. This assumption has been confirmed by the computations. The used method permits to obtain the stellar longitudes ϑ and polar distances β of the centres of the spots. The spots have been assumed to be circular and to have a uniform brightnees distribution. The radii R of the two spots, the fluxs from the unspotted surface Io and the flux densities from the spots, provided that the angle between the line of the sight and the rotation axis i and the limb darkening coefficient u are known, can also be obtained.

Period and lines width suggest $i \approx 90°$ for HD 170000. The assumed value was $i = 84°$. For $\lambda > 2000$ Å u has been taken from AL NAIMIY (1978) and for $\lambda < 2000$ Å $u = 0.9$ has been assumed. Because it is impossible to obtain β for the spots, if $i \approx 90°$, an equator symmetric model ($\beta = 90°$) has been adopted.

The differences between the parameters of the mean geometric models obtained from the different photometric observational sets (ANS , TD-1, 10 colour) are small and do not influence qualitatively the final results.
We accepted the mean geometrical model from the ANS observations, given by the following parameters:

spot 1	$\vartheta_1 = 36°$	$\beta_1 = 90°$	$R_1 = 70°$
spot 2	$\vartheta_2 = 251°$	$\beta_2 = 90°$	$R_2 = 80°$.

The vertical lines in Fig.1 and 2 illustrate the phase positions of the spots.

These parameters were used in the analysis of TD-1 scans. The fluxes of the two spots and of the rest surface were separated and plotted with 20 Å wavelength steps in Fig.3. The upper part of Fig.3 shows the spectral energy distribution of the "unspotted" star HD 170000, of spot 1 (star with the spectrum of spot 1), and of the comparison star HD 207971 (selected by JAMAR et al.,1978) fitted at λ 2740 Å. In the lower part of Fig.3 the comparison of the "unspotted" star and of spot 2 is given.

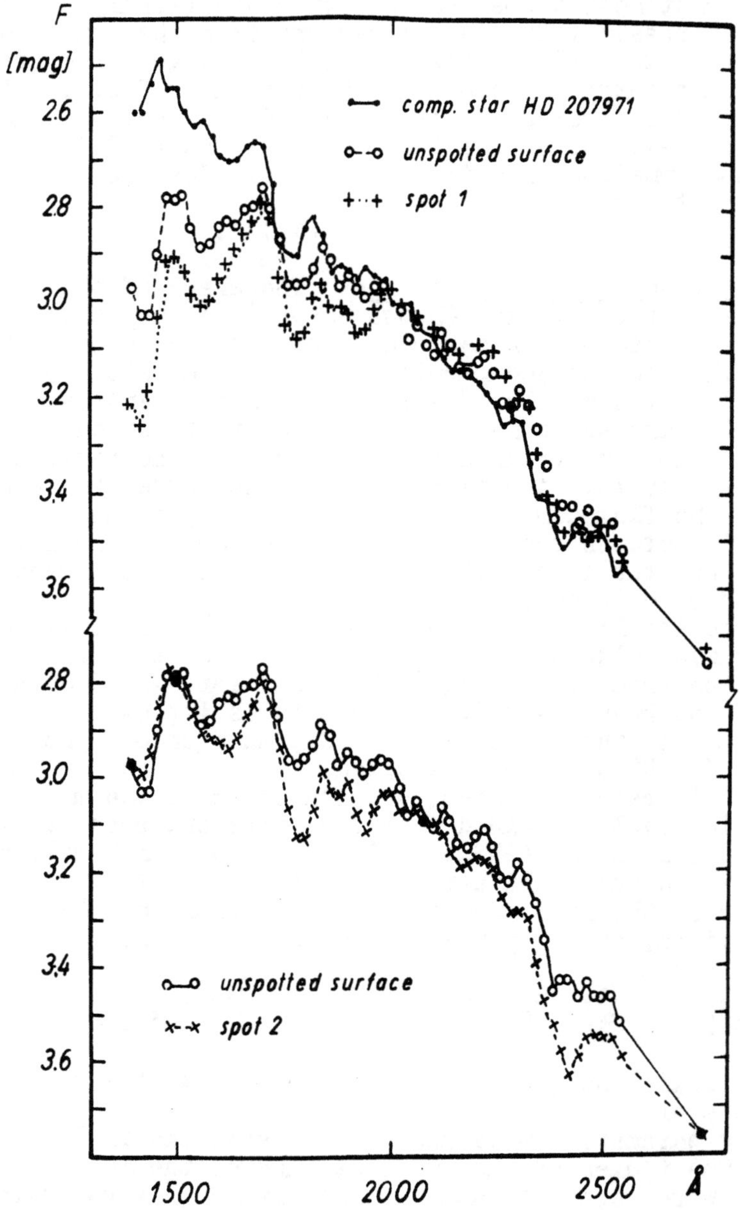

Fig.3. The spectral energy distribution of the unspotted star HD 170000 in comparison with those of spot 1 and the normal star HD 207971 (upper part) and with spot 2 (lower part).

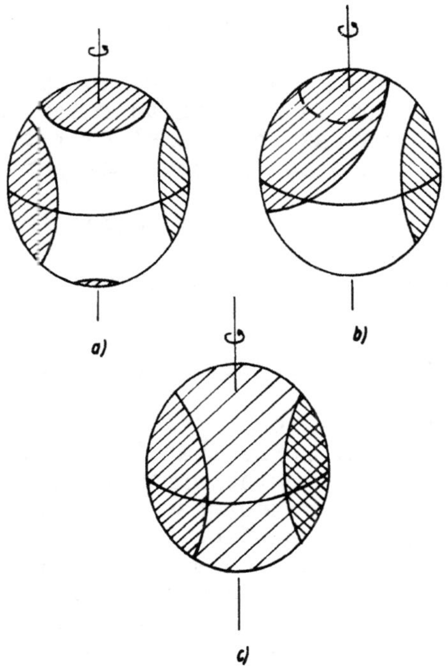

Fig.4. Schemes of the discussed models.

Discussions of the results

1) The spectral contrast distribution of the two spots (brightness relative to the unspotted surface) is quite different (Fig. 3). Spot 1 shows two spectral absorption features at λ 1420 Å and λ 1560 Å superposed by continuous darkening beginning at λ 1700 Å and a triple feature between λ 1740 Å and λ 2000 Å. The spot 2 spectrum shows a dark feature at λ 1600 Å, the same triple feature between λ 1740 Å and λ 2000 Å as spot 1, and a broad dark feature centred on λ 2400 Å. Possible sources of opacities in the UV spectrum of HD 170000 have been discussed by JAMAR (1977,1978).

2) The comparison with HD 207971 shows that the unspotted surface of HD 170000 is quite peculiar. Its spectral energy distribution is strongly correlated with that of spot 1.

3) The interpretation is clear for the accepted model. But this model is the simplest one of all possible models which are able to explain the observations. In fact, the variable part of the flux is related to rotational nonsymmetric structures, whereas rotational symmetric structures give a contribution to the constant part of the flux.
We will discuss together with the assumed model two other possible models (Fig.4).

a) The star has additional polar caps or belts with the spectral properties of spot 1. To produce the observed spectral energy distribution they must cover about 66% of the stellar disk. That is impossible because the diameter of spot 1 is about 70°.

b) The spot includes the rotation pole. Such a model fits the observations somewhat worse than the equator symmetric one, but still lies in the error range. However, this model can slightly decrease but not eliminate the peculiarity of the rest surface.

c) The most probable interpretation seems to be that HD 170000 is a star with a peculiar atmosphere. Spot 1 is a region with the same peculiarity as the rest surface, but strengthened. Spot 2 correlated with the positive magnetic pole (see Fig.2) has additional peculiarities with respect to spot 1 and the rest surface.

Although the analysis was carried out till now only for one star, the results indicate, that the simple conception often used (peculiar spots on the surface of a normal star) should be replaced by a more complex one.

References

Al Naimiy, H.M. (1978). Astrophys. Space Sci. 53, 181.

Jamar, C. (1977). Astron. Astrophys. 56, 413.

Jamar, C. (1978). Astron. Astrophys. 70, 379.

Jamar, C., Macau - Hercot, D., Praderie, F. (1978). Astron. Astrophys. 63, 155.

Landstreet, J.D. and Borra, E.F. (1977). Astrophys. J. 212, L43.

Musielok, B., Lange, D., Schöneich, W., Hildebrandt, G., Želwanowa, E., Hempelmann, A., Salmanov, G. (1980). Astron. Nachr. 301, 71.

COOL MAGNETIC CP STARS

Saul J. Adelman[1]
Code 681, Laboratory for Astronomy and Solar Physics[2]
NASA Goddard Space Flight Center
Greenbelt, MD 20771 USA

Charles R. Cowley[3]
Department of Astronomy
University of Michigan
Ann Arbor, MI 48109 USA

ABSTRACT. We review progress in research on the cool magnetic CP stars especially during the past four years. Studies of energy distributions and abundances are analyzed to reveal their systematics. Model atmospheres do not yet accurately describe the average energy distributions much less the variability with the rotational periods of these stars.

1. INTRODUCTION

In this review we are concerned with magnetic Cp stars cooler than the silicon stars: the SrCrEu stars and their variants. These objects have anomalous chemical compositions and magnetic fields which are not homogeneously distributed over their stellar photospheres. Radiative diffusion theory suggests that at a given location on the surface the abundances are functions of optical or physical depth. Wolff (1973) gave a general introduction to these stars, with extensive literature references.

2. OPTICAL ENERGY DISTRIBUTIONS

Following the discovery by Kodaira (1969), spectrophotometric studies of Adelman and Pyper (e.g. Pyper and Adelman 1985) and photometry by Maitzen and his collaborators (e.g. Maitzen 1976) have shown that the presence of broad, continuum features near $\lambda 4200$ and $\lambda 5200$ are signatures of the magnetic Ap star phenomena in the optical region. In some stars these features are variable on rotational timescales. For the most part the energy distributions of these stars, especially the cool Ap stars, remain anomalous over their entire rotational periods.

---

[1] Permanent Address: Department of Physics, The Citadel, Charleston, SC 29409 USA
[2] NRC-NASA Research Associate
[3] This paper was written while on sabbatical at the Institute for Astronomy, Vienna, Austria

C. R. Cowley et al (eds.), Upper Main Sequence Stars with Anomalous Abundances, 305–313.

There has been considerable interest in the origin and structure of these broad, continuum features. Maitzen and Seggewiss (1980) concluded that the λ5200 feature was composed of two main components with a different temperature behavior. There is a narrow, rather deep feature centered near λ5175 which shows its greatest strength in the hottest Ap stars and a broad, rather shallow feature which peaks at somewhat longer wavelengths and at a lower temperature. This pattern is consistent with results by Adelman and Pyper (1986) who find, besides the deep feature, subminima in some stars near λ5000 and λ5556. Subminima near λ5000 can also been seen in the work by Maitzen and Muthsam (1980) who find that they can synthesize the broad feature in the coolest Ap stars, but cannot model the narrow component in hotter Ap stars. The 4d-4f transitions in Fe II (Johansson and Cowley 1984) fall in this region. Since they are not in the Kurucz and Peytremann (1975) compilation, it would be of interest to re-synthesize this region by including Kurucz's (1981) Fe II gf-values.

North (1980) demonstrated that for the hottest cool Ap stars Δ(V1-G) and Δa are correlated with the surface magnetic field if it is not too large. This suggests that magnetic fields play some role in the formation of the λ5200 feature, but not necessarily a dominant one.

Carpenter (1985) constructed magnetic, line-blanketed model atmospheres with slightly distorted dipolar magnetic fields and a solar composition. The Zeeman splitting of the contributing atomic lines was taken into account in computing the opacity distribution function. As the magnetic effects are latitude dependent, the emergent spectrum varies with viewing inclination. A magnetic 15000 K model matches the fluxes of a 15200 K non-magnetic model with the same abundances in the optical and also displays a 10% flux discrepancy in the ultraviolet. As most magnetic Ap stars have larger flux deficiencies, this paper provides useful guidance in interpretation and is consistent with the usual interpretation that the abundances of magnetic Ap stars are greater than solar.

The structure of the atmosphere of a magnetic star is surely not homogeneous and probably not plane parallel, yet most of our analytical work thus far has been based on this assumption. The errors made by the use of such crude modeling depend on the application of the models. The magnetic structure affects both continuum and line features. The predicted fluxes in the optical region do not greatly depend on whether a magnetic field is or is not present (Carpenter 1985) or whether the metallicities are solar, three-times, or ten-times solar (comparison of results given by Kurucz 1979). But observationally the models are inadequate in the optical and more so in the ultraviolet.

Examples of the vicissitudes of attempts to fit optical continua to calculated models are given by Pyper and Adelman (1985) and references therein. The optical energy distributions of some Ap stars, which are often cool Ap stars, can be only crudely fit with predictions based on normal stellar abundances. At best in this temperature region the predictions can fit only those values not affected by the Ap star anomalies such as the broad, continuum features. This raises the question of whether observations at lower resolution, i.e. intermediate and broad-band photometry, can be calibrated to yield temperatures, gravities, and metallicities. Finding the true continuum between lines in the optical region is not too difficult for the silicon stars, but for the cooler Ap stars it can be quite difficult shortward of λ4500. The ultraviolet spectra for both silicon and cool Ap stars are heavily line blanketed.

3. PROBLEMS WITH ABUNDANCE ANALYSES

In his survey of the cool Ap stars Adelman (1973) gave tables of the errors to be expected in the abundances of various elements as a result of the uncertainties in T_{eff}. These errors vary considerably, depending both on the element and stage of ionization. There will also be similar errors if the surface gravity is not correct.

Lanz (1984) found that the magnetic Ap stars have larger Bolometric corrections than normal stars with the same Paschen continuum slope. The mean difference was 0.16 ± 0.07 mag. Further it was not possible in general to infer the Bolometric correction of peculiar stars with only the Geneva photometric system as the flux deficiency is related to the temperature, surface magnetic field, and chemical abundances (see also Adelman 1985).

Hauck and North (1982) have found that in the T_{eff} vs. B2-G diagram, Ap stars lie between the normal star class III and V relations. The interpretation of such a plot may not be unique, but one possibility is that the appropriate values of log g are slightly less than those of comparable normal stars. This is consistent with the effects of the magnetic fields. In fact analyses of HgMn stars (Adelman 1984a) and of the cool Ap stars (Adelman 1984b; Table 1) suggest log g = 3.5 ± 0.5 dex for CP stars. When one compares the Balmer line profiles of magnetic Ap stars with those predicted by model atmospheres, one has to be aware that the models may have been calculated with the solar He/H ratio rather than a smaller value appropriate for such stars. This effect can amount to about 0.10 dex in log g.

Errors which result from neglect of concentration of the elements into patches (departures from plane-parallel geometry) are somewhat harder to assess. We expect that calculations such as those carried out by Khokhlova and her colleagues (c.f. Piskunov and Khokhlova 1984) for the silicon stars could be applied to the cooler Ap stars. Sadakane's (1976) study of 73 Dra showed surprisingly little dependence of the overall abundances on magnetic phase. Nevertheless stars such as HR 465 must have abundance variations over the photosphere amounting to several orders of magnitude.

Adelman (1973)'s survey as well as more recent abundance analyses (c.f. Muthsam and Cowley 1984) have used a simple algorithm to account for Zeeman broadening, by simply adding a term to the Doppler width which behaves, analytically, like a microturbulence term. The usefulness of this approximation has been recently exploited by Ryabchikova and Piskunov (1984). Detailed calculations, for example, by Hardorp, Shore, and Witmann (1976) show this more explicitly.

Clearly, for a given line, there will be some microturbulence that approximates rather well the curve-of-growth behavior of Zeeman broadening, but it is not a simple matter to select the proper values, accurately, for a large number of lines. The analysis of high resolution high signal-to-noise data of some sharp-lined relatively non-variable cool Ap star with spectral synthesis techniques should illuminate the problems of this simple approximation. Magnetic null lines and lines with very simple Zeeman patterns will provide important checks on such analyses (Shore and Adelman 1974).

Saturation effects of various kinds are avoided by the use of truly weak lines although the problems due to errors in effective temperature and surface gravity remain, as do those of surface inhomogeneities (abundance patches). Still, we expect that the errors due to patches will be larger, usually, for stronger lines than for weak ones. In the case of the magnetic stars, few studies have had the benefit of sufficiently high quality material (low noise and high dispersion) to use truly weak lines, so the discussion below will apply to material that is subject to all of the sources of error that we have discussed.

4. UPDATE OF THE ABUBDANCE SURVEY OF ADELMAN

Since the largest survey of abundances of cool, magnetic stars is still that of Adelman (1973), we shall attempt an adjustment of his results. A start on the reanalysis of individual stars has been made by still assuming that the observed hemisphere is homogeneous. Then the values of the optical energy distributions unaffected by the broad, continuum features and the Hγ profiles were compared with the predictions of Kurucz's (1979) model atmospheres. The final choice of model was made by demanding that the values of log Fe/H from Fe I and Fe II lines be the same and the model metallicity was selected to be consistent with the adopted value of log Fe/H.

In HD 8441 for which H_s = 0.0 km s^{-1}, Adelman (1984b) found ξ = 0.0 km s^{-1}. But an unpublished re-investigation using Fe II gf values from a new critical compilation by Martin, Fuhr, and Wiese (1986) and better estimates of the line broadening for Fe I and Fe II suggests ξ = 0.7 km s^{-1} which is about 1/3 of that for a normal star of comparable temperature. Analyses following this prescription, especially with low noise data and metal enhanced model atmospheres, should be able to produce more realistic abundances than in the literature. But it is non-ideal due to the neglect of surface inhomogeneities and the failure to model the ultraviolet correctly. We hope that the relative abundances of some elements will be accurate (e.g. the rare earths).

Effective temperatures, surface gravities, and iron abundances derived consistent with the study of Adelman (1984b) are given for five of the hottest cool Ap stars of the 21 Ap stars of Adelman (1973) in Table 1. The values for HD 8441 are similar to those of Ryabchikova and Ptitsyn (1985).

Table 1: Cool Ap Star Parameters

Star	Adelman (1973) T_{eff}	This Review Paper T_{eff}	log g	log Fe/H	model
HD 8441	10150	9325	3.40	-3.82	3x
HD 50169	10100	9550	3.60	-3.26	10x
HD 89069	9900	9425	3.95	-3.34	10x
HD 111133	10550	9500	3.40	-3.45	10x
HD 192678	10350	9300	3.80	-3.14	10x

There is a mean systematic shift of 800 K in effective temperature between the two sets of results. Log g = 3.63 ± 0.24 instead of 4.0 as assumed previously. The new values are preliminary. If the microturbulences in the other stars are like that of HD 8441, the use of the new critically compiled Fe II gf values and better damping constants will result in small changes in the final effective temperatures, surface gravities, and log Fe/H values (of order 25 K, 0.1 dex, and 0.05 dex, respectively). The model atmospheres are now fully line-blanketed rather than hydrogen-line blanketed. There have also been improvements in the gf-values of Fe I and other atomic species since 1973. For these five stars, the new analyses presented here suggest that the average value of [Fe/H] is +0.93 rather than +1.22 as published. This is a reduction of a factor of two. Our expectation is that HD 5797, the most iron-rich star of Adelman (1973) will have its mean iron abundance reduced to about 20 times solar, which is comparable with the Muthsam-Cowley (1984) value for iron in HR 6870. There may well be regions on the surface with greater than the mean enhancement. A substantially enhanced iron abundance may prove to be a difficulty for the diffusion theory, but current uncertainties in both the theory and the observational determination makes a clear statement impossible at this time.

At the cool end of the Ap star sequence, van Dijk et al. (1978) found that the ultraviolet flux discrepancies seen in the hotter Ap stars disappeared. This result was confirmed by Adelman (1985a), who re-examined the data from the ANS, IUE, OAO-2, and TD-1 satellites as well as optical spectrophotometry and derived both optical and ultraviolet region color temperatures. The large amount of line blanketing complicates the process of estimating temperatures. For the coolest Ap stars such as 10 Aql, the change in T_{eff} from the values given by Adelman (1973) is probably much less than for the stars included in Table 1.

Table 2 gives some indication of the order of the overall changes in abundances now thought to be appropriate to Adelman's (1973) survey. This table of mean abundances does not necessarily represent our best estimate for a 'typical' cool magnetic Ap star. Indeed, the star-to-star abundance variations are such that a set of mean abundances, even if they were based on accurate determinations, might never actually apply to an individual star. We note in presenting these new results that an approximate correction method has been used which probably yields slightly different results than a more careful star-by-star analysis would. Nevertheless our procedure should indicate the proper magnitudes of the corrections for the mean abundances.

The procedure adopted to give an estimate of the corrected mean abundance is as follows. The old mean abundances were corrected for a temperature change of 600K and allowance was made for changes in gf values by comparing the old and new analyses of HD 8441. Table 2 contains the results for elements between magnesium and yttrium wherever possible along with solar values and the mean values from a study of 10 normal B and A stars (Adelman 1986) which use the same gf values as the most recent study of HD 8441. We did not have lines in HD 8441 to guide us for some other elements while for the singly-ionized rare earths one should carefully re-evaluate the presence of each species before performing re-analyses. As a guide to the changes we have included the change in log N/H for HD 8441.

Table 2: Abundance Values

Element	cool Ap star mean Adelman (1973) [N/H]	cool Ap star mean This Review [N/H]	Change in HD 8441 log N/H	The Sun log N/H	Ten B & A Stars log N/H
Mg	-0.08 ± 0.19	-0.10	-0.13	-4.38	-4.29
Si	+0.51 ± 0.35	+0.07	-0.22	-4.37	-4.40
Ca	+0.68 ± 0.43	+0.46	-0.46	-5.66	-5.71
Sc	+0.64 ± 0.41	+0.02	-0.75	-8.96	-9.16
Ti	+0.79 ± 0.37	-0.32	-0.73	-7.02	-6.97
V	+0.53 ± 0.24	-0.08	-0.40	-7.79	-7.40
Cr	+2.02 ± 0.51	+1.00	-0.62	-5.88	-5.92
Mn	+1.60 ± 0.38	+1.59	-0.18	-7.16	...
Fe	+1.06 ± 0.33	+0.71	-0.42	-4.37	-4.34
Co	+1.47 ± 0.56	+1.56	+0.11	-7.08	...
Ni	+0.33 ± 0.40	+0.95	...	-6.70	-6.57
Sr	+1.90 ± 0.80	+1.98	-0.01	-9.10	-8.88
Y	+0.28 ± 0.27	+0.39	-1.00	-9.76	...

Despite the 600 K mean temperature change for some elements the mean overabundance has not changed by more than 0.3 dex: magnesium, calcium, manganese, colbalt, strontium, and yttrium although some of these values depend on only one line. For yttrium there has been a substantial change in the adopted gf-values which results in the large change in the absolute abunance for yttrium. Silicon, scandium, and vanadium are now normal instead of being overabundant. The mean titanium anomaly has changed by a factor of 10. Titanium is now underabundant by a factor of 2. The quality of the Ti II-gf values has been improved substantially in the past decade. Chromium is overabundant by a factor of 10 instead of 100. The quality of the optical region Cr II-gf values is poor and systematic errors are probable. The mean iron abundance is now 5 times solar. It is consistent with the identification of high-level transitions of Fe II in the cool Ap stars β CrB and γ Equ (Johansson and Cowley 1984). Nickel may be 10 times overabundant on the average, but as for Cr II, Ni II-gf values are of poor quality. The large overabundance of cobalt is not surprising in light of the 10^3 times solar abundance in HR 5049 (Dworetsky et al. 1980), but it is uncertain in the cool Ap stars as it depends only on a weak line or two. Table 2 also contains the mean results for normal B and A stars. The differences from solar give some indication of how well the zero-points of the abundance anomalies are known. The largest differences are for scandium, vanadium, and strontium.

5. SUMMARY OF RARE EARTH ABUNDANCES

Cowley (1984) discussed the rare earth elements in stellar spectra so the present review will be limited to general comments and to a brief discussion of very recent work.

A point of major interest is whether observations of the lanthanides show a definite deviation from a pattern that might be expected from nuclear processing. Our attention has been focused primarily on deviations from the odd-even effect, known as the Oddo-Harkins rule in geochemistry. Clear deviations from this pattern are difficult to establish. Still in a few cases the evidence for deviations is very strong. In particular, there is an apparent odd-Z anomaly at europium in a few stars (e.g. 73 Dra, β CrB) which remains even after correction for hyperfine structure in Eu II (Hartoog et al. 1974). In addition for β CrB and HR 7575 the neodymium and samarium abundances appear depressed with respect to their light-lanthanide congeners. Such 'holes' possibly represent differentiation from some more primitive, possibly nuclear pattern.

Cowley (1980) pointed out that the strengths of the lanthanide lines in the Am and coolest, magnetic Ap stars were comparable, and that this implied their abundances must be similar. It is frequently thought that the magnetic Ap stars have abundances of the rare earths that are one or two orders of magnitude larger than those of the Am stars. This appears certain for some of the extreme magnetic Ap's, but the strong possibility exists that the abundance excesses of the lanthanides and of yttrium have been overestimated because of inadequate models or deviations from LTE. The observed, second spectra of the lanthanides, come from elements that are overwhelmingly doubly-ionized in the hotter magnetic stars.

Magazzu (private communication) carefully studied six lanthanides in the two cool Ap stars γ Equ and 10 Aql, and the classical Am star 32 Aqr. His work is based on carefully chosen weak (≤40 mÅ) lines. The oscillator strengths were obtained from Meggers, Corliss, and Scribner's (1975) intensities using a formula similar to the one suggested by Cowley and Corliss (1983) with the exception of those for Eu II, where the Biemont et al. (1982) values were used. The abundance differences for La, Ce, Nd, Sm, Eu, and Gd in these stars are rather similar and there is no indication that the two magnetic stars are richer in the light and intermediate lanthanides than the Am star.

Cowley predicted on the basis of wavelength coincidence statistics that the Nd/Sm ratio in γ Equ and 10 Aql would be found to be different. Magazzu's work shows the effect, but only if one neglects the uncertainties, which are of order 0.2 dex. Possibly, the result found by coincidence statistics could be accounted for by an overall factor of two difference in the relative abundances [Nd/Sm] (γ Equ - 10 Aql) ≃ 0.3 dex, which was obtained by Magazzu. Since the differences seem rather clear in the raw coincidence data, further work should be undertaken to clarify this question of differentiation amoung the lanthanides. It is especially important that the stellar spectra be obtained at low noise levels and high resolution.

We do not report the 'mean abundances' for the lanthanides. In γ Equ and 10 Aql, Magazzu finds the abundances of the odd-Z lanthanides La and Eu to be less than those of the even-Z elements Ce, Nd, Sm, and Gd. The difference is 0.7 - 1.0 dex. Thus these two magnetic Ap stars clearly do not resemble Adelman's (1973) rare-earth mean pattern. The origin of that particular pattern, with the abundances increasing steadily from La to Eu, involves a variety of factors that depend on the individual stars and cannot be discussed here. It is possible that 73 Dra (Sadakane 1976) has a rare-earth pattern resembling Adelman's (1973) mean. However, apart from Eu II, the lanthanide lines in 73 Dra are considerably weaker than those in typical cool magnetic Ap stars.

ACKNOWLEDGEMENTS. We thank A. Magazzu for permission to discuss his results prior to publication. SJA's attendance at this colloquium was made possible by a grant from The Citadel Development Foundation.

REFERENCES

Adelman, S. J.: 1973, Astophys. J. 183, 95
Adelman, S. J.: 1984a, Astron. Astrophys. Suppl. 58, 585
Adelman, S. J.: 1984b, Astron. Astrophys. 141, 362
Adelman, S. J.: 1985, Pub. Astron. Soc. Pacific 97, 970
Adelman, S. J.: 1986, Astron. Astrophys. Suppl., in press
Adelman, S. J., Pyper, D. M.: 1986, in preparation
Biemont, E., Karner, C., Meyer, G., Trager, F., zu Putlitz, G.: 1982, Astron. Astrophys. 107,166
Carpenter, K. 1985,: Astrophys. J. 289, 660
Cowley, C. R.: 1980, Vistas in Astron. 24, 245
Cowley, C. R., Corliss, C. H.: 1983, Monthly Notices Roy. Astron. Soc. 203, 651
Dworetsky, M. M., Trueman, M. R. G., Stickland, D. J.: 1980, Astron. Astrophys. 85, 138
Hardorp, J., Shore, S. N., Witmann, A: 1976, Physics of Ap-Stars, Weiss, W. W., Jenkner, H., Wood, H. J., eds., University of Vienna Observatory, p. 419
Hartoog, M. R., Cowley, C. R., Adelman, S. J.: 1974, Astrophys. J. 187, 551
Hauck, B., North, P.: 1982, Astron. Astrophys. 114, 231

Johansson, S., Cowley, C. R.: 1984, Astron. Astrophys. 139, 243
Kodaira, K.: 1969, Astrophys. J. (Letters) 157, L59
Kurucz, R. L.: 1979, Astrophys. J. Suppl. 40, 1
Kurucz, R. L.: 1981, SAO Special Report 390
Kurucz, R. L., Peytremann, E.: 1975, SAO Special Report 362
Lanz, T. 1984 : Astron. Astrophys. 139, 161
Maitzen, H. M.: 1976, Astron. Astrophys. 51, 223
Maitzen, H. M., Muthsum, H.: 1980, Aston. Astrophys. 83, 334
Maitzen, H. M., Seggewiss, W.: 1980, Astron. Astrophys. 83, 328
Martin, G. A., Fuhr, J. R., Wiese, W. L.: 1986, in preparation
Meggers, W. F., Corliss, C. H., Scribner, B. F.: 1975, NBS Monograph 145
Muthsam, H., Cowley, C. R.: 1984, Astron. Astrophys. 130, 348
North, P.: 1980, Astron. Astrophys. 82, 230
Piskunov, N. E., Khokhlova, V. L.: 1984, 6th Sci. Conf. Subcommission 4 Multilateral Cooperation Acad. Sci. Socialist Countries, 20
Pyper, D. M., Adelman, S. J.: 1985, Astron. Astrophys. Suppl. 59, 369
Ryabchikova, T. A., Piskunov, N. E.: 1984, 6th Sci.Conf. Subcommision 4 Multilateral Cooperation Acad. Sci. Socialist Countries, 27
Ryabchikova, T. A., Ptitsyn, D. A.: 1986, Upper Main Sequence Stars With Anomalous Abundances, Cowley, C. R., Megessier, C., eds. (this conference)
Sadakane, K.: 1976, Pub. Astron. Soc. Japan 28, 469
Shore, S. N., Adelman, S. J.: 1974, Astrophys. J. 191, 165
van Dijk, W., Kerssies, A., Hammerschlag-Hensberge, G., Wesslies, P. R.: 1978, Astron. Astrophys. Suppl. 66, 187
Wolff, S. C.: 1983, The A-Stars: Problems and Perspectives, NASA SP-463

DISCUSSION (Adelman and Cowley)

KHOKHLOVA: How many stars in your sample of 21 are variable?
ADELMAN: If we went to the 0.5% level, probably all of them would be variable. At present, variability is definitely known for about 2/3 of the sample. The rest are probably either long period variables or stars seen pole-on.
KHOKHLOVA: I should make one other remark. If you are observing variable stars which have abundance patches, then you can not determine the actual microturbulent velocity parameter.
ADELMAN: In a strict sense, I agree with you; spots and patches complicate the analyses. This survey is a first-order approximation, and by means of suitable observations of the variability one could improve on the assumption that the hemisphere being studied is homogeneous.

One other point is worth mentioning. Are there any Ap stars which have normal (solar abundances) regions on their surfaces? This question can not be answered from the abundance studies, but the spectrophotometry shows that some of these stars approach normal flux distributions at certain phases: 63 And, CU Vir, and 56 Ari do this.
SCHÖNEICH: I would like to say something about ultraviolet observations of cool Ap stars with the ANS satellite. For each star, we have only a few observations, but the high accuracy of the ANS photometry and the well known periods permit some conclusions about the UV variability of these stars. Some years ago, Musielok noted that HD 188041 is one of two stars which has the "null wavelength region" in the visible; ANS photometry has confirmed this. The UV varies in antiphase with the red, and is in phase with the blue part of the spectrum. The amplitude in the 1550 Å band is greater than 0.3 mag. Also, 73 Dra varies at 4200 Å in antiphase with the other bands in the visible, while the UV variations are in phase with the 4200 Å band. The amplitude of the variation at 1550 Å is greater than 0.5 mag. The amplitude-wavelength relation for HD 71886 is very similar to that of 73 Dra, but the amplitude at 1550 Å is more than 0.75 mag. That means that a dark spot must cover half the star's disc to produce the observed amplitude. In view of this, it is surprising that you found a decreasing UV flux deficiency for the coolest Ap stars.
ADELMAN: The result that the UV flux deficiency decreases with decreasing effective temperature is statistical in nature, so there will be individual exceptions to this result. My impression is that 73 Dra is not as cool as the coolest Ap stars.
DWORETSKY: You mentioned hyperfine structure. Although one can calculate relative displacements within levels for hyperfine shifts, and relative strengths, one has to have laboratory data to determine absolute shifts and hence the structure of patterns. So, if I want to adjust, say, a Mn curve-of-growth for hyperfine structure, how do I go about it?
ADELMAN: With extreme difficulty! For certain spectra, like Pr II, the presence of hyperfine components is obvious from atomic spectroscopy. In some cases measurement of the stellar line with high S/N can help to tell about its presence. But, that's about all the help I can give . . .

C. R. Cowley et al (eds.), Upper Main Sequence Stars with Anomalous Abundances, 315–317.

there are no magic formulae.

MICHAUD: You have suggested log g ≈ 3.5 for Ap stars, and it is often suggested that Am-Fm stars also have similarly small values. Considering the small amount of time evolutionary models spend with such low gravities, isn't it likely that such small gravities indicate errors in the analysis?

ADELMAN: The gravities are obtained by fitting Hγ profiles from spectrograms to the predictions of model atmospheres. The latter are dependent on the model, on the theory of hydrogen line broadening, and on the He/H ratio. Of course, corrections must be made for scattered light. Errors in the He/H ratio can cause errors of order 0.1 dex. The VCS (Vidal, Cooper and Smith) hydrogen line broadening theory appears to be adequate for normal B and A stars and matches the line shape in the wings and line shoulders. The models for Hg-Mn stars are not too bad, and I am inclined to believe log g values which average 3.5 dex. For the magnetic Ap stars, the models are much worse. However, other investigators (e.g., Ryabchikova and Ptitsyn in a poster paper at this Colloquium) have also found such values. To check for photographic errors, it would be desirable to obtain Hγ profiles using a Reticon with high S/N and sufficient resolution to see the metal lines.

STĘPIEŃ: I have a short comment on Dr. Schöneich's question about the disappearance of the flux deficiency in the cooler Ap stars. I am not surprised to see it, because the cooler Ap stars have much less flux in the UV, hence less flux is available for redistribution. A 0.5-mag variation in the UV for a hot star involves much more energy redistribution than a 0.5-mag variation in a cool star.

SADAKANE: What is the most important factor causing your effective temperatures for cool Ap stars to be reduced by more than 1000 K?

ADELMAN: The main reason that effective temperatures are substantially lower in this paper, compared to those given in my thesis, is that I am now trying to fit the Balmer jump region as well as the Paschen continuum. In the original study, I demanded that I obtain the same log (Fe/H) value from Fe I and Fe II lines. The temperatures so obtained therefore depended on the gf-values. The use of spectrophotometric data and Hγ profiles, in addition to this condition, makes the values of log g and T_{eff} less dependent on gf-values.

ROMANOV: Did you determine the abundances of elements for 73 Dra and β CrB?

ADELMAN: I have not analysed 73 Dra; it is not part of this sample. However, β CrB is in the sample, but it is cooler than the five stars for which I fit continua and Hγ profiles. There are difficulties in fitting the Balmer jump and Paschen continuum simultaneously for that star. I decided to analyse the hotter stars first, because finding matches between predicted and observed continua is easier for them.

ALECIAN: You mentioned one star with a microturbulence velocity parameter ξ = 0.7 km/s, which is a rather low value. What is the precision of this determination, and if the precision is poor, why did you not simply adopt ξ = 0? Information about turbulence is very important for diffusion calculations.

ADELMAN: For normal and Hg-Mn stars, the errors in ξ are 0.4 km/s in the best cases, if one uses Fe I and Fe II lines. For HD 8441, one has to increase this value to allow for the uncertainty of the model, to about 0.6 km/s. I believe this star has a small, but non-zero, microturbulence which is about 1/3 that of normal stars of similar temperature. Microturbulences determined with similar gf-values and damping parameters for Fe I and Fe II lines have a minimum of 0.0 km/s near 13000 K, and increase monotonically for both hotter and cooler effective temperatures. My values are smaller than those of other analyses. I do not understand the temperature dependence, especially as I have come to regard ξ as a "fudge" factor. When model atmospheres with a better representation of the line opacity become available, I plan to repeat some of these normal star analyses to see how ξ changes. It is important to know whether it is an artefact of the analysis technique.
DWORETSKY: Microturbulence is the number that we put into the equations to get the answer we thought of originally.
ADELMAN: A fudge factor!

ABUNDANCE ANALYSIS OF THREE Ap STARS: HD2453, HD8441, AND HD192913

T. A. Ryabchikova and D. A. Ptitsyn
The Astronomical Council
of the U.S.S.R. Academy of Sciences
Pyatnitskaya Str. 48, Moscow
109017 U.S.S.R.

ABSTRACT. Using 9 Å/mm-dispersion spectra of two Ap stars (HD2453 and HD8441) from the list by Adelman (1973) and the SiCr star HD192913 the abundances of 19 elements are obtained by the model atmosphere technique. The curve-of-growth method is used to estimate the surface magnetic field from Fe, Cr, and Ti lines.

One of the most extensive study of chemical composition of Ap stars was carried out by Adelman (1973). Now, more than 10 years later, the improved oscillator strengths and more refined methods of analysis of model atmosphere parameters are available, that makes it desirable to reinvestigate abundances of elements in these stars. Recently Adelman (1984) has redetermined the chemical composition of HD8441 and obtained quite different results. In this paper we present results of chemical analysis of two stars from the Adelman's list (HD2453 and HD8441) and the SiCr-type star HD192913.

For each star several 9 Å/mm-dispersion spectra were taken with the 2-m telescope of the National Astronomical Observatory of the Bulgarian Academy of Sciences and the 2-m telescope of the Ondrejov Astronomical Observatory of the Czechoslovakian Academy of Sciences. Observations were carried out in the frames of the Multilateral cooperation of Academies of Sciences of Socialist Countries. The spectra were traced with the microdensitometer 3CS Joyce Loebl and processed using the programme described by Piskunov et al. (1984). As far as possible unblended lines of 19 elements were chosen and their averaged equivalent widths were used to obtain abundances of elements by the model atmosphere technique.

Line intensities were calculated by means of the computer programme written by N.E.Piskunov at the Astronomical

C. R. Cowley et al (eds.), Upper Main Sequence Stars with Anomalous Abundances, 319–322.

Council of the USSR Academy of Sciences. The model atmospheres were taken from Kurucz et al. (1974). Europium abundances were corrected for hyperfine structure according to the paper by Landi Degl'Innocenti (1975).

The effective temperature T_e was determined from photoelectric photometry data. The final value of T_e was adopted taking into account available estimations of T_e based on continuum energy distribution. The surface gravity log g was obtained using Hβ and Hγ profiles. The theoretical curves of growth for Fe, Cr, and Ti were fitted to the observed ones by the least-mean-square method with 3 free parameters: microturbulent velocity ξ_t, abundance of the element log N, and surface magnetic field H_s. The broadening due to H_s was treated the same way as the Doppler broadening (Ryabchikova and Piskunov, 1984). The final mean values of T_e, log g, ξ_t, and H_s are given in Table I.

Derived logarithmic abundances are presented in Table II and in Fig.1. The scale corresponds to log N = 12 for hydrogen. The figures in parentheses indicate the number of

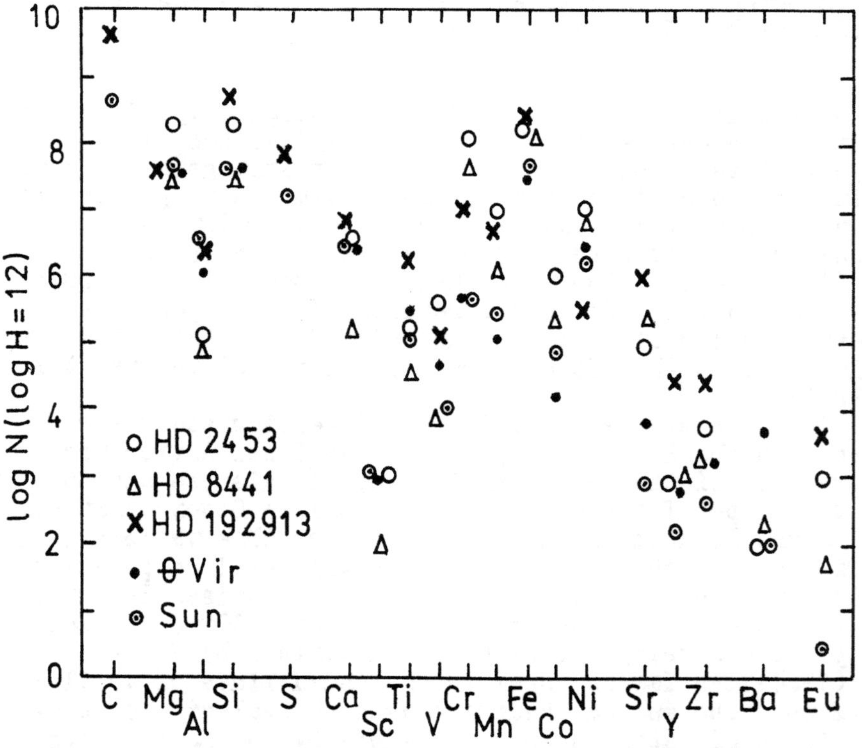

Fig.1. Elemental abundances of three Ap stars, normal star θ Vir, and the Sun.

Table I. Atmospheric parameters

HD number	T_e,K	log g	ξ_t,km/s	H_s,kGauss
2453	9000	3.75	2	2.54
8441	9200	3.50	0	0.56
192913	11000	3.50	1	1.80

Table II. Derived abundances

Element	HD2453	HD192913	HD8441	HD8441*	θ Vir	Sun
CII	...	9.72(2)	...	8.22	...	8.69
MgI	8.67(1)	...	7.43(2)	7.22	7.46(2)	7.58
MgII	7.95(1)	7.50(4)	7.29(4)	7.00	7.45(3)	
AlI	5.09(1)	6.37(1)	4.92(1)	...	6.02(2)	6.47
SiII	8.31(2)	8.66(8)	7.48(2)	7.02	7.58(5)	7.55
SII	...	7.78(2):	...	...	...	7.21
CaI	6.60(3)	6.90(1):	5.18(1)	5.63	6.45(3)	6.36
CaII	...	6.68(1)	...	4.70	...	
ScII	3.13(3)	...	2.12(1)	2.18	2.98(6)	3.1
TiII	5.24(11)	6.16(16)	4.59(15)	4.47	5.48(32)	5.02
VII	5.62(2)	5.13(5)	3.90(2)	3.74	4.70(12)	4.0
CrI	8.72(16)	6.94(2)	7.97(15)	7.60	5.43(3)	5.67
CrII	7.42(24)	6.97(14)	7.47(29)	7.69	5.91(18)	
MnI	6.89(6)	6.88(2)	5.90(3)	6.04:	5.13(2)	5.45
MnII	7.01(7)	6.60(9)	6.37(8)	6.20	5.02(2)	
FeI	8.28(14)	8.32(10)	8.00(16)	8.18	7.43(36)	7.67
FeII	8.31(13)	8.57(20)	8.22(15)	8.19	7.56(25)	
CoI	5.95(2)	...	5.45(1):	5.31:	4.18(1)	4.92
NiI	6.93(1)	...	...	5.87:	6.40(1)	6.25
NiII	7.26(4)	5.47(2)	6.80(4)	...	6.49(5)	
SrII	4.98(3)	5.98(3)	5.39(3)	4.66	3.84(4)	2.9
YII	3.00(3)	4.52(3)	3.13(3)	2.71	2.79(2)	2.24
ZrII	3.78(6)	4.50(2)	3.33(3)	...	3.15(3)	2.56
BaII	2.13(1)	...	2.30(1)	...	3.72(1)	2.13
EuII	3.12(5)	3.73(4)	1.76(3)	2.68	...	0.51

*from Adelman (1984)

lines used in the abundance analysis. For comparison are also given the chemical composition of normal star θ Vir (T_e=9300 K, log g=3.5, ξ_t =0 km/s) obtained by T.A.Ryabchikova and the solar abundances accoraing to Grevesse (1984).

Conclusions that can be made from Table II and Fig.1 are as follows. For HD8441 our results are in a good accord with those obtained by Adelman (1984). In the stars under study the abundances of practically all elements except Al (and light iron-peak elements in HD8441) are greater than or in some cases close to normal values. The overall patterns of relative abundances in all three peculiar stars are quite similar. There is a similarity in odd-even effect and in mean relative abundances of different groups of elements. The abundances of heavy elements (Sr, Y, Zr, and possibly Ba) in the standart star θ Vir appear to exceed solar values by 0.5 - 1 dex. Among the stars under study the highest content of metals (with the exception of some iron-group elements) is observed in the hottest star, HD192913, and the lowest one in the star with the weakest (practically negligible) magnetic field, HD8441. Cr, Mn, and Eu reveal the largest excesses (up to ~2 dex). It is worthwhile to note that the abundance of Eu corrected for the hyperfine structure turns out to be not so high for Ap stars as it is often believed: the excesses over the solar values are 1 - 3 dex. The variations of iron content in peculiar stars is remarkably small (of the order of the errors of analysis) as compared to other elements. The iron excess relative to standart abundance amounts to ~0.6 dex in all stars. Overdeficiency of some odd elements, Al and Y in particular, is observed. Their abundance ratio to neighbouring even elements for peculiar star is higher than in standart distribution.

We are grateful to colleagues from Ondřejov Observatory for placing some spectra of studied stars to our disposal and to the colleagues from Bulgarian National Observatory for help in observations.

REFERENCES

Adelman,S.J. 1973, Ap.J. Suppl., 26, 1.
Adelman,S.J. 1984, Astr. Ap., 141, 362.
Grevesse,N. 1984, Phys. Scripta, T8, 49.
Kurucz,R.L., Peytremann,E., and Avrett,E.H. 1974, Blanketed Model Atmospheres for Early-Type Stars (Washington: Smithsonian Inst.).
Landi Degl'Innocenti,E. 1975, Astr. Ap., 45, 269.
Piskunov,N.E., Ptitsyn,D.A., Ryabchikova,T.A., and Khokhlova,V.L. 1984, Nauchnye Informatsii, Astr. Sovet Akademii Nauk SSSR, 54, 45.
Ryabchikova,T.A., and Piskunov,N.E. 1984, in: Magnetic Stars, Proc. Conf. held in Riga, April 10-12, 1984, ed. V.Khokhlova, D.Ptitsyn, and O.Lielausis (Salaspils: Latvian Phys. Inst.), p.59.

ABSORPTION LINES OF CaII AND H IN THE NEAR IR REGION OF THE MAGNETIC STAR HD I52I07

T.N.Kuznetsova
The USSR Academy of Sciences
Central Astronomical observatory at Pulkovo
I96I40 Leningrad
USSR

ABSTRACT. A behaviour of H lines P_{12}- P_{18} and CaII triplet of the star HD I52I07 is studied in the range λλ 8400 - 8800 Å. An analysis of the obtained data enabled us to suspect the presence of a relationship between the variation of equivalent widths of the lines of the Paschen series of Hydrogen CaII lines and that of the star's magnetic field.

The star HD I52I07 (A2p - A4p, $4^m.86$, He +830 + I430± ± 300 gauss, B-V = $0^m.08$, U - B = $0^m.05$) is an Ap star with a variable magnetic field of constant polarity. It is a relatively cool Ap star of the Sr - Cr -Eu type of the spectral pecularity with rather broad lines in the spectrum (w ~ 0.4 Å) and amplitude of light variations $0^m.0I5$. The period of its variation has been found from photometric and spectroscopic observations of the K CaII line and equals $3^d.8575$ (Wolf and Preston, I978). No essential spectral variations were found in the range λλ 3600-4800 Å. Thus it was decided to observe the members of the Paschen series of Hydrogen, formed in the higher layers of the atmosphere of the star, and analyse variations of profiles and equivalent widths of the Hydrogen lines of the Paschen series P_{12}- P_{18} and CaII triplet. The spectrograms were obtained with the diffraction spectrograph of the 50-inch reflector of the Crimian astrophysical Observatory with dispersion 240 Å/mm in the range λλ 6I00-8800 Å. Table I gives data on the obtained spectrograms.

The spectrogram measurement were carried out with the microdensitometer Joyse Loebl with a I00 times magnification. The readings were done at a ~ I.3 Å interval

C. R. Cowley et al (eds.), Upper Main Sequence Stars with Anomalous Abundances, 323–326.

line profile plots were constructed by a routine method.

Table I

n	J.D. 2440000+	Phase	Dispersion $\mathring{A}\,mm^{-1}$	Plates
I	784.8062	0.472	2.40	Kodak-IN
2	784.8785	0.49I	"	"
3	785.8639	0.746	"	"
4	785.9222	0.76I	"	"
5	785.9965	0.78I	"	""

Figure I gives average equivalent widths of all the Hydrogen lines observed both free of blending with CaII lines (P_{12}, P_{14}, P_{17}, ...) and anomalously enhanced P_{13}, P_{15}, P_{16} which are physical blends of CaII lines with Hydrogen lines. The curve of the variation of the residual intensity of P_{12} - P_{19} lines free of blending with the CaII lines is also plotted. Detection of CaII lines from blends with Hydrogen is a difficult task (Bychkov et al., I978), because that we have only done approximate calculations. Table 2 gives mean values of equivalent widths of CaII triplet lines obtained by reducting of

Table 2

line	$\overline{W}_\lambda$(Å)	line	$\overline{\varphi}$ 0.48I	$\overline{\varphi}$ 0.763	mean	σ
CaII(8662)	0.I0	CaII(8662)	-0.65	0.87	0.3	±0.9
CaII(8542)	0.66	CaII(8542)	0.85	0.40	0.6	±0.4
CaII(8498)	I.I3	CaII(8498)	I.35	0.83	I.0	±0.4

average equivalent widths of P_{13}, P_{15}, P_{16} (interpolated by the curve in Figure I) from the equivalent widths of the corresponding blends. Similar calculations have been done for every observations. Values W_λ CaII = W_λ (P_m+CaII) - W_λ P_m (Table 2) are given for nightly mean phases of observations. In addition the values of the CaII (8662) equivalent width were obtained from the formula
W_λ CaII(8662) = w_λ (P_{13} + CaII) - 0.5($w_\lambda P_{12}$ + w_λ P_{14}) (Table 3). The total contribution of all CaII lines to Hydrogen blends (Table 3) can be calculated from the ratios of equivalent widths of the blended lines P_{13}+ CaII, P_{15} +CaII P_{16} +CaII to equivalent widths of nonblended lines P_{12} and P_{14} from the formula (Polosukhina et al., I978).

$$A = \left(W_\lambda(P_{13}+CaII) + W_\lambda(P_{15}+CaII) + W_\lambda(P_{16}+CaII)\right) / W_\lambda P_{12}$$

$$A' = \left(W_\lambda(P_{13}+CaII) + W_\lambda(P_{15}+CaII) + W_\lambda(P_{16}+CaII)\right) / W_\lambda P_{14}$$

$$B = 2\left(W_\lambda(P_{13}+CaII) + W_\lambda(P_{15}+CaII) + W_\lambda(P_{16}+CaII)\right) / 3\left(W_\lambda P_{12} + W_\lambda P_{14}\right)$$

Table 3

W_λ \ $\bar{\varphi}$	0.48I	0.763	mean	σ
$W_\lambda P_{12}$	8.0	9.5	8.9	±0.9
$W_\lambda P_{13}$	5.6	6.I	5.9	0.4
$W_\lambda P_{14}$	3.8	4.2	4.0	0.5
$W_\lambda P_{15}$	2.4	2.9	2.7	0.4
$W_\lambda P_{16}$	I.3	I.8	I.6	0.3
$W_\lambda P_{17}$	0.5	0.8	0.7	0.3
$W_\lambda P_{18}$	0.4	0.7	0.6	±0.2
CaII(8662)	-0.9	0.2	-0.2	0.9
A	I.35	I.37	I.36	0.I8
A'	2.89	3.I3	3.0I	0.22
B	0.6I	0.63	0.62	0.06

The electron concentration in the stellar atmosphere can be estimated using the Inglis-Teller limit log ne = 23.26-7.5 log nm where n_m - the number of the last observed line in the Hydrogen series. For each nightly averaged phase we have mean n_m and log n_e :

Table 4

$\bar{\varphi}$	0.48I	0.763	mean ± σ
n_m	I9.0	2I.0	20.0 ± I.5
log n_e	I3.66	I3.34	I3.47 ± 0.23

The analysis of the obtained results enables us to suspect an existance of correlation between variations of equivalent widths of absorption lines of the Paschen series of Hydrogen (P_{12} - P_{18}) with the phase during the time of observation the magnetic field and index CI (Wolf and Preston,I978),while the CaII lines vary reversly.The electron density changes little,within the error. The contribution of CaII lines in Hydrogen blends,i.e. A parameter did not in fact change during the observations. The dependence of this value (A) on the spectral class (Bychkov et al.,I978) gives an earlier then A2p-A4p spectral class. The spectral class of the star from the Paschen series and CaII triplet proves to be also somewhat earlier.

References

Wolff,S.C.,Preston,G.W.: 1978, P.A.S.P.,90,406.
Bychkov,V.D.,Vitrichenko E.A.,Scherbakov A.G.,:1978, Izv.Krymskoj Astroph.Obs., 58, 81.
Polosukhina N.S.,Shcherbakov,A.G.,Malanushenko V.P: Astrophisics, 1978, 15, 1, 85.
Butler,H.E.:Contr. Dunsink Obs.,1951,1.

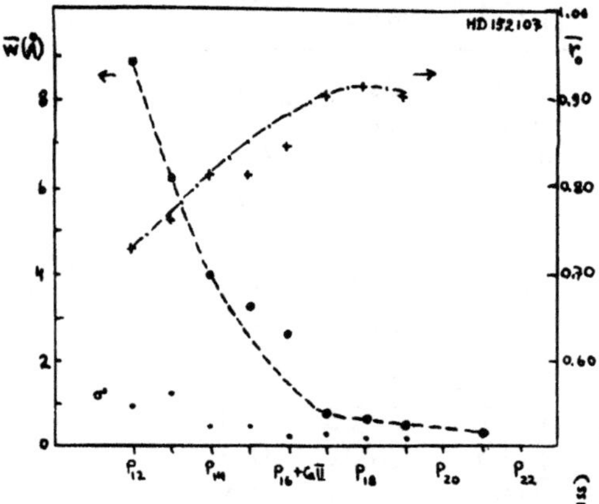

Fig.1. Dependence of mean $\overline{W}_\lambda$(Å) and mean r on values of m in Paschen series. The σ error for $\overline{W}_\lambda$(Å) measurements is indicated.

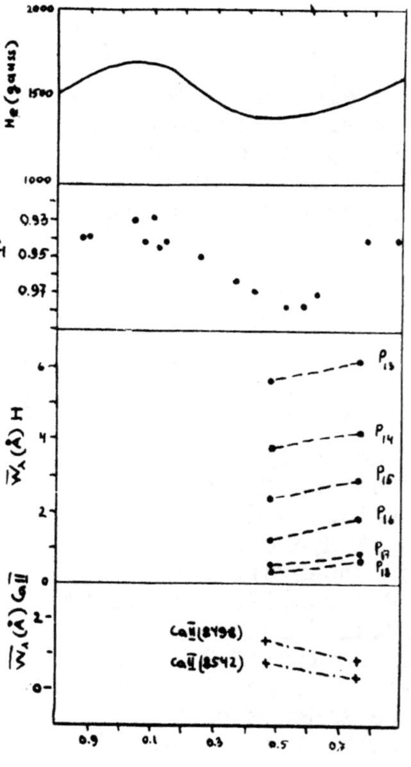

Fig.2. Behaviour of $\overline{W}_\lambda$ (Å) of P_{13} –P_{18} and CaII lines, He and C_1 with the phase.

A SEARCH FOR HEAVY ELEMENTS IN THE ULTRAVIOLET SPECTRA OF Ap STARS

A. B. Severny and L. S. Lyubimkov
Crimean Astrophysical Observatory
334413, p/o Nauchny, Crimea
USSR

ABSTRACT. The earlier attempts to detect the lines of heavy elements (Pb, U, Th and others) in the spectra of Ap-stars from the ground-based observations are briefly discussed. UV observations of Ap-stars with the ultraviolet telescope aboard ASTRON space station are described. The reduction of these data is considered. Three spectral regions containing the lines PbIIλ2203.53, UII λ2556.19 and ThII λ2368.05 were analyzed by the method of synthetic spectra. It has been shown that relatively cool Ap-star 73 Dra (T_{eff}=8150 K) has a great overabundance ($\geq 10^3 \varepsilon_\odot$) of heavy elements U, Th, Pb and W.

1. EARLIER IDENTIFICATIONS OF HEAVY ELEMENTS' LINES FROM THE GROUND-BASED OBSERVATIONS

Attempts have been made to determine the abundance of lead from visible spectra of stars all turned to be inconclusive. The line λ 5042.5 suspected by Burbidges (1955) as Pb-line in the spectrum of α^2CVn turned out to be wrong (Cohen et al. (1959)), as the other lines of Pb were absent in the spectrum of this star. The attempts of Guthrie (1972) to do the same with the line λ 4157.81 were unreliable because he arbitrarily assumed Os abundance to be 10^3. Besides, this line is strongly blended with the lines of FeII and MgI, and the estimates were based on the curve-of-growth analysis with rather arbitrary assumptions about the atmospheric structure (the osmium abundance is also

C. R. Cowley et al (eds.), Upper Main Sequence Stars with Anomalous Abundances, 327–340.

very uncertain). The Wegner and Petford (1974) estimates based on three weak, high excitation lines in Pzybilski star HD 10165, was "very shaky" as the authors concluded, because of the unreliability of the oscillator strengths.

Studying Os lines in 73 Dra, Guthrie (1969) suspected that UII-lines ($\lambda\lambda$4241.67 and 4543.63) were present. There appeared several other papers on the possible presence of U-lines in the visible spectrum of this star. Jashek and Malaroda (1970) found seven Pt-lines, four AuI-lines and five UII-lines. These authors were the first to note that the presence of these elements is difficult to be reconciled with the theory of Fowler et al. (1967) speculating that Ap-stars evolve after the red-giant phase: the presence of these elements (formed via r-process) on the surface of the stars requires the material of the star to have, at some time, been subjected to conditions that exist in stellar interiors only (these elements may also be formed at the explosion of a companion star).

Soon after that Brandi and Jashek (1970) found some traces of OsI, OsII, PtII and UII in five out of twelve stars. In particular, they called attention to a strong line λ 3860 identified as UII. This identification was confirmed by Hardorp and Shore (1971). The latters were able to identify except λ 3860, also $\lambda\lambda$4091 and 4242 lines as UII-lines in the spectrum of βCr B. Later Jashek and Brandi (1972) reported the identification of ThII, UII and AmII in the spectrum of the Ap-star HD 25354 (Sp-variable with $P=3\overset{d}{.}9$). They were also first to report the detection of a WI-line.

Hartoog and Cowley (1972) reported the identification of UII in the spectrum of HR 465 and a little later Cowley and Hartoog (1972) reported the U-abundance in this star about 10^6 times of the solar system abundance (at a significance level 3.8σ). The presence of UII-lines in βCrB was once again confirmed by Adelman and Shore (1973) at 4 sigma level. The line λ4019.13 of ThII may be present in stars HD 18078, 81009 and 165474. The UII-line λ 3859.58 is possibly present in 20 out of 21 star studied by Adelman (1973), while UII-line λ 4241.67 may be present in six of them. Adelman believes, that UII identification is not definite - more accurate measures of λλ 3860 and 4242 lines are needed. Seventy four elements have been identified for five Ap-stars (β CrB, 73 Dra, 78 Vir, HR 4816, HR 4072) using wavelength coincidence method (Cowley, Hartoog and Cowley, 1974). These authors concluded that only one heavy element, Pt, is identified with high confidence in HR 4072, while UII may be present in 73 Dra and β CrB; OsII may be present in β CrB. However the identification of UII is based mainly on λ 3859.58 line which is obviously present in the spectrum, while the λ 4241.67 line yeilds no definite indications. As for the other stars, there was no clear evidence for the presence of UII or ThII lines.

As far as the ultraviolet spectra of Ap-stars is concerned, there are still very few investigations aimed at the search for heavy elements (for instance, see the review of Castelli et al., 1983). Our investigations are based on the UV spectra of Ap-stars obtained with ultraviolet telescope aboard ASTRON space station.

2. "ASTRON" OBSERVATIONS IN THE ULTRAVIOLET

Space station ASTRON is equipped with 80-cm Cassegrain reflecting telescope having 8-m equivalent focal length and the X-ray sensors. The station was launched on highly elliptical orbit with apogee $200x10^3$ km and perigee $2x10^3$ km. Such an orbit is advantageous for avoiding the radiation belts influence on the photoelectric recordings of stellar spectra which are made near the apogee. Being launched in March 1983, ASTRON is still in operation.

The spectra are obtained with the concave toroidal grating and recorded by photoelectric scanning along the Rowland circle. Star pointing is realized by a slight inclination of secondary convex (hyperbolic) mirror as an element of guiding system. There exist two guiding systems: 1) by using the eccentricity of position of star image in question on the slit jaws and for weaker stars 2) the off-set system, where other bright stars in the vicinity of the star in question are used ($\leqslant$ 10' from a given star). More detailed description of ASTRON space telescope is given by Boyarchuk et al. (1984).

TABLE I. SPECTRAL LINES IN UV REGION

Element	Line	Z	A	$\log \varepsilon_c$	$\log \varepsilon_\odot$
PbII	2203.53	82	204,206,207,208	1.78	1.9
ThII	2368.05	90	232	0.7	0.2
UII	2556.19	92	238,235	0.0	<0.6

In the present work based on the analysis of ASTRON observations a search for heavy elements (lead, uranium and thorium) was carried out for several Ap-stars. The observed spectra were obtained by scanning the limited spectral ranges ± 2.5 Å wide centered on the wavelengths given in Table I, where Z and A are the atomic number and atomic weight, $\log \varepsilon_c$ is the cosmic abundance according to Allen (1973), $\log \varepsilon_\odot$ is the solar abundance after Hauge and Engvold (1977). The $\log \varepsilon$ values are given in the scale where the hydrogen abundance $\log \varepsilon(\mathrm{H})$ equals to 12.00.

Eleven Ap-stars were observed: $\varkappa$ Cnc, 15 Cnc, 78 Vir, ι Lib, 73 Dra, 9 ω Oph, 21 Per, 41 Tau, 21 Aql, 53 Tau and β CrB. It was shown recently by Adelman (1984) that 21 Aql is a normal star of spectral type B7 IV. Only three Ap-stars of the remaining ten stars have been subjected to a detailed analysis so far: 73 Dra(T_{eff}=8150 K), 9ω Oph (T_{eff}=9500 K) and $\varkappa$Cnc(T_{eff}=13600 K). The normal star λ UMa was investigated also to examine the oscillator strengths of lines in the spectral intervals under consideration.

The profile of each 5 Å interval was recorded with a small entrance slit aperture equal to 1" (accuracy of pointing being $\pm 0\overset{''}{.}3$) and narrow exit slit (resolution equal to 0.4 Å). The intensity was determined in relative units of the total wideband (28 Å-wide), relative to the intensity in the first ("comparison") channel, to remove possible influences of instabilities in the position of the star on the entrance aperture *). At each scanner position (fixed wavelength) 180 separate estimates were obtained (that corresponds to $1^m.83$ integration time). The dark current was recorded during the intervals between scans ($\sim 2^m$). (The total time required for one spectrum registration is a little more than 3 hours corresponding to six successive scans). In order to reduce more than 180x16=2880 measures (each section of the spectrum is 5 Å wide) a reasonable data reduction program should be provided for our computer, which is adopted to magnetic tapes obtained from the tracking station.

Clipping procedure is roughly as follows: all counts are reduced for dark current (d.c.) and all values of count numbers N_2 and N_1 smaller than fluctuations in d.c. ($\simeq \sqrt{d.c.}$) are eliminated. So after the first reduction we have the number of points M_1 with the mean value A_1 and r.m.s. deviation σ_0. After

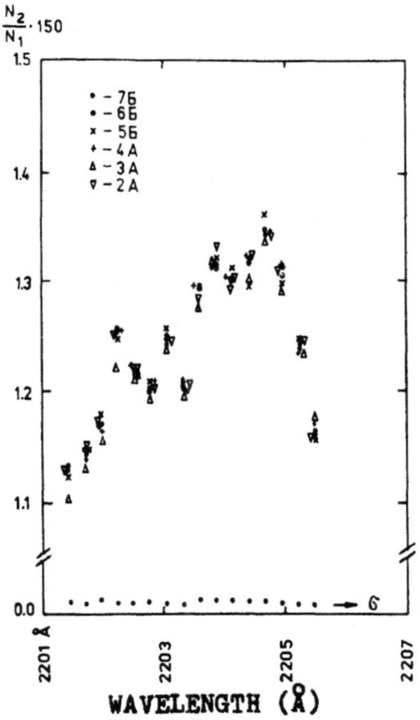

Figure 1

An example of using different data reduction programs to ASTRON observations in PbII section. M_1 and N_2 are number of counts in the first and second channels.

*) I.E. the average of $\Sigma(N_2/N_1)$ or $\Sigma N_2/\Sigma N_1$ was determined for each scanner position; here N1 and N2-numbers of counts in the first and in the second channels.

that all the points deviated from the mean $\geq 5\sigma_0$ are thrown away. This gives a new data set M_2 with the mean value A_2 and r.m.s. deviation σ_1. Then again all points with r.m.s. deviations $\geq 4\sigma_1$ are eliminated,and so on until we reach data sets M_5 and M_6 after throwing away the points beyond the $1\sigma_4$ and $1/2\sigma_5$ levels. The final data contains about 1500 points, which corresponds to about 60% of the original number.

From Fig. 1 it is clear that different programs of reduction yeild virtually identical results.

Further analysis are based on the comparison of the observed spectra in the considered UV regions with the synthetic spectra calculated for the same section of spectrum. The computer program SYNTHEL was applied in which the star rotation and the instrumental profile are taken into account. The list of spectral lines was selected from the table of Kurucz and Peytremann (1975). For the most UV lines this table is the only source of oscillator strengths gf. It is known that for some lines the semiempirical gf-values of Kurucz and Peytremann may contain significant errors. For the refinement of gf-values we analyzed at first ASTRON spectra of the normal star λ UMa with a known chemical composition. The adopted values of the effective temperature T_{eff}, surface gravity g and rotational velocity $v \sin i$ for λ UMa and for three Ap-stars investigated in detail are given in Table II. To determine the parameters T_{eff} and log g, we used

TABLE II. EFFECTIVE TEMPERATURES, SURFACE GRAVITIES AND ROTATIONAL VELOCITIES OF FOUR INVESTIGATED STARS

Star	Type	T_{eff}	log g	$v \sin i$ (km/s)
λ UMa	A2 IV	9300	3.7	45
73 Dra	Sr-Cr-Eu	8150	3.6	9
9 ω Oph	Sr-(Cr)	9500	4.0	35
$\varkappa$ Cnc	Mn-Hg	13600	3.6	6

the following features of visible spectrum: Balmer line profiles, $[c_1]$-index in uvby-system, and correspondence in iron abundance between FeI and FeII lines (ionization equilibrium).

The synthetic spectra were convoluted with rotation and with instrumental profile of ASTRON spectrometer (resolution is 0.4 Å). To establish the position of the observed continuum we calibrated from synthetic spectrum a rest intensity of some strong blends in the considered spectral sections (other than blends containing PbII, ThII or UII lines). The influence of scattered light on the observed profiles is negligible, less than 2% according to special investigation of A.A. Boyarchuk.

At first we consider the results of lead abundance determinations from the PbII-line λ 2203.53. This line has an excitation potential of lower level χ = 1.73 eV and oscillator strength log gf = -0.15 (Kunisz and Migdalek, 1974). It is blended with the MnII-line λ 2203.52 (χ = 4.5 eV) and with highly excited lines NiII λ 2203.47 (χ = 7.0 eV) and CrIII λ 2203.27 (χ=7.9eV). Two latter lines are not important for relatively cool Ap-stars (e.g. for 73 Dra), but in the spectra of hotter stars the contribution of NiII λ 2203.47 may be more considerable. The role of NiII line proved to be especially significant in the case of $\varkappa$ Cnc. The preliminary analysis of PbII-section for this star was carried out with nickel relative abundance [Ni] = $\log \varepsilon(Ni)_* - \log \varepsilon(Ni)_\odot$ = -1.0 obtained by Leckrone (1981) from UV spectrum of $\varkappa$ Cnc. For such Ni deficiency the lead overabundance [Pb] = 1.8 in $\varkappa$ Cnc has been found (Boyarchuk et al., 1984). But from the study of visible spectrum of this star follows, that Ni abundance is more close to solar value (Aller, 1970; Heacox, 1979). Our calculations showed that in this case the uncertainty in Ni abundance proves more important than variation in Pb abundance. Therefore it is difficult to obtain a reliable estimation of the lead content in this high temperature star.

In case of cooler Ap-star 9 ω Oph a synthetic spectrum is more sensitive to Pb abundance. A preliminary analysis of the visual spectrum shows that [Fe] = 0.3, [Mn] = 1.2 and [Cr] = 1.2. For these values and at normal Pb abundance we obtain a good agreement between the observed and synthetic spectra of this star if nickel abundance is normal, too. From the line WII λ2204.48 being in the same spectral section, we derived for 9ωOph tungsten overabundance [W] = 1.2.

Of three investigated Ap-stars, 73 Dra is the coolest. At first we used for it the model atmosphere of Sadakane (1976) with parameters T_{eff} = 8900 K and log g = 3.9. But later its visual spectrum was totally reanalyzed (Lyubimkov, in press), that allowed to conclude that the effective temperature of 73 Dra is considerably lower (see Table II). Such temperature decrease for this star has especially great importance for U abundance determination (see below).

Figure 2.

Comparison of the observed and computed spectra of 73 Dra in the section of PbII λ2203.53 and WII λ2204.48 lines. Thin line corresponds to synthetic spectrum convoluted with rotation, dotted and dashed indicate convolution also with instrumental profile (for normal and enhanced abundance of heavy elements analyzed : Pb and W).

ASTRON spectrum ———

Synthetic spectrum:
log ε(Pb)=5.4, log ε(W)=3.6 – – –
log ε(Pb)=1.9, log ε(W)=0.8 ....

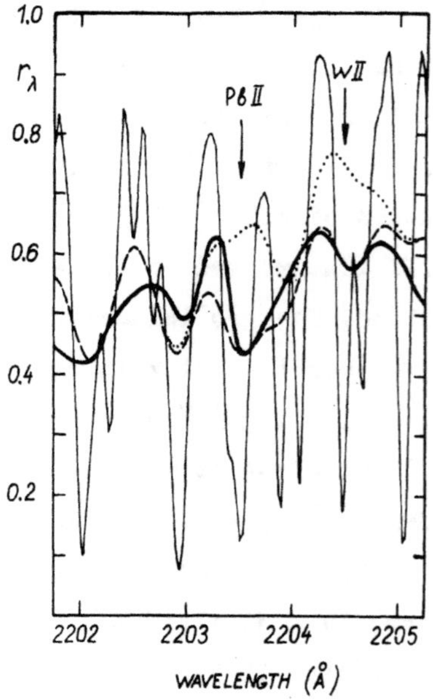

On Figure 2 the comparison of the computed and observed spectra of 73 Dra in PbII-section is shown. Calculations were made for a chemical composition found from visual region. If we take a normal Pb and W abundances (dotted line) a great dis-

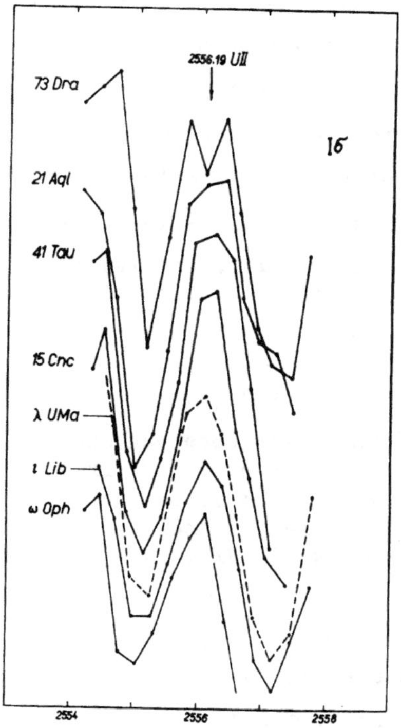

Figure 3.

ASTRON spectra of some program stars in the section of the line UII λ2556.19. The intensities are given in relative units; the curves of different stars are shifted relative to each other.

crepancy arises between theoretical and observed spectra. A good fit to ASTRON spectrum may be obtained only for considerable Pb and W overabundance : log ε(Pb) = 5.4 and log ε(W) = 3.6, i.e. [Pb] = 3.5 and [W] = 2.8 (dashed line).

Now let us consider ASTRON observations of UII-section. Of eleven Ap-stars only for 73 Dra a clear depression exceeding 3 sigma is visible at the place of UII λ2556.19 (Figure 3). This star has been observed twice - in 1983 and 1984 at the same phase φ = 0.77, and the depression appeared to be on both spectra. The line UII λ2556.19 is blended with the lines TiII λ 2555.98 and MnII λ2556.57. Relative contribution of the two lines depends on adopted effective temperature T_{eff}. Using Sadakane (1976) model atmosphere with T_{eff} = 8900 K, we have found enormously high uranium overabundance [U] > 5.0 (the oscillator strength log gf = -1.14 was adopted for the line UII λ2556.19 according to Corliss, 1976). But from the improved model atmosphere with lower temperature T_{eff} = 8150 K we obtained more moderate U overabundance.

Figure 4 shows that a good fit to the observed spectrum of 73 Dra is reached for log ε (U) = 4.4, i.e. for [U] $\geq$ 3.8. It is interesting to note that for normal U-abundance the depression on 2556.2 Å is absolutely absent in the computed spectra (dotted line in Figure 4).

Figure 4.

Comparison of the observed and computed spectra of 73 Dra in the section of the line UII λ 2556.19 (additional explanations see in Fig. 2).

ASTRON spectrum ———

Synthetic spectrum:
log ε (U) = 4.4 – – –
log ε (U) = 0.6 ·······

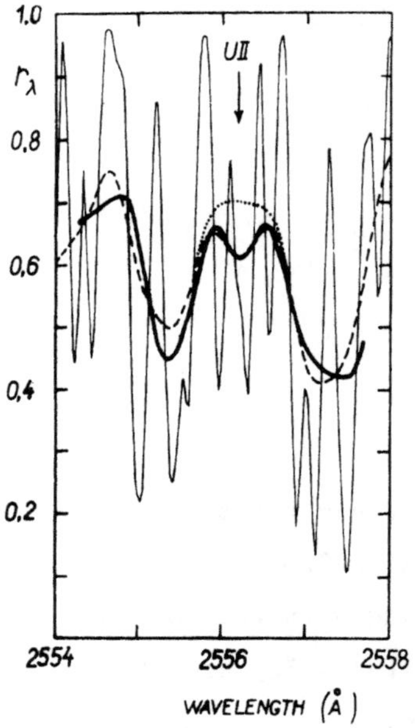

In order to confirm the value of log ε(U) derived from ASTRON spectrum, we consider also the line UII λ3859.58 in the visible region of the spectrum. The oscillator strength of this line is log gf = – 0.62 according to Voigt (1975). It is important to note that for both UII λ3859.58 and UII λ2556.19 the adopted gf-values correspond to the same scale of Voigt. The visual spectra of 73 Dra in the region considered were obtained by Cowley et al. (1977) and recently by Iliev (unpublished) at the Bulgarian National Astronomical Observatory (BNAO). In the latter case the phase φ = 0.88 was close to the phase of

ASTRON observations. By comparing the BNAO and synthetic spectra in the region of UII λ3859.58 (see Figure 5), the value of log ε (U) = 3.9 has been determined, which differed by 0.5 from the ASTRON value log ε (U) = 4.4. This difference corresponds just to ASTRON error of observations σ, which is indicated in Figure 3.

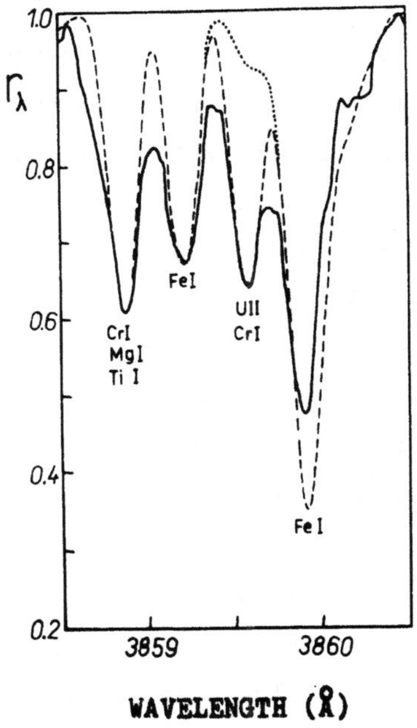

Figure 5.

Comparison of the observed and computed spectra of 73 Dra in the section of the line UII λ3859.58 (additional explanation see in Fig. 2).

BNAO spectrum ———

Synthetic spectrum:
log ε(U) = 3.9 – – –
log ε(U) = 0.6 ·······

Two DAO spectra (Cowley et al., 1977) obtained in two different phases have been also analyzed.

TABLE III. URANIUM ABUNDANCE IN 73 Dra DERIVED FROM VISIBLE AND UV-SPECTRA

Phase	UII-line	Spectrum	log ε (U)
0.19	3859.58	DAO	3.6
0.54	3859.58	DAO	3.2
0.77	2556.19	ASTRON	4.4
0.88	3859.58	BNAO	3.9

Table III shows that DAO spectra yeild lower values of log ε(U) than found from BNAO or ASTRON spectra, especially at the phase φ = 0.54. At this phase the extraordinary weakness of FeI lines has beenalso marked by Cowley et al. (1977). The difference in U abundances according to the data obtained at several observatories may be connected with real UII-line variations depending on the phase φ (such periodical variations of some lines in the spectrum of 73 Dra are well known). Further observations of UII λ 3859.58 are needed to solve this problem.

We conclude that great overabundance of uranium follows from visible and ultraviolet data on 73 Dra. As can be seen from Table III, the mean value is log ε (U) = 3.8, i.e. [U]$\geqslant$3.2. That is in good agreement with the overabundance of lead [Pb] = 3.5 and tungsten [W] = 2.8 found by us for 73 Dra from PbII-section.

Relatively cool Ap-stars (such as 73 Dra) are more suitable for a detection of the line UII λ2556.19. Our calculations showed that for the relatively hot peculiar star 53 Tau (T_{eff}= 12000 K) a marked depression at λ 2556.19 may appear only for great overabundance of uranium [U]>4.0.

Finally we mention the results concerning ASTRON observation of ThII-section. Clear depression at the place of ThII λ 2368.05 line is visible for 73 Dra. Preliminary analysis suggests that the thorium abundance in this star is also high, $\geqslant 10^3$ of the solar value. Therefore we can conclude that the investigations of Ap-star 73 Dra based on ASTRON ultraviolet spectra indicate high ($\geqslant 10^3 \varepsilon_\odot$) content of heavy elements, such as uranium, thorium, lead and tungsten .

REFERENCES

Adelman S.J., 1973. Astrophys. J. Suppl., 26, 1.
Adelman S.J., 1984. Mon. Not. R. Astr. Soc., 206, 637.
Adelman S.J. and Shore S.N., 1973. Astrophys. J., 183, 121.

Allen C.W., 1973. Astrophysical Quantities (third edition), the Athlone Press, London.

Aller M.F., 1970. Astron. Astrophys., 6, 67.

Boyarchuk A.A., Gershberg R.E., Granitskii L.V., Kovtunenko V.M., Khodzhayants Yu.M., Pronik V.I., Severny A.B., Krmoyan M.N., Tovmasyan G.M., Cruveilier P., Courtes G., Hua S.T., 1984. Sov. Astron. Lett., 10 (2), 67.

Brandi E. and Jashek M., 1970. Publ. Astr. Soc. Pacific, 82, 847.

Burbidge G.R. and Burbidge E.M., 1955. Astrophys. J. Suppl., 1, 431.

Castelli F., Faraggiana R., Catalano F.A., 1983. Mem. Astr. Soc. Italiana, 54, 585.

Cohen J.C., Deitsch A.J. and Greenstein J.L., 1959. Astrophys. J., 156, 629.

Corliss C.H., 1976. J. Res. Nat. Bur. Stand., A 80, 429.

Cowley C.R., Aikman G.C.L., Fisher W.A., 1977. Publ. Dominion Astrophys. Obs., 15, 37.

Cowley C.R. and Hartoog M.R., 1972. Astrophys. J. Lett., 178, L9.

Cowley C.R., Hartoog M.R., and Cowley A.P., 1974. Astrophys. J., 194, 343.

Fowler W.A., Burbidge E.M., Burbidge G.R., Hoyle F., 1967. Magnetic and Related Stars (ed. Cameron R.C.), Mono Book Corp., Baltimore, 233.

Guthrie B.N.G., 1969. Observatory, 89, 224.

Guthrie B.N.G., 1972. Astrophys. Space Sci., 15, 214.

Hardorp G. and Shore S.N., 1971. Publ. Astr. Soc. Pacific, 83,605.

Hartoog M.R. and Cowley C.R., 1972. Bull. AAS, 4, 311.

Hauge Ø., Engvold O., 1977. Inst. Theor. Astrophys. Report No49, Blindern-Oslo.

Heacox W.D., 1979. Astrophys. J.Suppl., 41, 675.

Jashek M., Brandi E., 1972. Astron. Astrophys., 20, 233.

Jashek M., Malaroda S., 1970. Nature, 225, 246.

Kunisz M.D., Migdalek J., 1974. Acta Phys. Polonica, A 45, 715.

Kurucz R.L., Peytremann E., 1975. Smithsonian Astrophys. Obs. Spec. Report, No 362, 1.

Leckrone D.S., 1981. Astrophys. J., 250, 687.

Lyubimkov L.S. (in press). Izvestia Krimsk. Astrophys. Obs., 75.

Sadakane K., 1976. Publ. Astr. Soc. Japan, 28, 469.

Voigt P.A., 1975. Phys. Rev., A 11, 1845.

Wegner G. and Petford A.D., 1974. Mon. Not. R. Astr. Soc., 168, 537.

Discussion appears after the following paper.

THE INVESTIGATIONS OF VARIATIONS IN THE DEPRESSION λ 4200, λ 5200 OF THE MAGNETIC STARS 53CAM AND BETA CrB.

V.I.BURNASHEV, V.P.MALANUSHENKO, N.S.POLOSUKHINA.
Crimean Astrophysical Observatory
Crimea, Nauchny, 334413
U.S.S.R.

INTRODUCTION.

During many recent years, from 1973, the search for short-term variations in the spectrum of Ap-stars has been carried out at the Crimean Astrophysical Observatory. Some results are presented in the poster dedicated to 53CAM obser-vasions. Spectral observations of the magnetic star 53CAM with time resolution of about 1 min were supplemented by the narrow-band photometrical observations within the framework of the cooperativ program of the magnetic stars observations, started in 1979.

I. OBSERVATIONS.

Narrow-band photometric observations are realized with aid of spectrophotometer. The band width is $\Delta\lambda = 30$ Å centered at $\lambda_c = 4220$ Å and $\lambda_c = 5220$ Å.

II. OBJECTS.

53CAM (A2p, $6^m.02$) and Beta CrB (F0p, $3^m.47$) was observed by the differential method (comparison stars are HD 65301 and γ CrB). Each estimation of the brightness $\Delta m = m_c - m_v$ was $\Delta t = 2$ min long.

III.A. RESULTS OF 53CAM OBSERVATIONS.

The analisys of observations (16 nights, total N=849) show:
1) The variations are complex, which can be described by a sum of three sinusoids with periods $P_1 = 20.1400$, $P_2 = 27.5828$, $P_3 = 79.2381$, reaining in phase during more than a year. (Edition Sov. Let. Astron. J.v.9, 286, 1983).
2) The amplitudes of each of three oscillations vary with

C. R. Cowley et al (eds.), Upper Main Sequence Stars with Anomalous Abundances, 341–344.

the stars rotation phase.
3) In the framework of the cooperative program "R.V." of 53CAM the following results were obtained other observers:
a) narrow-band photometry (Kuvshinov, Plachinda, Sov. Lett. Astron.J., vol.6, 368, 1980) in the core of H and K CaII lines shows variations with the time 20-30 minutes.
b) photometrical observations of Panov K.P. (1982, Comm. from Konkoly Obs., 83, 185) in the spectral region U,B,V show the variations with the periods: P_1 = 36, P_2 = 23, P_3 = 18.7, P_4 = 16, P_5 = 11 min. (3 nights, N=345).
c) photometrical observations of Zverko J. (1982, Bull. Astron. Inst. Czechosl. 33, 314) in λ_0 = 5260 Å, $\Delta\lambda$ = 190 Å during 4 nights with the total number of observations 535.

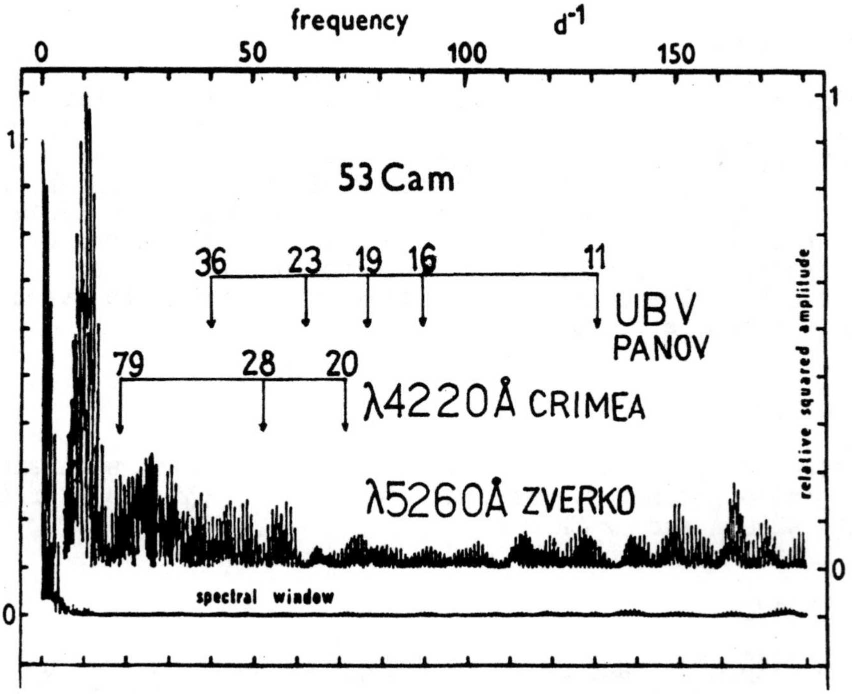

Figure I. The results of different observers: I) the power spectrum and the function of spectral window from data of Zverko; 2) the arrows show the positions of most significant peaks of power spectrum from Crimean data and the data of Panov. The numbers indicate periods in minutes.

The picture shows that in all probability the multiperiodicity exists and the most probable periods are P_1 = 20, P_2 = = 28 minutes.

Unfortunatly there was no an observational night common for all the participants of the program. Evidently the lack of observations does not permit us to draw any reliable conclusion.

III.B. RESULTS OF BETA CrB (will be published in IBVS).

The observations of BETA CrB were carried out during ten nights from 18.05 till 6.06.1984, (total N=551). It has been found that the flux variations do exist and could be described by a superposition of three sinusoidal oscilations with the periods P_3 = 196.9, P_2 =158.3 and P_1 = 58.6 min. The power spectrum and function of spectral window were obtained by the method of Diming. They are presented in fig.2.

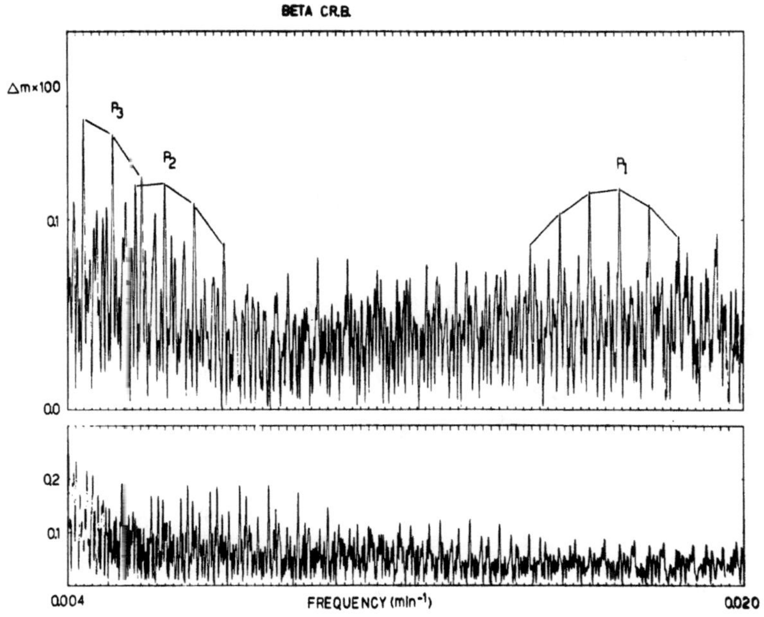

Figure 2. The power spectrum for the data on BETA CrB. P_1 ,P_2 ,P_3 — are the most significant periods.

The absense of any significant peaks in the function of spectral window for periods around P_1 ,P_2 ,P_3 and the presence of conjugate periods determined by the time gaps of siderial day confirm the reliability of the obtained results.

CONCLUSION.

There are two types of brightness variations of 53CAM in the depression λ 4200 and BETA CrB in the depression λ5200, namely:

a) Complex multi-periodical variations of Δm during one night, which might be caused by pulsations of stars with complex structure of atmosphere and high photosphere and magnetic activity of star.

b) Variations of Δ m with the phase of star rotation is well pronounced and might be explained in terms of oblique rotator model.

For 53CAM and BETA CrB the average values of Δ m for one night and magnetic field variations with the period of star rotations are compared in fig. 3.

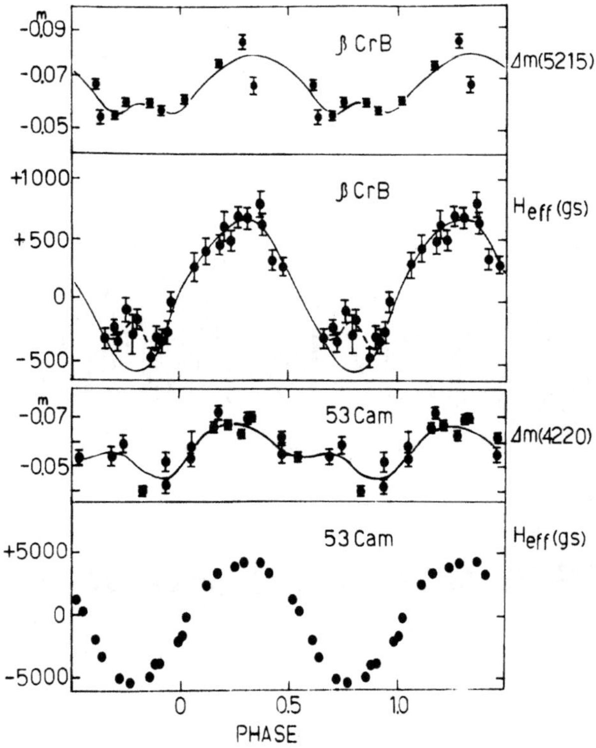

Figure 3. The variations of the depression λ 4200 for 53CAM and the depression λ 5200 for BETA CrB with the phase of rotation, and comparison with the magnetic field variations.

The picture shows a good correlation and indicates that maximum strengh of the depression λ 4200 for 53CAM and the depression λ 5200 for BETA CrB is related to the regions of magnetic poles.

DISCUSSION (Severny and Lyubimkov)

SADAKANE: I have six high dispersion IUE spectra of 73 Dra taken in 1978 and 1979. My analysis of the 2556 Å region does not confirm the U II line at 2556.19 Å. Also, two other lines of U II at 2562.94 Å and 2565.41 Å can not be confirmed in 73 Dra. The same thing occurs in the 1979 spectrum of HR 465 = HD 9996, in which strong U II lines close to 3859 Å are confirmed. Possibly, the U II lines in 73 Dra are strongly variable.

SEVERNY: Yes. [Inaudible] summed spectra from IUE and also from Earth for about the same phase. They did not find such a discrepancy with our result, they found traces of this line quite surely. It may be that the abundance depends on the phase because it is a variable star.

ROMANOV: Have you an interpretation of the variation of the abundance of uranium for 73 Dra?

SEVERNY: I prefer to avoid any conclusions or interpretations, because the investigation of the whole set of stars for heavy element abundances is not complete.

ADELMAN: I have looked at 21 Aquilae both in the optical and the ultraviolet. It is not a peculiar A star, but it is a rather sharp-lined slightly reddened B7 or B8 IV or V star.

SEVERNY: Ultraviolet spectra of this star are not yet analysed by us.

COWLEY: In HR 465, we found that we could interpret the line at 3859 Å as chromium at phases when the star was Cr-strong. There are Cr-weak phases in that star, and we [Cowley, Aikman & Fisher, Publs. DAO, 15, p37, 1977] believe in the uranium at those phases. In 73 Dra, the 3859 Å line can be very strong, but that star also has very strong Cr. Resolution of this question depends on having atomic data in the ultraviolet and more resolution, perhaps on the Hubble Space Telescope. Overabundance factors of 100 to 1000 are not unexpected, in view of the overabundances of the lanthanide rare earths. It's one thing to say that the actinides could also be overabundant by those factors, but quite a different thing to prove it! I'm not sure that we have sufficiently good data yet to make a definite conclusion.

SEVERNY: We have also looked at your published line profile, and it agrees quite well with the spectra taken at the Bulgarian national observatory.

C. R. Cowley et al (eds.), Upper Main Sequence Stars with Anomalous Abundances, 345.

The age of ε UMa

S. Hubrig
Zentralinstitut für Astrophysik der AdW der DDR
Potsdam, DDR

Abstract

A critical rediscussion of the luminosity confirms that ε UMa is brighter than expected for a star on the main sequence, in our determination by $0\overset{m}{.}6$.

ε UMa is a well known typical Ap star. The peculiarity refers mainly to Fe, Cr, Ti, Ca (Engin, 1975, Rice et al., 1981). Luminosity and spectrum vary with P = $5\overset{d}{.}0887$ (Guthnick, 1931, Provin, 1953). But the effective magnetic field strength is small: $H_{eff} < 110$ Gauss (Borra, Landstreet, 1980), $-300 \div +800$ Gauss (Glagolevski et al., 1981). What is the reason for this? Are the metallic lines to much broadened by other effects, so that a magnetic field cannot be measured precisely? Are the surface inhomogenities taken into accont in the correct manner? If so, why does the more homogeneously distributed hydrogen give nearly the same result, that means very small effective magnetic field strength? Because no clear answer could be given, another question arose, the question whether ε UMa is in a somewhat other state of evolution than most magnetic Ap stars. Therefore a critical rediscussion of the position of ε UMa in the HR diagram was made.
The visual absolute magnitude M_V of ε UMa is based on parallaxe determinations, predominantly those derived under the assumption that ε UMa belongs to the nucleus of the UMa moving cluster.
The following values are given:

$\pi = 0\overset{''}{.}042 \pm 0\overset{''}{.}005$ (Rasmunson, N., 1921 Lund Medd. Ser.II No 26)

$\pi = 0\overset{''}{.}043 \pm 0\overset{''}{.}004$ (Roman, N.G., 1949 Astrophys, Journ. 110, 205)

$\pi = 0\overset{''}{.}040 \pm 0\overset{''}{.}001$ (Wielen, R., 1977, Astron.Rechen-Inst. Heidelberg, No 116)

C. R. Cowley et al (eds.), Upper Main Sequence Stars with Anomalous Abundances, 347–350.

$\pi = 0''.041 \pm 0''.001$ (nucleus stars, in Eggen, O.J., 1984, Astron. J. 89, No 9, 1350)

The trigonometric parallaxe of ε UMa is beyond the limit of large significance, the following values are published:

$\pi_{trig.} = 0''.008 \pm 0''.010$ (Jenkins, L.F., Gen.Cat. of trig. stars paral. 1952)
$\pi_{trig.} = 0''.044 \pm 0''.027$ (Flint, A.S., 1919, Publ.Washburn Obs. XIII, p. I)

To the nucleus of the UMa moving cluster belong six stars from the FK 4 - ε UMa included - with very precisely known absolute proper motions. For these stars a space velocity V = 17.99 ± 0.10 km/sec turns out with the components in the directions of opposite to the galactic center u = -14.4 ± 0.06 km/sec, galactic rotation v = +1.3 ± 0.08 km/sec and the North Galactic Pole w = -10.7 ± 0.08 km/sec using Eggens determination of the convergent point (A= 308°.43; D=-38° 8'.2) derived from 31 stars. Another determination of the space velocity of the nucleus of the UMa moving cluster was made by Wielen (1977) with the result V = 16.0 ± 0.1 km/sec using the determination of A = 301°.92, D = -31°.22 from 20 stars. Wielen noticed that this cluster has the smallest velocity dispersion (0.1 km/sec) measured in a stellar system, indicating the strong gravitational binding. - Thus, the difference in V between the two sets of approximately 2 km/sec is much larger than the internal error.
The radial velocity V_r, which has to be expected for ε UMa as a member of the UMa moving clusters is V_r = -12.6 ± 0.07 km/sec in case using the data given by Eggen or V_r = -9.3 ± 0.06 km/sec using the data given by Wielen. The measured mean radial velocities show a large scattering. The following values in km/sec are given using metallic lines:

- 9	± 1	(Vogel, H.C., 1903, Astron. Nach. 163, 145)
- 7.1	± 0.46	(Baker, R.H., 1908, Publ.Allegheny Obs.1, 23)
- 8		(Lick) } Moor's Catalogue of Radial Velocities, 1932
- 5.4		(Mc Donald) } Moor's Catalogue of Radial Velocities, 1932
- 7.6	± 0.8	(Harper, W.G., 1937, Publ.Dom.Astrophys. Obs. 7, 1)
- 10		(Swensson, J.W., 1944, Astrophys. Journ. 99, 258)
- 7.1	± 1	(Woszczyk, A., Jasinski, M., 1980, Acta Astron. 30, No. 3, 331)
- 9.8	± 1	(Tektunali, H.G., 1981, Astrophys. Space Sci. 77, 41)
- 9	± 1	(Rice, J. et al., 1981, 23 Liege Coll. 265)

- 8.5 ± 1.3 (Hubrig, S., 1977, 1978)
- 9.4 ± 0.4 (Hubrig, S., 1978 - 1984)

The mean of all these determinations is V_r = -8.3 ± 0.4 km/sec. For the hydrogen lines Abt and Snowden (1973) give V_r = -11.´ ± 1.43 km/sec and Tektunali (1981), V_r = -10.0 ± 0.9 km/sec.
Our values have been obtained by two different methods of measurements:
V_r = -8.5 km/sec is derived from 38 spectra with dispersion 8 A/mm in such a way that for those lines which are split probably by inhomogeneities the radial velocity of each component was measured and the average afterwards computed. Contrary V_r = -9.4 km/sec was obtained using the wings of the whole line, in this case measuring 21 spectra with a dispersion of 4 A/mm.
If the radial velocity of ε UMa is -12.6 ± 0.07 km/sec as postulated by Eggens investigations then the systematic deviation from the mean measured radial velocity of -8.3 ± 0.4 km/sec need an explanation. One possibility would be the existence of a companion to ε UMa. The search for a period in the variation of the mean radial velocity had no success: the rotation period was recovered only. Even the values of proper motion do not exhibit any variation. Another possibility for the explanation of the deviation of the observed radial velocity value from the expected would be a contraction of the stars atmosphere; a possibility which is not discussed further. Finally such a difference could be produced by a special geometry of inhomogeneities of the elements. If all explanations for the discrepancy between observed and predicted radial velocity must be rejected, either a small correction to Eggens values of A, D or V is necessary or the membership of ε UMa to the nucleus of the UMa cluster must be inquired. We exclude the last mentioned possibility regarding that using Wielens values the predicted V_r agrees very well with the observed. Thus - as generally assumed - we conclude that ε UMa belongs to the nucleus of the UMa moving cluster and find from its parallaxe the visual absolute magnitude $M_V = -0.^m3 \pm 0.^m05$. On the main sequence such a luminosity corresponds to a B8 star with T_{eff} = 11900°K using the values given in Landolt-Bornstein, 1982.
From spectroscopic investigations and model calculations the following values T_{eff} are published:

T_{eff} = 9400° K (Schild, R. et al. 1971, ApJ 166, 95)
9600° K (Cucchiaro, A. et al. 1978, Astron. Astrophys. Suppl 33, 15)
10000° K (log g = 3.5)(Durrant, C.J., 1970, MNRAS 147, 75)
10200° K (Wolf, S.C. et al. 1968, ApJ 152, 871)
9900° K (Glagolevski, Ju. W., 1966, Astron. J. russ. 43, 73)
9500° K (log g = 3.5)(Glagolevski, Ju.W. et al. 1982, Astrophys. Issled russ. 15, 14)
9300° K (log g = 3.2)(Engin, S., 1975, IAU Coll. No 32, 623)
9985° K (log g = 3.5)(Tektunali, H.G., 1981, Astrophys. Space Sci. 77, 41)

Thus after rediscussion we come to the same conclusion given by Glagolevski et al. that the star is brighter than would be expected if it would be on the main sequence. The measured gravitational acceleration log g = 3.5 is in accordance with this assumption. The fact that ε UMa is the brightest member of the nucleus of the UMa moving cluster fits to a such a position in the HR diagram.
Whether ε UMa is in a somewhat other state of evolution than most magnetic Ap stars needs further investigations.

References:

Abt, H.A., Snowden, M.S., 1973, ApJ. Suppl. 25, No 215, 137.
Borra, E.F., Landstreet, J.D., 1980, ApJ. Suppl. 42, No 3, 421.
Guthnick, D., 1931, Sitz. preuss. Akad. Wiss. Berlin No 27, 618.
Provin, S., 1953, ApJ. 118, 489.

LINE SPECTRUM VARIATIONS IN THE AP STAR HD 51418

I.Kh.Iliev and I.S.Barzova
National Astronomical Observatory
Bulgarian Academy of Sciences
Smolyan, BG-4700
Bulgaria

ABSTRACT. Line spectrum variations in Ap star HD 51418 were studied. The intensities of the Eu II lines vary by about three times. Radial velocity measurements show remarkable variability of the Eu II lines with an amplitude of 16 km/s, but the Sr II and Fe II lines do not show significant changes.

1. INTRODUCTION

The star HD 51418 has been classified as a member of the Eu-Sr-Cr subgroup with one of the largest known V-magnitude variation (Gulliver et al., (1972). Gulliver and Winzer (1973) have found the spectrum of HD 51418 to be as a whole quite crowded. The maximum of the Sr II lines is shifted by about 0.35P from Eu II maximum. In contrast to many others Ap stars variations of Cr II are in phase with the rare-earths. Jones et al. (1973) pointed out the presence of lines of heavy rare-earths such as holmium and dysprosyum. The amplitude of the velocity curve from Eu II lines is somewhat larger than 10 km/s. All authors have pointed out that furter observations near the rare-earths maximum would be desirable.

2. OBSERVATIONAL DATA AND RESULTS

Thirteen spectrograms of HD 51418 were obtained with the Coude-spectrograph of the two-meter telescope of the Bulgarian National Astronomical Observatory. The plates are on IIaO emulsion and have a dispersion of 9 A/mm. The phases of observations were calculated using the ephemeris given by Gulliver and Winzer (1973).

Thirty six least blended lines of EU II, Dy II, Fe II

C. R. Cowley et al (eds.), Upper Main Sequence Stars with Anomalous Abundances, 351–354.

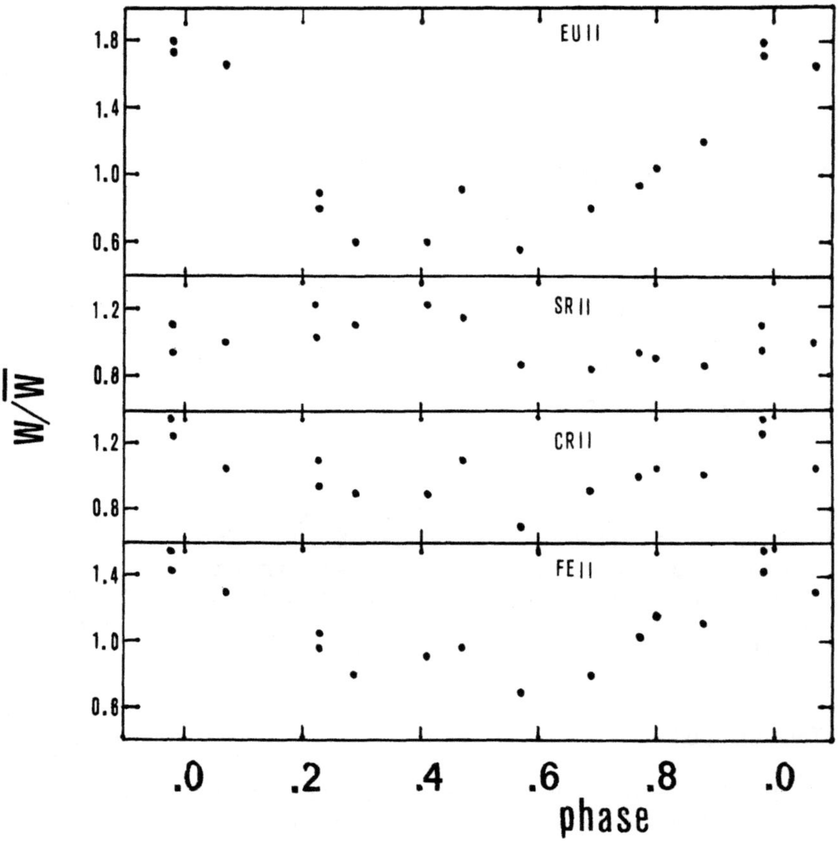

Fig.I. Equivalent widths variations for the Eu II, Sr II, Cr II and Fe II lines.

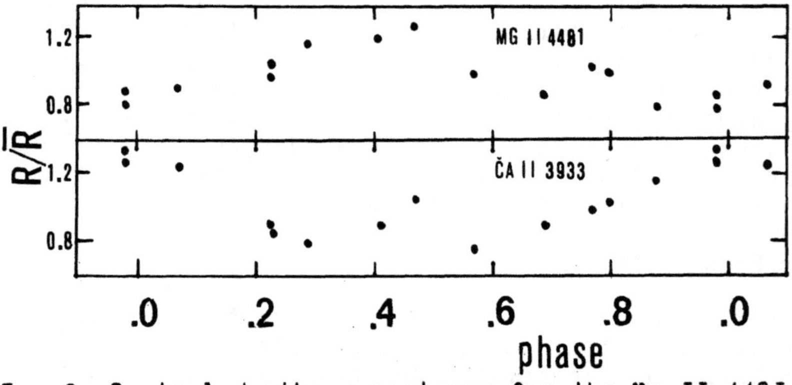

Fig.2. Central depths variations for the Mg II 448I and Ca II 3933 lines.

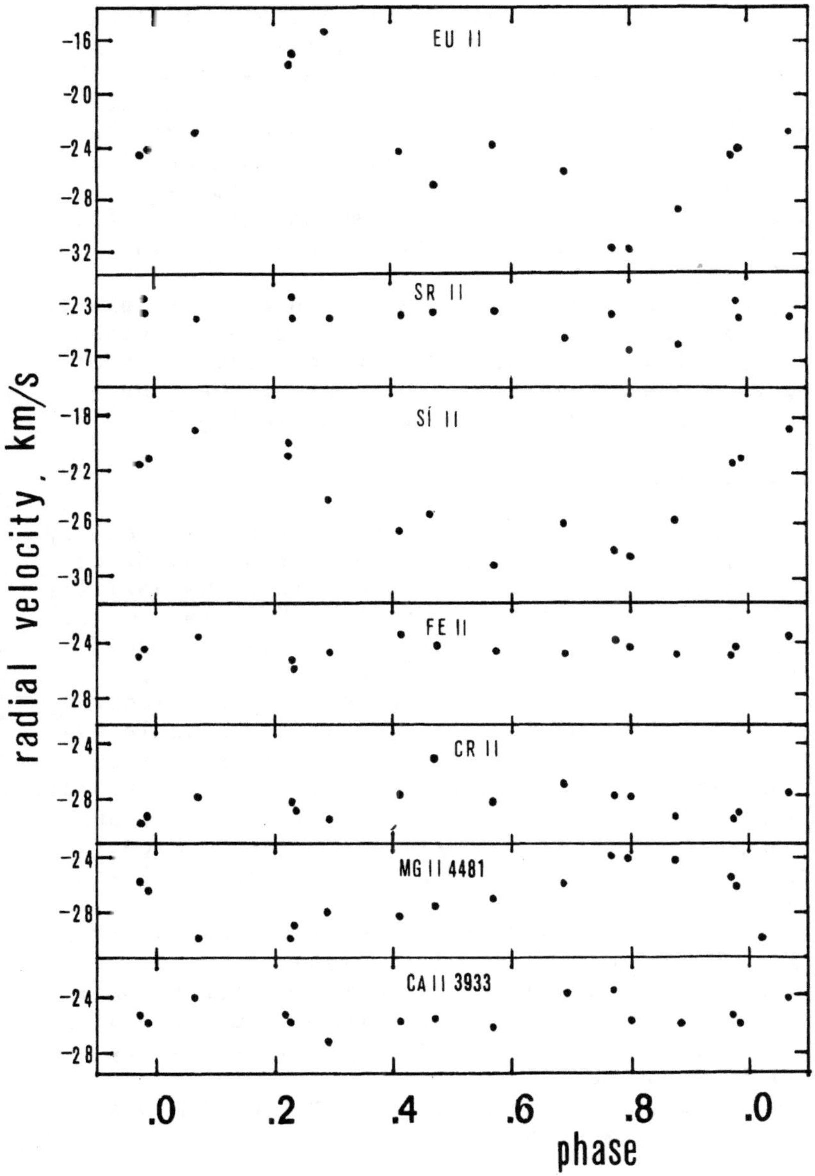

Fig.3. Radial velocity variations for Eu II, Sr II, Si II, Fe II, Cr II, Mg II 4481 and Ca II 3933

Cr II, Sr II, Si II, Ca II and Mg II were measured for equivalent widths. The typical error in the equivalent widths W of a single line are about I5 percent, but the errors in the mean equivalent widths $\overline{W}$ averaged over the period and over all the lines of the same element are close to 8-IO percent. With the exception of Ca and Mg for the rest elements from 2 to 7 lines were used. The ratio $W/\overline{W}$ for Eu II, Sr II, Fe II and Cr II is ploted versus phase in fig.I. The central depths ratio $R/\overline{R}$ for the Ca II 3933 and Mg II 448I lines versus phase is shown in fig.2.

For twenty seven lines of several elements radial velocities were derived. The typical error of a individual line are about 2-3 km/s. The radial velocities for each element were averaged in the same manner as the line intensities. The velocity curves are shown in fig.3.

3. DISCUSSION AND CONCLUSIONS

As shown in fig.I, both the phase shift of Sr II lines by about 0.3P-0.4P and the unusual variations of the Cr II in phase with the Eu II lines are confirmed. The variation of Dy II lines are not shown in fig.I but are similar to those of the Eu II. The intensities of these rare-earths vary by about three times. According to fig.2 Mg II 448I is the only line with antiphase variations.

Radial velocity measurements exhibit appreciable variations for the Eu II lines with an amplitude of about I6 km/s against IO km/s obtained from Jones et al. (I973). The amplitude of the velocity curve for the Si II lines is also large and reaches IO km/s. There are no significant radial velocity variations of the Fe II, Sr II and Ca II 3933 lines.

As a whole the obtained pattern of the line spectrum variability of the Ap star HD 5I4I8 seems to reflect a complex structure of the surface inhomogenities in this star. Probably there exist two different kinds of distribution of the elements - rare-earths and Si in spots, Fe, Cr and Sr in belts.

We wish to thank Dr. I.M.Kopylov for the helpfull discussion during the Crimean colloquium.

REFERENCES

Gulliver,A.F.,McRae,D.A.,Percy,J.R.,Winzer,J.E.: I972, Journ.Roy.Astron.Soc.Canada, 66, 72

Gulliver,A.F.,Winzer,J.E.: I973, Astrophys.J., I83, 70I

Jones,T.J.,Wolff,S.C.,Bonsack,W.K.: I973, Astrophys.J., I90, 579

β CrB - a Rosetta stone?

L. Oetken
Zentralinstitut für Astrophysik der AdW der DDR
Potsdam, DDR

Abstract

Combining the information from the speckle interferometric and spectroscopic binary β CrB a mass of 1.82 solar masses and an absolute visual magnitude $M_v = 1^m.42$ is found, indicating that the star may be in that state of evolution, when the stellar core has shrinked after hydrogen exhaustion in it and the energy generation mainly comes from the envelope. The question whether all magnetic Ap stars are in that special evolutionary state is revived.

Adelman et al. (1981) called β CrB a peculiar Rosetta stone. Today, from another point of view, I ask whether β CrB has the properties of that famous stone found some hundred years ago in northern Egypt near the town Rosetta. This stone was covered with entire different symbols arranged in three columnes; hieroglyphes, demotique symbols and greek letters. Each column obviously described the same contents. Thus the stone could be used as dictionary to interpret the hieroglyphes and generally to learn what has happened more than 2000 years ago. Can β CrB play the same role for the determination of the evolutionary state of magnetic (CP 2) stars?

β CrB is one of the best investigated magnetic stars, since it is bright and has very sharp absorption lines. But differing from most magnetic stars β CrB belongs to a binary system of such a kind that it can be observed by spectroscopic as well as by speckle interferometric methods. Thus mass and luminosity can be derived without the difficulties which otherwise come from the peculiarities in the spectra of magnetic stars. Since the components of the binary are not closely connected, the evolution of the spectroscopic observed primary will not be influenced by that of the companion. Moreover, β CrB may be a member of the Hyades supercluster, giving further

C. R. Cowley et al (eds.), Upper Main Sequence Stars with Anomalous Abundances, 355–358.

informations about the parallaxe. Therefore, because the fundamental stellar parameters can be determined on entire different ways, β CrB may be as valuable for the understanding of the physics of magnetic stars as the stone of Rosetta for the interpretation of the hydroglyphes.
The spectroscopic binary was extensively investigated by Neubauer (1944) and the elements were corrected including later measurements by Oetken and Orwert (1984) with the result given in the following tbl.

	Neubauer	Oetken, Orwert
P =	$3833\overset{d}{.}58 \pm 0.36 = 10\overset{a}{.}496$	$3873\overset{d}{.}0 = 10\overset{a}{.}604$
γ =	-20.19 km/sec	-21.4 km/sec
K =	9.19 ± 0.23 km/sec	9.95 km/sec
e =	0.406 ± 0.025	0.512
ω =	$185\overset{\circ}{.}4 \pm 2.2$	$187\overset{\circ}{.}7$
T_0 =	$242\ 8971\overset{d}{.}3 \pm 18.09$	$242\ 5156\overset{d}{.}541$

The visual and speckle-interferometric binary elements are based on measurements of different groups (Couteau, Labeyrie, Bonneau et al., Mc Alister et al., Morgan et al., Balega and Ryadchenko). Our preliminary elements and those derived by Tokovinin (1984) are given in the following tbl.

	Oetken, Orwert	Tokovinin
P =	$10\overset{a}{.}496$	$10\overset{a}{.}27$
e =	0.511	0.524 ± 0.006
i =	69°	$111\overset{\circ}{.}1 \pm 0.9$
Ω =	147°	$148\overset{\circ}{.}2 \pm 0.5$
ω =	$183\overset{\circ}{.}9$	$181\overset{\circ}{.}3 \pm 0.7$
T_0 =	1980.48	1980.506 ± 0.014

For the determination of the parallaxe, measurements were carried out during 1918-27 at the Allegheny- and during 1928-30 at the Mc Cormick-Observatory, giving the weighted mean of $0\overset{''}{.}031 \pm 0.008$. We used the now available speckle interferometric measurements to correct the extensively published equations of condition and performed a new least square solution to determine parallaxe and proper motion of β CrB. By this procedure the parallaxe found from observations at the Allegheny Observatory was reduced from 0.034 to 0.031, that from the Mc Cormick Observatory of 0.014 did not change.

Combining all informations the most probable parameters of the primary β CrB A are:

$$\mathfrak{M}_A = 1.82\,\mathfrak{M}_\odot \qquad M_V = 1\overset{m}{.}42$$

with a possible tendency to a somewhat larger mass.

From the mass-luminosity relation for main sequence stars, one would expect $M_V = 2\overset{m}{.}23$. Thus β CrB is brighter by $0\overset{m}{.}81$ than a main sequence star of the same mass. The difference decreases for a larger mass of β CrB, but the estimated external accuracy make it unprobable that the discrepancy fully disappears. Assuming the given values to be correct, β CrB would be near to that evolutionary state when the stellar core has shrinked after hydrogen is exhausted in it and the energy generation mainly comes from the envelope, which begins to expand.

If β CrB could be compared with the stone of Rosetta, this would mean that all magnetic stars may be in that particular stage of evolution. Then, moreover, it is suggested that in all stars with masses characteristic for A and late B-types magnetic fields are generated during this very active phase of stellar evolution. The condition may be favourable for magnetic field excitation in a dynamo process as long as this phase holds. A very complicated field may be excited which penetrates through the envelope thereby perfering the lowest multipoles, i.e. dipoles. To explain the observations the time for penetration has to be small compared to the time while the just mentioned phase of stellar evolution lasts. Estimations show that it may be possible. Moreover, under the given assumptions, one would expect that the number $N(A_p)$ of A_p (B_p) stars to the number $N(A)$ of normal A(B) stars corresponds to the ratio of times $\Delta t_1/\Delta t_2$ in stellar evolution: Δt_1 being the time the star spends from its start of the ZAMS till the core begins to shrink, while Δt_2 represents the following time till the star reaches its maximum luminosity. From Ibens (1985) calculations results $\Delta t_1/\Delta t_2 = 11$ % resp. 9.4 % for models with $\mathfrak{M} = 2{,}25\,\mathfrak{M}_\odot$ resp. $\mathfrak{M} = 3\,\mathfrak{M}_\odot$. The observed ratio $N(A_p)/N(A)$ is of the same order, i.e. ~10 %.

Concluding we state that the question whether β CrB can help to interpret the A_p phenomenon like the Rosetta stone is still open, although it seems a very promising idea. Firstly it is necessary to secure the position in the HR-diagram especially by increasing the accuracy of Δm between the binary components. Secondly the assumption that all CP2 stars have passed that active phase of stellar evolution just after hydrogen exhaustion in the core needs further investigation.

References:

Adelman, S.J., Boytim, B.A., Pyper, D.M., Shore, N.S.: 1981 Liège 23 Coll. Internat. d'Astrophys. 109.
Balega, J.J., Ryadchenko, V.P.: 1984 Pisma R. Astron. Jour. 10, 229.
Eggen, O.J.: 1960 Month. Not. 120, 540.
Iben, I.: 1985 Quart. Jour. Roy. Astr. Soc. 26, 1.
Neubauer, F.J.: 1944 Astrophys. Jour. 99, 134.
Oetken, L., Orwert, R.: 1984 Astron. Nach. 305, 315.

SOME STRUCTURAL FEATURES OF MAGNETIC FIELDS OF THE CHEMICALLY PECULIAR STARS β CrB AND α^2 CVn.

I.I.Romanyuk
Special Astrophysical Observatory
Stavropolskij Kraj
Niznij Arkhyz 357147
USSR

In spectra of chemically peculiar (CP) stars the weak lines originate at very different levels on the both sides of the Balmer jump (λ_B= 3646 A): $\tau \approx 0.01$ at $\lambda < 3646$ Å and $\tau \approx 0.5 - 1.0$, when $\lambda > 3646$ Å (Khokhlova,1978).

To estimate the radial gradient of the magnetic field, we have observed two CP-stars β CrB and α^2 CVn from 1979 to 1984 on the 6-meter telescope with an achromatic circular polarization analyzer (Glagolevskij et al.,1978). Compared with classical Zeeman analyzers,whose working wavelength band is only 300 Å, the achromatic analyzer covers a wide spectral region,enabling many more lines to be measured.

A combination of camera and grating was chosen so as record the 3300 - 4000 Å region all on the same plate at the 6.7 A/mm dispersion. Since the lines used to measure the field on either side of λ 3646 Å are exposed simultaneously on a single plate, one avoids many of the systematic errors arising from various positional,photometric and polarization effects, all these should influence the short and long wave ends of the spectrogramm identically.

Results of measurements.

1. Phase curves for the magnetic field variation of the CP-star β CrB in two spectral regions are given in Fig.1. The best agreement between our observations and Wolff's data (1978) we see in case of phase shift of our observations by 0.1 of the period. Probably,the cause of all these disagreements is that the β CrB is a binary star (Balega et al.,1984). We confirm the phase shift of the extrema of positive polarity by 0.15,discovered by Wolff.
2. For the star α^2 CVn we can see another picture in the properties of magnetic fields wich are measured using lines shortward (B_e^{3500}) and longward (B_e^{3800}) of the Balmer jump. The value of the magnetic field and its

C. R. Cowley et al (eds.), Upper Main Sequence Stars with Anomalous Abundances, 359–363.

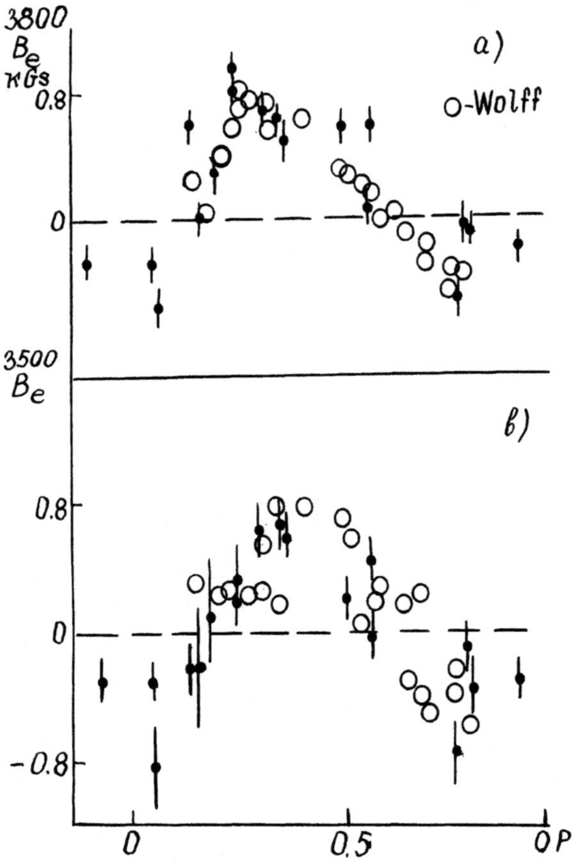

Fig.1

variation amplitude are weaker in the spectral region with $\lambda < 3646$ Å. This can be strengthening into the interior of $\alpha^2 CVn$ atmosphere at a rate of about 1 gauss/km (Fig.2). The field measurements, carried out separately from the iron lines (Fig.3a,b) and the chromium lines (Fig.4a,b) indicate a weakening of a strength of the magnetic field shortward of the Balmer jump.

A very large difference appears at the measurements of chromium lines, and at the same time, the magnetic field measurements from the titanium lines show similar variations from the different sides of Balmer jump. Neither instrumental nor measurement errors can be the cause of these differences. A comparison with the maps of iron, chromium and titanium (Fig.3c,4c,5c, respectively) in the surface of the $\alpha^2 CVn$ (Pavlova and Khokhlova, 1984) shows that, probably, there is a connection between the position of a spot of anomalous chemical composition relative to

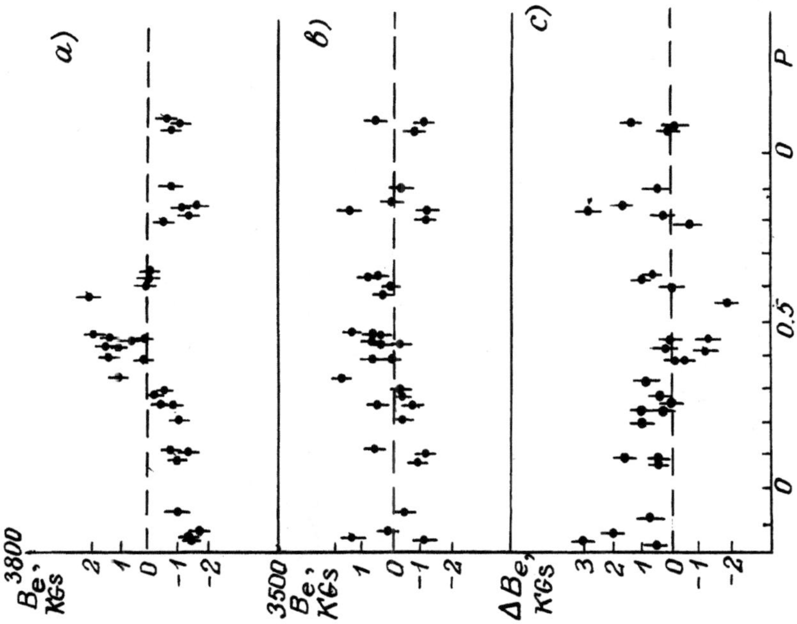
a)
3800
B_e, KGs
3500
B_e, KGs
в)
ΔB_e, KGs
c)
0
0.5
0
P

Fig.2

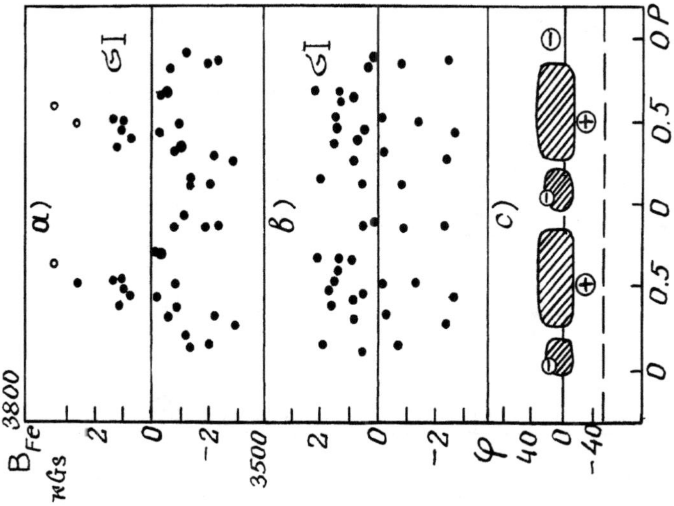
B_{Fe}
кGs
3800
a)
σI
3500
в)
σI
φ
c)
0
0.5
0
0.5
0P

Fig.3

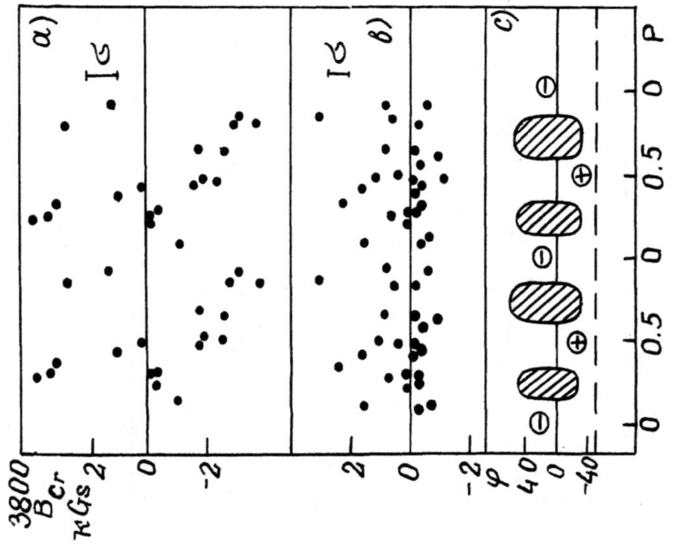

Fig.4

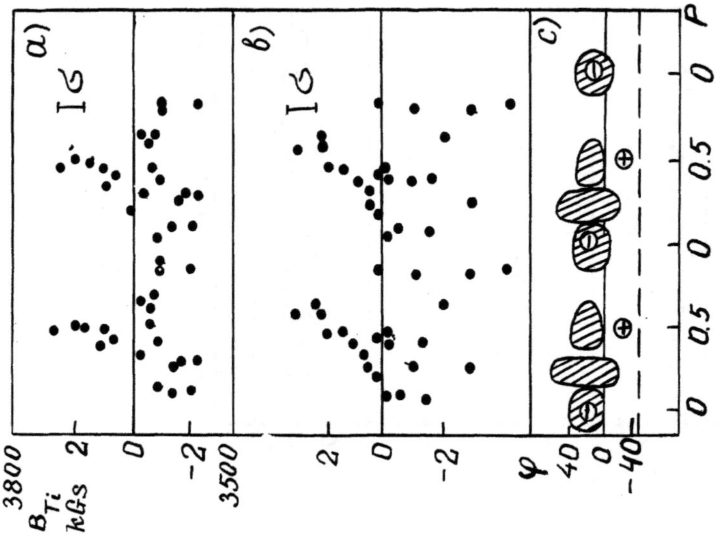

Fig.5

the pole of dipolar magnetic field and the value of radial gradient;when the spot coincides with the pole of dipolar magnetic field,there is no radial gradient,and if there are no spots on the poles,then the differences appear in the value of the magnetic field on both sides of Balmer jump. If we observe the radial gradient of magnetic field and since the values measured are averaged over the whole stellar surface,appreciably larger local gradients might well exist enougth to trigger the mechanism of magnetic element differentiation (Babcock,1963).

More detailed discussion of those problems are presented in other author's papers (Romanyuk,1984,1986).

REFERENCES

Babcock H.W.,Astrophys.J.,1963,137,p.690.
Balega Yu.,Bonneau D.,Foy R.,Astron.Astrophys.Suppl.Ser., 1984,57,p.31.
Glagolevskij Yu.V.,Naidenov I.D.,Romanyuk I.I.,Chunakova N.M.,Chuntonov G.A.,Soobsh.SAO Akad.Nauk SSSR,1978, 24,p.61.
Khokhlova V.L. Pisma Astron.Zh.,1978,4,p.228.
Pavlova V.M.,Khokhlova V.L.,Pisma Astron.Zh.,1984,10,377.
Romanyuk I.I.,Pisma Astron.Zh.,1984,10,p.443.
Romanyuk I.I.,Izv.SAO Akad.Nauk SSSR,1986,22,(in press)
Wolff S.C.,Publ.Astron.Soc.Pasif.,1978,90,p.412.

β LYRAE - A BINARY SYSTEM WITH ANOMALOUS ABUNDANCES AND MAGNETIC FIELD

M.Yu.Skul'skij
Polytechnical Institute, Lvov, SU-290013, U.S.S.R

ABSTRACT. Study of dependence of chemical abundance on conditions of excitation within the limits which could exist in the atmosphere of bright component have led to conclusion that by no modification of physical parameters a "normal" abundance of chemical elements in the atmosphere can be obtained. The variable effective magnetic field with mean value (-1350±50) Gauss is discovered.

The β Lyrae, a close binary system is at the final stage of active mass transfer (case B mass exchange) from the bright B8III component with the mass of 3,8 $M_{\odot}$ to the faint A5III component with the mass of 14,6 $M_{\odot}$ (Skul'skij, 1975a; Burnashev and Skul'skij, 1978). The rate of mass transfer is more than 10^{-5} $M_{\odot}$ per year. The atmospheric lines of bright component and emission-absorption lines of the circumstellar gas structures are seen in the visible β Lyrae spectrum range. For the first time the chemical abundances of the bright component atmosphere was determined by Boyarchuk (1959). In particular, the excess of He and the deficit of H were found. Hack and Job (1965) has decreased the ratio He/H taking higher temperature excitation. This uncertainty is due to the simultaneous presence of high excitation lines of NII, CII, SiIII and more extensive group of low excitation lines FeI, FeII, TiII, CrI, CrII etc., that requires special examination.

The quantitative analyses of the bright component atmosphere for ten main phases of the orbital period was

C. R. Cowley et al (eds.), Upper Main Sequence Stars with Anomalous Abundances, 365–368.

carried out by the curve of growth method (Skul'skij, 1975b). It was found that the excitation temperature defined in all phases is approximately about the same. But microturbulent velocity is substantially variable: from 5,5 km s^{-1} at the elongations to 18 km s^{-1} at the main minimum phase. It evidences that the shape of bright component is not spherical. The study of abundances at the second minimum where the faint component is not observed has confirmed the considerable deviation from the solar chemical composition. In particular, the excess of He, N, C, S, Si were obtained under all reasonable assumptions on excitation conditions (10000 K $\leq T \leq$ 14000 K, $1 \leq \lg p_e \leq 4$). The chemical composition is close to the "normal" one only if the lines of high excitation potential arise at the high temperature and low electron pressure and the FeI, FeII, TiII, CrII lines in the more dense layers at lower temperature. For example the plot of the value

$$[X] = \lg \frac{N_{el}}{N_{Fe}} \Big/ _{\beta\,Lyr} - \lg \frac{N_{el}}{N_{Fe}} \Big/ _{\odot}$$

versus exitation potential ε_i for visible lines of chemical elements is shown in Fig.1 for two cases of the abundance examinations of iron: the first case deals with the mean value of lg N for FeI and FeII (without FeIII lines) and the second one relates to the group of high excitation lines at the presence of FeIII in the spectrum (marked by the asterixes). In both cases $\lg p_e$ is 3. Such a stratification cannot be understood within the assumption of thermodynamical equilibrium. It seems that the atmosphere of bright component has a peculiar structure due to the active mass loss. Besides, Leushin and Snezhko (1980), suggest that matter of this component has been affected in

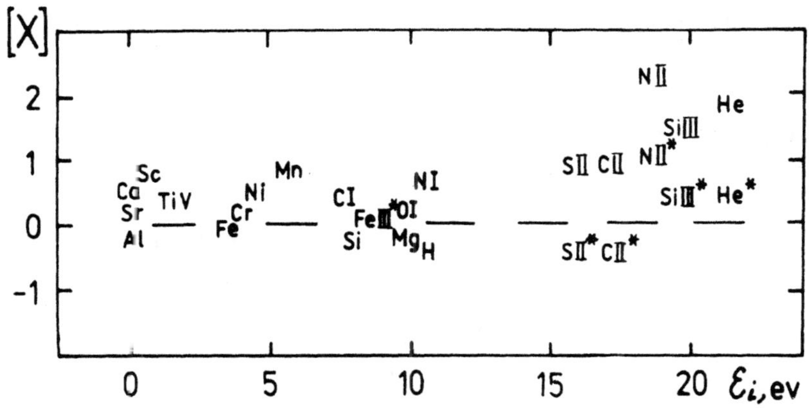

Fig.1. The [X] -index versus excitation potential.

hydrogen burning through CNO-cycle.

In this connection it is interesting that the magnetic field on the surface of the bright component of β Lyrae was discovered (Skul'skij, 1982; Skul'skij, 1985). The measurements of Zeeman splitting of ionized metal lines have been done using 250 spectrograms with dispersion of 9 Å/mm taken with the 6-m telescope in 1980-1984. The effective magnetic field H_e changes with the phase according to the dotted curve (Fig.2) obtained using the approximation of sinusoidal function and taking into account each observation's mean error. The mean value of H_e is (-1350±50) Gauss. Changing together with the orbital period phase with the amplitude of about 500 Gauss it reaches the maximum value near 0.87 P. Moreover, using the absolute spectrophotometric data by Burnashev and Skul'skij (1978) the depression of the continuum at λ 5200 has been found. The depth variation of the depression correlates with the magnetic field changes (Skul'skij, 1985).

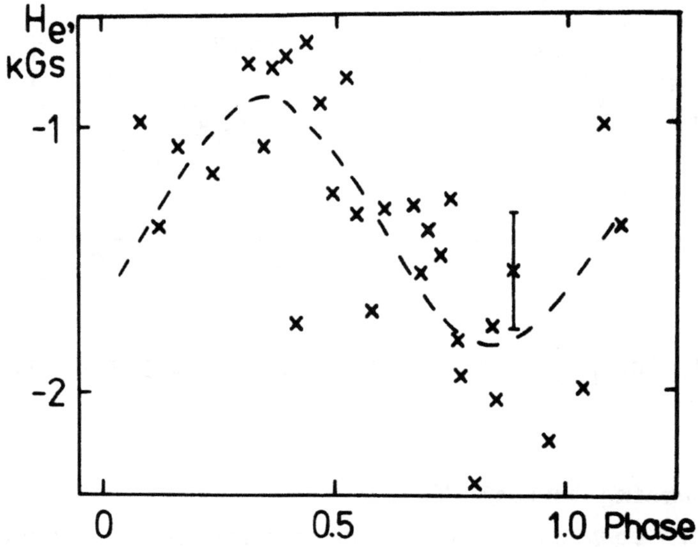

Fig.2. Variation of the effective magnetic field, H_e with the orbital phase. A typical error bar (2σ) is shown for one of the crosses.

References

Boyarchuk A.A. 1959, Sov. astron. J. 36, 5.

Burnashev V.I., Skul'skij M.Yu. 1978, Izv. Krimsk. Obs. 58, 64.

Hack M., Job F. 1965, Zs. Astrophys., B 62, 203.

Leushin V.V., Znezhko L.I. 1980, Pis'ma to Astron. Zh., 6, 171.

Skul'skij M.Yu. 1975a, Sov. Astron.J., 52, 710.

Skul'skij M.Yu. 1975b, In "Problemy cosmicheskoy fiziki", Kiev. University, 10, 160.

Skul'skij M.Yu. 1982, Pis'ma to Astron. Zh., 8, 238.

Skul'skij M.Yu. 1985, Pis'ma to Astron. Zh., 11, 51.

SPECTROSCOPIC ANALYSES OF HOT MAIN SEQUENCE STARS

Kozo Sadakane
Astronomical Institute
Osaka kyoiku University
Tennoji-ku, Osaka, Japan 543

ABSTRACT: Spectroscopic studies of normal O and early B type stars in the visual region are discussed. Present status of UV spectroscopic analyses of hot normal stars is reviewed. Discussions on a few practical problems in analyses of UV spectra are presented.

1. Introduction

Spectroscopic studies of photospheres of non-magnetic, normal and chemically peculiar (CP) O and B type stars have been relatively inactive in recent years. Attention has been mainly focused on outer atmospheres and active phenomena such as mass loss of these hot stars. This is in contrast with the recent activity in studies of late B type (mainly Hg-Mn type) CP stars in both visual and UV regions. Instead of summarizing a few results on hot CP stars, I will concentrate here on spectroscopic analyses of hot normal stars. Abundances of only a dozen or so elements have been well determined observationally for O and early B type normal stars. Elements heavier than Ca (except for Fe) have not been studied in them. Data of chemical composition of hot stars yield information on the state of recent nuclear processing in the Galaxy because the ages of these objects are only a fraction of the solar lifetime. Studies of normal stars are also necessary because they can be used as reference standards in analyses of CP stars. Detailed studies of normal stars are indispensable especially in the UV region where many spectral lines of heavy elements not observable in the visual region can be found. In view of the present day incompleteness of atomic data such as laboratory lists of atomic and ionic lines or transition probabilities, we have to carry out comparative studies of CP stars using normal stars of correponding temperatures as standards. Spectroscopic study of hot stars in the UV region is now in its early stage and much more efforts should be devoted to normal stars.

I will first summarize some recent analyses of O and early B type stars made in the visual region. Next, I will review recent spectroscopic works in the UV region. Several practical problems such as the damping constants and the microturbulence for UV metallic lines will be discussed. Finally, I will compare the results obtained in the UV and the visual regions and discuss the consistency of the current scheme of spectroscopic analyses.

C. R. Cowley et al (eds.), Upper Main Sequence Stars with Anomalous Abundances, 369–380.

2. SPECTROSCOPIC ANALYSES IN THE VISUAL REGION

Present status of both theoretical and observational studies of atmospheres of B type stars is summarized in Underhill (1982a). Discussions on the calibrations of basic physical quantities such as mass, radius, colors, effective temperature, and bolomecric correction are given in this article. Spectroscopic studies of O type stars have been extensively carried out by Conti and his associates (e.g., Garmany, Conti, and Massey 1980). A review article on O type stars will be given in a forthcoming book by Underhill and Divan (to be published).

Detailed spectroscopic analyses of 15 O and early B type stars in the visual region published before 1971 are reviewed in Scholz (1972). We can see on his Table 1 that abundances of elements heavier than Si are not known observationally in O type stars. In addition, abundances of P, S, Ar, Ca, and Fe are obtained in B type stars. Effective temperatures and surface gravities are determined mainly from ratios of line strengths (ionization equilibria) such as O II/O III or Si III/Si IV in these early studies. It has been known that ratios of line strengths in three different stages of ionization for the same element or in two stages of ionizaton for several different elements do not necessarily lead to the same temperature. The effective temperatures deduced from ionization equilibria are generally higher than the values obtained from the size of the Balmer jump or from the slope of the Paschen continuum. Sometimes we find discrepancies in the adopted effective temperatures by a few thousand degree in different analyses of the same star.

Table 1 lists the results of more recent analyses of O and B type normal stars. The B2 IV star γ Peg was studied by Peters (1976) and the B5 IV star τ Her was studied by Adelman (1977). Determinations of T_{eff} and log g in these studies are mainly based on the flux distributions and the profiles of the Balmer lines. They found that abundances are normal in these two stars except for slight overabundances of Ne, Cl, and Ar in γ Peg. Five O type stars are analyzed by Kudritzki and his associates and they determined effective temperatures, surface gravities and He abundances. Abundances of He in these hot stars are normal except for ζ Pup in which He is slightly overabundant. It is to be noted here that abundances of such elements as Cr or Mn which are important in physics of CP stars are not known observationally in hot O and B stars. This is because of the intrinsic weakness of absorption lines of singly or doubly ionized ions of these elements in the visual spectra of hot stars.

Now, I will briefly discuss the importance of re-analyses of sharp lined normal stars with modern techniques. These sharp lined stars can be used as reference standards in studies of CP stars of corresponding temperatures. A new systematic survey seems necessary because of the following reasons.

Table 1. Recent Analyses of O and B Type Stars

Star	HD/HDE	Author	T_{eff}	log g	Abundance
γ Peg	886	Peters 1976	21,500	3.7	14 elements
τ Her	147394	Adelman 1977	14,500	3.5	8 elements
	93250	Kudritzki 1980	52,500	3.95	He
ζ Pup	66811	Kudritzki et al. 1983	42,000	3.5	He
	93128	Simon et al. 1983	48,000	3.85	He
	93129A		45,000	3.60	
	303308		45,500	3.90	

(1) Very high signal to noise ratio (S/N) spectroscopic observations are now possible with modern electronic devices such as Reticon or CCD. A combination of these devices with a high resolution spectrogragh enables us to analyse profiles of even faint lines with high accuracy. For example, Smith (1981) detected extremely faint (equivalent width: $\lesssim 1 \sim 3$ mÅ) lines due to P II, Ar II, or Cl II near 4130 Å in the spectrum of ι Her using a Reticon detector system. Quantitative analyses of He I lines by Heasley and Wolff (1981) and Heasley, Wolff, and Timothy (1982) and of the H_α line by Heasley and Wolff (1983) in B type stars well illustrate the advantages of these new tools.

(2) Reliable data of transition probabilities are now available for many atomic and ionic lines. Although astrophysicists even today suffer from the lack of atomic data as discussed in the previous session of this colloquium, the situation has significantly improved in these ten years. Data used in older analyses should be replaced with the more reliable ones. It is important to incorporate the latest atomic data into analyses of both normal and CP stars.

(3) The effective temperature scale has been revised. Various methods of determining the effective temperatures of hot stars are discussed in Böhm-Vitense (1981) and in Underhill (1982a). Underhill (1982b) determined effective temperatures of 24 O3 to O5 stars and found no correlation between the effective temperature and spectral type or luminosity class for these hot stars. Recent revisions of the temperature scale depend on extensive high quality photometric observations from UV to IR regions. As discused by Code (1984), the calibration of UV flux measurements is critically important for O and B stars because these stars radiate most energy in this region. An uncertainty of at least ± 1000 K is expected even in the latest determinations of effective temperatures for early B type stars.

Table 2. Non-LTE Calculations

Ion	Ref.	T_{eff} (10^3 K)	log g	ξ_t (km s^{-1})
He I, II	1	30 - 50	3.3, 4.0, 4.5	
He I	2	15 - 27.5	2.5, 3.0, 4.0	
	3	15 - 27.5	2.5, 3.0, 4.0	
	4	15 - 27.5	2.5, 3.0, 4.0	
	5	15 - 27.5	3.0, 4.0	
Be II	6	10 - 15	4.0	0
B II	7	10 - 15	4.0	0
C II, III	8	25 - 35	3.5, 4.0, 4.5	
C III	9	30 - 55	3.3, 3.5, 4.0	10
N II	10	20 - 32.5	4.0	0, 5
N III	11	32.5 - 50	3.3, 4.0	5, 10
O I	12	7.5 - 17.5	1.0, 2.5, 4.0	0, 5
Ne I	13	15 - 22.5	3.0, 4.0	0, 4
Mg I, II	14	10 - 15	4.0	0
Mg II	15	15 - 40	4.0	0, 4
	16	15 - 35	2.5, 3.0, 4.0	0 - 15
Si II, III, IV	17	15 - 35	2.5, 3.0, 4.0	0 - 15
	18	15 - 35	3.0 - 4.5	0
Ca II	19	15 - 27.5	2.5, 3.0, 4.0	0, 4
	20	10 - 15	4.0	0
Sr II	20	10 - 15	4.0	0
Ba II	14	10 - 15	4.0	0

References: (1) Auer and Mihalas 1972, (2) Auer and Mihalas 1973a, (3) Mihalas et al. 1974, (4) Mihalas et al. 1975, (5) Dufton and McKeith 1980, (6) Boesgaard et al. 1982, (7) Borsenberger et al. 1979, (8) York 1980, (9) Sakhibullin and Solov'eva 1983, (10) Dufton and Hibbert 1981, (11) Mihalas and Hummer 1973, (12) Baschek et al. 1977, (13) Auer and Mihalas 1973b, (14) Borsenberger et al. 1984, (15) Mihalas 1972, (16) Snijders and Lamers 1975, (17) Kamp 1976, (18) Kamp 1978, (19) Mihalas 1973, (20) Borsenberger et al. 1981.

(4) Non-LTE computations of line profiles and strengths for many ions are now available. Published non-LTE predictions and thier range of applicability in T_{eff} and log g are summarized in Table 2. We can see on this table that detailed non-LTE computations are mainly carried out for light ions in the high temperature range. Kamp (1982) compared observed and predicted strengths of C II, and III and also Si II, III and IV lines in the UV region using high dispersion data obtained with the IUE satellite. He found reasonably good agreements but noted that theoretical predictions not always give satisfactory explanations of very strong and very weak lines. Quantitative comparisons of predictions with high S/N observations of normal stars are necessary before we apply them in studies of CP stars.

An important feature found in hot stars is the mass loss phenomenon. Mass loss effects are detected from asymmetric profiles of strong resonance lines such as C IV, N V, and Si IV in the UV region. Snow and Morton (1976) found that stars brighter than M_{bol} = -6.0 mag generally show mass loss effects. According to Gathier, Lamers, and Snow (1981), six stars among the 15 O and B stars listed in Table 1 of Scholz (1972) are losing mass. Lamers and Rogerson (1978) and Hamann (1981) analyzed the profiles of UV resonance lines in the B0 V star τ Sco and determined a mass loss rate of log $\dot{M} = -8.9 \pm 0.5\ M_{\odot}$/yr in this star. Smith and Karp (1978) found slight asymmetries in the photospheric lines of τ Sco and suggested that a subphotospheric, convective velocity field is responsible for these asymmetries.

3. ANALYSIS OF HIGH RESOLUTION UV SPECTRA

Now, we shall turn our attention to the observations made in the ultraviolet region below 3000 Å. Studies of high resolution UV spectra of various stars began in 1972 with the OAO-3 (Copernicus) satellite. The IUE satellite continues its activity since 1978 and a large amount of spectroscopic data are now available. The Soviet astrophysical space station Astron began its observation in 1983 (Boyarchuk et al. 1984). The importance of the UV region lies in the anticipation that we may find new infomation not yet discovered in the visual region. Because spectroscopic analyses in the UV region began recently, we do not have sufficient practical experience in the region yet.

The initial works to be carried out in a new wavelength region is to register the observed spectral features and to try thier identifications using available line lists. Table 3 summarizes these pioneering works on hot normal stars from high resolution spectroscopic data obtained with the Copernicus and the IUE satellites. B type supergiants are not included in this table. We can see on this table that no line identification list in the region between 1430 Å and 2000 Å has been published for B type main sequence stars so far. This is because of the restriction of the spctrogragh of the Copernicus satellite. Line lists in this spectral region for B type stars are urgently needed.

Table 3. Analysis of High Dispersion UV Spectra

Star	HD	Sp. T.	Range (Å)	Ref.
			Line List	
ζ Pup....	66811	O4f	923 - 3205	1
	49798	O6p	1175 - 2935	2
3 O-type Stars			1150 - 2000	3
42 Ori....	37018	B1 V	1040 - 1424, 2061 - 2903	4
γ Peg....	886	B2 IV	1000 - 1436, 2000 - 2990	5
ι Her....	160762	B3 IV	999 - 1467	6
τ Her....	147394	B5 IV	1025 - 1425, 2023 - 2959	7
ζ Dra....	155763	B6 III	1035 - 1425, 2000 - 3000	7
α Lyr....	172167	A0 V	1100 - 1460, 2000 - 3000	8
			2746 - 2881	9
			Atlas	
τ Sco....	149438	B0 V	949 - 1420, 1418 - 1560	10
ι Her....	160762	B3 IV	999 - 1422, 1418 - 1467	6
60 O and B Stars			1000 - 1450	11

References: (1) Morton and Underhill 1977, (2) Bruhweiler et al. 1981, (3) Dean and Bruhweiler 1985, (4) Johnson et al. 1977, (5) Hill and Adelman 1978, (6) Upson and Rogerson 1980, (7) Underhill and Adelman 1977, (8) Faraggiana et al. 1976, (9) Michelson 1981, (10) Rogerson and Upson 1977, and (11) Snow and Jenkins 1977.

Note: Data used in (9) are obtained with a balloon-borne spectrograph.

Discussions on difficulties in the analyses of UV spectra of hot stars have been given in the previous session and they are not repeated here. I think it is important to check, at the beginning, whether we can have consistent results with those obtained in the visual region for some well studied stars. This process has to be done quantitatively before we carry out analyses of some lines of unexplored elements in the UV region. I will discuss a few practical problems in the following.

(1) Damping Constants

It has been customary to assume ten times the classical damping constant in analyses of metallic lines of B and A type stars in the visual region when no experimental data are available. The choice of a damping constant raises no serious problem in the visual region because metallic lines are generally weak in this region. In contrast, many strong metallic lines are overlapping in the UV region even in the spectrum of an early B type star.

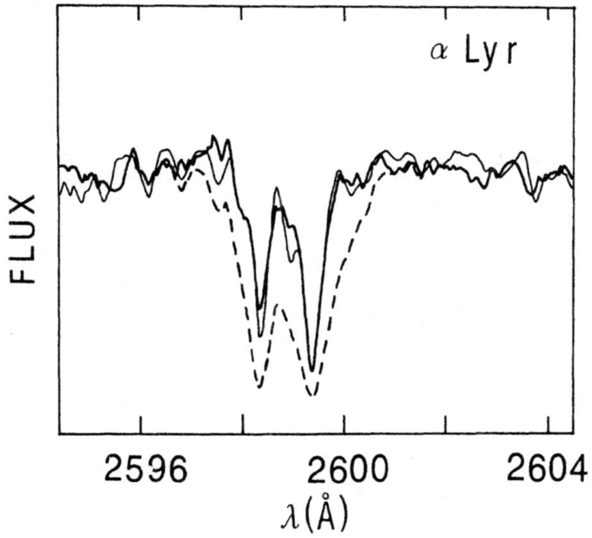

Fig. 1. Spectrum of α Lyr around 2600 Å. Bold line: Observed spectrum, thin line: computed spectrum with classical damping, and dashed line: computed spectrum with ten times the classical damping.

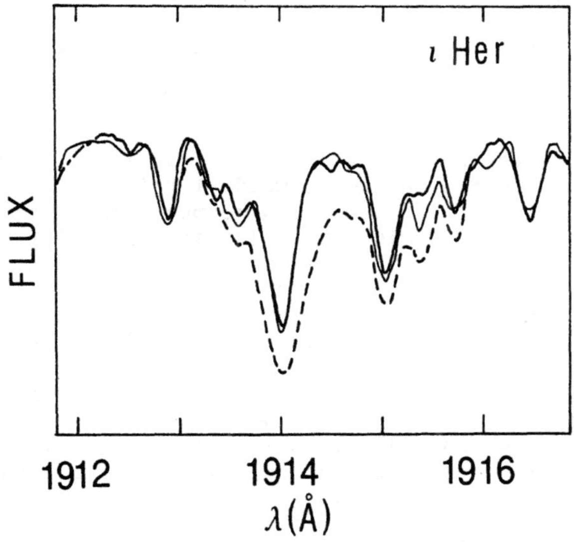

Fig. 2. Spectrum of ι Her around 1910 Å. Symbols are the same as in Fig. 1.

Thus, it is critically important to use correct damping constants to reproduce the observed spectrum in the UV region. Peytremann (1972) found that the damping by electron collisions for UV lines is always much smaller than either the radiative or the classical damping except for lines arising from high-excitation levels. According to him, a damping constant equal to 10 times the classical value is always much too large for the UV lines. Since then, however, no quantitative evaluation of the empirical damping constant in the UV region has been published.

Preliminary analyses of some strong lines of Fe II and Fe III in α Lyr (A0 V) and ι Her (B3 IV), respectively, are carried out using a spectrum systhesis technique. Six LWR (α Lyr) and three SWP (ι Her) images obtained with the IUE high dispersion spectrographs are co-added to make up the observational data. Figure 1 shows the region around 2600 Å in α Lyr where two strong resonance lines of Fe II are contained. I tried to reproduce the observed spectrum with a line blanketed model atmosphere (T_{eff} = 9650 K and log g = 3.95) and with the same abundances obtained in the visual region. Abundances of α Lyr obtained by Sadakane and Nishimura (1981) are used. 114 lines which are taken from the lists of Kurucz and Peytremann (1975) and Kurucz (1981) are included in the computation between 2595 Å and 2605 Å. Transition probabilities of the Fe II resonance lines are taken from the new compilation by Martin et al. (1986). Also assumed are a microturbulence velocity of 2 km/s (discussed below) and a rotational velocity, v sin i, of 23 km/s. Then, the damping constant is changed as a free parameter until a good fit is obtained. I obtained a satisfactory agreement with the observed profile using the classical damping constant as shown in Figure 1. If we assume a damping constant equal to 10 times the classical value, we are forced to assume a much smaller (more than 1.0 dex) abundance of Fe to account for the resonance lines.

Figure 2 shows the region around 1910 A in ι Her where many lines of doubly ionized Fe are crowded. I tried to reproduce the observed spectrum with a model atmosphere (T_{eff} = 17,800 K and log g = 3.8) and assuming the solar abundances of metals. A microturbulence velocity of 2 km/s and a v sin i value of 10 km/s are assumed. 80 absorption lines (mostly of Fe III) are included in the region between 1910 Å and 1918 Å. Transition probabilities are taken from Kurucz and Peytremann (1975) except for the Fe III multiplet 51 lines. Data for the multiplet are taken from Fuhr et al. (1981). Again, a good fit is obtained when the classcal damping constant is used. We cannot account for the profiles of the strong line (Fe III multiplet 34 at 1914.06 Å) and other weak Fe III lines simultaneously with a single value of the Fe abundance if we assume 10 times the classical value. Thus, I conclude that the actual damping constant for Fe II and Fe III lines originating from low excited levels is very close to the classical value.

(2) Microturbulence

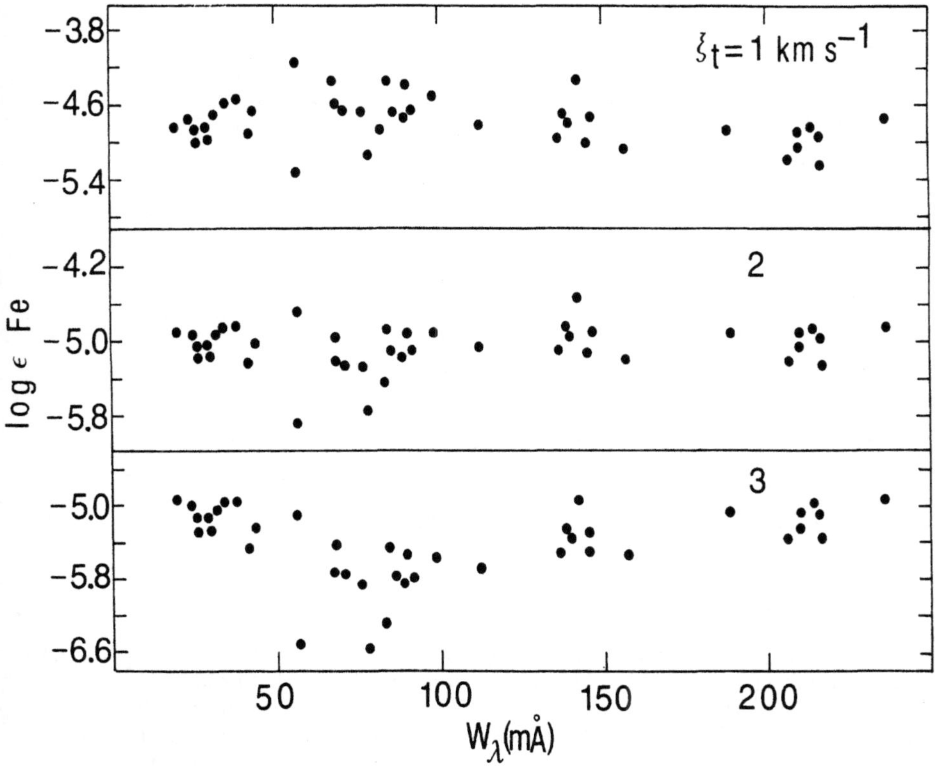

Fig. 3. Determination of the microturbulence ξ_t in the UV region (from 2300 to 3000 Å) from Fe II lines in α Lyr.

Castelli and Faraggiana (1979) noted that a depth-dependent microturbulence which decreases toward the surface has to be introduced in order to account for UV Fe II lines of α Lyr. I re-examined Fe II lines in α Lyr between 2300 Å and 3000 Å using high dispersion IUE data. Unblended lines are carefully selected using the two line lists (Kurucz and Peytremann 1975 and Kurucz 1981). Equivalent widths of 46 selected Fe II lines are measured. Transition probabilities (log gf values) for these lines are taken from the new compilation of Martin et al. (1986). Abundance of Fe is computed for each of the Fe II line using the same model atmosphere of α Lyr noted above. The microturbulence is changed as a free parameter in the range of 0 to 4 km/s. Figure 3 shows the dependence of the obtained abundance of Fe on measured equivalent widths for three microturbulences. The abundance is nearly independent on the equivalent width when a microturbulence of 2 ± 0.5 km/s is used. This value is the same as that obtained by Sadakane and Nishimura (1981) from the visual and the near UV (3250 Å to 3600 Å) spectrum of α Lyr. Thus it is apparent that we do not need to introduce a depth dependency in the microturbulence of α Lyr.

Table 4. Abundances of Iron Peak Elements in Vega

Ion	UV1[a]	N	UV2[b]	N	Visual	N
Cr I					-6.65	3
Cr II....	-6.99	7	-7.00	14	-6.81	15
Mn I					-6.87	3
Mn II....	-6.77	9	-6.81	4		
Fe I					-5.09	28
Fe II....	-5.06	46	-5.02	4	-5.10	19
Co II....	-7.45	7	-7.20	2		
Ni I			-5.94	11		
Ni II....	-6.04	8	-6.19	1	-6.43	1

a: 2100 Å to 3000 Å, b: 3250 Å to 3600 Å.

(3) Abundances

The abundance of Fe from the UV lines (log Fe/H = -5.0 ± 0.25) agrees with the result of Sadakane and Nishimura (1981). Iron is definitely underabundant in α Lyr when compared with the solar abundance of log Fe/H = -4.37 (Simmons and Blackwell 1982). Of course, it is to be noted that the above results depend on the adopted set of transition probabilities and the errors in the measurements. The gf values used here are revisions of the data of Fuhr et al. (1981) which for the most part resolve difficulties in the values of Fe II in the visual region.

I compared the equivalent widths of 22 apparently unblended Fe II lines given in Castelli and Faraggiana (1979) with my measurements and found that their values are systematically (by about 25 %) smaller than my results. This difference is most probably due to their underestimations of the height of the continuum level. I searched on the spectrum of α Lyr for line free windows by comparing with the synthesized spectrum. Usually, a few such windows can be found in each 10 Å interval between 2000 Å and 3000 Å. The highest points in these windows are connected by segments of straight lines and they are used as the continuum level. In addition, I measured equivalent widths of selected lines of Cr II, Mn II, Co II and Ni II in the UV spectrum of α Lyr. Abundances of these elements derived from UV lines are also in satisfactory agreements with those obtained in the visual region (Table 4). The RMS errors in the obtained abundances are ± 0.25 dex for Fe II and around ± 0.3 dex for other ions. A full description of the analysis of the UV spectrum of α Lyr will be published in a separate paper (Sadakane, Nishimura, and Hirata 1985). The importance of careful selections of lines and also a careful determination of the continuum level can not be overemphasized in analyses of stellar UV spectra.

4. CONCLUDING REMARKS

(1) The importance of normal stars in studies of CP stars is re-emphasized. Detailed analyses of normal stars in the visual region are still necessary. These analyses should be based on recent developements of both observational techniques and theoretical studies.

(2) Quantitative analyses of high resolution UV spectra of hot stars are now in the early stage of developement. Information on abundances of many unexplored elements will be obtained from UV data. Careful analyses of UV spectra of normal stars as reference standards are critically important in this region.

I would like to thank Dr. J. R. Fuhr for providing unpublished data of log gf values (Martin et al. 1986) and WDC-A at NASA/GSFC for providing released IUE data. Helpfull comments from Drs. J. Jugaku and M. Takada-Hidai are gratefully acknowledged.

REFERENCES

Adelman, S. J. 1977, M. N. R. A. S. 181, 667.
Auer, L.H., and Mihalas, D. 1972, Ap. J. Suppl., 24, 193.
———. 1973a, Ap. J. Suppl., 25, 433.
———. 1973b, Ap. J., 184, 151.
Baschek, B., Scholz, M., and Sedlmayr, E. 1977, Astr. Ap., 55, 375.
Boesgaard, A. M., Heacox, W. D., Wolff, S. C., Borsenberger, J., and Praderie, F. 1982, Ap. J., 259, 723.
Böhm-Vitense, E. 1981, Ann. Rev. Astr. Ap., 19, 295.
Borsenberger, J., Michaud, G., and Praderie, F. 1979, Astr. Ap., 76, 287.
———. 1981, Ap. J., 243, 533.
———. 1984, Astr. Ap., 139, 147.
Boyarchuk, A. A., et al. 1984, Pis'ma Astron. Zh., 10, 163, (Sov. Astr. Lett., 10, 67).
Bruhweiler, F. C., Kondo, Y., and McCluskey, G. E. 1981, Ap. J. Suppl., 46, 255.
Castelli, F., and Faraggiana, R. 1979, Astr. Ap., 79, 174.
Code, A. D. 1985, in IAU Symposium 111, Calibration of Fundamental Stellar Quantities, ed. D. S. Hayes, L. E. Pasinetti and A. G. D. Philip (Dordrecht: Reidel)
Dean, C. A., and Bruhweiler, F. C. 1985, Ap. J. Suppl., 57, 133.
Dufton, P. L., and Hibbert, A. 1981, Astr. Ap., 95, 24.
Dufton, P. L., and McKeith, C. D. 1980, Astr. Ap., 81, 8.
Faraggiana, R., Hack, M., and Leckrone, D. S. 1976, Ap. J. Suppl., 32, 501.
Fuhr, J. R., Martin, G. A., Wiese, W. L., and Younger, S.M. 1981, J. Phys. Chem. Ref. Data, 10, 305.
Garmany, C. D., Conti, P. S., and Massey, P. 1980, Ap. J., 242, 1063.
Gathier, R., Lamers, H. J. G. L. M., and Snow, T. P. 1981, Ap. J., 247, 173.
Hamann, W. -R. 1981, Astr. Ap., 100, 169.

Heasley, J. N., and Wolff, S. C. 1981, Ap. J., 245, 977.
———. 1983, Ap. J., 269, 634.
Heasley, J. N., Wolff, S. C., and Timothy, J. G. 1982, Ap. J., 262, 663.
Hill, I., and Adelman, S. J. 1978, Ap. J. Suppl., 37, 265.
Johnson, H. M., Snow, T. P., Gehrz, R. D., and, Hackwell, J. A. 1977, P. A. S. Pacific, 89, 165.
Kamp, L. W. 1976, Statistical Equilibrium Calculations for Silicon in Early Type Model Stellar Atmospheres, NASA TR-R455.
———. 1978, Ap. J. Suppl., 36, 143.
———. 1982, Ap. J. Suppl., 48, 415.
Kudritzki, R. -P. 1980, Astr. Ap., 85, 174.
Kudritzki, R. P., Simon, K. P., and Hamann, W. -R. 1983, Astr. Ap., 118, 245.
Kurucz, R. L. 1981, Smithsonian Ap. Obs. Spec. Rept., No. 390.
Kurucz, R. L., and Peytremann, E. 1975, Smithsonian Ap. Obs. Spec. Rept., No. 362.
Lamers, H. J. G. L. M., and Rogerson, J. B. 1978, Astr. Ap., 66, 417.
Martin, G. A., Fuhr, J. R., and Wiese, W. L. 1986, Atomic Transition Probabilities - Scandium through Nickel (A Critical Data Compilation), to be published.
Michelson, E. 1981, M. N. R. A. S. 197, 57.
Mihalas, D. 1972, Ap. J., 177, 115.
———. 1973, Ap. J., 179, 209.
Mihalas, D., Barnard, A. J., Cooper, J., and Smith, E. W. 1974, Ap. J., 190, 315.
———. 1975, Ap. J., 197, 139.
Mihalas, D., and Hummer, D. G. 1973, Ap. J., 179, 827.
Morton, D. C., and Underhill, A. B. 1977, Ap. J. Suppl., 33, 83.
Peters, G. J. 1976, Ap. J. Suppl., 30, 551.
Peytremann, E. 1972, Astr. Ap., 17, 76.
Rogerson, J. B., and Upson, W. L. 1977, Ap. J. Suppl., 35, 37.
Sadakane, K., and Nishimura, M. 1981, P. A. S. Japan, 33, 189.
Sadakane, K., Nishimura, M., and Hirata, R. 1985, submitted to P. A. S. Japan.
Sakhibullin, N. A. and Solov'eva, L. I. 1983, Tr. Kazan Gorod. Astro. Obs. Vyp., 48, 27.
Scholz, M. 1972, Vistas in Astron., 14, 53.
Simmons, G. J., and Blackwell, D. E. 1982. Astr. Ap., 112, 209.
Simon, K. P., Jonas, G., Kudritzki, R. P., and Rahe, J. 1983, Astr. Ap., 125, 34.
Smith, M. A. 1981, Ap. J., 246, 905.
Smith, M. A., and Karp, A. H. 1978, Ap. J., 219, 522.
Snijders, M. A. J., and Lamers, H. J. G. L. M. 1975, Astr. Ap., 41, 245.
Snow, T. P., and Jenkins, E. B. 1977, Ap. J. Suppl., 33, 269.
Snow, T. P., and Morton, D. C. 1976, Ap. J. Suppl., 32, 429.
Underhill, A. B. 1982a, in B Stars With and Without Emission Lines, ed. A. B. Underhill and V. Doazan (NASA SP-456), p. 3.
———. 1982b, Ap. J., 263, 741.
Underhill, A. B., and Adelman, S. J. 1977, Ap. J. Suppl., 34, 309.
Upson, W. L., and Rogerson, J. B. 1980, Ap. J. Suppl., 42, 175.
York, L. A. 1980, Ph. D. thesis, Boston University.

Discussion appears after the following paper.

DIFFUSION PROCESSES AND CHEMICAL PECULIARITIES IN MAGNETIC STARS

G. Alecian
(UA 173 - CNRS)
DAF, Observatoire de Paris-Meudon
92195 Meudon Principal Cedex
France

ABSTRACT. This review tries to expose some essential aspects of our current understanding of diffusion processes in magnetic CP stars. Many problems remain to be explained due to the fact that the presence of strong magnetic fields increases the number of free parameters in "diffusion models". More observational constraints are needed before one can attempt modelling every aspects of magnetic CP stars.

1. SOME GENERALITIES CONCERNING DIFFUSION OF ELEMENTS AND MAGNETIC CP STARS

Many papers, since the initial work of Michaud (1970), have exposed how elements may diffuse in stars (see the review papers by Michaud, 1980, Vauclair, Vauclair, 1982, and Alecian, Vauclair, 1983). In this section we shall try to show very briefly, why diffusion appears to be a powerful mechanism in explaining Chemically peculiar (CP) stars even if many problems remain to be solved.

1.1 How diffusion answers the global constraints deduced from the observations

From the whole set of informations given by the observations, one may work out a list of global minimum constraints which must be satisfied by any theoretical model for magnetic CP stars. We list them as follows:

(i) the anomalies affect the outer layers of the CP stars;
(ii) the anomalies must appear in a much smaller time than the stellar life time on the Main Sequence;
(iii) the phenomenon occurs in a well defined interval of T_{eff};
(iv) in this interval, there are several categories of CP depending on T_{eff};
(v) the peculiarities of magnetic CP are different from non-magnetic CP , for the same T_{eff};
(vi) in each category, there is an important diversity of abundance

C. R. Cowley et al (eds.), Upper Main Sequence Stars with Anomalous Abundances, 381–389.

peculiarities from star to star.

The diffusion processes are considered to be efficient in CP stars, because these constraints are satisfied respectively as follows:

(i) diffusion is efficient in the outer layers (more than in deeper layers);
(ii) diffusion may build anomalies in times smaller than 10^{4} years in stable atmospheres;
(iii) the effects of diffusion are destroyed if macroscopic motions are too strong. For instance, strong turbulence, convection or meridional circulation mix the stellar material and chemical differentiation cannot be established. On the other hand,strong mass losses carry away the atmospheric material faster than diffusion builds stratifications. Now, mixing and mass loss seem too be minimum in the range of T_{eff} where CP stars are found;
(iv) diffusion is very sensitive to T_{eff} and log g (through the radiative acceleration g^{rad} and the hydrodynamical situation which prevails), therefore, different abundance anomalies are obtained according to the spectral type;
(v) diffusion is very sensitive to the magnetic field in the atmosphere (above $\tau_{5000} \simeq 10^{-1}$);
(vi) All the parameters needed to describe completely a given star and which are important in how diffusion occurs, are not accurately known (essentially hydrodynamical ones). On the other hand, diffusion is a time dependent process: a given star may have several phases of peculiarities.

In the following sections of the present review, we shall put emphasis on two of these last points: magnetic field and time-dependent diffusion.

We do not consider that any other process than diffusion must be excluded in explaining the CP phenomenon, but all the processes which have been invoked till now, have appeared to be inadequate (see the comments of Bonsack, 1981). As for an example, the accretion process (which certainly exits more or less) cannot be considered as an alternative with respect to diffusion (Michaud, 1976): if mixing motions are too strong for diffusion, they are also too strong for accretion. And, for stable outer layers, the accreted material must diffuse as well as the original one. In this case, diffusion is much more efficient than accretion and well correlated with T_{eff} while accretion is not. Finally, accretion appears to be a secondary effect compared to diffusion.

1.2 A possible scenario for magnetic Ap stars.

According to the theoretical works on diffusion in CP stars during the last fifteen years, a possible scenario is given in figure 1. This scenario explains how a star may become an Ap star. Of course this scenario is very schematic and more sophisticated ones could be imagined.

The arrival on the main sequence is probably associated with some braking

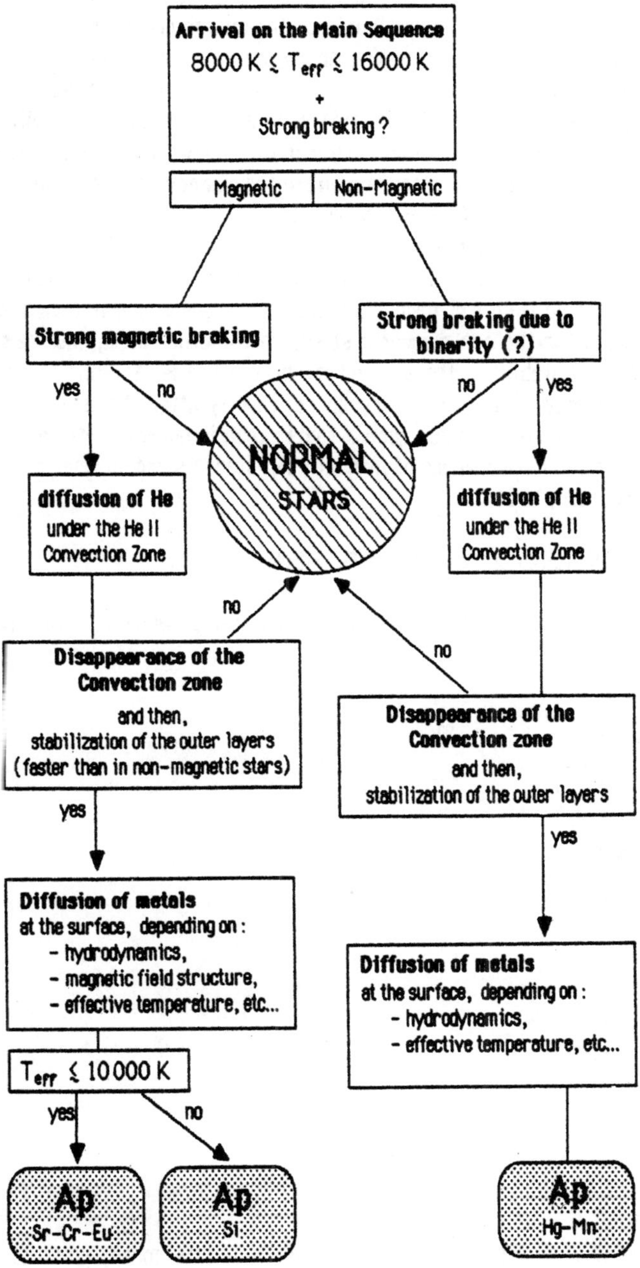

Figure 1
A possible scenario for Ap stars

mechanism which will help the star to be more stable than others having the same mass. This former step is rather badly known, but once the mixing processes occurring at the bottom of He II ionization-convection-zone, are practically smoothed out, helium may settle down by diffusion. When helium becomes underabundant in the outer envelope this leads to the disappearance of the convection zone and then, to the stabilization of atmospheric mixing motions. After that, efficient diffusion processes can start in the atmosphere, the detailed behavior depending on hydrodynamics, magnetic field structure and intensity, and also, on the effective temperature.

This scenario is almost the same for magnetic (Sr-Cr-Eu; Si) and non-magnetic (Hg-Mn) Ap stars except for the braking mechanisms. The rotational velocity is probably reduced by the magnetic field for magnetic stars and by something like tidal effects in the case of binaries for the non-magnetic stars. On the other hand, the presence of a strong magnetic field stabilizes the atmosphere before the He settling is complete. In that case, diffusion may go on in the atmosphere, shortly after the arrival on the Main Sequence and therefore, peculiarities appear faster than in non-magnetic Ap stars. This explains why the ratio of the number of magnetic Ap stars vs non-magnetic, is higher in young clusters (Michaud, 1981).

2. DIFFUSION IN PRESENCE OF A MAGNETIC FIELD

When magnetic field is present, the diffusion velocity may be written as follows:

$$v_{Di} = \frac{D_i}{1 + \omega_i^2 t_i^2} \left[\ldots + \frac{m_i}{kT} (g_i^{rad} - g) + \ldots \right], \quad (1)$$

with $\omega_i = \frac{Z e H}{m_i c}$,

and :

$$v_D \simeq \frac{\sum_i N_i v_{Di}}{\sum_i N_i} , \quad (2)$$

$i = 0, 1, 2, \ldots$

where D_i is the diffusion coefficient of the ion i, m_i its mass, g_i^{rad} the radiative acceleration, g the gravity, t_i the collision time, ω_i the gyro-frequency, Z e the ion charge, H the intensity of the horizontal component of the magnetic field, k, T and c are respectively the Boltzman constant, the local temperature and the light velocity. The velocity v_D (expression (2)) is a ponderated mean value which approximates, in the case of horizontal or zero magnetic field, or high densities (optically thick medium), the average velocity of the whole element (see the detailed study by Montmerle and Michaud (1976)).

In expression (1), we have only shown the terms which are the most important ones in usual cases, for more complete expressions see the review paper by Alecian and Vauclair (1983). As shown in expression (1) horizontal magnetic field affects diffusion through $\omega_i^2 t_i^2$ term: for zero magnetic field, neutral atoms, or high densities (which implies very small t_i) this term cancels out.

In CP stars, the presence of magnetic field may have important consequences on how diffusion occurs for some elements. Silicon provides a typical example of such a case. This element is often found overabundant in magnetic stars (about 10 to 10^2 with respect to solar abundance) while it is found normal or slightly overabundant in non-magnetic stars (Hg-Mn). This behavior has been first explained by Vauclair et al (1979) and studied after in more details by Alecian and Vauclair (1981). In the case of zero magnetic field (or when magnetic field lines are vertical), silicon is very weakly supported by the radiation field: Si II and Si III settle down while Si I goes toward the surface, the mean velocity v_D is slightly positive (element moves upward) and a weak overabundance (a factor 10) may be obtained at the stellar surface. Now, if there is a strong horizontal magnetic field (stronger than 5000 Gauss), the downward motion of the ions is impeded above $\tau_{5000} \simeq 10^{-1}$, and there, the negative contribution of the ions becomes weaker in v_D. This leads to higher overabundances (about a factor 10 to 10^2 can be obtained by diffusion in the atmosphere).

Other elements may also have special behavior in the presence of magnetic field. For instance boron, which has a strong positive v_D, can be overabundant only if there is a magnetic field (Borsenberger et al, 1981). Generally speaking, the less an element is abundant with respect to hydrogen the more it is pushed up by the radiation field, the detailed behavior depending on its atomic properties. Elements like boron, which are strongly pushed through the stellar outer layers by the radiation field, will reach the surface whether the star is magnetic or not, and then strong horizontal magnetic lines will force them to accumulate high in the atmosphere, before they leave the star.

Some other elements may also accumulate in the "line forming" region before reaching the place where magnetic field can trap them. In that case, these elements must show the same kind of overabundances than in non magnetic stars.

Actually, the situation is much more complex in real cases, since the magnetic field lines may have various angles with respect to the horizontal and in this case, the mean diffusion velocity v_D is no more vertical but it is the sum of a vertical and horizontal component (cf. Alecian and Vauclair, 1981). In spite of these difficulties, some predictions are possible. For instance Vauclair et al (1979) and, Alecian and Vauclair (1981) have predicted that silicon is preferentially overabundant at places where the magnetic lines are horizontal. However, horizontal diffusion (horizontal component of v_D) may change this feature (Mégessier, 1984), we shall discuss that problem in the next section.

All these studies have been made in the framework of the "oblique rotator model": the diffusion processes in the magnetic field, will produce patches and rings according to the geometry of magnetic lines. Michaud et al (1981) have studied this in detail.

3. TIME-DEPENDENT DIFFUSION

Expression (1) allows the computation of the diffusion velocity at a given place in the medium and for a given concentration of the considered element. Starting from an homogeneous concentration of that element throughout the star, diffusion may create abundance stratifications if mixing motions are not too strong. These stratifications may be interpreted as abundance anomalies by the spectroscopists, as far as the line forming region is concerned. Strictly speaking, the detailed computation giving a qualitative and quantitative description of these stratifications, is required in order to make any theoretical prediction on what kind of abundance peculiarities may be observed.

Abundance stratifications are obtained by solving the continuity equation:

$$\partial_t N + \boldsymbol{\nabla}.(N\mathbf{v}) = 0 \ , \qquad (3)$$

where:

$$N = N(\mathbf{r},t) \ , \qquad (4)$$

and :

$$\mathbf{v} = \mathbf{v}_D(N(\mathbf{r},t),\mathbf{r}) + [\ \mathbf{V}\ ] \ , \qquad (5)$$

where $\mathbf{V}$ represents some eventual hydrodynamical velocity like a stellar wind for instance (turbulence must be included in v_D (Vauclair et al, 1978)). Actually, equation (3) is a time-dependent non-linear partial differential equation which must be solved numerically.

3.1 Vertical stratification

Even in the case of magnetic Ap, a star has essentially a spherical symmetry and the gradients involved in $\mathbf{v}_D$ are essentially vertical. Therefore, diffusion in stars may be assumed to be a vertical one-dimensional process. However, this simplification is far from enough for solving easily equation (3). Fortunately, in many cases a "zero order" approximation allows a rough estimate of the overabundances that can be expected at the stellar surface. This approximation consists in comparing the radiative acceleration g^{rad} to gravity g throughout the star, and in assuming that equation (3) will reach a stationary solution with $N = N^*$ such as $g^{rad} = g$ everywhere in the star. Now, due to the non-linearities in equation (3), the existence of such a stable stationary solution is not certain.

This problem has been recently studied by Alecian and Grappin (1984) who considered the time-dependent diffusion of one ion in stellar envelopes. They have shown that stable solutions like $N \simeq N^*$ are possible in optically thick regions, but also abundance stratifications may go through transient states very different from the stationary solution before reaching it (in the case of zero mass loss). If such transient states affect upper optically thin regions, this may produce ups and downs of observable abundances during the stellar lifetime. On the other hand, these authors have also speculated on a possible unstable scenario in building abundance stratifications in outer optically thin stellar regions. If such transients states or unstable stratifications occur in Ap stars atmospheres, this means that there may be several phases of peculiarities for some elements in a given star. This can explain the wide diversity of abundance anomalies observed in magnetic and non-magnetic Ap stars because a given star may go through different phases of peculiarity.

3.2 Horizontal diffusion

We have shown in section 2 that, due to the presence of magnetic field, an horizontal component of the diffusion velocity shows up. Generally, the effect of this drift velocity is negligible comparing to the vertical process. But one can suppose that, when the vertical stratification approaches stationarity (with formation of rings and patches), horizontal diffusion must be taken into account. On another way, as pointed out by Michaud et al (1981), a slight turbulence may add also an horizontal component to the diffusion velocity in the case of horizontal concentration gradients. So in this case rings and patches become wider with time.

A quantitative study of horizontal diffusion has been made for silicon by Mégessier (1984). Using the radiative acceleration of Alecian and Vauclair (1981), Mégessier has computed the drift velocity of silicon assuming a dipolar configuration of the magnetic field. She has found that, after silicon accumulates in a ring at the magnetic equator (horizontal field lines), silicon migrates toward the magnetic pole and then, sinks (see her communication to this colloquium for some new observational results). Such a process will take about 10^8 years to be achieved (to be compared with 10^4 years for the vertical process). Now, long times as 10^8 years raise the problem of the permanence of the magnetic structures at the surface of Ap stars. Indeed, observational confirmation of silicon horizontal diffusion would imply that the magnetic structures remain unchanged during about 10^8 years (whereas the diffusion time of surface magnetic lines is precisely of the order of 10^8 years for structures having a size equal to the stellar radius).

4. SOME IMPORTANT PROBLEMS CONCERNING MAGNETIC CP STARS AND RELATED TO THE DIFFUSION PROCESSES

4.1 Rapid variations of cool Ap stars

The discovery by Kurtz (1982) of rapid variations (in the frequency range of 4 to 15 minutes) in cool Ap stars, brings a new qualitative knowledge about these stars. According to him, these variations seem to be linked to the presence of strong magnetic field; but why only few high overtone modes are excited and how they are related to the magnetic field was not understood.

Very recently, Dolez et al (1985) have considered the problem of the rapid variations in Ap stars in the framework of the diffusion processes. They have found that the helium stratification due to the combined effect of diffusion and a local mass loss at the magnetic pole, favours the existence of the oscillations found by Kurtz.

4.2 The possible existence of corona

There is not, at the present time, any observational evidence of corona around CP stars. The X-ray observations are near the detection limit (Golub et al, 1983) and cannot be considered as reliable.

The detection of corona around CP might pose a problem as far as its existence would imply some violent atmospheric activities which prevent any chemical separation in the photosphere. Recently, Havnes and Goertz (1984) have made a theoretical study on magnetospheres of early-type CP. According to this study (the first one in this subject), the heating of an eventual corona, would be achieved by the interaction of mass loss with magnetic field and rotation.

4.3 The radiative accelerations

The calculation of accurate radiative accelerations is very important in carrying out computations on diffusion. Several problems make this task difficult. For instance, many efficient atomic transitions absorb photons in UV, where the radiation flux is badly known. On the other hand, until now, these computations have neglected the Zeeman splitting which can increase the radiative accelerations up to a factor two in some cases (Borsenberger et al, 1981). Note that this last effect may act in optically thick regions, i.e. before the $\omega_i^2 t_i^2$ term becomes efficient in equation (1).

4.4 The observational constraints

For the moment the observational constraints are not strong enough to decide what is the real aspect of the magnetic Ap stars' surface. The measurement of magnetic field in the lines of some elements would be helpful: for instance to decide if silicon is accumulated in rings rather than spots (Michaud et al, 1981). Of course, this information might also be obtained by the mapping methods using accurate analysis of spectral line variations (see Khokhlova, this colloquium), however, the lack of uniqueness in the results of these methods (mainly due to the observational uncertainties), preclude for the moment the drawing of firm conclusion on the shape of the surface inhomogeneities.

Precise data on the line profiles and curves of growth should give also useful informations about vertical stratifications (Borsenberger et al 1981, Alecian, 1982).

5. A TIME-DEPENDENT CONCLUSION

Much more observational data for individual stars (mainly on magnetic fields), more constraints are needed before undertaking further detailed theoretical investigations on diffusion processes in the case of magnetic CP stars, because the existence of strong magnetic fields adds too many free parameters.

The general features of magnetic CP stars are well understood if one accepts that diffusion can act in the outer stellar regions. However the detailed features remain to be explained Models for individual magnetic stars will be probably built up next to the ones for non-magnetic Hg-Mn stars which have more simple properties and for which a parameter-free model can apply.

REFERENCES

Alecian, G., 1982, Astron. Astrophys., **107**, 61.
Alecian, G., Grappin, R., 1984, Astron. Astrophys., **140**, 159.
Alecian, G., Vauclair, S., 1981, Astron. Astrophys., **101**, 16,
Alecian, G., Vauclair, S., 1983, Fundamentals of Cosmic Physics, **8**, 369, Gordon and Breach Science Publishers Ltd.
Bonsack, W. K., 1981, in "Upper Main Sequence Chemically Peculiar Stars", 23 rd Liège Int. Astrophys. Symp., Université de Liège, ed. P.Renson, p.345.
Borsenberger, J., Michaud, G., Praderie, F., 1981, Ap.J., **243**, 533.
Dolez, N., Gough, D., Vauclair, S., 1985, in preparation (private communication).
Golub, L., Harnden, F.R.Jr., Maeson, C.W., Rosner, R., Vaiana, G.S., Cash, W., Snow, T.P.Jr., 1983, Ap.J., **271**, 264.
Havnes, O., Goertz, C.K., 1984, Astron. Astrophys., **138**, 421.
Kurtz, D.W., 1982, Monthly Notices Roy. Astron. Soc., **200**, 807.
Mégessier, C., 1984, Astron. Astrophys., **138**, 267.
Michaud, G. 1970, Ap.J., **160**, 641.
Michaud, G., 1976, in Physics of Ap Stars, IAU coll. N°32, ed. W.W.Weiss, H. Jenker and H. J. Wood, Vienna.
Michaud, G., 1980, Astron.J., **85**, 589.
Michaud, G, 1981, in "Upper Main Sequence Chemically Peculiar Stars", 23 rd Liège Int. Astrophys. Symp., Université de Liège, ed. P.Renson, p.355.
Michaud, G., Mégessier, C., Charland, Y., 1981, Astron. Astrophys., **103**, 244.
Michaud, G., Charland, Y., Vauclair, S., Vauclair, G., 1976, **210**, 447.
Montmerle, T., Michaud, G., 1976, Ap.J. Suppl., **31**, 489.
Vauclair, S., Vauclair, G., 1982, Ann. Rev. Astron. Astrophys., **20**, 37.
Vauclair, S., Hardorp, J., Peterson, D.M., 1979, Ap.J., **227**, 526.
Vauclair, S., Vauclair, G., Schatzman,E., Michaud, G., 1978, Ap.J., **223**, 567.

DISCUSSION (Sadakane)

SILANT'EV: What are the lowest values of magnetic field one can measure from the Zeeman effect in these hot stars?

SADAKANE: Perhaps the next speaker, Dr. Dworetsky, can answer that, as I do not know the detailed answer.

DWORETSKY: [barely audible] About 20 gauss.

ADELMAN: I want to comment on the assumption that the damping constant is about 10 times classical in the optical region, as you have advocated in the ultraviolet. We should abandon that assumption. Better approximations will change the microturbulence ξ by about 0.5 km/s.

Also, could you tell me where you obtained the information about the instrumental profile for IUE high resolution spectra?

SADAKANE: Data on the instrumental profiles for IUE were privately communicated by Dr. Y. Kondo of NASA.

COWLEY: Would you comment on the reality of the Fe/H and metal/H abundance fluctuations in these stars, and in particular, the extent to which we might consider α Lyr (Vega) as a possible borderline case of the λ Boo phenomenon?

SADAKANE: I think there is a diversity in metal abundances, especially iron, of about 1.0 dex among superficially normal late B and early A stars. Vega itself may be related to λ Boo stars, although it does not satisfy all of the classification criteria for them.

SEVERNY: In one of your tables you seemed to have a difference in chemical composition of α Lyr derived from the visible and UV spectra. What could be the cause of this? Different spectral resolution?

SADAKANE: My results show there are no significant differences in abundances of Cr, Mn, Fe, Co and Ni between the UV and visual regions. Small differences of about 0.3 dex are mainly due to the poor quality of some of the gf values.

POLOSUKHINA: How accurate are the radial velocities?

SADAKANE: We do not really measure true radial velocity from IUE spectra. I reduce my data by measuring the wavelengths of Fe II lines which can be identified, and adjust the observed wavelengths to the laboratory scale. The accuracy depends on the quality, signal-to-noise ratio and v sin i. In the case of α Lyr, experience shows that radial velocities (i.e., wavelengths) can be determined to ±0.05 Å when I use carefully selected lines of Fe II.

MENDOZA: For the solar abundances you gave in Table 4, what was the wavelength interval used?

SADAKANE: The solar abundances are based on observations in the visual region of the solar spectrum.

C. R. Cowley et al (eds.), Upper Main Sequence Stars with Anomalous Abundances, 391–393.

DISCUSSION (Alecian)

DOLGINOV: In your paper, you use the normal diffusion coefficient in the magnetic field, which is proportional to $1/H^2$. This form is valid only if you have very quiet conditions, with laminar flows and no currents, etc. However, if there is even a small amount of turbulence or a current, then the diffusion coefficient should be proportional to $1/H$ (the so-called Bohm diffusion). Have you considered this possibility?

My second question is similar to the one I asked Mégessier. How is it possible to keep the elements localized for a long time by the field, bearing in mind that an interchange instability exists because the system is like one in which the heavy medium is above the light one?

My third question is, the observations of spots on the stars (by Khokhlova and others) seems to contradict the predictions of the magnetic diffusion separation theory as presented. Can you explain these discrepancies?

ALECIAN: May I begin by answering the third question? At present, computations concerning element diffusion in the presence of a magnetic field at the surface of an Ap star are done with a very simple model, for example assuming dipole or quadrupole fields, but not allowing for the fact that the real magnetic structures are probably more complicated. Also, I do not think that the results on the distribution of patches of elements are precise enough to decide whether or not elements are accumulated in patches or in rings. Elements accumulated in rings may diffuse along the oblique field lines.

About your second point, I agree that this instability may exist, but we have not computed exactly for what overabundance of Si this may occur. At present, we are only considering simple situations, and in the future more complex physical situations will probably be considered.

MÉGESSIER: May I comment on the second question? The situation for Si is not static, but dynamic, and it is not at all like what happens in a glass containing two liquids of different density, one on top of the other. Here, you have a permanent motion, a sort of flow. As far as the observations of patchy distributions of elements is concerned, it should be remembered what Vera Khokhlova said about not being able to determine what is going on more than 45° from the sub-solar point. It is difficult to say that there are really contradictions between theory and observations.

ALECIAN: I am slightly embarrassed about the first question, because I do not know the answer!

MICHAUD: Could I start with the other questions first? On the second point, we make the calculations in the simplest possible way, and compare with the observations. It is possible that additional turbulence is important. We shall see when we compare with the observations.

The third point was the reason for my question to Dr. Khokhlova on her review paper. I asked if she could exclude rings and she said no. Michaud, Mégessier and Charland (**Astron. Astrophys.**, 103, p. 244, 1981)) showed that existing observations could be fitted as well by rings as by spots.

The first point I considered a number of years ago. I now forget the reason we decided to neglect the Bohm formula, as you call it, but it had to do with different physical situations.

KHOKHLOVA: You said that for vertical diffusion the characteristic time is about 10^4 years and for horizontal diffusion it is about 10^8 years. So, when a star starts from the zero-age main sequence, after 10^4 years anomalies appear, but distributed homogeneously. After that 10^8 years are required to form spots by horizontal diffusion. This means that stars in young clusters or associations of age about 10^6 years should not be spotty.

MÉGESSIER: It's a bit too schematic. You have to take into account the geometry of the magnetic field. You can't have a regular overabundance everywhere on the surface. You have it near the magnetic equator because where the lines are parallel to the surface Si II is prevented from sinking.

KHOKHLOVA: If the appearance of spots is really connected with the magnetic field, then complex chemical structures force us to think about more complicated magnetic fields than the simple dipole usually assumed.

MICHAUD: The 10^8 years mentioned was needed to get Si back to the poles. The appearance of other spots or rings can be much faster than that, perhaps on the order of 10^5 or 10^6 years. That is the reason for the observational test which was tried for Si.

DWORETSKY: At least one of the three problems you mentioned, the X-rays, apparently should not be a problem after all. I will discuss this in my review paper tomorrow. The X-ray observations of A stars are now interpreted as being due to two causes. For single late B or A stars, the emission follows the scaling law for soft X-rays which is obeyed by the earlier B stars. Stronger X-ray emission is only seen from A stars with binary companions, and the new interpretation is that it is the later type companions which give the high X-ray fluxes. Whether the star is normal or chemically peculiar seems to make no difference; it is only whether it is double that seems to be important.

CHEMICAL AND TEMPERATURE INHOMOGENEITIES ON STELLAR SURFACES AS A RESULT OF AN INSTABILITY

A. Z. Dolginov

A. F. Ioffe Physical-Technical Institute
Academy of Sciences of the USSR
194021 Leningrad K-21, USSR

Observations show that chemical anomalies are distributed inhomogeneously on Ap star surfaces. The most elaborated explanation of the observations is based on the fact that different ions and atoms are affected by radiative forces of different strengths and, hence, have different diffusion velocities. The diffusion across the magnetic field is a factor of $(1+\nu_H^2/\nu_C^2)^{-1}$ slower than along the field (ν_H is the gyrofrequency and ν_C is the collision frequency of the ions). It leads to the increasing of the heavy ion number density in regions occupied by magnetic traps.

However, such an explanation meets a number of difficulties: a) the conditions $\nu_H^2 > \nu_C^2$ holds for the most of ion species only in regions where the optical depth is less than 10^{-2} if the field H exceeds 10^5 gs. Although little is known on the depth of the region occupied by the chemical anomalies, there are some indirect indications that it is larger than 10^{-2}; b) The observed dipole field has a value 10^3–10^4 gs and does not form traps corresponding to the observed chemical spots which have very complicated configurations; c) the magnetic trap is imperfect. The separation process in the field is assumed to be produced by the diffusion which needs a long time. However, ions can escape from the trap together with the surrounding hydrogen plasma because of various plasma instablished which take much shorter time; d) observed space distribution of rare elements and also of Fe, Cr, Ti contradicts the predictions of the magnetic separation hypothesis (cf. V. Khohlova IAU Coll. No. 90, this volume).

Thus, new ideas are required to avoid the above difficulties. Dolginov (1984, the workshop paper "Magnetic star" Riga April 1984) has pointed out that some kind of instability can form and support local chemical anomalies. Suppose that in the stellar photosphere ($\tau<1$) there exists even a very small initial excess of atoms or ions which can be easily excited by impacts with surrounding plasma electrons. The energy taken from the electrons is converted into ion excitation energy and then into the energy of radiation, which leaves the star. This leads to the cooling of the surrounding plasma. The cooling results in the shift of the hydrogen ionization equilibrium. The emissivity of neutral hydrogen atoms is much larger than that of free protons and electrons. The pressure inside the cool region is smaller than outside, which leads to the compression of the region. The local cooling leads also to the predominant diffusion of heavy atoms into this region. All these processes result in the cooling continuation. This means that we deal with some instability of the thermal type.

At larger depths thes inverse scenario is possible. If the collisional deactivation of atoms excited by the radiation is large, then the plasma heating occurs instead of the cooling. In this case the energy of stellar radiation is partially recaptured by atoms and then transmitted to the plasma. The realization of the first or second regime depends on the relation between the radiative and collisional transition probabilities. Under the LTE condition and in the case when the radiation has the same temperature as the matter the both processes completely compensate each other and there occurs no cooling or heating. However, the equilibrium is never complete

C. R. Cowley et al (eds.), Upper Main Sequence Stars with Anomalous Abundances, 395–396.

in the surface layer.

The local cooling (heating) leads to more intensive horizontal and vertical diffusion of the heavy atoms inward (or outward) the considered region because of the radiative pressure.

Let us impose the following model assumptions: a) the medium consist of hydrogen and a small admixture of the heavy element 10^{-3}; b) spectral lines occupy about 10 - 20% of the continuous spectra; c) the oscillator strengths correspond to the silicon atom transitions; d) in the semi-infinite medium there exists a finite region near the surface with a vertical size $\tau << 1$, and horizontal one $\tau >> 1$, where the number density of the heavy atoms is larger than outside the region.

We need the temperature of this region as a function of the heavy element number density using the obtained temperature gradient, we can determine the diffusion velocity and, hence, the separation rate. To solve this problem quantitatively we have used the energy transfer equation, the continuity equation for particles, and the kinetic equation to describe the atomic level population. The solution of the above equations for the considered model leads to the following conclusions (see A.Z. Dolginov, A.V. Kljachkin Pisma v Ast.J. in press): a) in the region above $\tau \leq 0.1$ the cooling leads to $\simeq$10% temperature decrease; b) below this region there occurs the corresponding heating; c) the separation rate is the same order of magnitude as in the existing theories of the chemical separation by the radiative pressure; d) unlike these theories our approach can explain, not only global but also local chemical anomalies (spots); e) the magnetic field may play an important role suppressing the motions which may destroy the chemical spots;

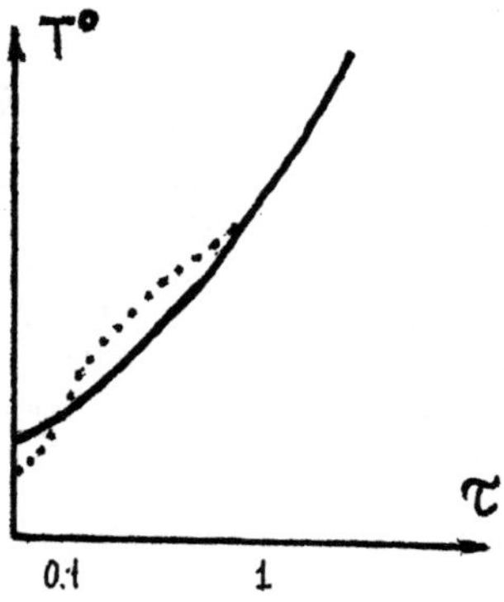

Fig.1

— without heavy element

.... with heavy element

f) one should take into account the real level structures of the heavy atoms for precise quantitative predictions, but even the present model shows the important role of this instability in the chemical separation prossess on Ap star surfaces; g) Temperature spots are well known on late-type stars and on the sun. The instability we have considered may play an important role for these cases as well.

NON-MAGNETIC INTERMEDIATE-TEMPERATURE STARS: A REVIEW

M. M. Dworetsky
Department of Physics and Astronomy
University College London
Gower Street, London WC1E 6BT
England

ABSTRACT. The literature concerning spectroscopic studies, abundances, photometry, and systematics of intermediate-temperature (9500 K < T_{eff} < 16000 K) normal, Hg-Mn, and related stars is reviewed. The review is restricted mainly to papers published since the last international meeting on chemically peculiar stars at Liège in June 1981, and is intended to be complete to 31 December 1984.

1. INTRODUCTION

It would be logical to start by defining carefully what I mean by the phrase "non-magnetic intermediate-temperature" stars. By non-magnetic I mean those upper main-sequence peculiar and normal stars which, as a group, have observed magnetic fields which are not significantly different from zero. We should bear in mind that the existence of undetected very weak fields is by no means excluded as a possibility. By intermediate-temperature I mean, somewhat loosely, those stars with 9500 K < T_{eff} < 16000 K, i.e., A0 - B5 stars, but out of necessity I will allow myself the liberty of relaxing the lower limit slightly in order to include a few specific objects.

To produce a complete review, I found it necessary to make some incursions into areas which are being reviewed at this Colloquium by Dr. Leckrone (Space Observations). I intend to make no distinction in principle between observations from space and those from the ground (referred to as 'optical' or 'visible' in this review). The review will encompass normal stars, Hg-Mn stars, related objects, and metal-weak λ-Boötis stars. Previous reviews covering earlier work include those by Wolff and Wolff (1976) on Hg-Mn stars, and those by Bonsack and Wolff (1980) and Hack (1981) on the more general problems of the observed properties of chemically peculiar stars. Brief discussions of recent results concerning peculiar stars in open clusters, X-ray observations, and microturbulence are also given.

C. R. Cowley et al (eds.), Upper Main Sequence Stars with Anomalous Abundances, 397–416.

2. NORMAL STARS: ABUNDANCES

2.1. General Remarks

Observers of normal stars have been exceptionally busy during the past few years; some rather remarkable results have emerged which are upsetting preconceived ideas. From the point of view of the student of chemically peculiar stars, normal A and B stars serve as comparison objects. However, they are of increasing interest in themselves, partly because the spread of abundances being deduced is surprising large.

The normal star to which all abundance determinations are ultimately compared is the Sun. Grevesse (1984) gives a critical compilation of solar photospheric abundances, and a comparison with the recent critical review of C1 carbonaceous chondrite abundances by Anders and Ebihara (1982). Grevesse's compilation is probably the best summary now available, and I recommend its use.

However, those of us who are stellar spectroscopists prefer to use normal comparison stars which have effective temperatures closer to those of the chemically peculiar stars we are analyzing. We are also well aware that the average normal late B or A0 main-sequence star has a high rotational velocity ($< v \sin i > = 150$ km s^{-1}) which renders it useless for detailed spectroscopic analysis. Those stars which have been studied during the review period all have small values of v sin i.

One very important point is that the analyses I summarize below were nearly all made using the new fully-line-blanketed LTE models of Kurucz (1979). Use of these models has significantly reduced the microturbulent velocities (v_t) required to produce abundances independent of equivalent width, when compared with results from unblanketed models.

In this review, logarithmic abundances are quoted on the scale log N(H) = 12.00. A few abundances will be quoted directly as ratios (e.g., Be, B).

2.2. Published Analyses of Normal A and B Stars

Adelman and Nasson (1980) studied θ Aql (B9.5 III), ν Cap (B9.5 V), and σ Aqr (A0 IVs) using 4.3 and 8.9 Å mm^{-1} spectrograms. The first two stars were reported to have essentially solar compositions, but σ Aqr may belong to a "hot" extension of the Am sequence.

Sadakane (1981) studied two superficially normal stars, 21 Peg (B9.5 V) and HR 7338 (A0 III), using 2.4 Å mm^{-1} spectrograms. He confirmed that 21 Peg has nearly solar abundances (actually it may have rather lower metal abundances if we accept Grevesse's compilation), but HR 7338 has remarkably low metal abundances, a conclusion not seriously affected by the presence of a binary companion.

Adelman (1984a) analyzed six slowly rotating normal mid- to late-B stars: π Cet (B7 V), 134 Tau (B9 IV), HR 2154 (B5 IV), HR 5780 (B6 IV), 21 Aql (B8 II-III), and ν Cap (again; see Adelman and Nasson 1980). Microturbulent velocities ranged from 0.0 to 1.8 km s^{-1}. He found essentially solar abundances in all six stars.

A few studies of specific elements in normal stars were conducted using ultraviolet spectra obtained with the Copernicus or International Ultraviolet Explorer (IUE) satellites. These studies were sometimes done as comparisons for chemically peculiar stars.

Boesgaard and Praderie (1981) analyzed the B II resonance line at 1362.46 Å in γ Gem (A0 IV) and found boron to be depleted by a factor 5-10 relative to other normal A and B stars studied earlier by Boesgaard and Heacox (1978). In that earlier study, a mean normal-star abundance relative to hydrogen of 1.4×10^{-10} (LTE) was found, although the scatter was considerable. Boesgaard and Praderie also found that B II was undetectable in α CMa. They considered non-LTE corrections which raised the average abundance estimate to 2×10^{-10}.

Boesgaard and Praderie (1981) also studied merged 3.4 Å mm^{-1} spectrograms of the Be II resonance line region at 3130-31 Å and found that beryllium in γ Gem is also depleted relative to the solar value ($Be/H = 1.3 \times 10^{-10}$) by at least a factor of 4, while it seems to be normal in α Lyr.

Sadakane *et al* (1983) studied the Al II and Al III resonance lines in seven normal stars for comparison with chemically peculiar stars. They concluded that aluminium was close to the solar abundance in all the stars except α Lyr (see below).

Adelman and Leckrone (1984) presented a discussion of IUE spectra of π Cet, 134 Tau, and ν Cap. They pointed out the importance of the new laboratory work on atomic spectra for UV astronomy. They found that, while 134 Tau and ν Cap have essentially normal abundances of manganese, π Cet has a Mn abundance at least 0.6 dex higher, based on their preliminary IUE analysis.

2.3. Vega (α Lyr)

This bright star has, in the recent past, often been used as a normal comparison standard for differential analyses. It has become increasingly clear during the review period that Vega is metal-weak compared to most other normal B and A stars. It is also surrounded by a dust shell (Aumann *et al*, 1984) which is probably composed of large grains; this shell may be a disk seen face-on, so that it is entirely possible that Vega is a rapid rotator seen nearly pole-on (assuming that the rotational angular momentum vector is perpendicular to the disk). There is also increasing evidence for the disturbing possibility that Vega, which is the primary astronomical spectrophotometric standard, is slightly variable on occasions, and possibly shows low-amplitude δ-Scuti behaviour (Fernie, 1981). To compound the likelihood that Vega is indeed doing odd things, Goraya and Singh (1983) reported Hα variability (a transient emission) on 1983 October 12 on a time scale of 1.5 hr. Charlton and Meyer (1985) were not able to confirm this behaviour on six nights in 1984 August. Further observations are clearly essential due to the transient nature of the phenomenon.

Several abundance analyses of Vega may now be considered. Dreiling and Bell (1980) obtained high signal-to-noise Reticon spectra and analyzed the abundances of Ti and Fe; for iron they calculated "astrophysical" gf-values based on the solar spectrum. Their LTE analysis gave v_t = 2.5 km s^{-1}, log N(Ti) = 4.7, log N(Fe) = 6.9 for Fe I and 7.1 for Fe II, depending on which set of oscillator strengths were used. An attempt to correct for non-LTE effects suggested that more nearly solar values would thereby be obtained, but this reviewer notes that other LTE results, e.g., those of Adelman, Sadakane, and their co-workers, should correspondingly be adjusted upwards, and this would produce embarrassing high abundances of Fe in many apparently normal stars. While Dreiling and Bell used observations of Vega in the region 4000 - 5000 Å, Sadakane and Nishimura (1981) compared results in the near UV (3250 - 3640 Å) with results in the visible (3900 - 4900 Å) to determine whether the same abundances and microturbulences would be obtained. They found v_t = 2.0 km s^{-1} in both regions, and log N(Fe) = 7.0 and 7.1 in the near UV and visible, respectively.

Most other metals also yielded low abundances in Vega when compared with the Sun, especially aluminium: Sadakane and Nishimura obtained log N(Al) = 5.5 in Vega, vs. 6.5 in the Sun and 6.14 in ν Cap (the A-star results are from analysis of the Al I resonance lines). The optical results, including those of Adelman (1984a) were confirmed in the UV analysis of the resonance lines of Al II and Al III by Sadakane, Takada, and Jugaku (1983) in several normal stars. They found that these two ions give essentially the same abundance results as the optical analyses of Al I. The fact that the resonance lines of all three ions give a very similar low abundance in Vega when analyzed by LTE techniques, while giving consistent solar or near-solar abundances in several other normal A and B stars, makes it difficult - if not impossible - to explain the observed deficiency in Vega by non-LTE effects.

Lambert, Roby, and Bell (1982) used high signal-to-noise Reticon observations in the visible and near infrared to study C I, N I, and O I in Vega. They concluded that the abundances of these elements are essentially the same as in the Sun.

Boesgaard and Praderie (1981) studied the Be II resonance line region in Vega with high signal-to-noise Reticon data (3100 - 3175 Å). Their analysis gave Be/H = 1×10^{-11}, a value close to the average for other normal stars.

Friere Ferrero *et al* (1983) analyzed Mg II h and k profiles in Vega based on Copernicus and balloon ultraviolet spectra. Although the unblanketed model atmospheres adopted by them may be open to criticism, they obtained log N(Mg) = 7.0, in agreement with the results of Sadakane and Nishimura (1979), who found that Mg was somewhat underabundant (by a factor of 4) in Vega compared with the Sun. Welsh *et al* (1983) published UV balloon spectra of Vega in the region 2000 - 3200 Å but did not carry out a detailed analysis.

3. Hg-Mn STARS: ABUNDANCES

3.1. Element by Element

The study of Hg-Mn stars has been undergoing the same kind of revolution as other stellar studies, due largely to the availability of ultraviolet spectra from the IUE satellite. In this section, I will deal first with abundance studies of individual elements (in order of atomic number).

3.1.1. Be and B

Sadakane and Jugaku (1981) found that boron is strongly enhanced in κ Cnc, HR 7361, and 20 Tau (Maia in the Pleiades, B7 III). The latter star may be related to the Hg-Mn stars; it is not a "classic" Hg-Mn star since Hg is not present and Mn only mildly enhanced. They found both B II 1362 Å and the B III 2065-67 Å doublet. Leckrone (1981) determined an overabundance of B of about 2.5 dex in κ Cnc, but the B II resonance line was absent in μ Lep, 46 Dra, ι CrB, HR 4072, and χ Lup. Sadakane and Jugaku also found the Be II 3130-31 Å lines in HR 6997, HR 7361, κ Cnc, and 112 Her, which indicated a large enhancement of beryllium.

Boesgaard *et al* (1982) used 6.7 Å mm^{-1} spectra taken at Mauna Kea to study 43 Hg-Mn stars and compare them with normal stars. They concluded that there is a trend for Be II lines to be enhanced when Mn II lines are enhanced. A temperature trend may be indicated: only half the cool Hg-Mn stars ($T_{eff} \simeq 11000$ K) studied have enhanced Be II, while all hot Hg-Mn stars in their sample have enhanced lines of this element. The overabundance factors range from 20 to 2 x 10^4.

3.1.2. Al

Sadakane, Takada, and Jugaku (1983) analyzed the Al II and Al III resonance lines in 22 Hg-Mn stars using IUE spectra. The cooler Hg-Mn stars have moderate deficiencies of aluminium (0.5 to 1.0 dex), while the hotter Hg-Mn stars have deficiencies of about 1.4 dex.

3.1.3. Cu

Jacobs and Dworetsky (1981) found that copper is enhanced in 10 of 11 Hg-Mn stars studied. A temperature trend of increasing abundance was found for all Hg-Mn stars except 112 Her, which, perhaps paradoxically, appears to have a strong deficiency of Cu. The maximum overabundance found was nearly a factor 10^3 (in HR 7361) compared to the Sun.

3.1.4. Ga

The remarkable gallium overabundance discovered many years ago in the hotter Hg-Mn stars at optical wavelengths has received spectacular confirmation through ultraviolet observations with IUE. Jacobs and Dworetsky (1981) confirmed the anomaly by observing the strong 1414 Å resonance line of Ga II in 7 out of 10 Hg-Mn stars, as well as two other Ga II lines. Takada and Jugaku (1981) presented broadly similar results in somewhat more detail for resonance lines of both Ga II and Ga III in several other stars. However, there is some disagreement between the two papers about the exact abundances giving rise to the observed lines;

Takada and Jugaku obtained lower Ga abundances by about 0.3 dex from Ga II and by about 0.5 dex from Ga III. These differences may reflect the somewhat preliminary nature of the analyses, rather than any profound physical problems or non-LTE effects. Jacobs (in preparation) has reconsidered the Ga abundance question and concludes that there are no significant differences between results from the two ions. The over-abundance of gallium, in the Hg-Mn stars which have it, is of the order of 10^3 - 10^4.

3.1.5. Pt

Dworetsky, Storey, and Jacobs (1984) presented theoretical oscillator strengths for strong Pt II transitions in the ultraviolet, together with a spectrum synthesis analysis of the abundance of Pt in three cool Hg-Mn stars, χ Lup, HR 4072, and HR 7775, based on IUE spectra. The least blended lines are 1777 Å and 2144 Å. The abundances found range from log N(Pt) = 6.0 in χ Lup to 6.6 in HR 7775. The solar abundance is 1.8, so the enhancement factor for platinum is of the order of 2×10^3 to 10^4. "Astrophysical" gf-values for Pt II lines in the optical region were also obtained.

3.1.6. Hg

Dworetsky (1980a) used published data to show that the oscillator strength of Hg II 3984 Å was obtainable from laboratory experiments. He also showed that the discussions of IUE observations of the Hg II resonance lines at 1942 Å and 1649 Å by Leckrone (1980) led to an abundance of mercury in agreement with those from 3984 Å and from the Hg I line at 4358 Å in the same stars, when LTE analyses were employed. Leckrone (1984) has reconsidered the entire problem in great detail, taking into account the well-known isotopic anomalies in Hg II and the observed splitting of the 3984 and 1942 Å lines (Guern *et al* 1976). The earlier work is confirmed: the results for LTE calculations now yield excellent agreement between Hg I and Hg II lines, when mercury is assumed to be homogeneously distributed in the atmosphere.

Mégessier, Michaud, and Weiler (1980) stated that they had found Hg in the first three stages of ionization in two Hg-Mn stars, μ Lep and χ Lup. Since these stars have very different T_{eff} (13000 and 11000 K, respectively), they argued that this could only happen if Hg were concentrated (by diffusion processes) in a thin layer at small optical depths, rather than homogeneously distributed. Leckrone (1984) showed that their identification of Hg II suffered from a wavelength calibration error. This suggests that their identification of Hg III may also be in error. Jacobs and Dworetsky (1981) searched for Hg III in a differential comparison of IUE spectra of two Mn stars (53 Tau and φ Her); φ Her has strong Hg II and 53 Tau, remarkably, has none. No credible evidence for Hg III was found nor was any found in the hot Hg-Mn star HR 7361, where it might have been expected.

Leckrone (1984) also discussed the enhancement factor for mercury in Hg-Mn stars. The five stars he studied all yielded abundances of the order of log N(Hg) = 6.0. The meteorite abundance given by Grevesse (1984) is log N(Hg) = 1.27, which is poorly determined, while a solar spectroscopic determination is probably not possible. Leckrone tentatively determined log N(Hg) = 2.1 from his analysis of the normal stars ν Cap and π Cet. Due to the weakness of the lines seen, this is likely to be an upper limit. If Leckrone's result is more representative of normal stellar compositions, the gross enhancement factor for Hg is typically 10^4 ; if the meteorite value is more correct, enhancements of 4 x 10^4 to 10^5 are representative. In the cool Hg-Mn stars, large enrichments of Hg^{204} are seen relative to other isotopes: the enhancement of this isotope may therefore reach values of 10^6 or more.

3.1.7. Bi

Jacobs and Dworetsky (1982) found that bismuth was extraordinarily strong in HR 7775 but absent in all other Hg-Mn and normal stars which they studied. They used spectrum synthesis to determine log N(Bi) = 6.7 to 6.8. The solar system abundance is from meteorites only: log N(Bi) = 0.7. Here, Bi is enhanced by a factor of 10^6 , which is the largest factor for any element so far obtained. Guthrie (1984) found a strong, broad line (W_λ = 36 mÅ) at 4259.41 Å in DAO coudé spectra of HR 7775 which does not occur in other Hg-Mn stars; it is almost certainly a Bi II feature.

3.2. Abundances in Individual Stars

Several abundance analyses were published during the review period, and are described briefly below.

3.2.1. α And

Derman (1982) published a new analysis of α And based on 7 and 12 Å mm^{-1} spectra obtained at Haute Provence by Hack and Stalio. This analysis, like many others, suffers from imprecision due to the high rotational velocity of the star. Derman points out that T_{eff} estimates for α And differ widely. His results confirm the long-held view that the chief anomalies are overabundances of P, Mn, Ga, Sr, Y, Zr, and Hg. Derman also reminds us that several previous investigators suspected that α And is surrounded by a circumstellar shell and that light and spectrum variations have been claimed by several authors.

Derman did not include mention of the paper by Rakos, Jenkner, and Wood (1981) in which further evidence of photometric and spectroscopic variations with P = 0.9636 days was presented, based on Copernicus data.

3.2.2. 33 Gem

Chunakova, Bychkov, and Glagolevskij (1981) give preliminary photographic magnetic field results for this star. The variation is roughly sinusoidal with a total range of 3 kGs and P = 3.099 days. Why is a magnetic Ap star discussed in this review? According to the authors, 33 Gem has Hg II 3984 Å, Mn II, and P II as well as features which correspond to those of the Si stars, and demonstrates that it is possible for Hg-Mn characteristics to appear occasionally in magnetic Ap stars.

3.2.3. HR 562 = HD 11905

This star has recently been analyzed by Ptitsyn and Ryabchikova (1986) using 8 Å mm^{-1} spectra obtained with the Bulgarian 2-m telescope. In most respects, the authors find HR 562 to have typical Hg-Mn abundances, except that the iron abundance is lower than solar by 0.9 dex! There is a suggestion that asymmetries in the line profiles may be due to a binary companion. HR 562 also seems to belong to the small subgroup of Hg-Mn and related objects with deficiencies of C and/or Si.

3.2.4 υ Her and 38 Dra

Adelman (1984b) gives a new analysis of υ Her based mainly on Mt. Wilson spectrograms. This star is a "classic" Hg-Mn type, with deficiencies of He, Mg, and Ni, roughly normal Na, Si, S, Ca, Sc, Cr, and Fe, and large enhancements of P, Mn, Ga, Sr, Y, and Hg.

Adelman also rediscusses the earlier analysis of the high velocity Hg-Mn star 38 Dra by Adelman and Sargent (1972). This star is also typical, although quantitatively different. Carbon was not detected, Ca is deficient, while Sc, Mn, and Sr are enhanced only by 1 dex and Ni is normal. Gallium was not observed, but Y and Zr are enhanced by about 2 dex. This is in contrast to υ Her: Adelman did not detect Zr II in that star. Again, Hg II 3984 Å is very strong. Adelman finds that Ba is enhanced in 38 Dra by about 2 dex over the solar value.

3.2.5. ν Cnc

Adelman, Young and Baldwin (1984) carried out a detailed analysis of the Hg-Mn star ν Cnc. This star is of interest because, with T_{eff} = 10375 K, log g = 3.60, it may be the coolest of the "classical" Hg-Mn stars, only a few hundred degrees hotter than Sirius. Besides the now-familiar anomalies associated with Hg-Mn stars, it has a strong Sc enhancement and a mild Ni enhancement, relative to solar abundances.

3.2.6. Guthrie's Survey

Guthrie (1984) performed a differential curve-of-growth analysis of 20 of the sharpest-lined Hg-Mn stars based on published equivalent widths, on 2.4 Å mm^{-1} DAO spectrograms loaned by Prof. C. R. Cowley, or on Edinburgh 6 Å mm^{-1} spectrograms. The comparison star was γ Gem (A0 IV). Among the more interesting conclusions are these:

1) Mg is very underabundant in some stars;
2) Sc has large star-to-star differences;
3) V may be deficient in at least some cases;
4) Ni is generally deficient;
5) the Pt II isotope shift in HR 1800 is nearly as large as that in χ Lup, and the Pt overabundance phenomenon is confirmed

as confined to the very narrow range of temperatures, 11000 K < T_{eff} < 12000 K;

6) the presence of Au and Bi is confirmed in HR 7775; it seems likely that Au is present in the atmospheres of HR 4072 and χ Lup primary stars as well.

4. RELATED STARS: SPECTRA AND ABUNDANCES

In this section I review publications on intermediate-temperature stars which are (or may be) related to the Hg-Mn phenomenon. Sirius (α CMa) is included because its T_{eff} is very close to the lower limits of the Hg-Mn range.

4.1. Sirius (α CMa)

Boesgaard and Praderie (1981) found that Sirius is strongly deficient in Be and B. For both elements, their analysis gave upper limits well below the normal-star abundances.

Bell and Dreiling (1981) analyzed Fe I, Fe II, and Ti II lines in LTE and concluded that log N(Ti) = 5.5, log N(Fe) = 8.0 and 8.1 from Fe I and Fe II, respectively.

Lambert, Roby, and Bell (1982) examined the C, N, O abundances in Sirius, based on visual and infrared Reticon spectra. Although they found that these elements had essentially solar abundances in Vega (see 2.3 above), in Sirius they found a strong C deficiency (-0.6 dex), a mild O deficiency (-0.3 dex), and a slight N excess (0.2 dex).

4.2. 20 Tau

This Pleiades star is Maia, HD 23408, B7 III. It had previously been classified as He-weak and has the smallest v sin i of any B star in the Pleiades cluster (45 km s^{-1}). Several analyses had been done previously; during the review period Mon, Hirata, and Sadakane (1981) published their new analysis. They find He to be very weak (-0.7 dex); Mg, Si, Ca, Ni to be underabundant (especially Si, -1.1 dex); C, Ti, Cr, Fe to be normal; and P, Mn to be overabundant (0.8 and 0.7 dex, respectively). It appears that this Pleiades star has many characteristics of mild Hg-Mn stars. This conclusion is strengthened by Sadakane and Jugaku (1981), who found a large enhancement of B in 20 Tau similar to that seen in κ Cnc and other Hg-Mn stars, and probable enhancement of Be. The Be enhancement in 20 Tau is, in all probability, rather marginal (about 1.0 dex), according to Boesgaard *et al* (1982). In addition, Sadakane, Takada, and Jugaku (1983) found that Al II and Al III lines in the ultraviolet spectrum indicate a deficiency of -1.2 dex, typical of Hg-Mn stars. The reader may care to note that, by 1982, several astronomers had routinely included 20 Tau in their samples of Hg-Mn stars for survey purposes.

4.3. HR 6000

Five papers on this important and most unusual star (= HD 144667) were published during the review period. The brief paper by Castelli, Cornachin, and Hack (1981) on the ultraviolet spectrum has been superseded by a more detailed analysis by Castelli *et al* (1984, 1985) and new optical spectroscopy is discussed by Andersen and Jaschek (1984) and Andersen, Jaschek, and Cowley (1984). A most important item concerning this remarkable object is the strong likelihood that it is extremely young: it is a visual common-proper-motion companion of the A7 IIIe star HR 5999 which is associated with a T Tauri group and a reflection nebula. If the connection is indeed proven, then this Ap star must have an age somewhat less than 10^6 yr. To the best of my knowledge no magnetic field observations of HR 6000 have been published. It remains to be seen whether or not HR 6000 has a magnetic field. The radial velocity is constant (Andersen and Jaschek, 1984).

The optical and ultraviolet results are in reasonable agreement about the main abundance anomalies. The authors quoted above agree that $T_{eff} \simeq 14000$ K, although opinions differ concerning surface gravity. (A new calibration of the Stromgren parameters c_0 and β by Moon and Dworetsky (1985) gives $T_{eff} = 14070$ K, $\log g = 4.28$, indicating that HR 6000 is on the zero-age main sequence.) The most striking anomalies are, along with the usual He weakness, large underabundances of Si and C. Lines of these elements are totally absent in the optical spectrum (limit 10 - 15 mÅ), while the ultraviolet results suggest underabundances of 2 - 3 dex. Underabundances of N and Al are also reported by Castelli *et al* (1985). The Fe II lines in the optical spectrum are very strong; P and Mn are mildly enhanced (1 dex). There is some evidence that Xe is enhanced. Although HR 6000 bears some resemblance to Hg-Mn stars, the optical and ultraviolet observations showed that Hg II lines are absent or very weak. The Ga II line at 1414 Å illustrated by Castelli *et al* (1985) is very weak when compared to the same line in Hg-Mn stars which have gallium anomalies, and may be identical to an unidentified feature at nearly the same wavelength which is also seen in normal stars (J. Jacobs, priv. comm.).

Castelli *et al* (1985) also report that resonance lines of C II, O I, Mg II, Si II, and Fe II appear to have circumstellar or interstellar components shifted by -15 km s^{-1}.

4.4. Superficially-Normal Late B Stars

Cowley (1980) presented 2.4 Å mm^{-1} spectroscopic observations of 13 sharp-lined late B stars. All but one had spectra which, at classification dispersions, showed no obvious line-strength anomalies. Cowley classed 46 Aql and 23 Cas as "mild" Hg-Mn stars on the basis of his high-dispersion spectra. For example, a weak Hg II line is visible in 46 Aql, and both stars have marked P II enhancements. The star 21 Peg may be a hot analogue of the Am stars. The case of 64 Ori is more complicated: this is a spectroscopic triple system consisting of a double, sharp-lined, short-period binary and a third component with broad lines. The rest of the sample studied seems to consist of relatively normal stars. Sadakane (1981) did a detailed study of 21 Peg and HR 7338 using Cowley's spectrograms and concluded that, in fact, 21 Peg has abundances within 0.3 dex of the solar values, while HR 7338 is clearly metal-deficient to a degree even more pronounced than Vega.

4.5. HOT Am STARS: o Peg and σ Aqr

Adelman, Young, and Baldwin have carried out analyses of these two hot Am stars (T_{eff} = 9625 K and 10125 K, respectively). They appear to have a significant He deficiency, while most of the metals are slightly enhanced. Also, Ni, Sr, Zr, and Ba are enhanced by about 1.0 dex.

5. STATISTICAL STUDIES OF LINE STRENGTHS

Cowley and Aikman (1980) studied wavelength coincidence statistics (WCS) and showed that, in many cases, WCS based on the measurement of high-dispersion spectra can give valid, though crude, estimates of photospheric abundances (they quote ± 0.5 dex). In other words, one can not only identify which elements are present by statistical means, but one can also obtain approximate quantitative information. They calibrated their statistical parameters with adopted model-atmosphere abundances of Cr, Mn, Fe, and Y for normal and chemically peculiar stars.

In a very important paper, Cowley, Sears, Aikman, and Sadakane (1982) studied WCS for a sample of 34 late B and early A stars. They pointed out that effects due to T_{eff} and v sin i could be allowed for quite well (see their Fig. 1), and that the results clearly suggest that Fe is underabundant in several stars in their sample (HR 562, HR 8226, Vega). Somewhat surprisingly, perhaps, the known metal-weak AO III star HR 7338 (Sadakane, 1981) did not show any indication from the statistical study that Fe was underabundant. The Fe lines may be enhanced because of the much lower surface gravity of this star. On the other hand, the recent study of HR 562 by Ptitsyn and Ryabchikova (1985) confirms the remarkable weakness of the Fe II spectrum. The enhancement of Fe II in Sirius shows up clearly in the WCS results.

Bord and Davidson (1982) applied WCS techniques to the long-wavelength IUE spectrum of κ Cnc (1850 - 3250 Å). As might be expected in a hot Hg-Mn star, they found clear evidence for Fe II, Mn II, and Cr II. Os II was also detected "possibly". A further study of the same spectrum by Davidson and Bord (1982) produced marginal evidence for the presence of Si II, Cl II, Ca II, Sr II, and Zr II.

Chjonacki, Cowley, and Bord (1984) applied the quantitative WCS technique developed by Cowley and Aikman to the ultraviolet IUE spectrum of κ Cnc in both the SWP and LWR regions (1223 - 2097 Å; 1969 - 3227 Å), and examined the results for Fe II, Fe III, Mn II, and Mn III. They concluded that Fe III and Mn III lines were clearly present, and that any departures from the Saha ionization equilibrium must be small. They also concluded that, within the admittedly large uncertainties (0.3 - 0.5 dex) of their results, the Fe and Mn abundances are equal.

Finally, Bord and Davidson (1985) have used WCS techniques to demonstrate that reasonably accurate stellar radial velocities can be obtained from IUE spectra of sharp-lined stars. The studied κ Cnc, ι CrB, α^2 CVn, and χ Lup. In the case of χ Lup, a double-lined spectroscopic binary, they obtained a clear detection of both the primary and the secondary spectra, even though the secondary star is approximately 1.5 mag fainter.

6. MAGNETIC FIELDS: UPPER LIMITS

It has been assumed implicitly throughout this review that normal late B stars and Hg-Mn stars do not, in fact, have measurable magnetic fields. During and shortly before the review period, two papers were published which formally confirm this assumption. Landstreet (1982) used photoelectric Zeeman techniques to search for longitudinal magnetic fields in 36 bright upper-main-sequence stars; of these, 9 are in or near the temperature range for this review. In no case did Landstreet record significant detection of a magnetic field. Among the best-observed cases is Vega (α Lyr) with $B = -9 \pm 19$ Gs; the median error for all of Landstreet's observations is ± 65 Gs. Borra and Landstreet (1980) previously used similar techniques to study 6 Hg-Mn stars. In no case was a significant field detected, with typical errors of ± 170 Gs. The bright Hg-Mn star α And was observed on 5 of 6 successive nights, with no significant detection of a magnetic field (average error ± 50 Gs). One other Hg-Mn star, ι CrB, was previously observed by Borra, Landstreet, and Vaughan (1973) with null results of the same precision.

7. LINE ASYMMETRIES

Rice and Wehlau (1982, 1984) have published the results of careful searches for line asymmetries of Fe II lines in the Hg-Mn stars ι CrB, φ Her, and κ Cnc, with negative results. Previously, Dworetsky (1980b) had shown that the asymmetries observed in ι CrB by Smith and Parsons (1976) were due to the binary nature of this star. Weak asymmetries of Hg-Mn line profiles might be expected on theoretical grounds (Michaud, 1978).

8. RADIAL VELOCITIES

Radial-velocity data for 96 Hg-Mn stars (or candidates which turn out to be normal) have been published by Stickland and Weatherby (1984). This compilation is based on 8.6 Å mm^{-1} plates taken at the Royal Greenwich Observatory. Several new orbits are proposed. Further discussion of the remaining literature on radial velocities and orbits is somewhat beyond the scope of this review.

9. SPECTROPHOTOMETRY

Adelman and Pyper (1983) presented additional spectrophotometry of 4 Hg-Mn stars and 3 other Ap stars to augment their earlier data for 11 Hg-Mn stars (Adelman and Pyper, 1979). These and other results are brought together and discussed by Adelman (1984c). In that paper, Adelman considers peculiarity indices centred on 3509 Å, 4200 Å, 5200 Å and 6300 Å. These are defined in terms of the spectrophotometric data given in the series of papers by himself and colleagues Pyper and White begun in 1979.

There are difficulties in comparing spectrophotometrically-defined indices with narrow-band photometric indices. For example, at different times Adelman has used three separate indices to study the 4200 Å feature (although one has been rejected as not very sensitive). He has also tried two different definitions of Δa, the 5200 Å parameter, and considers the use of an equivalent width measurement for this feature based on spectrophotometry.

There are structural differences in the 5200 Å feature in various Hg-Mn stars, best appreciated in Adelman's Fig. 11. It is still not clear what causes the 5200 Å feature to be strong only in some Hg-Mn stars and in nearly all magnetic Ap stars.

10. LAMBDA-BOÖTIS STARS

After many years of "non-status" λ Boo stars have once again attracted the attention of astronomers interested in chemical peculiarities. Hauck and Slettebak (1983) offer a spectroscopic definition: "λ Boo stars are A-F type stars with metallic lines which are too weak for their spectral types, when the latter are determined from the ratio of their K-line to Balmer-line strengths. They are distinguished from weak-lined Population II (horizontal branch) stars by the fact that they have normal space velocities and moderately large rotational velocities." In fact, the λ Boo phenomenon seems to extent to B9, as ε Sgr is listed by Hauck and Slettebak as a λ Boo star; thus the phenomenon extends to sufficiently hot stars that it may be included in this review.

Baschek *et al* (1984) have examined the ultraviolet spectra of λ Boo stars. There are indications of a normal carbon abundance; also, there is a strong broad unidentified feature at 1600 Å in λ Boo stars which is absent in normal stars. Paradoxically, they find that the λ Boo star 29 Cyg is only marginally of this type in the ultraviolet, while ρ Vir is a λ Boo star in the ultraviolet but fairly normal in the optical! Further characteristics are weak Al II and Mg II in the ultraviolet.

Morgan (1984) and Abt (1984) presented interesting discussions of a remarkable extreme metal-weak A star, HR 4049. All that can be seen in the optical spectrum 3900 - 4900 Å (at 39 Å mm^{-1}) are narrow H lines and a weak Ca II K-line; Morgan remarked that "no MK box...can accept the spectrum..." which has been assigned the type A0 Ib-IIpec. (The luminosity may be too high to fit the usual λ Boo type, but what is it?)

Abt (1984) also described his search for stars near A0 with weak Mg II lines. Only a very few have $(B - V) \lesssim 0.00$; one (20 Eri) is very blue $(B - V = -0.13)$. In the discussion after his paper, Abt pointed out that nothing was known about the binary frequency of these stars.

11. ANOMALOUS C IV AND Si IV LINES

Sadakane (1984) reported the startling news that high-resolution IUE spectra of 36 Lyn exhibits strong, broad resonance absorptions of

C IV and Si IV. Although 36 Lyn is a magnetic He-wk Bp star with Teff = 13,600 K and is not a Hg-Mn star, it seems worthwhile to call this remarkable phenomenon to the attention of astronomers. Sadakane concludes that these lines can not be photospheric and probably originate in a circumstellar corona or chromosphere.

Molaro *et al* (1983) found similar anomalous C IV and Si IV lines in the rapidly rotating A0 V star HD 119921 (HR 5174), which they find to be blue-shifted by nearly 70 km s^{-1} with a shortward absorption asymmetry. Variable broad C IV and Si IV resonance absorptions are often found in Be stars as well (e.g., Barker and Marlborough, 1985).

The possibility that 36 Lyn has a companion should not be ruled out, although there is no direct evidence for binarity. If it had a rapidly rotating companion whose expanding envelope surrounded both stars, an explanation for the anomalous lines might readily be imagined. A careful search for H_α emission in 36 Lyn might be worthwhile.

12. OPEN CLUSTERS

Klochkova (1983) examined the peculiar stars in the Pleiades at 9 Å mm^{-1}. She confirms that HD 23950 (HR 1185) is a Hg-Mn star (see Abt and Levato, 1978) although she did not comment on the shell-like nature of the spectrum which Abt and Levato mention (see below).

Mermilliod (1982, 1983) provides some remarkable insights into cluster stars. In the 1982 paper, he looked at the available data on blue stragglers and concluded that, in "middle-aged" clusters, the percentage of Ap's of all types among blue stragglers is very large. However, most of these seem to be He-wk and Si stars rather than Hg-Mn stars. In older clusters, blue stragglers are often Am stars.

In 1979, Abt published his very important summary of the occurrence of Ap stars in open clusters. He noted that a number of stars had characteristics of rapid rotators ("n" appended to spectral type) and simultaneously characteristics of slow rotators ("s" for sharp-lined). He suggested that these might be weak shell stars. A few of these are Ap stars of various types, including HD 23950, a Hg-Mn star.

In the 1983 paper, Mermilliod considered the "sn" stars in open clusters and associations. There are 17 cluster "sn" stars, and a further 12 in the Orion OB1 association. They have low values of v sin i, sharp lines of Ti II, Ca II, Si II, C II, and Fe II, and broad lines of He I, as if they had thin shells (i.e., Abt's original hypothesis). However, the "sn" stars often have He-weak or Hg-Mn characteristics; I have already mentioned the example of HD 23950; Maia (HD 23408; see 4.2) is another. Mermilliod considers that α Scl resembles "sn" stars if observed with resolutions comparable to that used by Abt. Mermilliod gives reasons why the "sn" stars are almost certainly intrinsic slow rotators and not pole-on, and argues that the "shell" hypothesis is an unlikely explanation of the appearance of the spectra. Discerning their true nature awaits better observations.

13. X-RAY OBSERVATIONS

Cash and Snow (1982) and Golub *et al* (1983) have published the results of observations of 51 A and late B stars with the Imaging Proportional Counter and (in a few cases) with the High Resolution Imager on the Einstein satellite. They attempted to detect 5 Hg-Mn stars, as well as several bright normal stars near A0 V. The essence of their conclusions may be stated as follows:

1) Single stars in this part of the HR diagram, whether normal or peculiar, generally emit soft X-rays (0.15 - 4.0 keV) in accordance with the scaling law for early-type stars, $\log (L_x/L_{bol}) = -6.9$.
2) A-star binaries often have enhanced X-ray emission. This usually comes from active young late-type companions, or, in the case of Sirius (α CMa) from a hot white dwarf companion.
3) The coronal temperatures deduced from single stars near A0 V are quite low, $\leqslant 10^6$ K. In the case of ϕ Her, the previously reported detection with HEAO 1 (Cash, Snow, and Charles 1979) was either an error or the temperature is very low.

14. BINARITY, STATISTICS, AND SYSTEMATICS

Guthrie (1981) compared binary and single Hg-Mn stars statistically in order to test two possibilities: firstly, whether the (unknown) braking mechanism is more effective in short-period binaries; secondly, whether the hypothesis of Guthrie and Napier (1980) has any observed effect, namely, that in SB2's mutual irradiation would inhibit diffusion and lead to different abundances.

The results showed a highly significant difference in that SB2's and shorter-period SB1's rotate much more slowly than longer-period SB1's and single Hg-Mn stars. The problem is that, although the former group has experienced more tidal braking than the latter, several SB2 stars are rotating more slowly than their synchronous velocities.

In the second investigation, Guthrie found only one significant pattern: Sc tends to be weak or absent in SB2's, and strong in single stars and longer-period SB1's.

Eggen (1984) has carried out an extensive analysis of the MK spectral classification, uvbyβ photometry, and space motions (U, V, W) of field A0 stars in the Yale Catalogue of Bright Stars. Of much interest to us is his discovery that, due to the standards used, the "Houk" sample (i.e., the new Michigan catalogues) fit MK standard photometric boxes far better than the Cowley *et al* (1969) sample, which has the A0/A1 border well within the standards' A0 box.

Eggen's discussion of membership in moving groups is interesting, although perhaps controversial. His Hyades and Sirius "supercluster" groups contain a few Ap stars, and yield a high number of blue stragglers of which several are Ap's of various types. Whether or not this is a result of including higher luminosity Ap stars in the A0 sample might be a subject for further investigation.

15. MICROTURBULENCE: SOME NEW RESULTS

The usual way in which microturbulence is determined in a fine analysis is to plot log (Abundance) vs. W_λ, and choose that value of v_t

which makes abundances independent of line strength. Dufton, Durrant, and Durrant (1981) showed that this procedure leads to systematic overestimates of v_t due to the non-linearity of the curve of growth, i.e., the random errors in W_λ give rise to an adopted v_t which is too large. The effect is particularly important for the light elements which they studied in B stars. Dworetsky (1982) showed that this effect can be reduced by using heavier elements like Fe. The fundamental problem remains, and a neat solution has been proposed by Magain (1984). The systematic error can be eliminated by plotting log (Abundance) vs. W_λ (synthetic) for an assumed abundance, model, and v_t. One should then choose v_t as before, but now, since the W_λ's are theoretical, they do not contain random observational errors.

The use of fully-line-blanketed model atmospheres clearly leads to lower required values of v_t, as predicted by Kurucz. Dworetsky, Storey and Jacobs (1984) found that the values of v_t obtained in analyses of HR 4072 and χ Lup using unblanketed models (2.0 km s^{-1}) were much higher than those obtained using blanketed models (0.9 km s^{-1}). Any published microturbulent velocity obtained with unblanketed models should be viewed with scepticism.

16. SUMMARY AND CONCLUSIONS

Dr Hack (1981), in her review paper for the Liège Colloquium, concluded with a discussion of what was known, what was unknown, and what needed to be studied more. To a remarkable extent, the points she raised then remain unanswered.

It is increasingly clear that the future for abundance determinations lies with high-quality observational material and detailed careful analysis. The virtually untouched problem of CNO abundances in late B and early A stars should be approached from both the optical (i.e., near infrared) and ultraviolet (for C, N) data. The optical study by Lambert, Roby, and Bell (1982) sets a demanding high signal-to-noise standard for future workers to emulate. There is still an enormous amount of work remaining to be done on the vast accumulation of IUE high-resolution spectra of both normal and chemically peculiar stars. Due to the extremely crowded nature of the spectra, it is important that only appropriate techniques such as spectrum synthesis be used.

The study of Hg-Mn and other peculiar stars in clusters needs to be taken a lot further. We urgently need to obtain high-resolution spectra of these stars, and normal stars in the same clusters, and analyze them carefully. It may be possible to identify age-dependent anomalies in this way. The observations can not be done photographically, because the stars are too faint; high-quantum-efficiency detectors must be used.

Photographic observations still play an important part in abundance analysis of A and B stars. The important remarks by Cowley and Adelman (1983) are relevant to their use; compared with only a few years ago, it is much more readily possible now to perform feats of numerical manipulation on digitized tracings to obtain co-added spectra of high signal-to-noise, and to use smoothing algorithms to obtain further improvements. I recommend that this paper be read by all workers in the field.

Although photographic spectroscopy is regarded as an obsolete technique in some astronomical circles, it remains largely true that very few high efficiency solid-state detectors are used with spectrographs which provide anything approaching the spectral resolution and coverage of the classical coudé spectrographs ($\lambda/\Delta\lambda = 10^5$ or better, over 1000 Å). One hopes that this situation will change, and there are encouraging signs in this direction.

We still do not know the nature of the braking mechanism which slows down Hg-Mn and related stars (presuming, of course, that these stars do not form as slow rotators; HR 6000 may be a case in point). Whatever it is, it seems to be exceedingly effective. Because it has now been demonstrated that the fields in these stars are non-existent or at least too weak to be detected at a noise level of 50 - 100 Gs, magnetic explanations seem increasingly unlikely.

It is becoming clearer that the Hg-Mn phenomenon concerns not only stars which are "fully formed" Hg-Mn objects, but that "marginally peculiar" stars with many similarities to them exist. Discovery of these objects is (by Dr Cowley's definition) virtually impossible from ordinary MK classification, unless one's suspicions are aroused by discrepant colours and He-types; one can readily see that selection processes will inhibit discovery of cooler members of this class (in fact, very few are known), because He lines would not be expected in any great strength anyways. Most often they are discovered accidentally during the course of other work. This promises to be an important and fruitful area for further investigations.

At the same time, it has also become clear that the range of metal abundances in "normal" late B and A0 stars is greater than expected. We now see quite plainly that Vega has normal (i.e., solar) CNO abundances, but is obviously metal-weak. On the other hand, Sirius, perhaps the prototype "early Am" star, has a "disturbed" CNO abundance pattern and a considerable metal enrichment. More and better detailed abundance analyses of normal stars near A0 are urgently needed; results of this quality for only two stars are woefully insufficient.

The revival of interest in the λ-Boötis stars is very welcome. We seem to know almost nothing about them. Here lies material for a dozen highly significant doctoral theses!

One new observational tool will become available, if all goes according to schedule, late in 1986. The Hubble Space Telescope High Resolution Spectrograph will be capable of producing spectra of resolution and signal-to-noise equal to the best available in the optical from coudé spectrographs. However, the amount of time expected to be available for stellar abundance studies will, regrettably, be small.

REFERENCES

Abt, H. A. (1979). Astrophys. J., 230, 485.
Abt, H. A. (1984). In The MK Process and Stellar Classification, ed. R. F. Garrison (David Dunlap Obs., Toronto), p. 340.
Abt, H. A., and Levato, H. (1978). Publs. Astron. Soc. Pacific, 90, 201.
Adelman, S. J. (1984a). Mon. Not. R. Astron. Soc., 206, 637.
Adelman, S. J. (1984b). Astron. Astrophys. Suppl., 58, 585.

Adelman, S. J. (1984c). Astron. Astrophys. Suppl., 55, 479.
Adelman, S. J., and Leckrone, D. S. (1984). Physica Scripta, T8, 25.
Adelman, S. J., and Nasson, M. A. (1980). Publs. Astron. Soc. Pacif., 92, 346.
Adelman, S. J., and Pyper, D. M. (1979). Astron. J., 84, 1603.
Adelman, S. J., and Pyper, D. M. (1983). Astron. Astrophys., 118, 313.
Adelman, S. J., and Sargent, W. L. W. (1972). Astrophys. J., 176, 671.
Adelman, S. J., Young, J. M., and Baldwin, H. E. (1984). Mon. Not. R. Astron. Soc., 206, 649.
Anders, E., and Ebihara, M. (1982). Geochim. Cosmochim. Acta, 46, 2363.
Andersen, J., and Jaschek, M. (1984). Astron. Astrophys. Suppl., 55, 469.
Andersen, J., Jaschek. M., and Cowley, C. R. (1984). Astron. Astrophys., 132, 354.
Aumann, H. H., Gillett, F. C., Beichman, C. A., de Jong, T., Houck, J. R., Low, F. J., Neugebauer, G., Walker, R. G., and Wesselius, P. R. (1984). Astrophys. J., 278, L23.
Barker, P. K., and Marlborough, J. M. (1985). Astrophys. J., 288, 329.
Baschek, B., Heck, A., Jaschek, C., Jaschek, M., Koppen, J., Scholz, M., and Wehrse, R. (1984). Astron. Astrophys., 131, 378.
Bell, R. A., and Dreiling, L. A. (1981). Astrophys. J., 248, 1031.
Boesgaard, A. M., and Heacox, W. D. (1978). Astrophys. J., 226, 888.
Boesgaard, A. M., Heacox, W. D., Wolff, S. C., Borsenberger, J., and Praderie, F. (1982). Astrophys. J., 259, 723.
Boesgaard, A. M. and Praderie, F. (1981). Astrophys. J., 245, 219.
Bonsack, W. K., and Wolff, S. C. (1980). Astron. J., 85, 599.
Bord, D. J., and Davidson, J. P. (1982). Astrophys. J., 258, 674.
Bord, D. J., and Davidson, J. P. (1985). Astron. Astrophys., 143, 461.
Borra, E. F., and Landstreet, J. D. (1980). Astrophys. J. Suppl., 42, 421.
Borra, E. F., Landstreet, J. D., and Vaughan, A. H. (1973). Astrophys. J., 185, L145.
Cash, W., and Snow, T. P. (1982). Astrophys. J., 263, L59.
Cash, W., Snow, T. P., and Charles, P. (1979). Astrophys. J., 232, L111.
Castelli, F., Cornachin, M., and Hack, M. (1981). In Upper Main Sequence CP Stars, 23rd Liège Astrophys. Colloq. (Univ. Liège), p. 149.
Castelli, F., Cornachin, M., Hack, M., and Morossi, C. (1984). Astron. Astrophys., 141, 223.
Castelli, F., Cornachin, M., Hack, M., and Morossi, C. (1985). Astron. Astrophys. Suppl., 59, 1.
Charlton, J. C., and Meyer, B. S. (1985). Publs. Astron. Soc. Pacif., 97, 60.
Chjonacki, G. T., Cowley, C. R., and Bord, D. J. (1984). Astrophys. J., 286, 736.
Chunakova, N. M., Bychkov, V. D., and Glagolevskij, Yu. V. (1981). Soob. Sp. Astrofiz. Obs., 31, 5.
Cowley, A., Cowley, C., Jaschek, M., and Jaschek, C. (1969). Astron. J., 74, 375.
Cowley, C. R. (1980). Publs. Astron. Soc. Pacif., 92, 159.
Cowley, C. R., and Adelman, S. J. (1983). Q. J. R. Astron. Soc., 24, 393.
Cowley, C. R., and Aikman, G. C. L. (1980). Astrophys. J., 242, 684.

Cowley, C. F., Sears, R. L., Aikman, G. C. L., and Sadakane, K. (1982). Astrophys. J., 254, 191.
Davidson, J. P., and Bord, D. J. (1982). Astron. Astrophys., 111, 362.
Derman, I. E. (1982). Astrophys. Space Sci., 88, 135.
Dreiling, L. A., and Bell, R. A. (1980). Astrophys. J., 241, 736.
Dufton, P. L., Durrant, A. C., and Durrant, C. J. (1981). Astron. Astrophys., 97, 10.
Dworetsky, M. M. (1980a). Astron. Astrophys., 84, 350.
Dworetsky, M. M. (1980b). Mon. Not. R. Astron. Soc., 191, 521.
Dworetsky, M. M. (1982). A Peculiar Newsletter, No. 9, p.3.
Dworetsky, M. M., Storey, P. J., and Jacobs, J. M. (1984). Physica Scripta, T3, 39.
Eggen, O. J. (1984). Astrophys. J. Suppl., 55, 597.
Fernie, J. D. (1981). Publs. Astron. Soc. Pacif., 93, 333.
Friere Ferrero, R., Gouttebroze, P., and Kondo, Y. (1983). Astron. Astrophys., 121, 59.
Golub, L., Harnden, F. R., Maxon, C. W., Rosner, R., Vaiana, G. S., Cash, W., and Snow, T. P. (1983). Astrophys. J., 271, 264.
Goraya, P. S., and Singh, M. (1983). Inf. Bull. Var. Stars No. 2455.
Grevesse, N. (1984). Physica Scripta, T8, 49.
Guern, Y., Bideau-Méhu, A., Abjean, R., and Johannin-Gilles, A. (1976). Physica Scripta, 14, 273.
Guthrie, B. N. G. (1981). In Upper Main Sequence CP Stars, 23rd Liège Astrophys. Colloq. (Univ. Liège), p. 189.
Guthrie, B. N. G. (1984), Mon. Not. R. Astr. Soc., 206, 85.
Guthrie, B. N. G., and Napier, W. M. (1980). Nature, 284, 536.
Hack, M. (1981). In Upper Main Sequence CP Stars, 23rd Liège Astrophys. Colloq. (Univ. Liège), p. 79.
Hauck, B., and Slettebak, A. (1983). Astron. Astrophys., 127, 231.
Jacobs, J. M., and Dworetsky, M. M. (1981). In Upper Main Sequence CP Stars, 23rd Liège Astrophys. Colloq. (Univ. Liège), P. 153.
Jacobs, J. M., and Dworetsky, M. M. (1982). Nature, 299, 535.
Klochkova, V. G. (1983). Soob. Sp. Astrofiz. Obs., 37, 73.
Kurucz, R. L. (1979). Astrophys. J. Suppl., 40, 1.
Lambert, D. L., Roby, S. W., and Bell, R.A. (1982). Astrophys. J., 254, 663.
Landstreet, J. D. (1982). Astrophys. J., 258, 639.
Leckrone, D. S. (1980). Highlights of Astronomy, 5, 277.
Leckrone, D. S. (1981). Astrophys. J., 250, 687.
Leckrone, D. S. (1984). Astrophys. J., 286, 725.
Magain, P. (1984). Astron. Astrophys., 134, 189.
Mégessier, C., Michaud, G., and Weiler, E. J. (1980). Astrophys. J., 239, 237.
Mermilliod, J.-C. (1982). Astron. Astrophys., 109, 37.
Mermilliod, J.-C. (1983). Astron. Astrophys., 128, 368.
Michaud, G. (1978). Astrophys. J., 161, 592.
Molaro, P., Morossi, C., Ramella, M., and Franco, M. (1983). Astron. Astrophys., 127, 413.
Mon, M., Hirata, R., and Sadakane, K. (1981). Publs. Astron. Soc. Japan, 33. 413.

Moon, T. T., and Dworetsky, M. M. (1985). Submitted to MNRAS.
Morgan, W. W. (1984). In The MK Process and Stellar Classification, ed. R. F. Garrison (David Dunlap Obs., Toronto), p. 18.
Ptitsyn, D. A., and Ryabchikova, T. A. (1986). Paper presented at IAU Coll. 90, Crimea (This volume).
Rakos, K. D., Jenkner, H., and Wood, J. (1981). Astron. Astrophys. Suppl., 43, 209.
Rice, J. B., and Wehlau, W. H. (1982). Astron. Astrophys., 106, 7.
Rice, J. B., and Wehlau, W. H. (1984). Astrophys. J., 278, 721.
Sadakane, K. (1981). Publs. Astron. Soc. Pacif., 93, 587.
Sadakane, K. (1984). Publs. Astron. Soc. Pacif., 96, 259.
Sadakane, K., and Jugaku, J. (1981). Publs. Astron. Soc. Pacif., 93, 60.
Sadakane, K., and Nishimura, M. (1979). Publs. Astron. Soc. Japan, 31, 481.
Sadakane, K., and Nishimura, M. (1981). Publs. Astron. Soc. Japan, 33, 189.
Sadakane, K., Takada, M., and Jugaku, J. (1983). Astrophys. J., 274, 261.
Smith, M. A., and Parsons, S. B. (1976). Astrophys. J., 205, 430.
Stickland, D. J., and Weatherby, J. (1984). Astron. Astrophys. Suppl., 57, 55.
Takada, M., and Jugaku, J. (1981). In Upper Main Sequence CP Stars, 23rd Liège Astrophys. Colloq. (Univ. Liège), p. 163.
Welsh, B. Y., Boksenberg, A., Anderson, B., and Towlson, W. A. (1983). Astron. Astrophys., 126, 335.
Wolff, S. C. and Wolff, R. J. (1976). In Physics of Ap-Stars, IAU Colloq. No. 32, ed. W. W. Weiss, H. Jenkner, and H. J. Wood (Universitätssternwarte Vienna), p. 503.

DISCUSSION (Dworetsky)

PTITSYN: I'd like to make a remark on the superficially normal late-B star HR 8226 = HD 204754, which according to the WCS analysis of Cowley, Sears, Aikman & Sadakane (1982) was expected to be iron-weak. We have done a detailed chemical analysis of this star using 9 Å/mm spectrograms. It turns out to have a fairly normal chemical composition. The abundance of Fe may be slightly low, but no more than 0.3 dex, which is within the errors of the analysis. We suspect that the reason for the discrepancy between our results and those of Cowley and his co-workers is that they adopted too low a value of v sin i, which we think is in the range 10-12 km/s, but we can not be certain of this.

DWORETSKY: [to Cowley] Do you want to reply?

COWLEY: I find no problem with this result. Our method should have statistical validity, but I would not make a strong claim in any individual case.

DWORETSKY: The possible slight weakness of Fe mentioned (0.3 dex) may have some bearing on this question. I agree with your remark about the statistical nature of the results of Cowley et al. (1982).

DROBYSHEVSKI: Are there any indications of possible duplicity of Vega, not only spectroscopic Doppler shifts, but possibly also in the X-ray emission?

DWORETSKY: I am not aware of any. The X-ray observations of Vega were very difficult to make. The Einstein Image Proportional Counter was, I recall, unable to detect Vega, and the High Resolution Imager had to be used, with a very long integration time. Vega has a very, very weak X-ray emission consistent with a fairly low temperature corona, about $5 \cdot 10^5$ K. This is consistent with the scaling law for single B and A stars which I mentioned in my review.

SEVERNY: I have a short comment about the magnetic field in Vega. If it exists at all, it may not be constant.

DWORETSKY: I thought all results were null so far, such as yours and those of Borra and Landstreet. Are you saying that there may be a real but variable field?

SEVERNY: There might be, but sometimes it can not be detected, within error limits of ±5 gauss.

DWORETSKY: We are at the limits of the statistical accuracy of the data, and perhaps we should not conclude that there is a field, but say only that it is proven that the field is no stronger than some upper limit.

SEVERNY: Yes, that is correct.

DWORETSKY: I would interpret the published results by saying that no field has been detected, within the errors of ±20 gauss or so. The existence of very weak general fields is not yet excluded.

ALECIAN: One must be very careful in interpreting the abundance determinations for Hg-Mn stars. If there are abundance stratifications in the atmosphere, the "classical" methods, which assume homogeneity, may fail. For example, a "cloud" of Mn with 10^5 overabundance located above $\tau_{5000} = 10^{-4}$ gives a curve of growth for Mn III lines which looks like the curve of growth given by uniformly distributed Mn with 10^2 overabundance and $\xi = 3$ km/s. This is an example: other kinds of

C. R. Cowley et al (eds.), Upper Main Sequence Stars with Anomalous Abundances, 417–419.

effects are possible.

DWORETSKY: I don't quite know how to reply. The use of the homogeneous model does not strictly imply that we believe every assumption made when the model is used. I believe that you yourself suggested a possible explanation a few years ago in unpublished work. You pointed out that small amounts of microturbulence can co-exist with diffusion. This turbulence can spread the thin cloud produced by diffusion into something resembling a homogeneous distribution. Perhaps this explains why homogeneous models always seem consistent with the data, within the errors of analysis. The stratification effects mimic other things, and are consequently difficult to detect.

ALECIAN: Have you determined the microturbulence from different elements in the same star, and do you get the same value of ξ for the different elements?

DWORETSKY: Yes, the values are the same within the errors, for Fe I, Fe II, and Y II, in the few stars I have investigated recently, using fully line blanketed Kurucz model atmospheres.

I would like to emphasize that great care has to be taken to eliminate the spurious results for microturbulence which result from improper treatment of statistical errors in the observed equivalent widths. Nowadays, one usually plots line strength against derived abundance, and adjusts the value of ξ to produce constant abundance independent of equivalent width. However, the equivalent widths are observed values, and their random errors can produce a systematic increase in the microturbulence derived. I discussed Magain's (1984) paper on this subject in my review, and recommend that all of us adopt his technique of substituting theoretical equivalent widths, for the guessed microturbulence and abundance, for the observed values. This should lead us to still more accurate microturbulence estimates. It is also very important to use fully blanketed model atmospheres, because Kurucz has shown that unblanketed models lead to overestimates of ξ, and I have confirmed that use of fully blanketed models reduces the value considerably, though not usually to zero.

HUBENÝ: I would like to return to the question of the disadvantages of Vega (α Lyr) as a spectrophotometric standard, and a consequent skipping of observations of this star in the Space Telescope programme mentioned by Dr. Adelman. I would like to see this reconsidered, because Vega, besides serving as a standard from the observational point of view, would be an excellent standard star from the point of view of theoretical modelling. For example, much more work has been done for Vega than for Sirius, detailed NLTE studies have already been performed, both for overall model atmospheres and detailed transfer solutions for individual atoms, and the lower metallicity makes NLTE modelling easier (blanketing is not so heavy), and finally the search for chromospheres and coronae could be continued. High resolution observations would help resolve one of the basic questions about A stars, the structure of the most superficial layers.

DWORETSKY: I am not a spokesman for the Space Telescope programme. However, some of us plan to apply to observe Vega if no one else wants to. In other words, although Vega is not a standard star for Space Telescope, it will probably be observed.

HUBENÝ: Concerning the discussion of λ Boo stars and Vega, I would like to show you a plot of the TD-1 observations of Vega, which here [shows a transparency] is compared to theoretical spectra based on NLTE calculations. One may recognize a feature near 1600 Å.

DWORETSKY: There is indeed some sort of depression there. Perhaps we should vote whether Vega is a λ Boo star or not! All in favour? [pause] All against? [pause] All not voting? [pause] Most people are not voting either way!

JOHANSSON: During this conference we have heard of three different broad depressions at 5200, 1400, and now at 1600 Å. Do you know of any similar feature at another wavelength in any star, for which there is a satisfactory explanation?

DWORETSKY: Not really, though there is also a 4200 Å feature, I think. Explanations, so far are very unsatisfactory for all the features. There has been some limited success in identifying part of the 5200 Å feature as Fe I lines, by Maitzen and Muthsam (Astron. Astrophys., 83, p. 334, 1980), but perhaps others can answer this question better than I can. What is astonishing, of course, is that we are seeing a strong depression in a metal-weak star. Very, very interesting.

ADELMAN: The Fe identification only works for 5200 Å in the coolest Ap stars, and you need a lot of iron to produce it.

THE ABUNDANCE OF GALLIUM IN B-TYPE CHEMICALLY PECULIAR STARS *

Masahide Takada-Hidai+
Research Institute of Civilization, Tokai University
1117 Kitakaname, Hiratsuka-shi, Kanagawa 259-12, Japan

Kozo Sadakane+
Astronomical Institute, Osaka Kyoiku University
Minamikawahori-cho, Tennoji-ku, Osaka 543, Japan

Jun Jugaku+
Tokyo Astronomical Observatory
University of Tokyo, Mitaka-shi, Tokyo 181, Japan

ABSTRACT. Quantitative analyses of the Ga abundances in 27 Hg-Mn stars, 11 Si (magnetic) stars, 8 He-weak stars, and 7 normal stars are made with the resonance lines of Ga II at 1414 Å and Ga III at 1495 Å in IUE spectra. The Ga overabundances are confirmed as a genuine anomaly in many peculiar stars. Only upper limits of the Ga abundance can be obtained for some stars. However these upper limits are much lower than those inferred from visual spectra. Among the 27 Hg-Mn stars, 17 stars are distributed in the range of 2.0 - 3.8 dex of overabundances of Ga. Ten other stars show upper limits less than 2.1 dex. Fifteen stars of 20 hotter Hg-Mn stars with $T_{eff} > 11000$ K show high overabundances in a narrow range of 2.6 - 3.8 dex, while, among 7 cooler Hg-Mn stars, the only star HR 7775 shows the same overabundance as in these 15 stars. Of 11 Si stars, 9 stars have Ga overabundances ranging from 1.9 to 3.2 dex. Gallium is overabundant in 5 out of 8 He-weak stars in the range of 1.7 - 3.2 dex, while in the three other stars the upper limits are less than 1.3 dex. The Ga abundances in normal stars are all upper limits which roughly correspond to the solar value.

* The full text of this paper will be submitted to the Astrophysical Journal.

\+ Guest Observer with the International Ultraviolet Explorer satellite.

C. R. Cowley et al (eds.), Upper Main Sequence Stars with Anomalous Abundances, 420.

GALLIUM OVERABUNDANCE IN THE Ap-Si STAR HD 25823

Marie-Christine ARTRU
Département d'Astrophysique Fondamentale
Observatoire de Meudon
F-92195 Meudon Principal Cedex - FRANCE

Rubens FREIRE FERRERO
Observatoire de Strasbourg
11, rue de l'Université
F-67000 Strasbourg - France

Abstract : An anomalous enhancement of the resonance line of Ga II at 1414.44 Å is observed in IUE spectra of the silicon star HD 25823. High-resolution spectrum of this star is compared to synthetic spectra calculated in the range 1406-1422 Å. The LTE abundance of gallium is evaluated to $\log N_{Ga} = 6.3 \pm 0.5$ in the scale where $\log N_H = 12$.

INTRODUCTION

The overabundance of Gallium is currently observed in the Mg-Mn stars. Heacox (1979) established abundances larger than 10^3 times the solar value for eight Hg-Mn stars, over a sample of 21 of these stars. Takada and Jugaku (1981) also deduced such overabundances from IUE ultra-violet observations of seven Hg-Mn stars. On the contrary, Gallium was generally considered as normal in magnetic Ap stars. An exceptional detection of the visible lines of Ga II has been reported in two Ap-Si stars HD 200311 (Adelman, 1974) and HD 25823 (Didelon, 1985).

It will be important to establish if the Gallium overabundance is, or not, a definite characteristic differentiating silicon and Hg-Mn stars of similar temperature : in the diffusion theory in Ap stars, an extra opacity at 1400 Å may influence the radiative forces pushing out Gallium ions in the stellar atmospheres (Michaud, private communication). The Ap-Si stars show a large depression at 1400 Å, tentatively attributed to Si II autoionization (Artru et al, 1981) and the resonance line of Ga II at

C. R. Cowley et al (eds.), Upper Main Sequence Stars with Anomalous Abundances, 421–424.

1414.44 Å occurs exactly in this region. For six Ap-Si stars (over a sample of 12) this line appears stronger than for the comparison normal star πCeti. We present here a quantitative analysis of the Ga II enhancement in HD 25823, the Ap-Si star where it is, by far, the most important.

OBSERVATIONAL DATA

The Ap-si star HD 25823 (41 Tau, m_v = 5.19) has an effective temperature of 13000 K (Lanz, 1984) and vsini = 21 km s^{-1}. We compared the high dispersion IUE spectra of HD 25823 (image # 14956) and HD 17081 (πCeti, image # 16256) chosen as comparison stars. We show on fig. 1 the two spectra with identified lines in the spectral range 1406-1422 Å where the Ga II resonance line appears.

SYNTHETIC SPECTRA IN THE SPECTRAL REGION OF THE Ga II RESONANCE LINE

For quantitative analysis of the Ga II resonance line ($3s^2\ ^1S_0$ - $3s3p\ ^1P_1$) at 1414.44 Å, we computed LTE synthetic spectra in the spectral range 1406 - 1422 Å. A special effort was done to actualize the atomic data on individual lines of this region. As basic wavelength reference we used the experimental data compiled by Kelly and Palumbo (1973). The gf values are taken from Kurucz and Peytremann (1975) or Kurucz (1981) when no other source was available. The main lack of data occurs for other ions such Mn II, Cr II and Ni II; especially Mn II has a line at 1414.40 Å blended with the Ga II one, but probably much weaker.

The gf value of the Ga II line is well established to 1.8 ± 0.1 from different recent determinations, either theoretical (Froese-Ficher and Hansen, 1978) either experimental (Andersen, 1979). The Stark broadening was deduced from an empirical estimation (Dimitrijevic, 1985) $\gamma/ne = 2.10^{-6} s^{-1} cm^{+3}$ at 13000K.

The LTE synthetic spectrum was calculated for a blanketed model (Kurucz et al. 1975) with effective temperature T_{eff} = 13000 K and gravity g = 10^4 CGS ; the microturbulence was fixed to 3 km/s and the spectrum was degraded to the IUE resolution of 0.2 Å which is slightly larger than

Fig. 1 - Observed IUE high-resolution spectra of the stars HD 25823 (Ap-Si) and HD 17081 (normal) in the range 1406-1422 Å

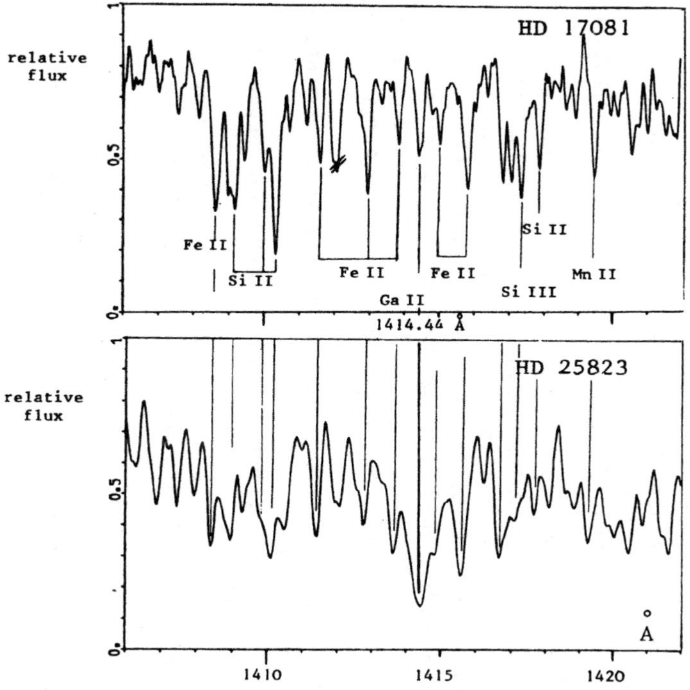

Fig. 2 - Calculated LTE synthetic spectra of a model atmosphere (13000 K, log g = 4) ;
a : with the solar abundances for all elements ;
b : with increased abundances of silicon (+1 dex) and gallium (+3.5 dex)

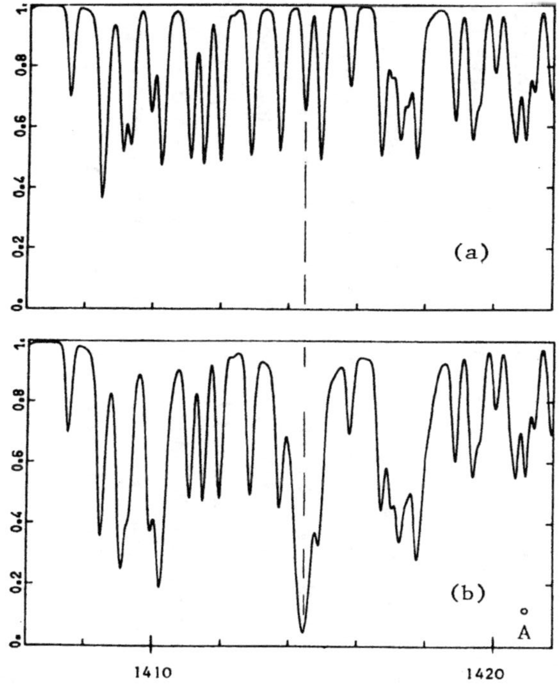

the broadening due to the rotationnal velocity of HD 25823. The element abundances were fixed to solar values excepted that silicon was increased by a factor of 10 and gallium was varied from 1 to 10^5 times the solar value.

The two synthetic spectra shown on fig. 2 were computed with Ga/H = 1 and Ga/H = 3. 10^3. Most of the observed individual lines are well reproduced in the synthetic spectrum, except in a few cases (1411.07 Ni II, 1415.50 Fe II and 1417.70 Si II). The best fit with the observed HD 25823 spectrum was obtained for an overabundance of gallium of 3000 times the solar value. We thus estimate log NGa/NH = 6.3 $\pm$ 0.5 (solar value = 2.8). Further synthetic spectra for other spectral regions, in particular those of Ga III lines, will be undertaken.

REFERENCES

- Adelman S.J., 1974, Ap. J. Suppl. Ser. 28, 51.
- Andersen, T., Eriksen, P., Poulsen, O., Ramanujam, P.S., 1979, Phys. Rev. A 20, 2621.
- Artru, M.C., Jamar, C., Petrini, D., Praderie, F., 1981, Astron. Astrophys. 96, 380.
- Didelon, D., 1985, to be published in Rev. Mex. de Astron. y. Astr.
- Dimitrijevic, M.S., Astron. Astrophys., 1985, 145, 439.
- Froese-Fisher, C., Hansen, J.E., 1978, Phys. Rev. A17, 1956.
- Heacox, W.D., 1978, Ap. J. Suppl. Ser. 41, 675.
- Johansson, S., 1978, Physica Scripta, Vol. 18, n° 4, 217.
- Kelly, R., L., Palumbo, L.J., 1973, Atomic and Ionic Emission Lines below 2000 Å, NRL Report 7599, Naval Research Laboratory, Washington.
- Kurucz, R.L., Peytremann, E., 1975, A table of Semiempirical gf Values, SAO, Sp Rep. n° 362.
- Kurucz J.L., 1981, Semiempirical Calculation of gf Values, Fe II, SAO Sp Rep. n° 390.
- Kurucz, R.L., Peytremann, E., Avrett, E.H., 1974, Blanketed Model Atmospheres for early-type Stars (Washington, Smithsonian Institution Press).
- Lanz, T., 1984, Astron. Astrophys. in press.
- Takada, M. Jugaku, J., 1981, in Upper Main Sequence CP Stars, 23rd, Liège Astrophys. Coll., Univ. de Liège, June 1981, p.163.

THE MOST IRON-DEFICIENT MANGANESE STAR HR562

D. A. Ptitsyn and T. A. Ryabchikova
The Astronomical Council
of the U.S.S.R. Academy of Sciences
Pyatnitskaya Str. 48, Moscow
109017 U.S.S.R.

ABSTRACT. Abundances of 11 elements are determined in the atmosphere of the manganese star HR562 = HD11905 (B8III, T_e=14000 K, log g=3.5). The deficit of iron of 0.9 dex is extreme among HgMn stars. The excess of mercury abundance amounts to 4.8 dex.

1. INTRODUCTION

In their search for iron variations in Population I late B and early A stars Cowley et al. (1982) have suspected some stars to be iron-deficient. One of these stars, HD204754, turned out to have normal chemical composition (Ptitsyn and Ryabchikova, 1985). In this paper we present results of preliminary chemical analysis of another star, HR562=HD11905 (B8III). Wolff and Wolff (1974) classified HR562 as a manganese star. Later Wolff and Preston (1978) found variations of radial velocities of hydrogen lines with a period of 5.1 days, that they considered as an indication of the binary nature of the star. They found no lines of secondary component and concluded that they might be very broad. The lines of the primary component seemed to be slightly weaken as if the radiation from the secondary component would contribute to the continuum.

2. OBSERVATIONS

One 9 Å/mm-dispersion spectrogram of HR562 was taken with the 2-m telescope of the National Astronomical Observatory of Bulgarian Academy of Sciences on September 9, 1982 (2^h09^m UT). This spectrogram was measured with the 3CS Joyce Loebl microdensitometer. About 100 lines were identified in the 3800-4700 Å range, nearly half of which belong to the MnII spectrum.

C. R. Cowley et al (eds.), Upper Main Sequence Stars with Anomalous Abundances, 425–428.

3. MODEL ATMOSPHERE

From Genéva photometry (Rufener, 1981) using the calibration of Cramer and Maeder (1979) we obtained T_e=13930 K. The $(b - y)_0$ index from Philip et al. (1976) with calibration of Relyea and Kurucz (1978) gave T_e=14100 K. Finally the effective temperature of 14000 K was adopted.

The surface gravity log g=3.5 was obtained from the Hγ profile. We couldn't estimate the microturbulent velocity by usual way from comparing strong and weak lines of a given element because the observed lines are weak or not numerous for all elements except Mn, while for Mn lines the scattering is large due to uncertainties in the oscillator strength values. The value v_t=2 km/s, typical of HgMn stars (Dworetsky, 1971), was adopted.

4. ABUNDANCES OF ELEMENTS

The chemical composition was determined by the model atmosphere technique. Theoretical intensities of lines were computed for the Kurucz et al. (1974) model atmosphere with T_e=14000 K and log g=3.5 by means of the programme written by N.E.Piskunov at the Astronomical Council of the USSR Academy of Sciences.

The chemical composition derived is given in Table I.

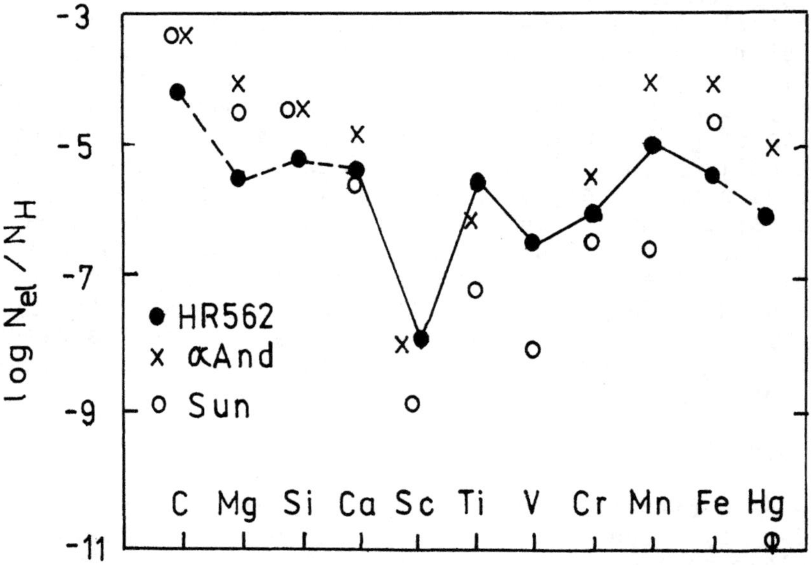

Fig.1. Abundances of HR562 with those of αAnd for comparison and solar abundances.

Table I. Abundances of HR562

Element	$\log\frac{N_{el}}{N_H}$	Number of lines	Source of gf-value
CII	-4.23	4	1
MgII	-5.58	1	2
SiII	-5.22	5	2
CaII	-5.36	1	2
ScII	-7.90	1	3
TiII	-6.48	12	4
VII	-6.49	4	5
CrII	-5.91	6	6
MnII	-4.92	29	7
FeII	-5.41	7	8
HgII	-6.06	1	9

Key to references:

1. Wiese,W.L., Smith,M.W., and Glennon, B.M. 1966, Atomic Transition Probabilities, v.1, NSRDS-NBS 4.
2. Wiese,W.L., Smith,M.W., and Miles, B.M. 1969, Atomic Transition Probabilities, v.2, NSRDS-NBS 22.
3. Wiese,W.L., and Fuhr,J.R. 1975, J. Phys. Chem. Ref. Data, 4, 263.
4. Roberts,J.R., Andersen,T., and Sørensen,G. 1973, Ap.J., 181, 567.
5. Roberts,J.R., Andersen,T., and Sørensen,G. 1973, Ap.J., 181, 587.
6. gf-Catalogue. Astr. Council, USSR Acad. of Sci. (unpublished).
7. Warner,B. 1967, Mem.R.A.S., 70, 165.
8. Groth,H.G. 1961, Z. Astrophys., 51, 231. Corrected by +0.1 dex.
9. Dworetsky,M.M. 1980, Astr. Ap., 84, 350.

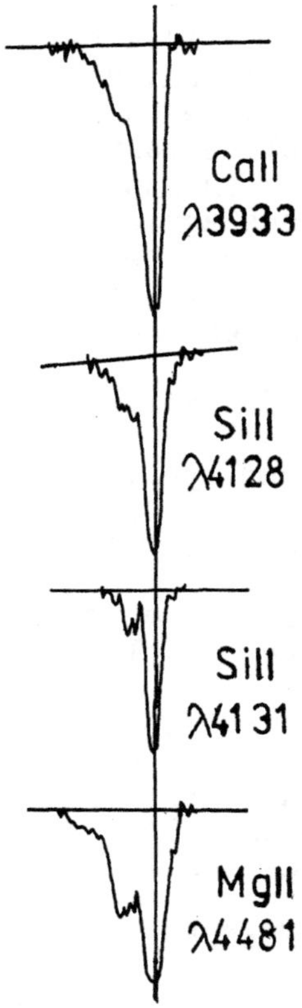

Fig.2. Line asymmetries.

The number of lines used in the analysis as well as the sources of oscillator strengths are also indicated. We didn't take into account a possible contribution to the continuum from the secondary component, but according to rough estimates it wouldn't effect essentially our results.

Logarithmic abundances relative to hydrogen for HR562, solar atmosphere and well studied manganese star αAnd are shown in Fig.1. With respect to solar abundances, C, Mg, Si are deficient by ~1 dex, Ca is normal, iron-peak elements with the exception of Fe itself are in large excess, Hg is by 4.8 dex enhanced. The content of Fe is 0.9 dex below normal, therefore HR562 turns out to be a HgMn star with extreme deficit of iron. So the range of iron abundance variations in these stars amounts to 1.5 dex, exceeding such variations in other types of CP stars. The comparison with α And, another HgMn star having nearly the same T_e and log g, reveals the similarity of relative abundances, the absolute values in HR562 being lower by ≳0.5 dex. The ratio Mn/Fe in HR562 is definitely greater than unity, that evidences against the nuclear origin of anomalies. High mercury abundance in HR562 gives one more example of manganese stars with weak Fe lines which have enhanced Hg. This emphasizes the uniqueness of the 53 Tau phenomenon.

5. LINE ASYMMETRY

In our spectrogram of HR562 the profiles of some lines appear to be asymmetric. Fig.2 illustrates this effect for some strong lines. Rough estimation of radial velocity from the displacement of blue component is in accord by its absolute value and sign with that predicted from ephemeris given by Wolff and Preston (1978).

6. ACKNOWLEDGEMENTS

We are grateful to the staff and director of the Bulgarian National Observatory for possibility to carry out observations and to Dr. V.Khokhlova for helpful discussion.

REFERENCES

Cowley,C.R., Sears,R.L., Aikman,G.C.L., and Sadakane,K. 1982, Ap.J., 254, 191.
Cramer,N., and Maeder,A. 1979, Astr. Ap., 78, 305.
Dworetsky,M.M. 1971, D.Ph.Thesis, Univ. of Calif.
Kurucz,R.L., Peytremann,E., and Avrett,E.H. 1974, Blanketed Model Atmospheres for Early-type Stars, (Washington: Smithsonian Inst.).
Philip,A.G.D., Miller,T.M., and Relyea,L.J. 1976, Dudley Obs. Rep., No.12.
Ptitsyn,D.A., and Ryabchikova,T.A. 1985, Soviet Astr. J., in press.
Relyea,L.J., and Kurucz,R.L. 1978, Ap.J. Suppl., 37, 45.
Rufener,F. 1981, Astr. Ap. Suppl., 45, 207.
Wolff,S.C., and Preston,G.W. 1978, Ap.J. Suppl., 37, 371.
Wolff,S.C., and Wolff,R.J. 1974, Ap.J., 194, 65.

THE LIGHT AND SPECTRUM VARIABLE CP2 STAR HR 6127

J. Žižňovský
Astronomical Institute
Slovak Academy of Sciences
059 60 Tatranská Lomnica
Czechoslovakia

ABSTRACT. Photometric and spectroscopic variability of HR 6127 was found. The value of the magnetic axis inclination was derived from the radial velocity variations of metallic lines. The star was classified as cool CP2 star.

1. INTRODUCTION

The studied star was alternatively classified as A2p Si, Sr (Cowley et al., 1969), normal A1 V (Bonsack, 1974), A2 Si 3955 (Floquet, 1975), early Am (Cowley, 1976), normal with strong Y II lines (Cowley et al., 1978) and as cool Ap star (Žižňovský, 1980b). Pirronello and Strazzula found its Sr, Y and Zr abundances to be typical for Am stars. Adelman and Pyper (1983) admit a mild Ap star classification. Possible light and spectral variability was reported by Žižňovský (1980a). Magnetic field of the value of 0.1 tesla can be expected from the Geneva photometry by Hauck and North (1982). Variable magnetic field of (-0.03 to -0.1) tesla was found by Glagolevskij et al. (1984).

2. OBSERVATIONS

Photometric observations of HR 6127 were performed with the 0.6 m telescope of the Skalnaté Pleso Observatory in the years 1981 - 1982. An intermediate-band filter centered to 526 nm was used, in order to detect possible 530 nm continuum depression variability. Variability of the star in the mentioned spectral region was expected, as observations made by Gettys and Schild (1977) indicated variable values of the A(53) peculiarity index for HR 6127.

The star was observed in 19 nights, BD + 54 1809 served as the comparison star. The amplitude of the light variations is 0.017 mag. The interval of (0.5 - 175) days

C. R. Cowley et al (eds.), Upper Main Sequence Stars with Anomalous Abundances, 429–432.

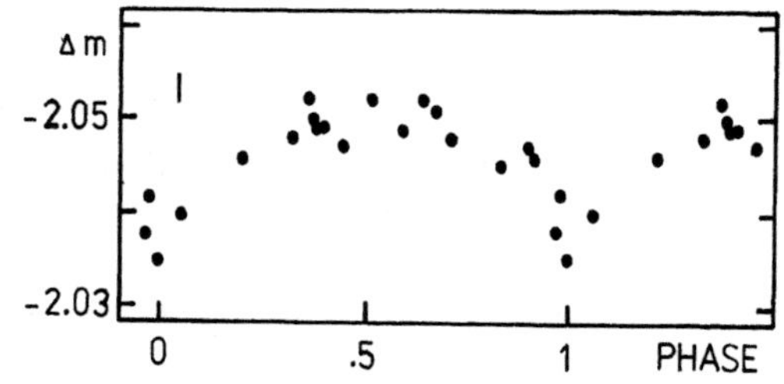

Figure 1. Light curve of HR 6127. The 3 value is marked in the upper left corner.

was searched for periodic variability, using a program written according to Morbey (1978). The best fitting yields to the following ephemeris:

$$\text{JD min} = 2\ 444\ 985.6031 + 2.144202 \times E.$$

The light curve is represented in Figure 1. Dots are nightly means, each dot represents 10 to 140 individual measurements.

The spectroscopic material consists of 30 spectrograms obtained in Coudé spectrographs of 2m telescopes of Ondřejov and Rozhen observatories in the years 1975 - 1981. The dispersion of the spectra is 0.85 and 0.42 nm/mm for Ondřejov and Rozhen spectra respectively. Some of the spectra were underexposed, these were excluded from further interpretation.

Equivalent widths of spectral lines were measured on intensity tracings, using the intensitometer described by Minarovjech et al. (1983). Radial velocities of the metallic lines were measured at the Ondřejov Observatory's comparator, excluding the Ca II K line, which was measured

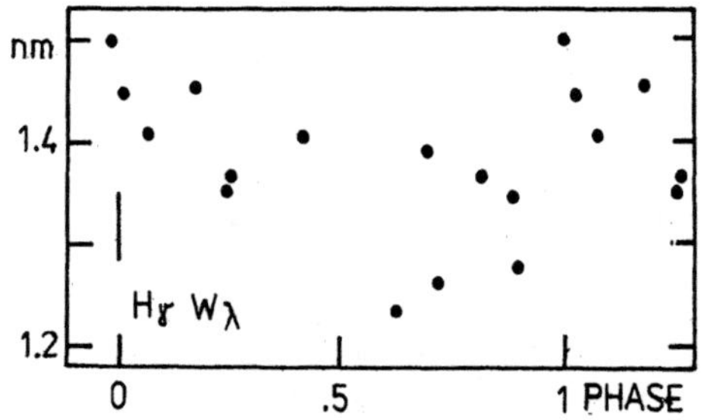

Figure 2. Variability of the hydrogen Hγ line equivalent width in the photometric period. The mean value of standard errors is marked at the left size of the figure.

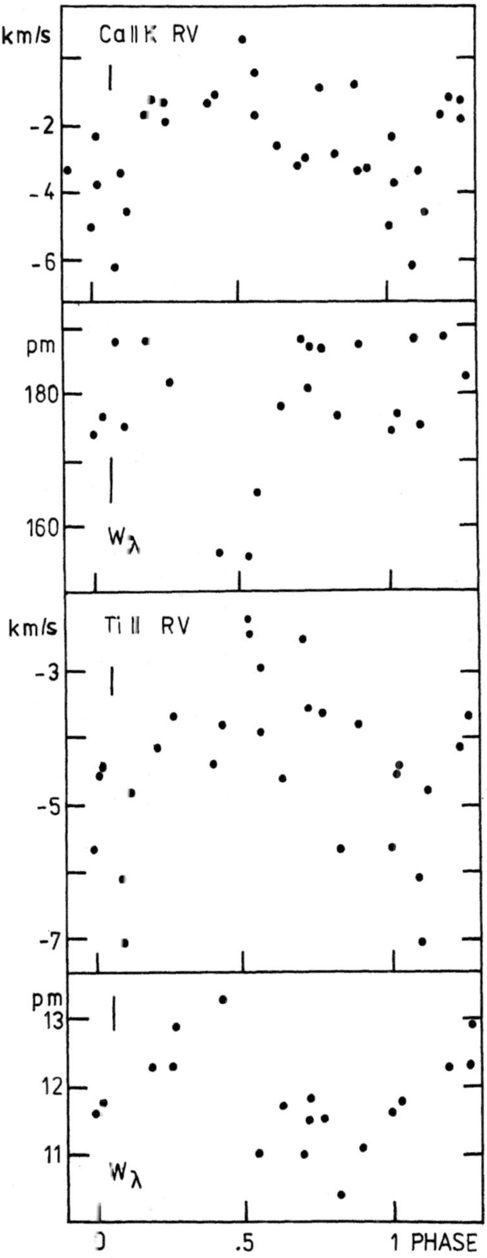

Figure 3. Radial velocity and equivalent width curves for Ca II and Ti II lines. Mean values of standard errors are marked at the left side for each data set.

at the TV-equipped comparator of the Skalnaté Pleso Observatory. Radial velocities and equivalent widths of spectral lines of Ca II K, Ti II, Sr II, Si II and Sc II and equivalent widths of Fe I and the Hγ line show periodic variability in the photometric period. The resulting curves for the Hγ, Ca II K and Ti II lines are represented in Figures 2 and 3. Radial velocities are mean values of 16 Ti II lines, equivalent widths are means of 5 Ti II lines measured on each plate.

3. DISCUSSION

The observed light and spectral variability of HR 6127 can be explained by the oblique rotator model. From the Mg II 448.1 nm line halfwidth we derived following value of the projected rotational velocity:
$V \sin i = 11.6$ km/s.
Adopting the value of $R = 2.5\ R_\odot$ for the radius of the star, $i = 11^\circ$ for the angle of inclination can accepted. The positions of Ca II K, Ti II and Sr II spectroscopic spots are probably closely correlated with the visible magnetic pole of the star. The amplitudes of the radial velocity variations then lead to an unusual value of β, the angle of the magnetic axis inclination, $\beta \sim 13^\circ$. The problem of the magnetic field configuration is to be

solved when magnetic field observations and their interpretation will be finished.

On the basis of presented light and spectral variability and the presence of magnetic field, HR 6127 is classified as cool CP2 star.

REFERENCES

Adelman, S.J., Pyper, D.M.: 1983, Astron. Astrophys. 118, 313
Bonsack, W.K.: 1974, Publ. Astron. Soc. Pacific 86, 408
Cowley, A., Cowley, C., Jaschek, M., Jaschek, C.: 1969, Astron. J. 74, 375
Cowley, C.R.: 1976, Astrophys. J. Suppl. 32, 631
Cowley, C.R., Elste, G.H., Urbanski, J.L.: 1978, Publ. Astron. Soc. Pacific 90, 536
Floquet, M.: 1975, Astron. Astrophys. Suppl. 21, 25
Gettys, J., Schild, R.E.: 1977, Publ. Astron. Soc. Pacific 89, 519
Glagolevskij, Yu.V., Romanyuk, I.I., Zverko, J., Žižňovský, J.: 1984, in Magnetic Stars, Salaspils, p. 22
Hauck, B., North, P.: 1982, Astron. Astrophys. 114, 23
Minarovjech, M., Rybanský, M., Žižňovský, J.; Zverko, J.: 1983, Bull. Astron. Inst. Czechosl. 34, 51
Morbey, C.L.: 1978, Publ. Dominion Astrophys. Obs. 15, 105
Pirronello, V., Strazzula, G.: 1981, Astron. Astrophys. 104, 80
Žižňovský, J.: 1980a, Bull. Astron. Inst. Czechosl. 31, 26
Žižňovský, J.: 1980b, Bull. Astron. Inst. Czechosl. 31, 300

THE ANALYSIS OF CHEMICAL COMPOSITION OF Am STAR ATMOSPHERES

A. A. Boyarchuk and I. S. Savanov
Crimean Astrophysical Observatory
334413, p/o Nauchny, Crimea
USSR

ABSTRACT. The Survey of principal physical characteristics of the atmospheres of Am stars has been carried out. The anomalies of chemical composition have been discussed. The deficiency in C, O, Mg, Ca, and Sc alongside the excesses of heavier-than-Ni elements, with respect to Fe has been pointed out. The hypothesis on the nature of the observed anomalies are briefly discussed.

1. INTRODUCTION

The "metallic-line" stars are the stars with enhanced lines of metals in their spectra compared to normal stars of the same spectral classes, determined by hydrogen and Ca II lines. They are indicated by the Am index. Naturally, when using common criteria of spectral classification for A stars, they belong to later spectral classes according to the metal lines. Calcium lines are very weak in the spectra of Am stars and therefore spectral classes determined by their intensities are consequently earlier than those determined by the hydrogen lines.

A number of hypotheses have been advanced to explain the anomalies of Am star spectra. Some of them suppose specific conditions of excitation in the atmospheres of Am stars, leading to the observed anomalies. Other hypotheses assume that we deal with real anomalies in the chemical composition of the atmospheres of Am stars.

By now more than 40 Am stars have been studied on the basis of spectrograms with dispersion better than 10A/mm. The analysis has been facilitated using model atmospheres. The majority of stars was analyzed by Smith (1971, 1973, 1974), Conti (1965, 1970), and Conti and Strom (1968), and in the Crimea by Lyubimkov and Savanov (1983). Many other astronomers analyzed no more than one star. As a result we obtained the material involving a great number of stars distributed over a wide range of effective temperatures thus making it possible to establish correlations and general laws concerning spectral anomalies of Am stars. However, it should be noted, that the data available for Am stars are imperfect. First, different authors used as standards various stars, that might lead to additional inhomogeneity of the material. Second, the most extended analyses of spectrograms have been carried out with obsolete values of oscillator strengths.

2. PHYSICAL CONDITIONS IN THE ATMOSPHERES OF Am STARS

By using model atmospheres it is possible to estimate the effective temperature, T_{eff}, the surface gravity g and the value of the turbulent velocity V_t. For this aim the following criteria should be used: hydrogen line profiles, estimates of ionization equilibrium and energy distributions in narrow spectral regions determined by scanner observations as well as narrow band photometry.

Figure 1 shows the position of Am stars on the diagram log(g)–T_{eff}. One can see that

C. R. Cowley et al (eds.), Upper Main Sequence Stars with Anomalous Abundances, 433–442.

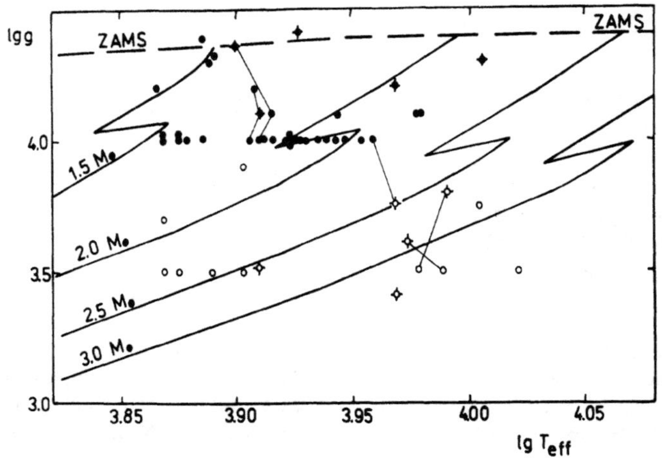

Figure 1. The location of Am stars on the diagram of log(g)-T_{eff}. Evolutionary tracks are from Mengel et al. (1979). Open circles correspond to the evolved stars—filled ones to the unevolved. Circles with crosses stand for the stars studied in the Crimea. Lines connect the results obtained by different authors for the same star.

Am stars have the effective temperatures in the range from 7400 to about 10000K, that fall within the range of spectral classes B9 - F0. The low temperature limit for Am stars is probably reliable. A special search for cool Am stars has been carried out (Smith 1973). In the spectra of F0 stars and cooler, Sc, Ca and Fe lines are intensive enough and the existence of anomalies between their intensities might be easily found. As far as the high temperature range of Am stars distribution is concerned, the situation is less reliable. The matter is that due to the increase of the level of ionization, the Sc. Ca, and metallic lines noticeably weaken and after that completely disappear. Probably in the future the UV data analysis of Sc III and Ca III lines would clarify this phenomenon, but spectral anomalies can be discovered only by quantitative analysis of high dispersion spectrograms. Thus Boyarchuk (1963) has found that the brightest star α CMa is an Am star and Smith (1974) showed that the star θ Leo, which was believed to be the standard for spectral classification is also an Am star.

The values of microturbulent velocities in the atmospheres of Am and A stars are almost the same. Such conclusions were drawn earlier by Baschek and Reimers (1969) and Smith (1971). They have also remarked, that for the stars with the effective temperature from 7600 to 9000K, the microturbulent velocity is higher than for the adjacent regions. Recently the assumption of modern values of oscillator strengths and blanketing and the account on the superfine structure resulted in the fact that the value of the turbulent velocity for the Sun considerably decreased from 2 km/s to 0.8 - 1.0 km/s. However, since the atmospheres of Am and A stars have been analyzed by the same method the conclusion that in the atmospheres of Am and A stars the turbulent velocities are equal, probably remains reliable. Modern model atmospheres being applied for Am stars has shown that to determine the relative intensities of lines and their profiles, no sophistication of the model with respect to the model of normal stars is needed. Particularly, low electron density on the surface of a star is not necessary.

Furthermore, according to modern methods of analysis, it might be concluded that the atmospheres of Am and A stars are similar by their physical parameters, and anomalous relations between line intensities observed in the spectra of Am stars shouldn't be ascribed to the effects of excitation.

Finally, we remark that the rotational velocity of Am stars is considerably lower than that of A stars. Metallicism may occur at the stars with v·sin(i) up to the value of 60 – 70 km/s (like in β Ari, cf. Mitton 1977).

3. CHEMICAL COMPOSITION OF THE ATMOSPHERES OF Am STARS.

As stated above, the anomalous relations of line intensities in Am star spectra shouldn't be ascribed to excitation potentials. Therefore we assume that these anomalies are connected with the anomalous abundances of Am stars.

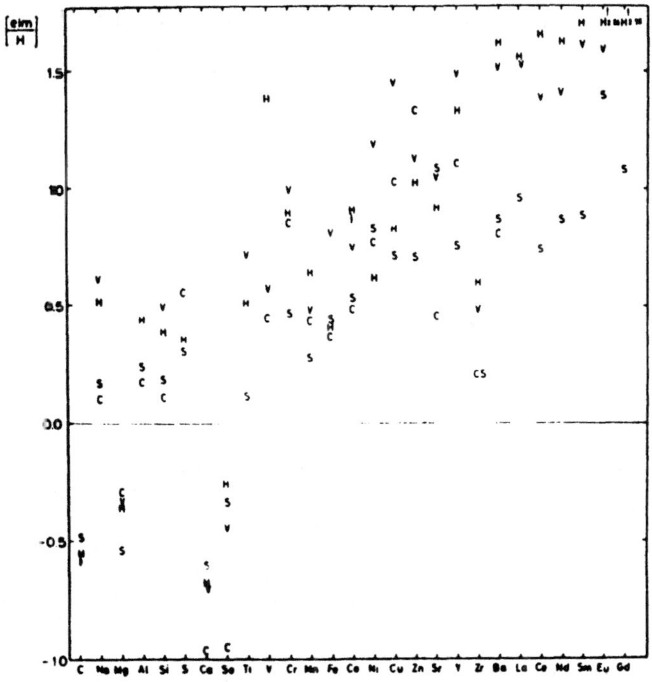

Figure 2. The abundance anomalies of the atmosphere of the star 63 Tau, determined by different authors. H stands for Hund (1972), V is for van't Veer-Menneret (1963), C is for Conti (1965), and S is for Smith (1976).

Hereafter, we use the following conventional notation:

$$[Fe/H] \equiv \log\,(N(Fe)/N(H))_{Am} - \log\,(N(Fe)/N(H))_{Std\ star}$$

So, we obtain the data on chemical compositions of the atmospheres available for more than 40 Am stars. Unfortunately, different authors used different values of oscillator strengths and different standard stars, that might led to the discrepancy of data. As an example, Figure 2 shows the values of element-to-hydrogen with respect to the atomic number, for 63 Tau. One can see that the discrepancy here is rather high. With due regard for the specific determination of the comparison of different elements, let us consider the abundance of different elements relative to iron, and separately the abundance of Fe relative to hydrogen. We readily assume, that such comparative methods would help to eliminate the contribution of errors. Figure 3 demonstrates relative abundances of elements in the atmospheres of Am stars.

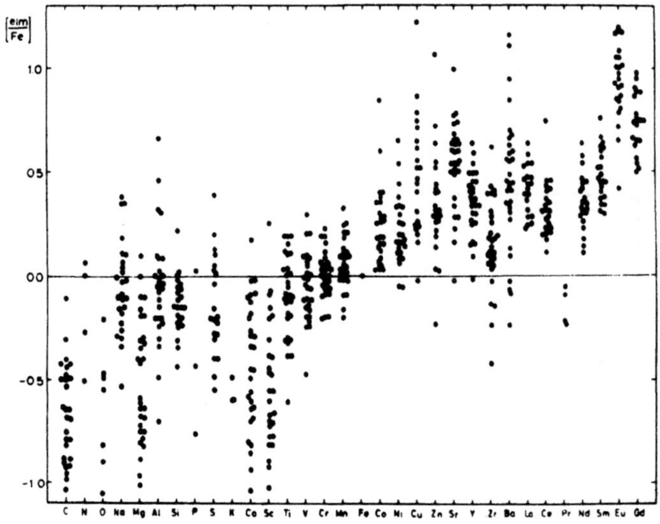

Figure 3. Relative abundance of the elements in the atmospheres of the Am stars (see the text).

According to Figure 3, one can draw the following conclusions:

1) While passing from light elements to heavier ones the excesses of elements increases.
2) The increase of overabundance of elements is not monotonic. Among light elements, the deficiency of C, O, Mg, and Ca, Sc is obvious, but at the same time Na, Al and probably Si and S are of normal content. Among heavier elements there is a great excess of Sr, Ba, Eu, and Gd together with moderate excesses of Zr, Ce, and Nd. More precise consideration of Eu content also evidences for its excess by 10 times (Hartoog and Cowley 1974).
3) The scatter is rather high for all elements than the elements of the Fe group. The abundance of these elements are determined by a great number of lines and the errors are not appreciable. But for other elements the scatter exceeds an order of magnitude: Ba, for instance. From several general considerations we proceed to the analysis of the results obtained for separate elements. Here we shall deal with the data obtained by high dispersion spectrograms with blanketing model atmospheres reduced to the same scale of oscillator strengths.

Figure 4 shows the abundance of iron in the atmospheres of Am stars with respect to that in the atmospheres of standard A stars. The iron abundance in the atmospheres of Am stars turns out to be thrice that in the normal star's atmospheres. There is a weak tendency in the atmospheres of hotter Am stars to have greater excess of Fe, than in cooler stars.

For some elements which are not presented in Fig. 3, we obtain some qualitative data.

Wallerstein and Conti (1969) haven't found any Li or Be lines in the spectra of Am stars, this fact testifies that there is no great excess of these elements. Boesgaard and Praderie

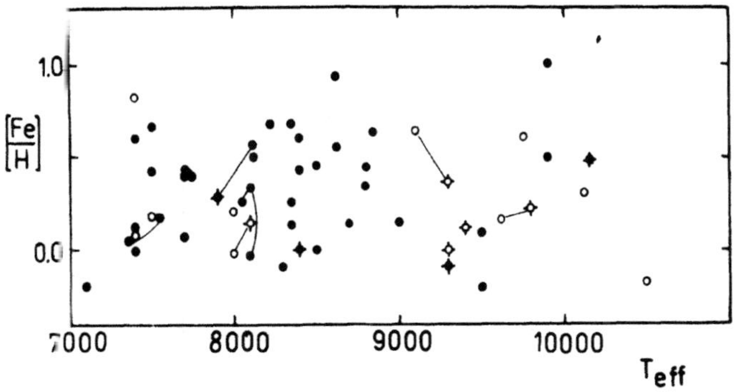

Figure 4. Abundances of iron as a function of the effective temperature. Signs are the same as in Fig. 1.

(1981) obtained later, that Be and B are greatly underabundant in the atmospheres of α CMa and γ Gem. Smith (1974) found that the He abundance in three stars, θ Leo, α GemA and α CMa is about twice as low. Boyarchuk and Show (1978) haven't found any Tc, Pt, and Hg in the ultraviolet spectra of α CMa obtained with the Copernicus satellite with 0.05 and 0.1A spectral resolution. Cowley, Aikman, and Hartoog (1976) haven't found any evidence for actinides. Lyubmikov and Savanov (1985) found strong excesses of Th in the atmospheres of Am stars: 15 Vul, 16 Ori, 63 Tau, and no excess in the atmosphere of 81 Tau.

It is interesting to consider the correlations between abundance anomalies of different elements. Figure 5 shows the comparison of the values of [Ca/Fe] and [Sc/Fe]. There is a good correlation between the anomalies of Ca and Sc abundances. Such a correlation also exists between [Ca/Fe] and [Mg/Fe]. However, the correlation between [C/Fe] and [Ca/Fe] is practically absent.

The dependence of [Ca/Fe] on the effective temperature is quite weak (see Figure 6), i.e. cooler stars may have more deficiencies of Ca. There is almost no correlation between [C/Fe] and T_{eff} values. Insofar, the anomalies of C are to some extent different than the ones of Mg, Ca, and Sc.

The anomalies of heavy elements probably also correlate with each other (with insignificant exceptions), but since the scatter is rather high, a definite conclusion is hard to draw. Following Figure 7, one can observe substantial correlations between [La/Fe] and [Nd/Fe]. On the other hand, according to Figure 8, there is no reliable correlation between [Zr/Fe] and [Ba/Fe].

But taking into account Figure 9, one cannot see a tendency of the value [Zr/Ba] to depend on the effective temperature.

There is no strong correlation between the value of the deficiency in light elements and the excess of heavy ones, e.g. between [C/Fe] and [Eu/Fe]. the lower the rotational velocity of the star, the stronger the anomalies of chemical composition. Kodaira (1976) found that

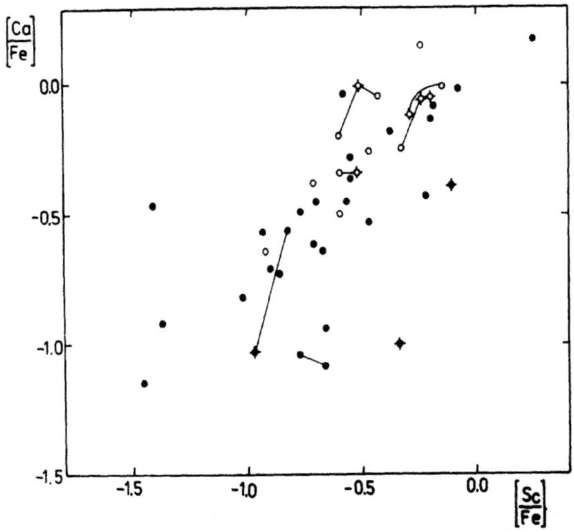

Figure 5. The comparison between the abundance anomalies of Ca and Sc. Signs are the same as in Fig. 1.

metallicity index $[m_1]$ decreases with the increase of v·sin(i). Correlations of the abundances with rotation were studied by Smith (1973). We consider the dependence of [V/H] on v·sin(i) for evolved and unevolved stars (a) and "hot" and "cool" Am stars (b), as shown in Figure 10.

Significant scatter on the presented figures involves difficulties for conclusions, and makes them less reliable. But, some qualitative conclusions are likely possible:

1) The peculiarity of chemical abundance of Am star atmospheres is the excess of heavy elements increasing on the average with the atomic weight. The deficiency of several light elements has been found.
2) In some cases the correlations between anomalies exist, in other cases they vanish. At present, due to insufficient accuracy, we are not apt to make any definite remarks.

To make these conclusions more well founded, more accurate and homogeneous data on the chemical composition of the atmospheres of Am stars are needed.

4. THE ORIGIN OF CHEMICAL COMPOSITION ANOMALIES

As it was stated above, the detailed analysis of the spectra of Am stars showed, that the anomalies in correlations between line intensities of different elements should not be ascribed to some specific conditions of excitation, and we deal with real anomalies of chemical composition.

The existing hypotheses on the nature of anomalies can be provisionally divided into two groups.

According to the first group, chemical anomalies are formed outside the stars, primarily in different types of accretion, where the separation or formation of chemical composition

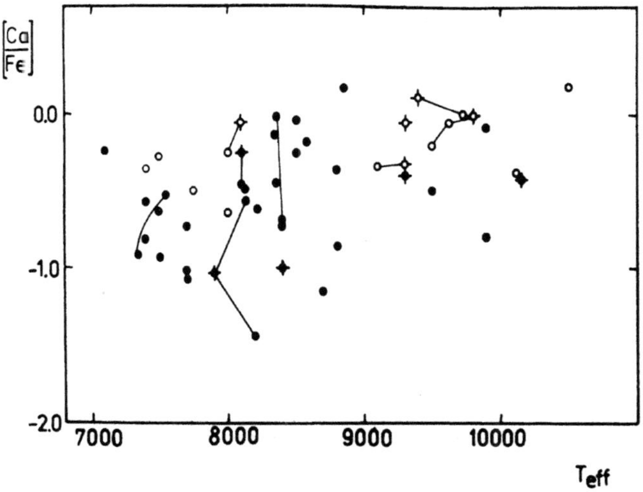

Figure 6. Abundance of calcium as a function of the effective temperature. Signs are the same as in Figure 1.

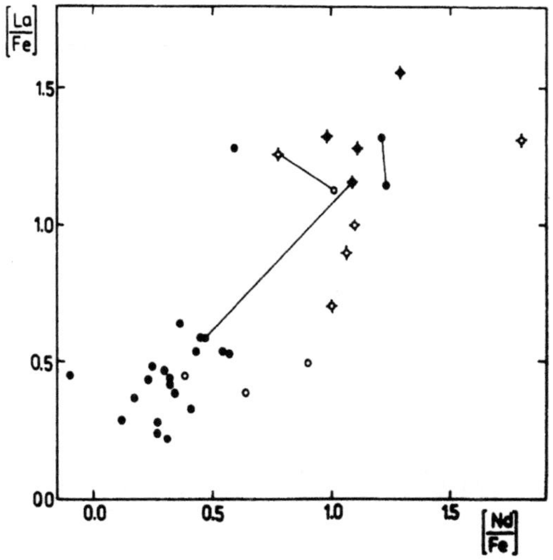

Figure 7. The comparison between the abundance anomalies of [La/Fe] and [Nd/Fe]. Signs are the same as in Figure 1.

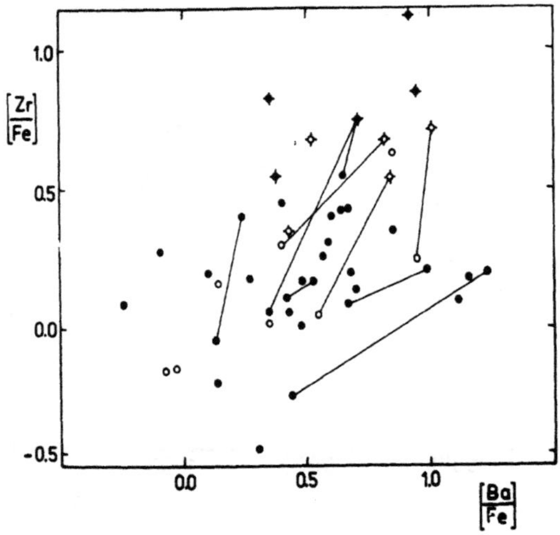

Figure 8. The comparison between the abundance anomalies of [Zr/Fe] and [Ba/Fe]. Signs are the same as in Figure 1.

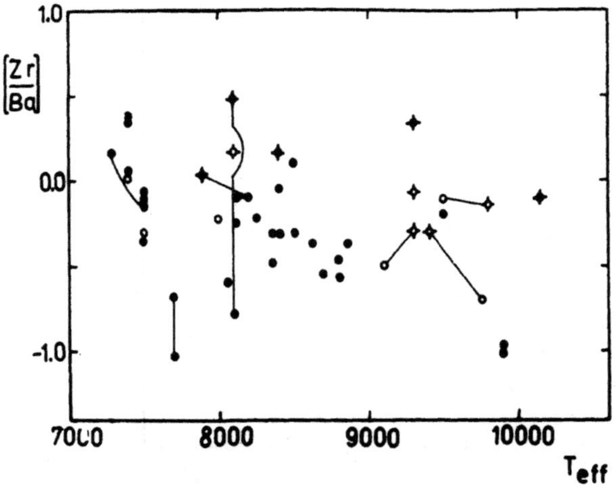

Figure 9. The dependence of [Zr/Ba] values on the effective temperature. Signs are the same as in Figure 1.

depends either on the magnetic field influence or by matter outflow in the binary system. It is noteworthy, that these initial hypotheses transform the problem of chemical composition in the

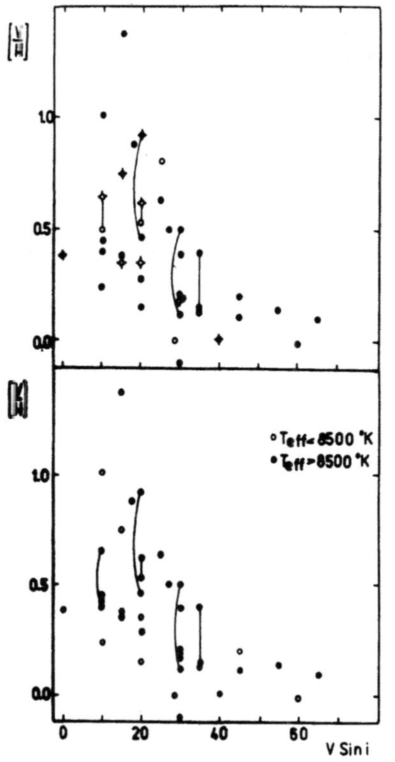

Figure 10. The comparison between [V/H] and rotational velocity for: evolved and unevolved stars (above), and "hot" and "cool" Am stars (below).

atmospheres of Am stars into two problems: formation of the elements outside the star, and matter transfer on the surface of the "metallic" star. The theories of this trend come across such difficulties are, for example, the interpretation of the presence of Am stars in young clusters and the absence of strong magnetic fields in the Am stars.

Other hypotheses ascribe the formation of chemical composition anomalies to the processes in the interior of stars, for instance, mixing processes, where the upper layers are enriched with heavy elements. In spite of numerous obstacles, this theory is advantageous to explain the presence of heavy elements in the atmospheres of "metallic" line stars. Probably the most promising is the hypothesis on diffusive separation of elements by radiation pressure. However, this hypothesis needs to consider not only the behavior of separate elements, but to extrapolate it to the abundances of different elements in total.

REFERENCES

Baschek, B., and Reimers, D. 1969, *Astron. Ap.*, **2**, 240.
Boesgaard, A. M., and Praderie, F. 1981, *Ap. J.*, **245**, 219.
Boyarchuk, A. A. 1963, *Izw. Crimean. Ap. Obs.*, **29**, 219.
Boyarchuk, A. A., and Snow, T. P. 1978, *Ap. J.*, **219**, 515.
Conti, P. S. 1970, *P. A. S. P.*, **82**, 781.

Conti, P. S., and Strom, S. E. 1968, *Ap. J.*, **152**, 483.
Cowley, C. R., Aikman, G. C. L., and Hartoog, M. R. 1976, *Ap. J.*, **206**, 196.
Cox, J. P, and Giuli, R. T. 1968, *Stellar Structure*, (New York: Gordon and Breach), vol. 2, p. 626.
Hartoog, M. R., and Cowley, C. R. 1974, *Ap. J.*, **187**, 551.
Hundt, E. 1972, *Astron. Ap.*, **21**, 214.
Kodaira, K. 1976, in *Physics of Ap Stars*, IAU Colloquium No. 32, ed. W. W. Weiss, H. Jenkner, and H. J. Wood, p. 675.
Kurucz, R. L. 1979, *Ap. J. Suppl.*, **40**, 1.
Mengel, J. G., Sweigert, A. V., Demarque, P., and Gross, P. G. 1979, *Ap. J. Suppl.*, **40**, 733.
Mitton, J. 1977, *Astron. Ap. Suppl.*, **27**, 35.
Lyubimkov, L. S., and Savanov, I. S. 1983, *Astrophysica*, **19**, 505.
Lyubimkov, L. S., and Savanov, I. S. 1985, *Astrophysica*, **22**, 63.
Smith, M. A. 1971, *Astron. Ap.*, **11**, 325.
Smith, M. A. 1973, *Ap. J. Suppl.*, **25**, 277.
Smith, M. A. 1974, *Ap. J.*, **189**, 101.
Van't Veer, C. 1963, *Ann. d'Ap*, **26**, 289.
Wallerstein, G., and Conti, P. S. 1969, *Ann. Rev. Astron. Ap.*, 7, 99.
Watson, W. D. 1971, *Astron. Ap.*, **13**, 263.

DISCUSSION (Boyarchuk and Savanov)

PTITSYN: I wonder, do you not think that one can conclude from your diagrams that there is a correlation between elements which are produced in the same nuclear process, and vice versa?
BOYARCHUK: I have not looked at correlations with nuclear or diffusion processes. I am not sure the results are accurate enough to compare with some theory.
COWLEY: I can comment on this briefly, based on the material I have looked at, which is mainly that published by Myron Smith. The major correlations are between each element and its neighbour. There is a general trend for heavy elements to be enhanced and for light elements to be depleted. This general effect produces the correlation that Dr. Ptitsyn mentioned. Since nuclear processes build up nuclei from lighter ones to heavier ones, this kind of a correlation is to be expected.
PTITSYN: I agree, with the exception of the Zr-Ba correlation.
MICHAUD: Do you find a correlation [of abundance] with rotational velocity only for vanadium, or does it appear also for other elements?
BOYARCHUK: They are similar for the iron group, but I only made plots for a few elements.
HENSBERGE: I was surprised to see a good correlation between the relative La and Nd abundances. Cowley pointed out that, at least for cool magnetic Ap stars, the Nd lines in the visual are badly blended by Cr lines. Is this also the case in Am stars?
BOYARCHUK: The data we presented is based on published analyses. I would have to have examined the spectra myself in order to answer your question.
COWLEY: I believe it is probably not affected by the Cr, because Cr is not extremely overabundant the way it is in some of the magnetic stars.
DROBYSHEVSKI: One should be cautious when interpreting anti-correlations between chemical anomalies and rotational velocities in Am stars. These correlations favour only two facts: the anomalies are concentrated in the outermost layers, and they must be continuously replenished. In this sense they agree with both the radiation pressure diffusion theory and the planetoidal hypothesis.
BURKHART: I am not convinced by the correlation between vanadium abundance and the rotational velocity of the star, because of the sin i factor in the figure. It appears to me that there are not enough points on the diagram, especially at large values of v sin i.
BOYARCHUK: On this diagram we have plotted log $\varepsilon(V)$ vs. v sin i. We think that the correlation is indicated by drawing an upper envelope over the points, because the scatter is due to the sin i factor. We realize that for any one star, the true rotational velocity can not be determined, but we think the sample is large enough to indicate that the correlation exists.
ADELMAN: May I make a plea to all observers to publish their equivalent widths and their $H\gamma$ profiles? This would permit one to intercompare the data of different observers for the same star.

Also, I want to mention something which seems to have been omitted. I am very interested in stars which lie between the coolest of the Hg-Mn

C. R. Cowley et al. (eds.), Upper Main Sequence Stars with Anomalous Abundances, 443–445.

stars and the hottest of the classical Am stars. It would be interesting to know whether or not the abundances have smooth trends between the two types.

DWORETSKY: I have comments as well. I agree with what Saul Adelman said about publishing all the equivalent widths and Hγ profiles; these should certainly be published. I only wish that we could persuade editors that this was a good idea. They usually want you to store these data in a vault someplace, or make them into a microfiche. The latter is at least a partially acceptable result, because they will exist, but they are then subject to accidental loss, etc. But, at least in principle, they are available.

There has been a lot of discussion of surface gravities of stars at this Colloquium. I would like to mention a paper in press [since published] by T. Moon and myself (MNRAS, 217, p. 305, 1985) concerning the calibration of the Strömgren c_0 and β indices in terms of T_{eff} and log g. We have checked E. G. Schmidt's theoretical β indices and the Relyea-Kurucz c_0 indices against eclipsing binaries of known gravity and Code's fundamental T_{eff} stars. We find that there is a discrepancy in log g of about 0.15-0.20, in the sense that the Kurucz profiles yield smaller gravities. We do not fully understand the reasons for this.

BOYARCHUK: The accuracy for determination of spectroscopic profiles is less than 0.15.

DWORETSKY: The precision with which any individual measurement can be made is quite good; that's not what I am talking about. There is a systematic deviation, and all the previous gravities are about 0.2 dex too small, on average. I wonder if anyone has a comment on that? We do not know the cause, but I am quite convinced that the discrepancy is real. It looks as if the Kurucz Balmer line profiles may somehow be at fault.

COWLEY: If we don't know the effective temperature of these objects to within, say 500°, I'm sure this affects the attempts to determine surface gravity by Hγ profile in some significant way. As you know, Lane and Lester suggest downward modifications of the order of 500° or even 900° for the Am stars, so surely this is a factor that should be considered.

DWORETSKY: It should be, but our method disagrees with the results of Lane and Lester. They found surface gravities which were very low for some stars, and we find that there is a strong correlation between the difference in gravity and the metallicity of the star as determined from the δm_1 index. The precise cause of the difference between Lane and Lester and ourselves is difficult to understand. It may have to do with the fact that they tried to fit the continuum in the visible and ultraviolet to model atmosphere fluxes, while we have looked at the Balmer jump and Hβ profile. We do find that Am stars in open clusters which should lie near the zero-age main sequence have a mean log g = 4.3 with our method. This is about the only evidence we have that our method is independent of the amount of line blanketing.

HUBENÝ: In spite of arguments against Kurucz's set of models or the LTE approach, there is at least one argument in favour of the applicability of LTE. Dreiling and Bell calculated models for Sirius and claimed that the spectroscopically and dynamically determined gravities agree quite well. This suggests the problem may lie in the treatment of opacities, or some error in the Kurucz fluxes, because the model for Sirius was calculated using Bell's own program.

DWORETSKY: It is interesting that Schmidt used other models also, and calculated theoretical β indices for the older Kurucz, Peytremann and Avrett models which used Griem theory. If we had used those instead, there would have been no discrepancy between the theory and the observed surface gravities. This is one reason that we suspect that the broadening theory itself is at fault, because the only large differences in predicted β's arose from the differences in broadening theory and model atmospheres.

HUBENÝ: Unfortunately, Kurucz only stated that he used the Vidal, Cooper and Smith theory, but that it was based on routines supplied by Peterson. It is not easy to check these results; maybe there is some error in the treatment of the atomic profile of the Balmer lines.

DWORETSKY: I have no comment on that at all!

COWLEY: Are there any other comments? We still have a few minutes.

DROBYSHEVSKI: When doing statistics among Am stars, one has to be mindful of two groups of Am star binaries, the young and the evolved. Otherwise, some important statistical correlation may be lost.

ADELMAN: I have found good agreement between the observed Hγ profiles in the line wings and shoulders for the predictions of VCS theory as calculated using the ATLAS-compatible program BALMER for normal B5 to A2 stars. I shudder to think what substantial errors in the hydrogen lines would mean for the abundance analyses.

DWORETSKY: I shudder as well. I hesitate to come before you and state that I have these doubts, but when the evidence is staring you in the face, you must state what you think is going on, even if it turns out to be wrong later. But, I think this is right!

LITHIUM IN Am AND δ Del STARS

C. BURKHART AND M. LUNEL
Observatoire de Lyon
69230-St Genis-Laval- FRANCE

M.F. COUPRY and C. Van 't VEER
Institut d'Astrophysique
98 bis Bd Arago
75014- Paris - FRANCE

SUMMARY. High-resolution observations of the doublet LiI-6707 A were made to study the abundance of lithium in a sample of Am, δ Del, and reference stars. It is found that Am and δ Del stars may be underabundant, normal or overabundant in lithium (normal Li abundance meaning log N_{Li} = 3.0 with log N_H = 12).

1. INTRODUCTION

The abundance of the light element Li is not yet known in Am and δ Del stars. The detection of the LiI doublet at 6707A is indeed very difficult for such hot stars with fairly broad lines unless Li is overabundant. Today, high-resolution and good signal-to-noise spectra enable equivalent width measurements of Li in the A star domain. A set of model atmospheres (Kurucz, 1979a,b) is used to calculate the equivalent width of LiI-6707 A as a function of Te in the weak-line limit. The stellar Te, which is of critical importance in the abundance determination, is determined from published high-dispersion abundance analyses or from ubvy,β photometry. The Li abundance follows from the measured λ6707 equivalent width.

2. OBSERVATIONS

Spectra of 9 Am, 4 δ Del and 6 F stars were obtained in the region $\lambda\lambda$6675-6725 with the Coudé echelle spectrometer of the European-South-Observatory 1.4m telescope. The detector was a cooled Reticon array of 1872 photodiodes. The linear dispersion was 1.9 A/mm and the spectral resolution R = $\lambda / \Delta\lambda$ was equal to 100 000. The signal-to-noise ratio of the spectra was generally greater than 200.

Some spectra are shown in Fig. 1; they are flat-field corrected and referred to a hand drawn continuum, the scale is

C. R. Cowley et al (eds.), Upper Main Sequence Stars with Anomalous Abundances, 447–450.

Fig 1.

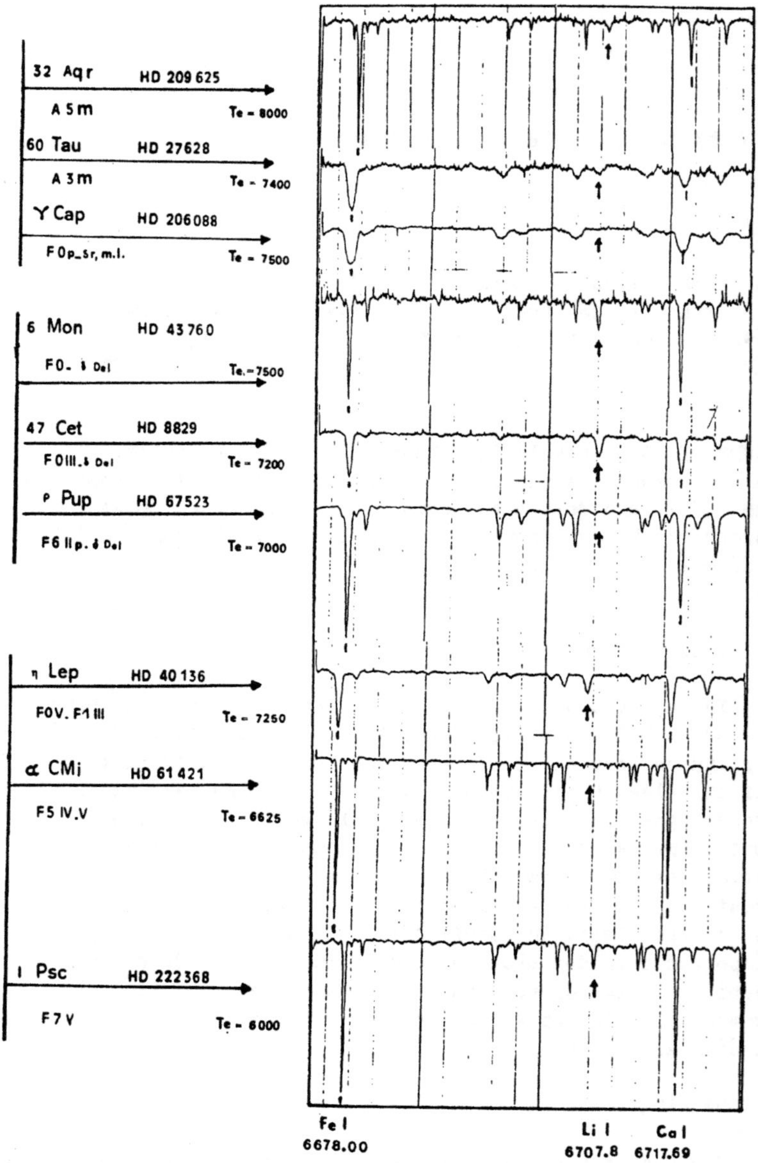
32 Aqr HD 209 625
A 5 m Te - 8000
60 Tau HD 27628
A 3 m Te - 7400
γ Cap HD 206088
F 0p_Sr, m.l. Te - 7500
6 Mon HD 43 760
F0 - δ Del Te - 7500
47 Cet HD 8829
F 0 III. δ Del Te - 7200
ρ Pup HD 67523
F6 II p. δ Del Te - 7000
η Lep HD 40 136
F0 V. F1 III Te - 7250
α CMi HD 61 421
F5 IV.V Te - 6625
ι Psc HD 222 368
F 7 V Te - 6000
Fe I
6678.00
Li I
6707.8
Ca I
6717.69

Fig 2.

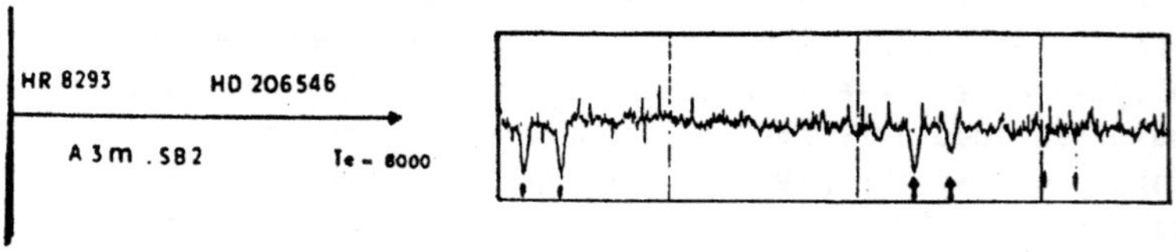
HR 8293 HD 206546
A 3 m . SB2 Te - 8000

the same for every spectrum. With a line profile fitting procedure (Cayrel et al., 1985) equivalent width measurements are possible with a high accuracy and lines as weak as 3 mA can be detected.

There are 3 double-line spectroscopic binary stars in our Am sample. Fig. 2 shows the spectrum of one of them, HR 8293. Besides the rather low quality of the spectrum compared to those of Fig. 1, we see the greater difficulty of continuum setting for a sb2 star than for a single one . The LiI lines of both components are as strong as those of much cooler stars like η Lep or ι Psc and, contrasting with the spectra of Fig.1, are the most prominent lines in the observed spectrum, except the FeI (6678 A) lines. Assuming that both components are just alike (Stickland, 1973) and taking into account the dilution effect, we get the intrinsic equivalent widths multiplying by 2 the equivalent widths as measured on the spectrum, thus leading to much more remarkable overabundances, in fact the highest ones in the star sample.

In our sample of stars, the ratio of the strength of the CaI-6717A line to that of the FeI-6678A line is nearly 1 for all the δ Del and F stars, but always less (sometimes much less) than 1 for all the Am stars (see figs. 1 and 2). This may be put together with the already known fact that the Am stars are underabundant in Ca and overabundant in Fe but that for δ Del or F stars the ratio of the abundances of Ca to Fe is the solar one.

3. LITHIUM ABUNDANCE

The derived Li abundances are plotted against Te in Fig.3 for the 3 groups of stars. The points with downward arrows represent upper limits on the Li abundance of a star where no Li is detected. The points with upward arrows represent lower limits on the Li abundance of 2 of the sb2 stars, this lower limit is obtained when we do not take into account the dilution effect. For the third sb2 star, HR 8293, we represent both the lower limits of each component and the true values if both components are identical stars with Te = 8000 K and if the dilution effect is considered.

In Am stars, Li may be overabundant (up to log N_{Li} = 3.9), normal, or underabundant (down to log N_{Li} < 2.0). The Li abundance dependence with temperature is definitely not simple : around 8000 K, Li abundance results are remarkable by their variety.

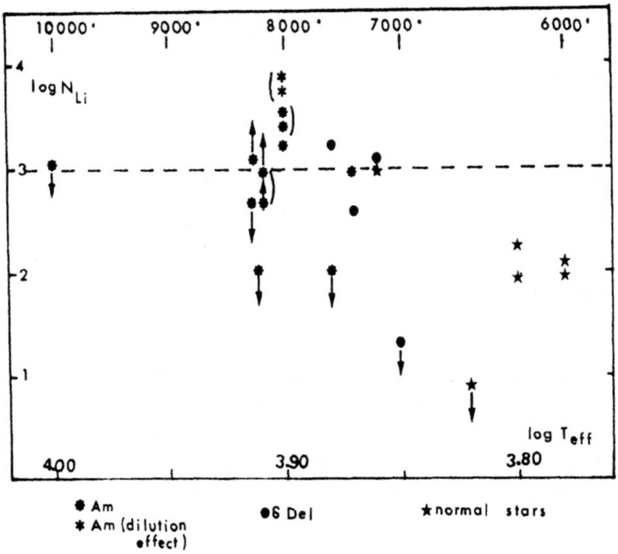

Fig 3

In δ Del stars, Li may be slightly overabundant, normal, or underabundant (down to log $N_{Li} < 1.30$). We cannot study any abundance trend with temperature because of the few stars observed.

There are not yet enough stars observed to conclude definitely, but those abundance results were expected for the Am stars in the frame of the microscopic diffusion hypothesis in presence of either a laminar meridional circulation or a turbulent diffusion by Vauclair et al (1978). Moreover, could the Li overabundant stars actually be explained by as simple a physical process as microscopic diffusion ?

References

Cayrel, R., Cayrel de Strobel, G., Campbell, B. : 1985, Astron. Astrophys. 146, 249.
Kurucz, R.L. : 1979a, Astrophys. J. Suppl. 40, 1.
Kurucz, R.L. : 1979b, Dudley Obs. Rep. No 14, 363
Stickland, D.J. : 1973, Monthly Notices Roy. Astron. Soc. 161, 193.
Vauclair, S., Vauclair, G., Schatzman, E., Michaud, G. : 1978, Astrophys. J. 223, 567

THE LI I 6708 FEATURE IN CP STARS*

R. Faraggiana
F. Castelli

Dipartimento di Astronomia
Osservatorio Astronomico
Via Tiepolo 11
I-34131 Trieste, Italy

M. Gerbaldi
Institut d'Astrophysique
98 bis, bd Arago
F-75014 Paris, France

M. Floquet
Lab. Associé CNRS n°337
Observatoire de Meudon
F-92190 Meudon, France

ABSTRACT Preliminary results are presented on the behaviour of the feature at 6708 Å ; the location of the observed feature coincides with the doublet of Li I 6707.761 and 6707.912 Å.

The observations were obtained in september 1983 with the Coudé Echelle Spectrometer (CES) of ESO using a Reticon as detector. A spectral range of about 50 Å centred at 6700 Å was covered with a resolution of 10. Raw data were reduced using the standard procedure with the ESO IHAP program package.

The analysis of the stellar spectra was performed with the synthetic spectrum method developed by one of us (F.C.).

The two coolest standard stars show a relatively strong Li feature, and the Li abundance required to reproduce the observed equivalent width is approximately coincident with the upper limit of the Li cosmic abundance, i.e. log ε=-9, for the HD 40136 and only slightly less for HD 739, i.e. log ε=-9.56.

The analysis of the spectra of CP stars was performed using a constant value of log g=4 and T_e values found in the literature.

Six stars of our sample do not show any detectable line at the predicted Li I wavelength. The spectra of three more stars, HD 3980, HD 25267, HD 220825, are very complex since they are SB2 ; the spectra have not yet been reduced. The spectra of the remaining six stars were compared with the computed ones.

The T_e of HD 15144 (Sr-Cr) is in the range 8400-8800 K. The Li I doublet is absent and an unidentified line is present at 6706.8 Å.

* Based on observations collected at the European Southern Observatory, La Silla, Chile.
To be published in Astronomy and Astrophysics.

C. R. Cowley et al (eds.), Upper Main Sequence Stars with Anomalous Abundances, 451–453.

HD 24712 (Sr-Cr-Eu) is one of the coolest Ap stars. Its T is in the range 7300-7500 K. Li I is absent, but the weak line at 6707.45 is present, and may be identified either as Fe I or Sm II.

HD 187474 (Cr-Si) is the hottest star of our sample with T_e 12000 K. The observed feature at the lithium position requires a Li abundance of log ε=-6.2, corresponding to more than 600 times the cosmic abundance.

HD 188041 and HD 201601, both classified as Sr-Cr-Eu stars, show a strong asymmetric Li I doublet. The Li abundance required to reproduce this observed feature are log ε=-7.1 and -7.99 for HD 188041 and HD 201601 respectively. Both stars show the same unidentified feature at 6707.0 Å as HD 15144.

HD 206088, classified either as an Am or as a Sr star, has a v sin i higher than the previous stars. The lines are broad and shallow and a weak Li I feature may be present.

CONCLUSION Among Cp stars belonging to the Sr-Cr-Eu subclass, the strength of the Li feature appears not to be correlated with the temperature of the stars. The same lack of correlation is found between the 6708 Å feature and the intensity of the magnetic field or the peculiarity parameter Δ (V1-G) of the Geneva photometric system.

In several stars a higher than cosmic abundance of Li is derived from the 6708 feature.

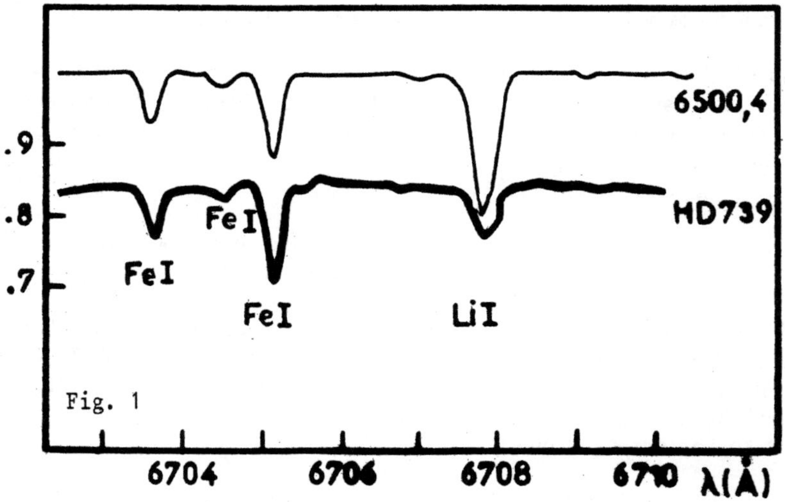

Fig. 1

The spectrum of the standard star HD 739 (F5 V) and the computed spectrum with 6500,4, solar abundances, but log ε_{Li}=-9 (cosmic abundance). The spectra are normalized to the continuum.

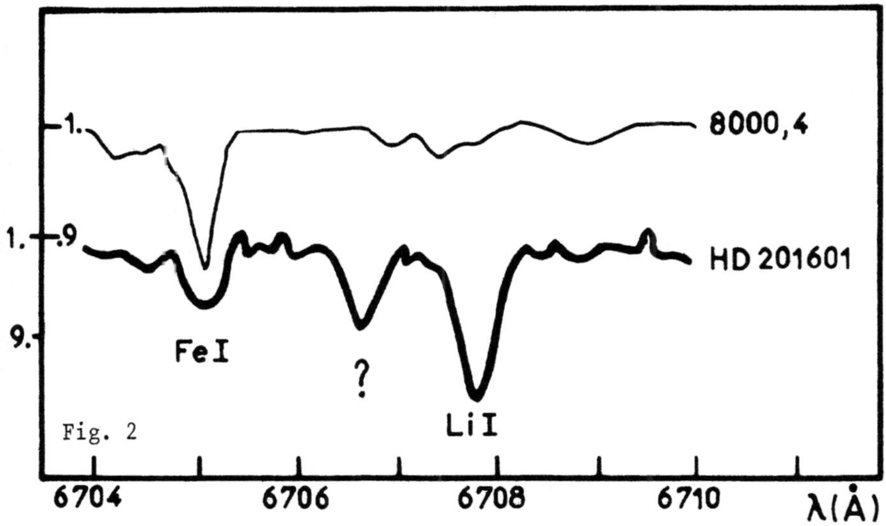

The spectrum of HD 201601 (γ Equ) and the computed spectrum with 8000,4 and enhanced metallic abundances (cosmic abundance for Li ; Cr and Sr 100 x ; Re 1000 x ; other elements 10 x). The spectra are normalized to the continuum.

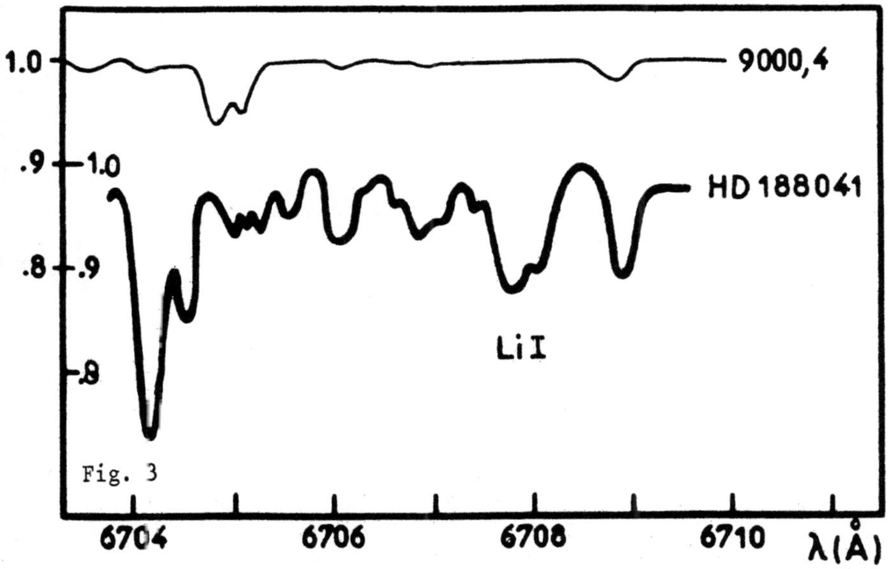

The spectrum of HD 188041 (HR 7575) and the computed one with 9000,4 and same overabundances as for HD 201601. The spectra are normalized to the continuum.

THE METALLIC-LINED STAR 32 AQUARII

Dursun Kocer, Cetin Bolcal
Kandilli Observatory of Bogazici University
Cengelkoy, Istanbul
Turkey

Saul J. Adelman[1,2,3,4]
Code 681, Laboratory for Astronomy and Solar Physics
NASA Goddard Space Flight Center
Greenbelt, MD 20771
United States of America

ABSTRACT. Prelimary results from an optical region study of the metallic-lined star 32 Aquarii indicate T_{eff} = 7600 K, log g = 3.10, ξ = 4.6 km s^{-1}, and an iron abundance about one-half solar.

1. INTRODUCTION

32 Aquarii (HD 209625 = HR 8410) is a sharp-lined single-lined spectroscopic binary (v sin i = 19 km s^{-1}) which is a member of the Ursa Major Stream (Hoffliet 1982). One of us (SJA) obtained 3 4.3 Å/mm well-widdened IIaO spectrograms of this star with the coude spectrograph of the 2.5-m telescope of Mt. Wilson Observatory. These plates were traced with the PDS microdensitometer at Kitt Peak National Observatory. This data was used to produce intensity vs. wavelength plots.

2. LINE IDENTIFICATIONS

The lines were identified on these tracings with the aid of standard sources, especially A Multiplet Table of Astrophysical Interest (Moore 1945) and Wavelengths and Transition Probabilities for Atoms and Ions, Part I (Reader and Corliss 1980). There are approximately 1500 lines in the wavelength interval λλ3820-4630. We have identified lines of the following atomic species: H I, Mg I, Mg II, Al I, Si I, Si II, Ca I, Ca II, Sc II, Ti I, Ti II, V I, V II, Cr I, Cr II, Mn I, Mn II, Fe

---

[1] NRC-NASA Research Associate
[2] On leave from the Department of Physics, The Citadel, Charleston, SC 29409 USA
[3] Guest Investigator, Mt. Wilson and Las Campanas Observatories
[4] Visiting Astronomer, Kitt Peak National Observatory

C. R. Cowley et al (eds.), Upper Main Sequence Stars with Anomalous Abundances, 455–457.

I, Fe II, Co I, Ni I, Ni II, Sr I, Sr II, Y II, Zr II, Ba II, La II, Ce II, Pr II, Nd II, Sm II, Eu II, Gd II, and Dy II.

3. INITIAL RESULTS

We have also compared the spectrophotometric measurements of Lane and Lester (1980) and the Hγ profile as derived from our optical region spectrograms with the predictions of Kurucz's (1979) fully-line blanketed solar composition model atmospheres. The best match occurs for T_{eff} = 7600 K and log g = 3.10 as shown in Figures 1 and 2. The use in our final analysis of the same sources of oscillator strengths and damping constants as the analyses of other B and A stars with similar spectroscopic material (Adelman 1984, 1985) will aid the determination of abundance anomalies and the comparison with HgMn and hot Am stars.

We have completed measuring the equivalent widths of the Fe I and Fe II lines. For each atomic species, the microturbulent velocity was found from the requirement that the derived abundances should not depend on the size of the equivalent widths. From 25 Fe II lines, we found ξ = 4.5 km s^{-1} and log Fe/H = -4.50 ± 0.26 while from 177 Fe I lines, ξ = 4.7 km s^{-1} and log Fe/H = -4.75 ± 0.19. The analysis of lines of other atomic species is in progress.

Fig. 1 The optical region energy distribution of 32 Aqr compared with the predictions of a 7600 K, log g = 3.10 solar composition model atmosphere.

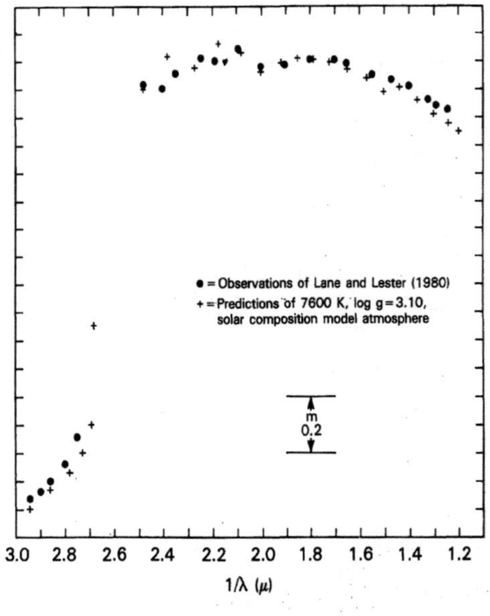

Fig. 2 The Hγ profile of 32 Aqr derived from measurements of the spectrograms compared with the predictions of a 7600 K solar composition model atmosphere.

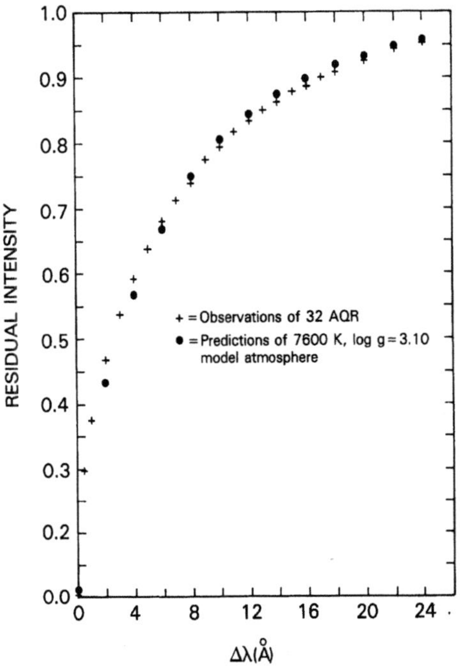

REFERENCES

Adelman, S. J.: 1984, Astron. Astrophys. 58, 585

Adelman, S. J.: 1985, Astron. Astrophys. Suppl., in press

Hoffleit, D.: 1982, The Bright Star Catalogue, 4th edition (Yale University Observatory: New Haven, CT)

Lane, M. C., Lester, J. B.: 1980, Astrophys. J. 238, 210

Kurucz, R. L.: 1979, Astrophys. J. Suppl. Ser. 40, 1

Moore, C. E.: 1945, A Multiplet Table of Astrophysical Interest (Princeton University Observatory: Princeton, NJ)

Reader, J, Corliss, C. H.: 1980, NSRDS-NBS 68, part I (US Government Printing Office: Washington, DC)

PARTICLE TRANSPORT IN NON-MAGNETIC STARS

G. Michaud
Département de Physique
Université de Montréal
C.P. 6128, Succ. A, Montréal
CANADA H3C 3J7

ABSTRACT. The observations of AmFm, λ Booti, HgMn and He rich stars that are explained without any arbitrary parameter by diffusion are briefly reviewed, followed by those observations that are not explained by this simple model. Mass loss is then shown to explain a large fraction of the observations that are not explained in the parameter free model. It seems to play a role in the λ Booti, AmFm, He rich and the hot horizontal branch stars. It is only of about 10^{-15} to 10^{-13} $M_\odot$/yr. Abundance anomalies then help to determine stellar hydrodynamics. It is finally suggested that recent observations of Li underabundances in F stars of the Hyades represent an extension of the AmFm star phenomenon.

1. THE PARAMETER FREE MODEL

Diffusion is a basic physical process and plays a role everywhere a more efficient transport process does not wipe out its effects. If one assumes that a star arrives on the main sequence with the convection zones as given by standard evolutionary models and one allows the chemical separation to go on unimpeded, one obtains what I call the parameter free model for the non magnetic stars. It is a simply defined stellar model that, as we shall see, leads to large abundance anomalies, indeed larger than observed in the peculiar stars considered here.

As an A or B star arrives on the main sequence it has a He II convection zone that extends roughly down to T = 50 000 K. It is joined by overshooting to the superficial hydrogen convection zone (Latour, Toomre and Zahn 1981) if the star is cool enough to have one (T_{eff} < 10 000 K). Diffusion then goes on directly below the He II convection zone (see Fig.1a). Helium settles gravitationnally since the radiative acceleration on it is smaller than gravity (Michaud et al. 1979). At the same time the heavy elements either diffuse into or out of the convection zone depending on wether the radiative acceleration is larger or smaller than gravity. Once the helium abundance has been reduced by a factor of about 3, the helium convection zone disappears (Vauclair, Vauclair and Pamjatnikh 1974). The chemical separation then starts occuring (see Fig. 1b) below the hydrogen convection zone if the star

C. R. Cowley et al (eds.), Upper Main Sequence Stars with Anomalous Abundances, 459–472.

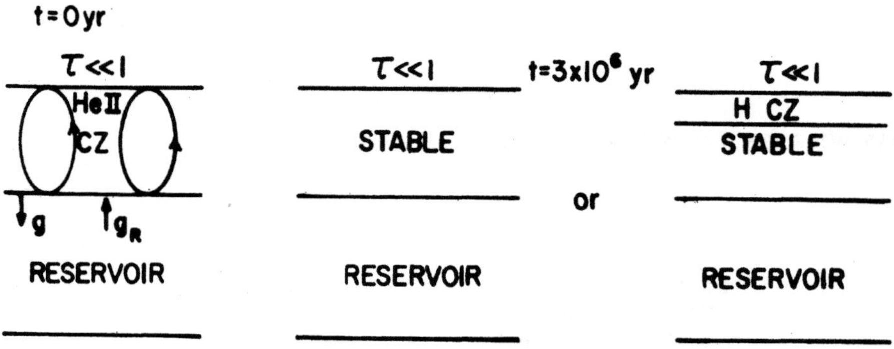

Figure 1. In the effective temperature range considered, stars form with a superficial He convection zone. When the He abundance has been sufficiently reduced it disappears and there remains a superficial H convection zone only if the star is cooler than 10 000 K.

has one or in the atmospere itself if the star does not have one. The starting conditions for the evolution of the abundances in that phase are those left over by the separation that occured below the He convection zone.

As I have already discussed in Liège (Michaud 1981), this very simple model, which is the only one that can be described without arbitrary parameters, is known to make assumptions that are not always valid. Additional phenomena are believed to be important in stellar interiors and, because of our ignorance of stellar hydrodynamics, they cannot all be described without arbitrary parameters. In Liège I have already discussed the potential effects of meridional circulation, of the tidal effects of a companion and of horizontal instabilities that could lead to horizontal inhomogeneities. In the review I made in Montréal (Michaud 1980), I have described the possible effects an inverted μ gradient may have in causing vertical inhomogeneities in stratified regions. Turbulence, accretion, a corona, mass loss are all effects that are potentially important but that we have to use arbitrary parameters to represent. Even if the HgMn and AmFm stars have no observational magnetic fields, a magnetic field below the observational limit may still have important effects on chemical separation.

We however adopt the methodological point of view that, while such phenomena may be important, one should first try to explain as many observations as possible without the use of any arbitrary parameter. Only because the parameter free model will be shown to explain a large fraction of the observations shall we investigate the possible complications. The importance of a given hydrodynamical process will be considered established only if one proces that can be described by, say, one arbitrary parameter can explain a substantial fraction of the observational facts the parameter free model does not explain. Mass loss will be seen below to be such a process.

More complete lists of references may be found in Vauclair and

Vauclair (1982) and Alecian and Vauclair (1983).

2. PHENOMENA EXPLAINED BY THE PARAMETER FREE MODEL

I here very briefly describe phenomena that are explained in a natural way by the parameter free model. References will be given to papers where a more complete description can be found.

1) The peculiarities appear in those stars that are likely to have the most stable atmospheres (Michaud 1970) so where the effects of diffusion are least likely to be wiped out by mass motion. They are slow rotators, are at an effective temperature where convection zones are either absent or very shallow and where mass loss is small, and have magnetic fields that may well stabilize the outer atmosphere.

2) The HgMn stars occur at $T_{eff} > 10\ 000$ K while the AmFm stars occur below. In the HgMn stars the separation then goes on in the atmosphere itself whereas it must go on below the hydrogen convection zone in the AmFm stars. This leads to the observed qualitative difference between the abundance anomalies of the two groups of stars (Michaud et al. 1976).

3) That the number of anomalous abundances be much larger on the magnetic than on the non magnetic stars is explained by the magnetic field making it impossible for the elements to leave the atmosphere where it is horizontal. An atomic species can become overabundant even if its radiative acceleration would push it into the interstellar medium since it is stopped by the horizontal magnetic field when it is ionized (Michaud, Mégessier and Charland 1981).

4)The observed near exclusion of the AmFm and the Delta-Scuti phenomena. If the abundance anomalies are caused by diffusion, helium has settled below the convection zone in the AmFm stars so that the driving mechanism of the pulsations has disappeared in the AmFm stars (Kurtz 1976, Baglin 1972, Michaud 1980).

5)The model correctly predicts which elements can be supported by radiation pressure and be overabundant and which elements are not supported and are underabundant. The maximum observed anomalies can further be explained by the migration of the elements from the envelope to the surface. The envelope of the anomalies is thus explained without arbitrary parameters (Michaud 1970, Watson 1971, Michaud et al. 1976 and Fig.4 of Michaud 1976).

6)Using the recent results of Tassoul and Tassoul (1982 and the following serie of papers) it can be shown that the anomalies appear in those stars rotating slowly enough for the meridional circulation velocity to be small enough for the chemical separation to go on unimpeded. This is true for both the HgMn stars (Michaud 1982) and the AmFm stars (Michaud, Tarasick, Charland and Pelletier 1983). In the latter case the tidal effects of the companion were also taken into account.

7)For the HgMn stars, detailed radiative calculations (mostly NLTE) have been carried in the atmosphere for He, Be, B, Mg, Si, Ca, Mn, Sr and Ba. In nearly all cases they reproduce, without arbitrary parameters, the envelope of the observed abundance anomalies (Michaud et al. 1981, Borsenberger, Michaud and Praderie 1979, 1981a, 1984, Alecian and Michaud 1981, Alecian and Vauclair 1981). These calculations were undertaken in order to allow a precise comparison of the predictions of the parameter free model with observations. Some problems have arisen. They constitute the facts that point to the need for additional hydrodynamic processes.

3. OBSERVATIONS THAT CONTRADICT THE PARAMETER FREE MODEL

There are a number of observations that cannot be explained by the parameter free model. I here list only those that appear best established.

1)It has long been clear to workers in the field that the large variations in the anomalies observed from star to star could not all be explained by the parameter free model. Some are explained by T_{eff} and log g differences. But this does not seem to be enough. It has been suggested at this conference that rotation may be an additional important parameter (see the evidence presented by Boyarchuck and the results of Didelon presented by Mégessier). An additional parameter (other than T_{eff} and log g) has to be involved and it might be related to rotation.

2)The observed overabundances in the AmFm stars are by factors of about 10 whereas diffusion would easily lead to overabundances by factors of 10^3 (Michaud et al 1976). The attempt to explain these lower abundances by turbulence has not been very successful (Vauclair, Vauclair and Michaud 1978) mainly because a small amount of turbulence wipes out the Ca and Sc underabundances without affecting significantly the overabundances of the heavy elements.

3)The existence, at the same temperature as the AmFm stars, of a group of stars that have <u>underabundances</u> of heavy elements, the λ Booti stars (Cowley et al. 1982). Their existence has been emphasized at this meeting.

4)The existence of the helium rich stars. Explanations in terms of diffusion based models have been suggested. However in the parameter free model, helium, not being supported by radiation pressure (Michaud et al. 1979), is underabundant. Only models involving mass loss and/or magnetic fields can lead to overabundances of He.

5)The existence of boron in some of the HgMn stars. According to the results of Borsenberger et al. (1979), boron is not only supported by radiation pressure in the atmosphere but is also pushed into the interstellar matter, so that no overabundance is expected to accumulate in the line forming region. Boron is however observed in a large fraction of the HgMn stars (Leckrone 1981, Sadakane and Jugaku 1981).

6)Even though the observed overabundance of Be is supported by the radiation pressure in the atmospheres of the HgMn stars and Be remains bound to the star (it is not pushed into the interstellar matter) it however accumulates higher than the line forming region unless some turbulence is present. The strength of its lines is not explained by the parameter free model even though the observed abundance is (Borsenberger et al. 1984, Borsenberger et al. 1981b).

4. ABUNDANCE ANOMALIES AND MASS LOSS

In my opinion the parameter free model explains a sufficient fraction of the observations for us to conclude that diffusion is the basic process responsible for the abundance anomalies. A model explaining the remaining facts would probably involve some of the hydrodynamical processes that I mentioned in Section 1. Whether it is possible to gain information on the hydrodynamical processes from abundance anomalies will depend on the number of observed facts that a simple and plausible phenomenon explains. Only if many observations are explained by a phenomenon parametrized by one arbitrary parameter should we accept that it plays a significant role in stars. Even if it were not possible to establish beyond reasonable doubt the importance of a given phenomenon however, merely knowing the potential effects of a plausible hydrodynamical process on abundance anomalies is interesting. Here I will restrict myself to a study of the potential effects of mass loss on abundance anomalies. How large does a mass loss need to be before it affects the anomalies? Does it need to be larger than in the OB stars or than in the Sun? It is clear that stellar atmospheres cannot be in hydrostatic equilibrium to infinity; the outer boundary condition must somewhere include dynamic processes. Here we assume that at the boundary there is homogeneous mass loss, that is the wind leaves with the chemical composition of matter at the top of the atmosphere. The wind is then parametrized by a single parameter, the mass loss rate.

The mass loss rate leads to a generalized outward flow of matter:

$$\frac{dM}{dt} = -4\pi R^2 N_H m_p v_w \qquad (1)$$

I assume that the mass loss rate does not substantially modify the stellar structure. This is justified for mass loss rates of 10^{-14} solar masses per year lasting 10^9 years, since only a very small fraction of the mass is involved. The conservation equation may then be written:

$$\nabla (c N_H(v_w + v_D)) = 0 \qquad (2)$$

where the diffusion velocity is given by its usual expression:

$$v_D = -D_{12} \left[\frac{\partial \ln c}{\partial r} + (g - g_R) \frac{m_p}{kT} - k_T \frac{\partial \ln T}{\partial r} \right] \qquad (3)$$

Calculations have been carried out for the AmFm stars, the λ Booti stars, the He rich stars and hot horizontal branch stars.

4.1 The AmFm Stars

The paramater free model for the AmFm stars already explains which elements are overabundant and which are underabundant as the result of the competition between the radiative and gravitational accelarations at the bottom of the hydrogen convection zone (Michaud et al. 1976). It also explains the binary period range, or equatorial rotation velocity range, over which the phenomenon is observed (Michaud et al. 1983). The predicted overabundances are however larger than observed (see Fig 3a of Michaud et al. 1983). Turbulence appears to be ruled out as the main cause of the reduction of the overabundances because, if it is strong enough to decrease sufficiently the overabundances of the heavy elements, it completely wipes out the underabundances of Ca and Sc (Vauclair et al. 1978).

Consider the effect of a small mass loss rate, say 10^{-15} $M_\odot$/y. As can be seen from Fig. 4 of Michaud et al. (1983), such a mass loss rate does not affect the underabundance of Ca. Similar calculations would similarly show that it does not affect those of Sc either. As can be seen from Fig. 3 of the same paper it however reduces the overabundance factor of, say, Eu from 1000 to 10. The exact overabundance factor becomes time dependant. The observed overabundances usually do not exceed factors of about 10 in AmFm stars (Preston 1974). A comparison with observations shows that the iron peak and the heavy elements are affected in a consistent manner. The observations then suggest the existence of mass loss rates of about 10^{-15} $M_\odot$/yr in AmFm stars. If the mass loss rate were larger by a factor of more than ten, the overabundances would be reduced sufficiently that this star would not be classified AmFm any more. If it were smaller by more than a factor of ten the overabundances would be larger than observed.

To understand the physical effects of mass loss in other objects to be discussed below, it is useful to understand how the mass loss modifies the separation. Consider where the separation occurs in the particularly simple case of helium (see Fig 1b of Michaud et al. 1983). The profile of the helium abundance is shown at the moment the helium abundance has been reduced sufficiently in the convection zone for the He convection zone to disappear. For a mass loss rate of 10^{-13} $M_\odot$/yr the separation occurs mainly at a depth where there is 10^{-5} of the stellar mass above. It occurs two orders of magnitude in mass closer to the surface if the mass loss rate is smaller by a factor of ten. It then occurs within a factor of ten of the bottom of the convection zone. This can be understood by calculating the diffusion flux as a function of depth. The diffusion flux of a given element is proportional to the product of the diffusion velocity by the density. Since the gravitational settling diffusion velocity is inversely proportional to density but directly proportional to $T^{1.5}$, the diffusion flux ends up increasing with depth as the temperature increases. Even if the diffusion flux is smaller than the mass loss flux just below the convection zone, it can still be larger than the mass loss flux deeper in the star. This happens in the cases considered here. Helium becomes underabundant in the atmosphere when arrives at the surface, the mass that was at the depth where the He separation flux was larger than the mass loss flux as the star formed. For the larger mass loss rate considered, this happens after 10^9 years. In the presence of mass loss, the separation can so occur only where the diffusion flux is larger than the mass loss flux and this happens deeper as the mass loss rate is increased. In what follows, other examples of this competition between mass loss rate and diffusion fluxes will be presented.

4.2 The λ Booti Stars.

The effect of increasing the mass loss rate is most dramatically emphasized by our recent calculations for the λ Booti stars. These are found at about the same effective temperature as the AmFm stars and are on or close to the main sequence (Hauck and Sletteback 1983). Instead of overabundances of heavy elements, they show underabundances of most heavy elements by factors of about three (Baschek and Searle 1969). Even if the details of the anomalies may not be so well known, there seems to be little doubt that there exists stars that have underabundances of heavy elements at the same temperature as the Am stars have overabundances. The success of diffusion based models in explaining the overabundances seems to have led to a non-status for the λ Booti stars as noted by Cowley et al. (1982).

To understand how it is possible for diffusion to lead to underabundances of heavy elements it is first necessary to realize how the radiative acceleration on heavy elements varies with depth in stellar atmospheres. In Fig. 2 is shown the radiative acceleration on Cr as a function of the mass above the point of interest. In this model $\log g = 4$, so that the radiative acceleration on Cr is clearly larger than gravity close to the convection zone but becomes equal to or smaller than gravity at a depth of about 10^{-3} of the total mass. If one adds the downward contribution of thermal diffusion the diffusion velocity is certainly downward there. From Fig. 2 of Michaud et al. (1983) one sees that a similar behavior occurs for most heavy elements. There is then a point in the star where the diffusion is downward when the star arrives on the main sequence. In the absence of mass loss this will lead to local underabundances of elements. If the mass loss rate is sufficient, the mass loss will expose these regions to view within the life of the star. Underabundances will be large if the mass loss rate flux sufficient to expose the matter is smaller than the diffusion flux as defined above for He in the case of the AmFm stars (see section 4.2).

Examples of results of detailed calculations are shown on Fig. 3 for Cr and Ti. For a mass loss rate of 10^{-13} $M_\odot$/yr, Cr becomes underabundant after about 10^8 years and remains underabundant for the rest of the life of the star. _If_ the mass loss rate is 3 times smaller, Cr remains mainly overabundant. _If_ the mass loss rate is ten times larger, the underabundances that appear on the surface within the stellar life time are of only 25%. There is then only a factor of about 3 around the optimum mass loss rate that allows underabundances of the size observed to materialize. Very similar results are also shown for Ti. The early overabundances are caused by the arrival on the surface of matter that was originally where diffusion was upwards.

In Table 1 are shown the underabundance factors that a mass loss rate of 10^{-13} $M_\odot$/yr leads to in a star with an effective temperature of 7800 K and $\log g = 4.4$. These are of the same order as those observed in λ Booti stars. Even though no effort was made to maximize the underabundance factors obtained, we have conducted calculations for different values of the mass loss rate, temperature and gravity so

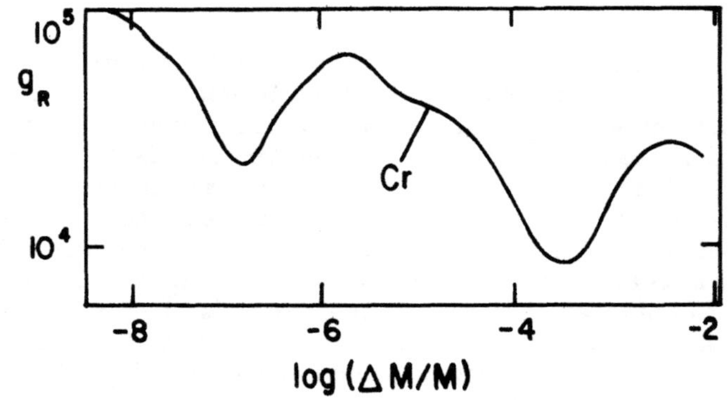

Figure 2. Radiative acceleration on Cr as a function of the mass above the point of interest.

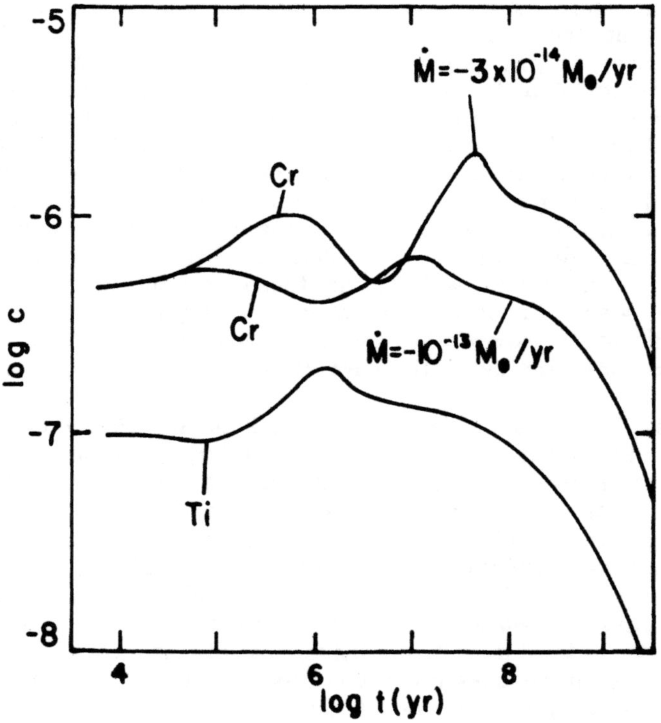

Figure 3. Time evolution of the abundance of Cr and of Ti. A solar abundance is assumed when the star formed.

that we can say that the underabundance factors cannot be much larger than those indicated. For a few individual elements they could be

TABLE 1
Underabundance Factors after 10^9 Years

Element	Abundance factor	Element	Abundance factor
C	0.41	Ti	0.24
Si	0.61	Cr	0.40
Ca	0.26	Mn	0.48

somewhat larger but they could not be systematically much larger than indicated (see Michaud and Charland, in preparation, for details). For instance, diffusion with mass loss could not explain generalized underabundances by factors of 100, in stars with lifetimes of 3 10^9 years and less.

4.3 The He Rich Stars

As briefly mentioned above, He is not supported by radiation pressure even in the early main sequence stars where it is observed to be overabundant (Michaud et al. 1979). In the absence of mass loss diffusion then leads to underabundances of He in the non-magnetic stars.

Vauclair (1975) has suggested that, if the mass loss rate is appropriate, it could lead to overabundances of helium in the atmospheres of the relatively hot stars observed to have such anomalies (Osmer and Peterson 1974). If the mass loss rate is appropriate, hydrogen will drag helium along and He will accumulate where the dragging is least effective that is where He is most in the form of neutral helium, since the diffusion coefficient is then some two orders of magnitude larger than when it is ionized. In the stars where it is observed to be overabundant, helium is least ionized in the atmosphere, so that is where it accumulates. This model is the only one based on diffusion and able to explain overabundances of He in the non-magnetic stars.

This model implicitly assumes that the helium flux continues to arrive in the atmosphere for the whole life of the star. This might not be the case if the mass loss rate were slow enough that the separation could be effective in the envelope, just as described above for the AmFm stars. We have (Michaud and Dupuis in preparation) evaluated how long it takes for the helium abundance to be reduced by a factor of three in the flux arriving in the atmosphere. It takes 1.5 10^9 years if the mass loss rate is 10^{-12} $M_\odot$/yr, a mass loss rate about equal to that needed to lead to the He overabundance. This is longer than the expected life time of those stars. This is then consistent with the assumptions of the model of Vauclair (1975) for the overabundance of He.

4.4 The Si Underabundance in the B and O Subdwarfs.

It has been revealed by IUE observations of O and B subdwarfs that they have underabundances of silicon by 4 or 5 orders of magnitude while having about normal N, underabundant C and ten times underabundant

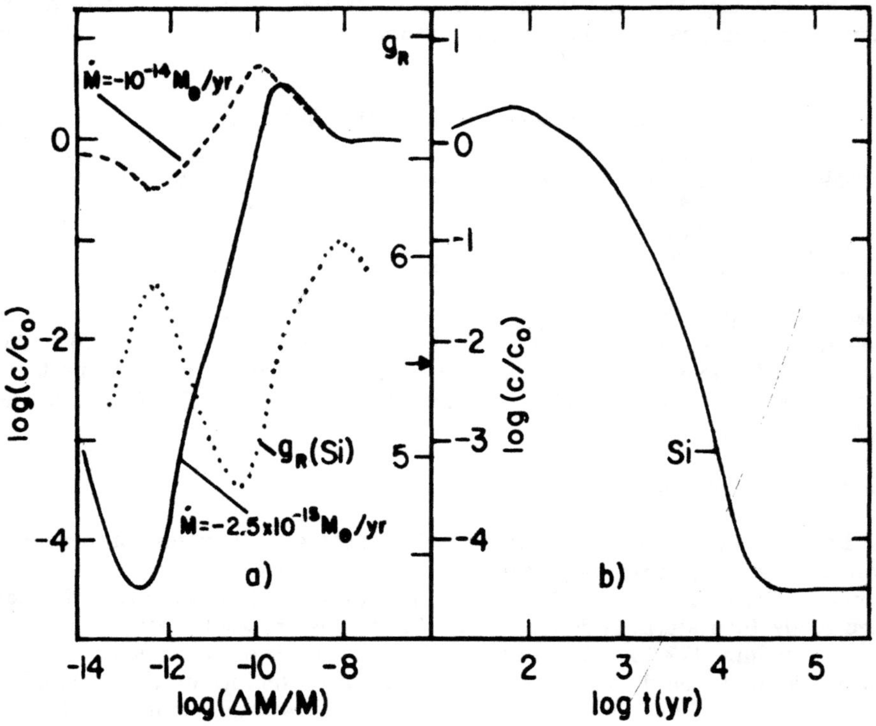

Figure 4. On part a) are shown the radiative acceleration (center scale) as a function of the mass above the point of interest, as well as profiles of the Si abundance (left hand scale) also as a function of depth. The time evolution of the Si abundance is shown on part b). Gravity is indicated by an horizontal arrow on the central scale.

helium. The C abundance varies considerably from star to star (Baschek et al. 1982, Heber et al. 1984, Lamontagne et al. 1985). These subdwarfs have log g = 5.5 and effective temperatures between 30 000 and 40 000 K. Their evolutionary status is not well known.

Explaining such objects by models based on nuclear physics seems extremely difficult: it is difficult to destroy helium by nuclear reactions (Vauclair and Reeves 1972) and probably impossible to destroy Si by such a large factor while not destroying N. Ever since the He underabundance was measured, diffusion was called upon to explain it. We here show how diffusion can also explain the Si underabundance but only if mass loss is also present.

In Fig. 4a is shown the radiative acceleration on Si in the envelope of a star with T_{eff} = 35 000 K and log g = 5.5, as is appropriate for the sdOs and sdBs. The radiative acceleration has a maximum in the atmosphere, where Si is supported by the radiative acceleration for a solar Si abundance (the radiative acceleration is larger than gravity). In the absence of mass loss diffusion is not expected to lead to underabundances of Si in the region where the lines form, contrary to the suggestion of Baschek et al. (1982) and of Heber et al. (1984). The radiative acceleration becomes smaller than gravity only much deeper than the line forming region: it is only deeper than the line forming region that Si is sufficiently in the rare gas configuration, that of Ne, for its radiative acceleration to be affected sufficiently. In the absence of mass loss Si is then expected to be supported and not to be very underabundant (Michaud et al. 1985).

In the presence of a small mass loss, the process described above for the λ Booti, stars occurs but much closer to the surface, where the radiative acceleration is smaller than gravity. As a function of time the abundance of Si then decreases in the atmosphere since it leaves by the wind but is not replenished by a flux from below. The results of detailed time dependant calculations are shown on Fig. 4b. For a mass loss rate of 2.5 10^{-15} $M_\odot$/yr, the Si abundance is reduced in the line forming region by more than four orders of magnitude after 10 000 yrs. In Fig. 4a is shown the space distribution of Si when it has about reached an equilibrium value. Its abundance goes down rapidly outward where the radiative acceleration is too small to support it. It reaches a minimum where the radiative acceleration has a maximum. This is because in the presence of an outward flux, the conservation of the Si flux leads to the abundance being smallest where the outward velocity is largest, that is where the radiative acceleration is largest (see Michaud et al. 1985 for details). If the mass loss rate is 4 times larger, the Si abundance is hardly reduced as can be seen from the equilibrium profile shown in Fig 4a. This qualitative difference in the solution is due to the sum of the diffusion and mass loss velocities changing sign in between. Whereas, in the absence of an abundance gradient, the net velocity is downward for the smaller of the two mass loss rates considered, it is upward for the larger mass loss. Mathematically there is a sign change in the differential equation and the nature of the solution changes. This occurs quite suddenly and so the underabundance of Si becomes a precise determination of the mass loss rate. If the mass loss rate is smaller, one expects that the time scale for the appearance of the anomaly will increase bacause it will take longer to empty the region where Si is supported, but then the underabundance should be larger. This remains to be checked by detailed calculations that are currently underway. The He, C and N abundances that such a mass loss rate leads to are also consistent with the observations (Michaud et al. 1985). We here only determine an upper limit to the mass loss rate, a lower limit is determined by the timescale of the sdB and sdO evolutionary phase. It is currently not known.

There is in our opinion no reason why stars should stop losing mass at the lowest currently detectable mass loss rate and so the hypothesis of a small mass loss rate is plausible. The existence of abundance anomalies puts very strong constraints as to how strong it can be.

5. LITHIUM UNDERABUNDANCES: EXTENSION OF THE AmFm PHENOMENON

Recent observations of Li in the F stars of the Hyades (Boesgaard and Tripicco 1985 preprint) have shown a most unexpected effective temperature dependance of the underabundance. It is sketched in Fig. 5. While the Li abundance is about normal in stars hotter than 7 000 K and in those around 6 200 K, it is underabundant by 2 orders of magnitude or more at 6 600 K with well defined intermediate values in between. Cooler than 6 000 K it shows the usual progressive reduction in the Li abundance.

It is relatively easy to see that diffusion is expected to lead to a dip in the Li abundance at about the observed temperature. On Fig. 5b is shown the effective temperature dependance of the depth of the He II convection zone and of the radiative acceleration on Li at the base of the He II convection zone. Slightly above 7 000 K, the convection zone is shallow and Li is supported by the radiation pressure in the hydrogenoid configuration. Below 7 000 K, the depth of the convection zone increases rapidly and Li is not supported any more at the bottom of the convection zone: it is completely ionized. Lithium then settles gravitationnally and becomes underabundant. As the temperature is further decreased, the depth of the convection zone continues to increase and the diffusion time scale increases, finally exceeding the age of the Hyades at an effective temperature of about 6 400 K (see Fig. 5 of Michaud et al. 1983). Gravitational settling then has no time to lead to underabundances in the cooler stars. The underabundances observed below 6 000 K are expected to be due to another process: the burning of the Li that has been transported to the Li burning region by turbulent diffusion (Vauclair et al. 1978).

Even though the parameter free model will always lead to the presence of such a dip, its properties depend on the details of the

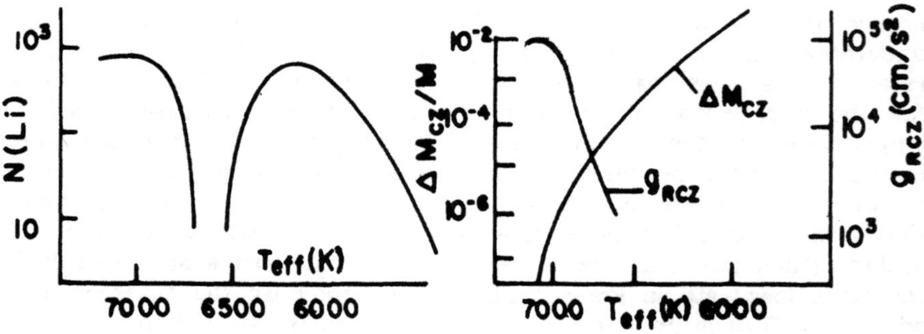

Figure 5. On part a) are sketched the results of Boesgaard and Tripicco (1985) for Li abundances (log N(H) = 12) in the Hyades. On part b) are shown, as a function of the effective temperature, both the depth of the convection zone (left hand scale) and the value of the radiative acceleration (right hand scale) at the bottom of the convection zone.

convection zones. In particular its exact location will depend on α, the value of the ratio of the mixing length and the pressure scale height. This is poorly known but we have used here $\alpha = 1.4$ as suggested by some evolutionary model calculations. Our argumentation has been based on static models but the time dependant calculations of the settling should really be carried out in evolutionary models that take into account the variation of the position of the bottom of the convection zone as the He abundance varies (Michaud, Fontaine and Beaudet 1984): if Li settles gravitationnally, so will helium. The dip will not be eliminated by changing α or taking evolution into account but its width will be changed and so will its depth. The precise position of the dip as well as its properties will turn out to depend on hydrodynamic properties which the Li underabundance will allow us to measure.

I would like to thank Dr Ann Boesgaard for communicating to me results ahead of publication.

REFERENCES

Alecian, G., and Michaud, G. 1981, Ap. J., 245, 226.
Alecian, G., and Vauclair, S. 1981, Astr. Ap., 101, 16.
Alecian, G., and Vauclair, S. 1983, Fundamentals Cosmic Phys., 8, 369.
Baglin, A. 1972, Astr. Ap., 19, 45.
Baschek, B., Kudritzki, R. P., Scholz, M., and Simon, K. P. 1982, Astr. Ap., 108, 387.
Borsenberger, J., Michaud, G., and Praderie, F. 1979, Astr. Ap., 76, 287.
______. 1981a, Ap. J., 243, 533.
______. 1984, Astr. Ap., 139, 147.
Borsenberger, J., Radiman, I., Michaud, G., and Praderie, F. 1981b, in Les étoiles de composition chimique anormale du début de la séquence principale, (Liège: Université de Liège), p.389.
Cowley, C. R., Sears, R. L., Aikman, G. C. L., and Sadakane, K. 1982, Ap. J., 254, 191.
Heber, U., Hamann, W.-R., Hunger, K., Kudritsky, R. P., Simon, K. P., and Méndez, R. H. 1984, Astr. Ap., 136, 331.
Kurtz, D. W. 1976, Ap. J. Suppl., 32, 651.
Lamontagne, R., Wesemael, F., Fontaine, G., and Sion, E. M. 1985, preprint.
Latour, J., Toomre, J., and Zahn, J.-P. 1981, Ap. J., 248, 1081.
Leckrone, D. S., 1981, Ap. J., 250, 687.
Michaud, G. 1970, Ap. J., 160, 641.
Michaud, G. 1976, in Physics of Ap Stars, IAU Colloquium No. 32, ed. W. W. Weiss, H. Jenkner, and H. J. Wood (Universitatsternwarte Wien, Vienna), p.351.
Michaud, G. 1980, A. J., 85, 589.
Michaud, G. 1981, in Les étoiles de composition chimique anormale du début de la séquence principale (Liège: Université de Liège), p.355.
Michaud, G. 1982, Ap. J., 258, 349.

Michaud, G., Bergeron, P., Wesemael, F., and Fontaine, G. 1985, Ap. J., in press.
Michaud, G., Charland, Y., Vauclair,S., and Vauclair G. 1976, Ap. J., 210, 447.
Michaud, G., Fontaine, G., and Beaudet, G. 1984, Ap. J., 282, 206.
Michaud, G., Mégessier, C., and Charland, Y. 1981, Astr. Ap., 103, 244.
Michaud, G., Montmerle, T., Cox, A.N., Magee N.H., Hodson, S.W., and Martel, A. 1979, Ap. J., 234, 206.
Michaud, G., Tarasick, D., Charland, Y., and Pelletier, C. 1983, Ap. J., 269, 239.
Osmer, P. S., and Peterson. D. 1974, Ap. J., 187, 117.
Preston, G. 1974, Ann. Rev. Astr. Ap., 12, 257.
Sadakane, K., and Jugaku, J. 1981, Pub. Astr. Soc. Pac., 93, 60.
Tassoul, J.-L., and Tassoul, M. 1982, Ap. J. Suppl., 49, 317.
Vauclair, G., Vauclair, S., and Michaud, G. 1978, Ap. J., 223, 920.
Vauclair, G., Vauclair, S., and Pamjatnikh, A. 1974, Astr. Ap., 31, 63.
Vauclair, S. 1975, Astr. Ap., 45, 233.
Vauclair, S., and Reeves, H. 1972, Astr. Ap., 18, 215.
Vauclair, S., and Vauclair, G. 1982, Ann. Rev. Astr. Ap., 20, 37.
Watson, W. D., 1971, Astr. Ap., 13, 263.

Discussion appears after the following paper.

THE PLANETOID-IMPACT HYPOTHESIS OF CP F, A, AND B STAR FORMATION: POSSIBILITIES AND PERSPECTIVES

E.M.Drobyshevski
A.F.Ioffe Physical-Technical Institute
Academy of Sciences of the USSR
194021 Leningrad, USSR

ABSTRACT. A possibility is analyzed of explaining the chemical anomalies of chemically peculiar (CP) F-A-B stars basing on the assumption of the formation of a large number of moonlike planetoids both in the course of separation of the components and in late stages of close binary evolution.

Primitive igneous differentiation of such planetoids results in their crust becoming deficient in Mg, Ca, Sc and enriched in Fe, Sr, Ba, and the Rare Earths. Infall of such planetoids or crust fragments ejected in their collisions with one another onto an A star makes it Am-type. The deficiency of some elements relative to normal abundance can be accounted for if one assumes that these elements present in the matter streaming from one binary component to another condense with subsequent rain-out into a component or formation of the planetoids.

The more diverse and complex anomalies (including the separation of isotopes) can be explained in the same context of the close binary evolution by invoking the ideas of <u>magnetic cosmochemistry</u> which considers the consequences of extremely nonequilibrium processes associated with the flow of magnetic-field generated electrical currents through a rarefied matter in space.

There have been many attempts to explain the CP phenomenon on F, A, and B stars within various versions of the accretion hypothesis. These usually involve the accretion of remnants of the protostellar cloud including the possibility of elemental separation in condensation, the formation of planetesimals, and even nuclear processes (Searle and Sargent 1967; Tomley et al. 1970; Dolginov 1975), as well as the effect of magnetic field on gas and dust accretion (Havnes and Conti 1971; Havnes 1975). Cowley (1977) has recognized broad possibilities inherent in the planetesimal-impact hypotheses, particularly if combined with other processes. However he pointed out

C. R. Cowley et al (eds.), Upper Main Sequence Stars with Anomalous Abundances, 473–476.

difficulties which can hardly be overcome within these hypotheses, namely: (1) it is unclear how infall of condensate onto a normal star could produce deficiency of some elements, and (2) it is unclear how to account for the strong isotopic shifts of Hg, Pt, He.

Our version of the accretion hypothesis for the CP phenomenon (Drobyshevski 1975a) provides a possibility to overcome these difficulties. It is predicated on the assumption of planetary systems being a by-product and/or limiting case of the evolution of binaries (Drobyshevski 1974a). In this context, numerous (up to 10^4) moonlike planets can form as matter flows from one component to another both in the course of their formation and separation as a result of rotational-exchange breakup of the rotating protostar (Drobyshevski 1974b) and in late stages of the close binaries evolution, when the primary becoming a red giant overflows the Roche lobe. There are grounds to believe that moonlike bodies can form not only in accretion disks (Drobyshevski 1975a) but within the primary itself as well, the conditions for fast condensation of planetoids being generally more favorable here than in classical protoplanetary disks (Drobyshevski 1975b). the most probable place for the formation of planetoids within the primary is the vicinity of the inner Lagrangian point, indeed, one can assume that conditions ensuring gas-dynamic suspension of growing planetoids can obtain here in the stream of outflowing matter. Part of the condensate can also rain-out into the primary thus forming a rocky core (Drobyshevski 1975a; Slattery 1978). Therefore the material streaming to the secondary should be depleted in refractory condensate (Sc_2O_3, ZrO_2, Al_2O_3, $CaTiO_3$, etc.) which can account not only for the deficiency of some elements but also for the frequently observed abundance differences between binary components which may be even equal mass and luminosity.

As a result of primitive igneous differentiation of material in initially molten moonlike bodies, the last-to-solidify melt forming their anorthosite crust becomes usually depleted in Mg, Al, Ca, and Sc while concentrating such elements with a large ionic radius as Sr, Y, Zr, Ba, and Rare Earth Elements, primarily Eu. A comparison of the composition of lunar highland rocks as measured in the Apollo and Luna 20 experiments with that of Am stars shows them to be practically identical (Fig.1) (Drobyshevski 1975b). A certain exclusion to this is the anomalies of iron and the geochemically related anomalies of Al and Ca which are due to the Moon's being generally deficient in iron. In more "normal" moonlike bodies the composition of the crust should agree closely with that of Am stars.

All Am stars are believed to be components of either young or evolved close binaries (Abt 1983; Drobyshevski

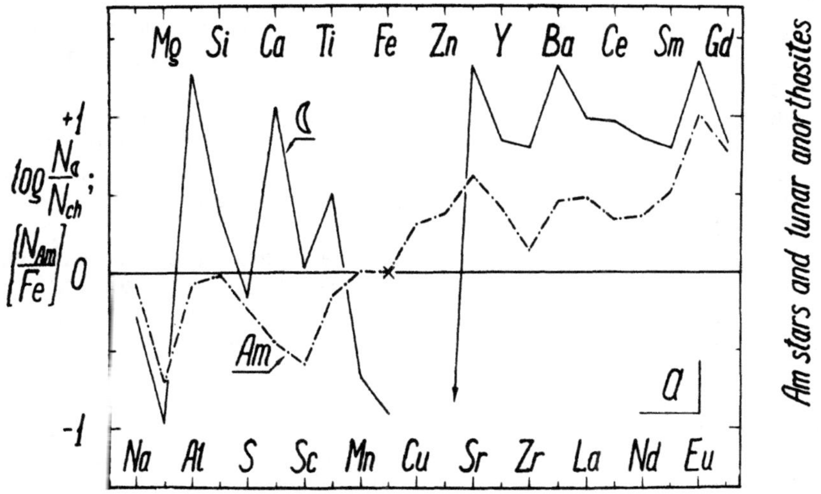

Fig.1. A comparison of the chemical composition anomalies of Am stars (normalized to "normal" stars and iron (Smith 1971)) with the chemical composition of the Moon's anorthosite crust (normalized to that of C1 chondrites).

1973b), so that in the context of the above ideas the presence in them of numerous moonlike planetoids would appear to be only natural. In rare collisions with one another the planetoids shed off fragments of their anorthosite crust. Infall of these fragments (and of the planetoids proper) onto the stellar components of the A class (with convective envelope masses of $\sim 10^{-10}-10^{-8} M_\odot$) makes them Am stars. The reasons for differences in the anomalies between even similar components and for correlations of some anomalies with star temperature are obvious. Thus the problem of the Am phenomenon may be considered solved.

The properties of Ap stars are more complex and more diverse. One may suggest that early Ap ("helium" and "silicon") stars possess primordial magnetic fields and are single. As for late Ap stars, their properties and the presence of a rather disordered magnetic fields on them can be readily accounted for by assuming them to be evolved binaries with a white dwarf-type companion (van den Heuvel 1968). Actually, they are second generation magnetic stars (first generation magnetic stars, where a weak field generated by outer convection builds up under the convective envelope due to the accretion of matter from the outside, are Am stars) (Drobyshevski 1973a). It is only natural to attribute the diversity of the properties of Ap stars to the presence of magnetic fields and their effect on the differentiation of matter and condensation of planetoids. (In particular, it is the effect of magnetic field which is usually held responsible for the maintenance of patchiness in the surface chemical composition.)

The transport of magnetic field by weakly ionized rarefied gas generates in the latter electric currents producing extremely nonequilibrium conditions (enhanced electronic temperatures, particle beaming, lasing effects, etc.) which are accompanied by exotic chemical reactions. This gives us grounds to define a new branch of science, namely, magnetic cosmochemistry. As evidenced by laboratory experiments (Basov et al. 1974; Thiemens and Heidenreich 1983; Letokhov 1983) under such conditions, in particular, efficient isotope separation can occur. The success of the planetoid-impact hypothesis in the interpretation of the Am phenomenon gives us grounds to believe that this as yet undeveloped science may offer promise for the explanation of some "nonstandard" anomalies on Ap stars. This approach is linked in more than one way with Alfvén's (1981) ideas on the role of electric currents in space.

REFERENCES

Abt,H.A.: 1983, Annu.Rev.Astron.Astrophys. 21, 343.
Alfvén,H.: 1981, "Cosmic Plasma", Reidel Publ.Co.
Basov,N.G., Belenov,E.M., Gavrilina,L.K., Isakov,V.L., Markin,E.P., Oraevski,A.N., Romanenko,V.I., and Ferapontov,N.B.: 1974, Pis'ma ZhETF 19, 336.
Cowley,C.R.: 1977, Astrophys.Space Sci. 51, 349.
Dolginov,A.Z.: 1975, in "Magnetic Ap-Stars" (ed. I.A. Aslanov), Elm, Baku, pp.147-150.
Drobyshevski,E.M.: 1973a, Astrophizika 9, 119.
Drobyshevski,E.M.: 1973b, Preprint PhTI-445, Leningrad.
Drobyshevski,E.M.: 1974a, Nature 250, 35.
Drobyshevski,E.M.: 1974b, Astron.Astrophys. 36, 409.
Drobyshevski,E.M.: 1975a, Astrophys.Space Sci. 35, 403.
Drobyshevski,E.M.: 1975b, Earth Planet.Sci.Letts. 25, 368.
Havnes,O.: 1974, Astron.Astrophys. 32, 161.
Havnes,O., and Conti,P.S.: 1971, Astron.Astrophys. 14, 1.
Letokhov,V.S.: 1983, "Non-Linear Selective Photoprocesses in Atoms and Molecules", Nauka, Moscow (in Russian).
Searle,L., and Sargent,W.L.W.: 1967, in "The Magnetic and Related Stars" (ed. R.C.Cameron), Mono Book Corp., Baltimore, pp.219-231.
Slattery,W.L.: 1978, Moon Planets 19, 443.
Smith,M.A.: 1971, Astron.Astrophys. 11, 325.
Thiemens,M.H., and Heidenreich III,J.E.: 1983, Science 219, 1073.
Tomley,L.J., Wallerstein,G., and Wolff,S.C.: 1970, Astron. Astrophys. 9, 380.
van den Heuvel,E.P.J.: 1968, Bull.Astron.Inst.Neth. 19, 326.

DISCUSSION (Michaud)

STĘPIEŃ: What is the mechanism for the uniform mass loss you have introduced for the stars in your computations?

MICHAUD: I treated it as an arbitrary parameter, because it is clear from the very large number of abundance anomalies that there must be an additional parameter which is important. For example, the λ Boo stars are more rapidly rotating than the Am-Fm stars, and it is possible that the hypothetical mass loss is related to the rotation rate. In the He-rich stars it may be driven by radiation pressure. In cooler stars, especially when convective zones are present, weak coronae are also possible. We can only consider an additional parameter, because the parameter free model explains many properties.

KROLL: Is the assumption of mass loss that starts at the top of the atmosphere not in contradiction with the assumption of no turbulence?

MICHAUD: If the process driving the mass loss is radiation pressure, then it need not cause any turbulence. If there is a convection zone in the atmosphere as in the Am-Fm stars, the existence of a turbulent corona need not cause problems. The separation in these stars only takes place below the convection zone. Even in the Sun, there is an apparent underabundance of He in the solar wind, by a factor 2.5, so that somehow the solar wind succeeds in carrying less He than (presumably) there is in the bulk of the Sun.

COWLEY: If one tries to enhance the abundances of heavy elements at the surfaces of stars by adding differentiated planet-like (or Earth-like) material, one has the problem of making Ca and Sc underabundant. Could the diffusion process for these two elements act rapidly enough to deplete them relative to the other heavy species that have been added to the atmosphere?

MICHAUD: This would, more or less, be the reverse of the mass loss process. Accretion rates comparable to the proposed mass loss rates would probably be required, perhaps 10^{-15} or 10^{-14} $M_\odot$ per year. But, since we know that stars lose mass down to the lowest rates we can measure, why should stars stop losing mass at the level we stop being able to measure? The reason to prefer winds to accretion is that it gives far less arbitrary results. We have no observational evidence of planetary accretion. Of course, by having accretion and later differentiating it on the star you could explain a lot of things!! However, the model already explains a large number of the properties of these objects.

DROBYSHEVSKI: Why do you think that the convective envelope mass for a star with $M \approx 1\ M_\odot$ will be only $M_c/M \sim 10^{-4}$? It is widely believed that for the Sun $M_c/M \sim 10^{-2}$, and that stars with $M \lesssim 1.5\ M_\odot$ (about F5) have $M_c/M \gtrsim 10^{-3}$; this ratio is much smaller ($\sim 10^{-8}$ to 10^{-10}) in the range 1.5-3.0 $M_\odot$.

MICHAUD: I agree that the depth of the convection zone in the Sun is likely to be of the order $M_c/M_\odot \sim 10^{-2}$. The only reason for the smaller values I gave was that I only had with me, in Paris, results from models with α=1.0 and no helium. The exact value of the convection zone at F5 is very sensitive to α.

C. R. Cowley et al (eds.), Upper Main Sequence Stars with Anomalous Abundances, 477–478.

DISCUSSION (Drobyshevski)

MICHAUD: I have two questions. First, what is the observational evidence for the existence of such planetesimals in binary systems? Second, how much mass transfer (in $M_{\odot}$) do you expect on to the Am or Ap star?

DROBYSHEVSKI: The main observational evidence is a very nice detailed correlation between abundances in the uppermost layers of an igneously differentiated body (e.g., the lunar crust) and abundances observed on Am stars. The mass expected to be transferred from one component to another in a close binary system containing A-type components is of order 1-2 $M_{\odot}$. There may be thousands of these moon-like bodies in such a system.

MICHAUD: There have been small companions observed by Wolff (**Ap. J.**, 222, p. 556, 1978) around B stars and all these stars turn out to be normal.

DROBYSHEVSKI: In some close binaries, especially those with hot components, there will be no planetary bodies. If the gas temperature is more than about 2000 K, condensation does not take place and planetoids would not form.

DOLGINOV: If the main difference between normal A stars and Ap stars is the absence or presence of the planetesimals, how can the strong helium deficit in Ap stars be explained?

DROBYSHEVSKI: The He anomalies may be explained in the planetoidal impact model if one takes into account the magnetocosmochemical processes in matter streaming from one close binary component on to the other, when this matter transports magnetic field. Electric-discharge-like phenomena take place leading, for instance, to the isotope separation effect. In such discharges H may be well ionized, whereas He, due to its high ionization potential, is not. In such a case the process of He-H separation may proceed quite effectively. Similar ideas were developed by H. Alvén.

GPSR Compliance

The European Union's (EU) General Product Safety Regulation (GPSR) is a set of rules that requires consumer products to be safe and our obligations to ensure this.

If you have any concerns about our products, you can contact us on ProductSafety@springernature.com

In case Publisher is established outside the EU, the EU authorized representative is:

Springer Nature Customer Service Center GmbH
Europaplatz 3
69115 Heidelberg, Germany

Batch number: 10815974

Printed by Printforce, the Netherlands